Topics in Applied Physics
Volume 113

Available online at
SpringerLink.com

Topics in Applied Physics is part of the SpringerLink service. For all customers with standing orders for Topics in Applied Physics we offer the full text in electronic form via SpringerLink free of charge. Please contact your librarian who can receive a password for free access to the full articles by registration at:

springerlink.com → Orders

If you do not have a standing order you can nevertheless browse through the table of contents of the volumes and the abstracts of each article at:

springerlink.com → Browse Publications

Topics in Applied Physics

Topics in Applied Physics is a well-established series of review books, each of which presents a comprehensive survey of a selected topic within the broad area of applied physics. Edited and written by leading research scientists in the field concerned, each volume contains review contributions covering the various aspects of the topic. Together these provide an overview of the state of the art in the respective field, extending from an introduction to the subject right up to the frontiers of contemporary research.

Topics in Applied Physics is addressed to all scientists at universities and in industry who wish to obtain an overview and to keep abreast of advances in applied physics. The series also provides easy but comprehensive access to the fields for newcomers starting research.

Contributions are specially commissioned. The Managing Editors are open to any suggestions for topics coming from the community of applied physicists no matter what the field and encourage prospective editors to approach them with ideas.

Stephen J. Paddison
Keith S. Promislow
Editors

Device and Materials Modeling in PEM Fuel Cells

Editors
Stephen J. Paddison
Department of Chemical
  & Biomolecular Engineering
University of Tennessee
Knoxville, TN USA 37996
spaddison@utk.edu

Keith S. Promislow
Department of Mathematics
Michigan State University
East Lansing, MI USA 48824
kpromisl@math.msu.edu

ISSN: 0303-4216
ISBN: 978-0-387-78690-2   e-ISBN: 978-0-387-78691-9
DOI 10.1007/978-0-387-78691-9

Library of Congress Control Number: 2008936882

© Springer Science + Business Media, LLC 2009
All rights reserved. This work may not be translated or copied in whole or in part without the written permission of the publisher (Springer Science + Business Media, LLC, 233 Spring Street, New York, NY 10013, USA), except for brief excerpts in connection with reviews or scholarly analysis. Use in connection with any form of information storage and retrieval, electronic adaptation, computer software, or by similar or dissimilar methodology now known or hereafter developed is forbidden.
The use in this publication of trade names, trademarks, service marks, and similar terms, even if they are not identified as such, is not to be taken as an expression of opinion as to whether or not they are subject to proprietary rights.

Printed on acid-free paper

springer.com

Contents

Part I Device Modeling

Section Preface 3
 Jean St-Pierre

1. Modeling of PEMFC Catalyst Layer Performance
   and Degradation 19
   Jeremy P. Meyers

2. Catalyst Layer Operation in PEM Fuel Cells: From Structural
   Pictures to Tractable Models 41
   B. Andreaus and M. Eikerling

3. Reactor Dynamics of PEM Fuel Cells 91
   Jay Benziger

4. Coupled Proton and Water Transport in Polymer Electrolyte
   Membranes 123
   J. Fimrite, B. Carnes, H. Struchtrup and N. Djilali

5. A Combination Model for Macroscopic Transport in
   Polymer-Electrolyte Membranes 157
   Adam Z. Weber and John Newman

6. Analytical Models of a Polymer Electrolyte Fuel Cell 199
   A. A. Kulikovsky

7. Phase Change and Hysteresis in PEMFCs 253
   Keith S. Promislow

8. Modeling of Two-Phase Flow and Catalytic Reaction Kinetics
   for DMFCs 297
   J. Fuhrmann and K. Gärtner

9. Thermal and Electrical Coupling in Stacks 317
   Brian Wetton

Part II Materials Modeling

Section Preface 341
    K. D. Kreuer

The Membrane

10. Proton Transport in Polymer Electrolyte Membranes Using
    Theory and Classical Molecular Dynamics 349
    A. A. Kornyshev and E. Spohr

11. Modeling the State of the Water in Polymer Electrolyte
    Membranes 365
    Reginald Paul

12. Proton Conduction in PEMs: Complexity, Cooperativity
    and Connectivity... 385
    S. J. Paddison

13. Atomistic Structural Modelling of Ionomer Membrane
    Morphology 413
    J. A. Elliott

14. Quantum Molecular Dynamic Simulation of Proton Conducting
    Materials.. 437
    G. Seifert, S. Hazebroucq and W. Münch

15. Morphology of Nafion Membranes: Microscopic
    and Mesoscopic Modeling 453
    Dmitry Galperin, Pavel G. Khalatur and Alexei R. Khokhlov

The Catalyst

16. Molecular-Level Modeling of Anode and Cathode
    Electrocatalysis for PEM Fuel Cells...................... 485
    Marc T.M. Koper

17. Reactivity of Bimetallic Nanoclusters Toward the Oxygen
    Reduction in Acid Medium 509
    Perla B. Balbuena, Yixuan Wang, Eduardo J. Lamas,
    Sergio R. Calvo, Luis A. Agapito and Jorge M. Seminario

18. Multi-Scale Modeling of CO Oxidation on Pt-Based
    Electrocatalysts... 533
    Chandra Saravanan, N. M. Markovic, M. Head-Gordon
    and P. N. Ross

19. Modeling Electrocatalytic Reaction Systems from First Principles 551
    Sally A. Wasileski, Christopher D. Taylor and Matthew Neurock

Index 575

Foreword

Computational studies on fuel cell-related issues are increasingly common. These studies range from engineering level models of fuel cell systems and stacks to molecular level, electronic structure calculations on the behavior of membranes and catalysts, and everything in between. This volume explores this range. It is appropriate to ask what, if anything, does this work tell us that we cannot deduce intuitively? Does the emperor have any clothes? In answering this question resolutely in the affirmative, I will also take the liberty to comment a bit on what makes the effort worthwhile to both the perpetrator(s) of the computational study (hereafter I will use the blanket terms modeler and model for both engineering and chemical physics contexts) and to the rest of the world. The requirements of utility are different in the two spheres. As with any activity, there is a range of quality of work within the modeling community.

So what constitutes a useful model? What are the best practices, serving both the needs of the promulgator and consumer? Some of the key components are covered below. First, let me provide a word on my 'credentials' for such commentary. I have participated in, and sometimes initiated, a continuous series of such efforts devoted to studies of PEMFC components and cells over the past 17 years. All that participation was from the experimental, qualitative side of the effort. I have been privileged to work with a series of excellent modelers and computational physical chemists, including Tom Springer, Lawrence Pratt, Stephen Paddison and, more recently, Matt Neurock and Vladimir Gurau. One conclusion from these efforts is that it is serious work to bring a serious computational study to fruition. There are few 'one size fits all' approaches that yield valuable results. Active coordination between experimentalists and modelers is difficult but ideal. Those researchers who publish many papers with a single modeling construct might delude themselves into thinking that their system provides a framework for a wide range of problems. However, such work often fails to deliver the punch line in terms of any deep insight because the reformulation in physically intuitive terms is sacrificed to the use of the solver.

My bias is that models are useful in that they isolate elements of the complex physics of reality and allow us to understand the contributions of different factors. This is difficult, if not impossible, in experimental studies, in which we usually work in a reductive manner. The most important thing to look for from models of chemical phenomena occurring in fuel cells is insight. Curve fitting is not modeling, but is possibly useful to system level engineers seeking simple algebraic expressions for transfer functions.

If I might, let me next briefly remind the reader of the obvious: A model is not reality, it is an (often gross) approximation to reality. Indeed, the very notion of modeling implies decisions on approximations and simplifications. Models are usually tuned to specific ranges or types of phenomena. They have their own internal logic and consistency and can tell us about trends to be expected from combinations of assumptions and parameters. Quantitative predictions are another matter. Occasionally, one hears a complaint (probably from the engineering side) that 'modeling does not yet get it right.' That implies that there is some all-encompassing, multiscale model to be had that will somehow bridge between messy, wet chemical items and mechanical engineering design needs. It is my contention that this is an overly ambitious role for modeling. The oft hoped-for hierarchical, multiscale models dreamt about in every workshop that I have attended on the subject may not be possible if it is to be based on stitching together models across a chain of ever longer length and time scales. One then also ends up with a hierarchy of assumptions. How these assumptions 'interact' can be quite complex. Assumptions that appear appropriate at one length scale may limit incorporation into a 'higher' level in the scheme. In many cases, it may be best to focus on specific phenomena of interest.

Much of the effort and often much of the value of a modeling exercise comes about from the process of formulating the model. In this process, the dissection of a complex phenomenon into component physics provides an instant level of insight even before any mathematics is done. It provides a framework for thought. The more clearly it is laid out for the reader, the better. There is a cautionary element to be added here. For the many of us who are visually oriented thinkers, the seduction of pictures may lead us down some blind alleys. This is particularly true for chemical schemes. Our pictures of interactions are often static, while processes of interest are of course *dynamic*. One example of this gap arises in thinking about proton transfers in hydrogen-bonded systems. The famous Bockris-Conway-Linden model scheme describing the Grotthus 'hopping' mechanism for proton transport in water was debunked long ago in its details. However, the mental picture of hand-off coupled to rotation is pervasive. More recently, evocative schemes showing lines of inter-molecularly hydrogen-bonded molecules are almost certainly grossly oversimplified and misleading.

We have difficulty formulating sufficiently detailed and accurate mental pictures of subtle processes. In water, for example, proton transfer takes place involving exchange of a proton between two primary water molecules, the

'Zundel' configuration, but an ineffable, fluctuating (I even use the word 'twinkling') network of weakly associated molecules likely supports this. Such weak and extended interactions rarely are referred to in the mental models or pictures provided to justify some new scheme proposed, for example, to deliver proton conduction in water-free environments. Yet it is the very weakness of the interactions that allows proton transfer to occur on a sufficiently rapid time scale. Germane to this discussion, it is the quality of such mental pictures and the correctness of the lessons drawn from them that will determine the quality of the resulting model. The latter factor cannot be over-emphasized.

A further key element of good modeling is validation of the model. Often, we will see a fit to a polarization curve from a fuel cell model. However, it is much more rare to see models pushed to provide simultaneous fits to multiple polarization curves under different conditions, or to other observables that might be tested as side-products. Rather than assert the validity of a model based on weak criteria, such as the fit to a single polarization curve, we should be trying to expose weaknesses and range of validity of the model. There is no shame in providing only a partial explanation to observed phenomena. Indeed, it is far more useful to the field to directly elucidate the weaknesses in one's own argument, indicating speculation where necessary. Many are the times when I find myself using published literature as a foil for discussion with my students or colleagues of how to construct a misleading argument.

One area in which a high degree of exactitude is possible, even the expected norm, is in quantum mechanics calculations. However, the proliferation of canned programs for various computational chemistry activities, such as those found in this volume, and their increasing use by non-specialists lead to another very important concern. The range of validity of a result needs to be carefully stated with such work. It is also difficult to know which methods are most appropriate for a given type of problem. For example, DFT calculations are frequently presented as a kind of gold standard of theory. Is the ubiquitous B3LYP functional combined with a 6-31G** basis set adequate to all problems? Is there, for example, a careful study that checks its accuracy for, say, depicting hydrogen-bonded structures? How do we assess such work? The combination mentioned above usually shows up in a methods section of a paper with little commentary on why we should believe that this approach is validated. Frankly, this is a form of obfuscation. At a time when these tools approach the level of routine for anyone willing to sign up to Gaussian or VASP's license agreements, it is most critical to ensure the correct use of various methods.

So, to sum up, in my view, the most important points helping the utility of your model for the rest of us are

1. Validation of results, especially of the rigorous and thorough-going variety.

2. Clear statements of the basic physics of the processes being modeled, the range and limitations of the model, and any assumptions.

Models are an excellent way to uncover aspects of physical reality. However, we cannot be too proud of our own creations, particularly of our pet hypotheses. I find this to be particularly true of modelers. After all, what else does the bedraggled theoretician have? Careful pruning of any inflated claims and self-analysis coupled to good communication will greatly enhance the contribution made via this exercise.

Thomas A. Zawodzinski Jr.
CWRU, Cleveland, Ohio

Book Overview

The design and operation of polymer electrolyte membrane (PEM) fuel cells requires the understanding and control of systems and processes which span fully 11 orders of magnitude in length scale and 17 orders of magnitude in time scales. In systems with widely dispersed, yet strongly interacting length and time scales, such as occurs in turbulent fluid flow as governed by the Navier Stokes equations, there has been only halting progress in developing simplified schemes which are amenable to analysis and computation. However, in the PEM setting the length and times scales naturally divide into groupings and in some instances the groups interact only weakly. In such a paradigm the groupings form steps on a ladder, each rung possessing its own governing equations, characteristic data, and observables. This dichotomy gives hope that an otherwise intractable problem will yield some of its secrets to analysis.

In a laddered paradigm, modeling-computational applications take information gathered from one length-scale grouping to make predictions about behavior at the next rung of the ladder. In PEM fuel cells, a key divide occurs around the one-micron length scale. On the supermicron length-scale porous and heterogeneous media can be averaged to give effective transport coefficients for diffusion, pressure-driven flow, and thermal conductivities. The ionomer membrane can be treated as a conductor with a specified resistance, dependent upon water content and perhaps temperature. This is the realm of the macrohomogenious models of the catalyst layer, and multiphase-flow models for the porous electrode. Piecing together systems of equations for electrodes, membrane, and flow fields yields unit cell models which can predict the transient impact of water-management strategies, reproduce AC impedance data, optimize flow field patterns, and model the interaction of the unit cell with a radiator or a humidifier; see the review articles by [1] and by [2] for excellent summaries. These unit cell models describe length scales from 10 μ meter up to 10's of centimeters and time scales from the millisecond to the half hour. Their purpose is to take the effective media characterizations of the constituent materials and to describe the response of individual cells so that optimal operational strategies can be developed, or to eliminate the slower time scales, so as to capture long-term loss of effective Pt surface area and membrane degradation.

Stack-level models lie on the next rung up on the ladder. For automotive or stationary applications unit cells are placed in stacks of 100–200. The individual cells are coupled, sharing current density and heat and influencing each other through oscillations in voltage and the sharing of fuel through common inlet headers. It is not possible to compute a couple model involving hundreds of detailed unit cells, the smallest length scales of the unit cell models must be averaged out or otherwise approximated. One approach is to replace a detailed model of a component of a unit cell, say the gas diffusion layer, with a single effective parameter for thermal conductivity or gas transport, ignoring two-phase effects and replacing the micron length scale with a millimeter length scale. Another approach is to simplify the detailed component models, removing complicating issues, and resolving the model asymptotically. This second approach amounts to deducing the effective parameters which describe the components from their governing equations, a procedure referred to as up-scaling. Stack models attempt to describe the propagation of anomalies from cell to cell, for example the influence of a large backing plate at the end of the stack or the impact of a "low cell" in the middle of a stack on its neighbors. Again the overall goal is to find operational strategies which optimize performance and longevity.

The unit cell and stack models, also called device models, lie above the micron cutoff, they take averaged material parameters and upscale them to make predictions on overall performance. Device models can be optimized over a range of material parameters, but they cannot explain how to construct the materials which have the desired parameters. The components of device models, Fick's and Maxwell-Stefan laws for diffusion, Darcy type laws for pressure-driven flows, together with saturation-capillary pressure relations, have a long history of successful modeling and validation. However the novel materials which comprise PEM fuel cells, and which give them their great efficiency, also behave far more subtly below the micron length scale than more traditional composite materials, such as rock, sand, or soils.

There are unresolved issues in device models. For example sorption isotherms, the waters per acid group measured at steady state under zero flux conditions as a function of ambient relative humidity and temperature, play a central role in modeling hydration levels in ionomer membranes. However, the sorption isotherms depend sensitively upon pretreatment of the membrane, see [3], and moreover there is no reliable data for hydration levels under the conditions typical of fuel cell operation in which significant water and ion flux pass through the membrane which is under mechanical constraints which impact its ability to swell.

The dependence of sorption isotherms upon pretreatment is an example of the hystereses which can bedevil models which rely on "effective parameter" reductions of subtle sub-scale processes. In delicate regimes, models at a particular length scale cannot self-validate against coarse-grained experimental data but must either be compared to data which resolve their smallest length scale or be supported by rigorous upscaling from sub-scale modes, or preferably both. It is not enough to duplicate gross device-level output, for example [4] show that one can accurately reproduce overall cell polarization curves with models based upon distinctly different loss mechanisms. The models produced the same cell power, but distribute the power production very differently through the cell.

The subtlety of the submicron behavior of ionomer materials impacts the modeling of key aspects of water management. It is common in PEM cell operation for the cathode to be wet (liquid water present) while the anode is dry (only water vapor present). In such a situation, at 80°C, sorption isotherms predict the cathode end of the membrane should hold almost double the water content of the anode. Diffusion-based models for water transport predict not only a smooth water concentration profile connecting the cathode side of the membrane to the anode side but also water fluxes of a magnitude that would either saturate the anode or dehydrate the cathode. Neither of these events are seen to occur within typical operational conditions. The diffusion-based models can be combined with, or replaced by, models of pressure-driven flow which tailor a capillary pressure to fit the desired outcome or which take a degenerative transport process in which some of the transport parameters become small or zero when bulk variables (pressure, concentration, activity) reach prescribed thresholds. However such models can only be considered as phenomenological until they have been justified by upscaling from sub-scale models or validated by data which resolve the length scales under question. Such experimental validation could be possible from in-situ MRI measurements of water distributions and fluxes on a 1–5 micron length scale which have been pioneered by [5].

The role of sub-micron, or materials modeling is both to give insight into the development of novel materials and to elucidate the mechanisms behind the averaged transport models of the super-micron scale. However much of the computational work in sub-micron level resides at the sub-nanometer level, while the ionomer membrane, catalyst layer, and, to a lesser extent, the gas diffusion layers of a PEM fuel cell all have rich structure on the nanometer-to-micron length scale. For many applications, this midscale is a missing rung on the modeling ladder. To focus the discussion, we consider the ionomer membrane, where the driving issue is to understand the mechanisms of water and ion transport, and particularly the development of membranes which retain their ionic conductivities at temperatures on the order of 120–140°C.

The fundamental difficulty with sub-micron modeling is the proliferation of length-scale niches, the ladder becomes a bit of a slide. At the level of 5–500 nanometers reside classical dynamic methods for atoms, which include many body interatomic potentials as well as the embedded atom method. At this length scale, Nafion looks like a crosslinked comb polymer in a polar solvent.

Given the chain statistics of the polymer branching and interaction energies, the self-consistent field theory calculates the phase separation into hydrophobic (backbone) and hydrophilic (pore) regions. On a cruder level, this can be simulated with a Ginzburg-Landau energy-minimizing approach; see the excellent introductory book by [6]. In the self-consistent field theory, one groups atoms into chains whose interactions are treated with techniques from statistical mechanics. The theory yields expressions for the free energy functional in terms of the polymer densities. Evaluating the free energy numerically requires several orders of magnitude more computation than typical at the super-micron scale. Approximations developed by linearizing around equilibrium states leads to expansions of the free energy whose variational derivatives yield Ginzburg-Landau-type partial differential equations for phase separation. However little work on this length scale has yet been done for the modeling of transport in ionomer membranes. The development of a compact and robust description of the phase-separation process within Nafion type materials would be a major advance.

Below the 5-nanometer length scale the pores of the phase-separated membrane come into resolution. The electric field generated by the charging of the double layers in the catalyst drives ions and water through the pores of the ionomer membrane, so the phase-separation process couples to the transport, as well as to the external mechanical stress imposed on fuel cells to maintain electrical contact and tighten mechanical seals. However, the coupling continues to even smaller length scales. Within the 2–4 nanometer hydrophilic regions the $-SO_3H$ acid groups of the pendant side-chains disassociate into tethered $-SO_3^-$ anions and free protons which interact strongly with the polar water. Classical double layer models for the interaction of cations with anions pose a Boltzmann distribution for the cations around the anions with an exponential decay which scales with the Debye length, typically 0.5–1 nanometer for the ionic molarities found in Nafion; see [7] for details. However, detailed electronic structure studies of the interactions of $-SO_3^-$ with water with protons, resolving below the angstrom level, show that the protons congregate not within the Debye length of the anions but within the bulk water in the center of the pore; see the survey article by [8]. This is an essential feature for which there is no current self-consistent non-atomistic models. It is the congregation of the protons within the bulk water at the pore center, where the waters are mobile and possess a hydrogen bond network conducive to hopping transport, that gives ionomer membranes their exceptional protonic conductivity at high hydration levels. Indeed at low hydration levels, below 3 waters per acid group, there is no bulk water within the membranes, protons reside among the waters of hydration surrounding the anions, and the ionic conductivity drops by two orders of magnitude.

At the nanoscale level and below, there are a wide variety of computational approaches, from orbital free density functional theories, which can handle on the order of 10^4 atoms, to tight-binding methodologies, electronic density functional theory, and finally ab initio quantum mechanics, which can handle no more than 10–100 atoms. The scale of computations at this level is enormous, with huge numbers of degrees of freedom arising from the electronic orbitals and time scales stretching from femtoseconds to picoseconds. However, without computations at this level, the mechanisms behind the fundamental properties of the ionomer membrane, its ionic conductivity, and water transport, have no first principles underpinning. An essential direction for future work is the robust coupling of angstrom-level models for reactivity to nanoscale models for polymer phase segregation.

References

[1] Wang, C.Y., Fundamental models for fuel cell engineering, *Chemical Reviews* **104** 4727–4765 (2004).
[2] Weber, A.Z., Newman, J., Modeling transport in polymer-electrolyte fuel cells, *Chemical Reviews* **104** 4679–4726 (2004).
[3] Hinatsu, J.T., Mizuhata, M., Takenaka, H., Water uptake of Perfluorosulfonic acid membranes from liquid water and from water vapor, *Journal of Electrochemical Society* **141** 1493–1498 (1994).
[4] Ju, H. and Wang, C.Y., Experimental validation of a PEM fuel cell model by current distribution data, *Journal of Electrochemical Society* **151** A1954–A1960 (2004).
[5] Ouriadov, A., MacGregor, R., and Balcom, B., Thin film MRI-high resolution depth imaging with a local surface coil and spin echo SPI, *Journal of Magnetic Resonance* **169** 174–186, (2004).
[6] Kawakatsu, T., *Statistical Physics of Polymers*, Springer, Hiedelberg (2004).
[7] Newman, J. and Thomas-Alyea, K., *Electrochemical Systems*, Wiley-Interscience, New York (2004).
[8] Kreuer, K.D., Paddison, S.J., Spohr, E., and Schuster, M., Transport in proton conductors for fuel-cell applications: simulations, elementary reactions, and phenomenology, *Chemical Review* **104** 4673–4678 (2004).

List of Contributors

Luis A. Agapito
Department of Chemical
  Engineering and Department of
  Electrical Engineering
Texas A&M University
College Station, TX 77843

B. Andreaus
Institute for Fuel Cell Innovation
National Research Council Canada
3250 East Mall, Vancouver, BC,
  V6T 1W5 Canada

Perla B. Balbuena
Department of Chemical
  Engineering
Texas A&M University
College Station, TX 77843

Jay Benziger
Department of Chemical
  Engineering
Princeton University
Princeton, NJ 08544

Sergio R. Calvo
Department of Chemical
  Engineering
Texas A&M University
College Station, TX 77843

B. Carnes
Institute for Integrated Energy
  Systems
University of Victoria
PO Box 3055 STN CSC
Victoria, BC, V8W 3P6, Canada

N. Djilali
Institute for Integrated Energy
  Systems
University of Victoria
PO Box 3055 STN CSC
Victoria, BC, V8W 3P6, Canada

M. Eikerling
Department of Chemistry
Simon Fraser University
Burnaby, V5A, 1S6, B.C., Canada

J. A. Elliott
Department of Materials Science
  and Metallurgy
University of Cambridge
Pembroke Street, Cambridge, CB2
  3QZ

J. Fimrite
Institute for Integrated Energy
  Systems
University of Victoria
PO Box 3055 STN CSC
Victoria, BC, V8W 3P6, Canada

xviii   List of Contributors

J. Fuhrmann

Dmitry Galperin
Physics Department
Moscow State University
Moscow 119899, Russia

K. Gärtner

S. Hazebroucq
Technische Universität Dresden
Institut für Physikalische Chemie
01069 Dresden, Germany

M. Head-Gordon
Department of Chemistry
University of California
Berkeley CA 94720

Pavel G. Khalatur
Institute of Organoelement
  Compounds, Russian Academy of
  Sciences, Moscow 117823, Russia
Department of Polymer Science,
  University of Ulm, Ulm D-89069,
  Germany

Alexei R. Khokhlov
Physics Department, Moscow State
  University, Moscow 119899,
  Russia
Institute of Organoelement
  Compounds, Russian Academy
  of Sciences, Moscow 117823,
  Russia
Department of Polymer Science,
  University of Ulm, Ulm D-89069,
  Germany

Marc T. M. Koper
Leiden Institute of Chemistry
Leiden University
PO Box 9502, 2300 RA Leiden, The
  Netherlands

A. A. Kornyshev
Department of Chemistry
Faculty of Physical Sciences
Imperial College London
London SW7 2AZ, UK

K. D. Kreuer
Max-Planck-Institut für
  Festkörperforschung
Heisenbergstr. 1, D-70569
  Stuttgart

A. A. Kulikovsky
Institute for Materials and Processes
  in Energy Systems (IWV–3)
Research Center "Jülich"
D–52425 Jülich, Germany

Eduardo J. Lamas
Department of Chemical
  Engineering
Texas A&M University
College Station, TX 77843

N. M. Markovic
Materials Sciences Division
Lawrence Berkeley National
  Laboratory
Berkeley CA 94720

Jeremy P. Meyers
UTC Fuel Cells

W. Münch
EnBW, Energie Baden-
  Wurttemberg AG
76131 Karlsruhe, Germany

Matthew Neurock
Department of Chemical
  Engineering
University of Virginia
Charlottesville, VA 22904

John Newman
Lawrence Berkeley National
  Laboratory and Department of
  Chemical Engineering
University of California
Berkeley, CA 94720-1462

S. J. Paddison
Department of Chemical &
  Biomolecular Engineering
University of Tennessee
Knoxville, TN USA 37996

Reginald Paul
Department of Chemistry
University of Calgary
2500 University Dr. N. W.,
  Calgary, Alberta, Canada
  T2N 1N4

Keith S. Promislow
Department of Mathematics
Michigan State University
East Lansing, MI USA 48824

P. N. Ross
Materials Sciences Division
Lawrence Berkeley National
  Laboratory
Berkeley CA 94720

Chandra Saravanan
Materials Sciences Division,
  Lawrence Berkeley National
  Laboratory, Berkeley
  CA 94720
Department of Chemistry,
  University of California, Berkeley
  CA 94720
Nanometrics Inc, 1550 Buckeye Dr,
  Milpitas CA 95035

G. Seifert
Technische Universität Dresden
Institut für Physikalische Chemie
01069 Dresden, Germany

Jorge M. Seminario
Department of Chemical
  Engineering and Department
  of Electrical Engineering
Texas A&M University
College Station, TX 77845

E. Spohr
Institute for Energy Process
  Technology (IWV-3)
Forschungszentrum Jülich
D-52425 Jülich, Germany

Jean St-Pierre
Future Fuels^{TM}
University of South Carolina
Columbia, SC 29208, USA

H. Struchtrup
Institute for Integrated Energy
  Systems
University of Victoria
PO Box 3055 STN CSC
Victoria, BC, V8W 3P6,
  Canada

Christopher D. Taylor
Department of Engineering
  Physics
University of Virginia
Charlottesville, VA 22904

Yixuan Wang
Department of Chemical
  Engineering
Texas A&M University
College Station, TX 77843

Sally A. Wasileski
Department of Chemistry
University of North Carolina
at Asheville
Asheville, NC 28804

Adam Z. Weber
Lawrence Berkeley National
  Laboratory and Department
  of Chemical Engineering

University of California
Berkeley, CA
94720-1462

Brian Wetton
Mathematics Department, UBC

Thomas A. Zawodzinski Jr.
Department of Chemical Engineering
Case Western Reserve University
Cleveland, Ohio

Part I
Device Modeling

Section Preface

JEAN ST-PIERRE

Introduction

PEMFCs are a potential power source alternative for a wide range of applications. Their development has progressed significantly during the last few years driven both by political will and the perceived market potential for fuel cell technology penetration into the economy. The U.S. president's Hydrogen Fuel Initiative: A Clean and Secure Energy Future was announced early in 2003 [1]. Fuel Cells Canada, a non-profit organization, promotes and supports the industry [2]. Fuel cell systems are available on a limited basis (continuous mass manufacturing is not yet a reality) from a number of developers [3]. Future increases in market penetration are reliant upon costly additional technology development and implementation of a fuel infrastructure. Since revenues from fuel cell production is not yet sufficient to keep the industry solvent, a cost-effective strategy is essential for future progress.

Mathematical modeling represents a potential approach, which has already gained significant recognition as evidenced by a number of recent reviews [4, 5, 6, 7] and a forthcoming book [8]. A recent survey article by Haraldsson and Wipke analyzes 7 competing commercial software packages and 32 literature models covering a period up to 2002 [4]. Yao et al's review is more extensive and conveniently summarizes many of the effective/constitutive relations upon which macro-scale models are built [5]. Wang's review focuses on computational fluid dynamics (CFD)-based models and includes work up to 2004 [6]. Cheddie and Munroe's review is more narrowly focused on the solution strategy, comparing and discussing both single-domain and multi-domain approaches [7]. Mathematical modeling requires both funding and time [4], for maximum benefit, a sustained, focused program is most appropriate.

In view of the significant modeling activities and number of active groups worldwide, it is useful to assess the current status to ensure future developments will address PEMFC areas of major concern. This is especially important considering that PEMFCs are at a critical juncture. Large capital amounts have already been spent during the last ten years with even larger sums likely required to achieve mass manufacture and develop a fuel infrastructure. In this communication, future modeling needs are discussed and supported by a focused literature search. A number of mathematical model evaluation criteria are also introduced and some of these are used to analyze recent model activities.

Model Evaluation Criteria

PEMFC Development Needs

PEMFC current development needs are relatively well documented and easily accessible by browsing the US Department of Energy web site [9–11]. The four major areas of PEMFC development needs are succinctly summarized as:

- Material cost reduction (catalysts, gas diffusion media, proton exchange membranes, bipolar plates)
- Degradation rate improvement (including applications)
- Freezing tolerance including start-up
- Water-management optimization within gas diffusion electrodes and flow field channels

Mathematical Model Development Advantages and Features

Advantages and features are regularly mentioned in introductory comments of model-related work [7]. These are summarized as follows:

- Cost savings associated with reduced experimental test plans and, faster design and analysis cycles
- Faster design and analysis cycles
- More focused and effective experimental test plans
- Potential to derive new experimental techniques
- Validation needed to build confidence in model predictions

Cost savings are realized after an initial investment in model-development activities. A faster design and analysis cycle is a potential strategic advantage which is closely tied to knowledge and fundamental understanding acceleration. Although modeling cannot replace experimentation, a focus change from design evaluation under expected operating conditions to model parameter evaluation under the same operating condition range has the potential to accelerate and reduce testing needs. Other features, such as code speed, code flexibility/adaptability (modular aspect), and documentation, are also important but are not discussed in detail due to the limited scope of the present communication.

Literature Search

Several lists of PEMFC mathematical models were generated from the basic search engine of the Journal of Power Sources [12] covering the 1994 to the present range. The first searches were performed to determine the yearly evolution of model appearances (search A). The second search was performed to extract more detailed information from the recent record (search B). More details are provided about these searches (completed on March 13, 2006) in Table 1. Searches A and B provided a number of papers addressing a range of topics that did not meet the initial objectives. Thus, papers addressing topics

TABLE 1 Literature survey search criteria summary.

Basic Search Field	Basic Search Field Content	
	Search A	Search B
Term(s)	Model or Modeling or Modeling or Simulation or Analysis or Prediction or Optimization or Calculation And PEMFC or PEM fuel cell or PEFC or Proton exchange membrane fuel cell or Polymer electrolyte membrane fuel cell or Polymer electrolyte fuel cell or Polymeric fuel cell or Polymeric membrane fuel cell or Direct methanol fuel cell or DMFC	
Within	Abstract	
Source	Subscribed journals*	Subscribed journals* including articles in press
Subject	All sciences	
Limit by document type	Article	
Dates	Selected calendar year to same selected calendar year. Selected calendar years spanned the 1994 to 2005 range	2005 to present

* Journal of Power Sources

such as hybrid technologies, regenerative systems, measurement methods, and fuels other than hydrogen, methanol, and reformate were not considered. Only device-level models were found as opposed to material-level models (at the molecular or group of molecules scale).

Present and Future Modeling Status

General Trends

Figure 1 illustrates the results of search A to which some of the results of search B were added (model published in 2006 or in press). For the case of PEMFC, it is clear that mathematical modeling has gained recognition in the last 2–3 years due to its multiple advantages. The trend does not show any sign of slowing down. However, the same situation does not exist for the direct methanol fuel cell (DMFC). Although there is an upward trend in the number of papers published, there is no evidence that a sudden surge of interest is currently developing. This is rather surprising, considering that the DMFC was considered the prime candidate for early release into commercial production for portable applications. Product release or commercialization dates were recently quoted, spanning the 2006–2007 range [13, 14]. By comparison, PEMFCs for automotive applications are not expected to be commercialized before 2020, being preceded by bus and government fleets during the 2010–2020 period [9].

Figure 2 illustrates the results of search B. Two major trends are evident. In the first instance, degradation and freezing do not seem to receive as much attention as required to push PEMFC technology to a commercial success. Several reasons could be invoked to explain this observation, including insufficient physical understanding of the phenomena to formulate meaningful models, significantly expensive and time-consuming experimentation, and lack of interest potentially founded on the perception that the field is not sufficiently fundamental. If these assumptions are correct, these barriers need

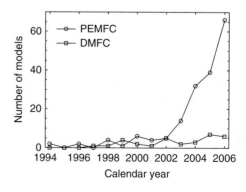

FIGURE 1. Evolution of the number of published PEMFC and DMFC mathematical models in the Journal of Power Sources. Two papers addressed both PEMFC and DMFC aspects (1 in 2003 and 1 in 2004). Therefore, these papers were recorded once in each category.

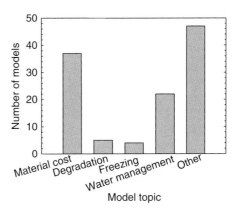

FIGURE 2. Topical distribution of PEMFC (including DMFC) mathematical models for the 2005 to the present period. 155 papers were examined, 3 of them having duplicate objectives, which were both recorded in the appropriate topical categories.

to be timely addressed. The second major observed trend relates to a significantly large group of researchers seemingly working on aspects that are not presently essential to PEMFC commercial success (the 'other' category in Figure 2). The existence of such a group could provide the necessary resources to intensify the research in the degradation and freezing areas assuming that some of the identified issues can be satisfactorily addressed.

Mathematical model development advantages were not discussed in the selected literature and the analysis was therefore dropped. A detailed analysis during the course of a specific model development and implementation phases would be beneficial to demonstrate the value of mathematical modeling.

Material Cost

Many catalyst layer models have appeared in the literature during the last few years [15, 16, 17, 18, 19, 20, 21]. This observation partly explains the complications associated with this topic. Still, much work remains to be completed since many effects have not yet been included, such as proton surface diffusion (outside the ionomer, [22,23]) and ionomer density (water content effect), which effectively and respectively increases/modifies the reactive surface area. The surface-sensitive nature of Pt catalysts on the oxygen reduction reaction rate [24] and electrochemical promotion (a catalytic effect, [25]) represent other examples which can also affect the reaction rate and surface area. All these effects are further compounded by the potential presence of liquid water which effectively modifies the reaction front, access to specific catalyst particles and surface properties.

Gas diffusion electrodes have been characterized with the objective to link structural parameters (such as permeability, fraction of hydrophobic pores, pore size distribution and volume, and catalyst layer thickness and composition [26, 27, 28]) to cell performance. Although this information is valuable to validate existing gas diffusion electrode models [29, 30, 31], the link between structural parameters and cell performance is not necessarily direct. Most gas diffusion electrode characterization methods are destructive. As a result, tests can only be completed with related samples (same material batch) or after cell operation (physical/chemical procedures need to be developed for obtaining the required part of the gas diffusion electrode which could affect its structure/composition). Therefore, errors may be introduced in derived correlations. Errors are also compounded by factors such as membrane/electrode assembly bonding and cell compression (affecting, for example, porosity), which are not necessarily taken into account during characterization. Development of in situ structural and compositional characterization methods (along the x, y, and z directions, Figure 3) would benefit both the optimization of gas diffusion electrodes and manufacturing quality control.

The search for new and cheaper membrane chemistries is partly fulfilled by the development of material level models, which is not part of the present review. Other membrane-related aspects such as degradation and water management are covered in succeeding sections.

Flow field models are available for conventional channel designs [32, 33] as well as for prototype designs, including foam [34] and fractal designs [35]. However, the turn-down ratio of such structures (the ability to function over a wide range of current densities) does not appear to have been addressed despite its importance. Flow fields are expected to provide low pressure drop (especially at high current densities), as well as effectively manage reactant distribution over the whole active surface area and liquid water removal (especially at low current densities). These conflicting requirements can be addressed by increasing reactant stoichiometry at low current densities, negatively impacting efficiency. Flow field designs able to access increasingly larger portions of the active surface as current density rises (geared flow field) represents an alternative. Design, modeling and optimization of such a flow field would positively impact cell operability and efficiency.

Degradation

Degradation can take many forms. In most cases, the structure and chemistry of the materials can severely be affected. Development of predictive models is therefore significantly more challenging due to the added time evolution of

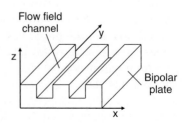

FIGURE 3. Coordinate system definition.

model parameters. Additionally, characterization of lifetime degradation is a time-consuming and costly activity presently resulting in limited access to validation data sets. This issue is further compounded by the absence of in situ structural and compositional characterization methods (as already mentioned in relation to material cost research needs), requiring test duplication to carry out destructive tests at scheduled intervals. For these reasons, this area offers many opportunities and even greater fundamental challenges for model development (Figure 2).

Degradation encompasses areas as diverse as bipolar plate corrosion affecting contact resistance by the appearance of surface films [36], catalyst particles agglomeration reducing the active surface area [37], platinum dissolution and subsequent movement in the membrane also affecting active surface area [38, 39], platinum oxidation modifying the oxygen reduction reaction rate [40], carbon oxidation leading to surface groups modifying component hydrophobicity, water management and mass transport [41, 42, 43], and contamination originating from reformate fuel species reducing anode active surface area [44, 45, 46]. Several cases were modeled, including platinum dissolution and its subsequent movement [38, 39] and CO [44, 45, 46] and CO_2 contamination [47]. A general degradation model has also been proposed [48]. Still, many aspects have remained unexplored. Whereas CO and CO_2 contaminants can reversibly adsorb on catalyst surfaces, cases such as irreversible adsorption, subsequent potential-dependent reactions, cathode contaminants, and multi-contaminant cases also need to be investigated. Membrane contamination, which can occur as a result of exposure to marine environments or winter deicing agents, has not been addressed despite the negative effects of cations on both membrane conductivity and mass transport properties (water content, electro-osmotic drag, [41]). The physical loss of polytetrafluoroethylene from the gas diffusion media surface [42] impacting water management in both flow field channels and porous media is poorly documented (models do not appear to have been proposed). Another potential explanation for the appearance of a mass transport effect relates to the need for stack assembly mechanical compression [49], which can result in porous material relaxation reducing porosity. The presence of carbon oxidation as a result of partial or complete fuel starvation requires reaction kinetic modifications to adequately predict cell performance (voltage and current distribution). Some exploratory work has recently been completed in this direction [50], but more is needed to capture materials structural and compositional changes and integrate the parallel reaction framework to unit cell performance models. Membrane degradation and more specifically the kinetics associated with side reactions (radical and peroxide formation and their subsequent reactions with the ionomer) is another example of a fundamental issue. Membrane degradation rate predictions based on fundamental knowledge are not currently possible since rate constants, intermediates transport properties and intermediates space distribution within the polymer are either unavailable or poorly understood. Many of these examples do not necessarily require tests in fuel cells to yield relevant results being much more amenable to bench type experimentation and consequently lower investment in both time and cost. A recent cell durability survey [51] could complement the cited literature and form a basis to initiate research activities in this critical area.

Freezing

Freezing can be viewed as a special case of degradation requiring hundreds of repetitive freeze storage and freeze start-ups to meet automotive requirements. Therefore, this area is subject to the same limitations of costly and time-consuming tests, and relative absence of in situ structural and compositional measurement methods. The result of this situation is a challenging and varied opportunity for fundamental research and discovery (Figure 2).

A recent review [52] suggests many of the modeling needs. Heat generation methods, an essential aspect of a successful freeze start, such as reactant starvation, also requires integration of an additional electrochemical reaction framework to the model (beyond oxygen reduction and hydrogen oxidation) to predict cell performance and heat management especially near the outlet where the starvation will have the most effect (hydrogen evolution will occur below a stoichiometry of 1). Ice damage is partly mitigated by purging the cell with a dry gas after operation. Purging also limits the appearance of mass transport losses with repetitive freeze cycles. Tracking water should therefore extend beyond the time during which the cell is operated (initial conditions influence the outcome of a freeze start). The link between the observed mass transport loss and water management is further reinforced by the results of heat treatment (operating the cell beyond its specified operating temperature) to recover performance. This observation suggests that freezing modifies water content, location or state within the membrane/electrode assembly. The characteristics associated with this water have not yet been clearly established. Ethylene glycol, the present automotive coolant, can be adapted to fuel cell systems. Degradation by auto-catalytic reactions (exposure to oxygen) is partly minimized by limiting the effect of the glycolic acid intermediate. Modeling of the several parallel and serial reaction pathways in combination with mitigation approaches could lead to coolant life predictive capabilities.

An end cell effect was recently predicted during a freeze start (end cells are cooler due to the thermal mass associated with the end plate assembly, [53]). This situation suggests that stack models offer advantages over single cell models. Relatively little activity has been reported in the stack modeling area. More experimental studies are required to define issues, assess degradation and further understanding for modeling purposes, especially for large numbers of thermal cycles and for complete systems, since the literature is relatively scarce [52, 53, 54, 55, 56]. Of particular concern is the need for property values below the water freezing point such as heat transfer through porous materials, water content and membrane conductivity (including model to correlate data), and coefficients of expansion (to predict stresses in stack assemblies). Ice formation, quantification and its localization within the fuel cell porous materials is an example of another fundamental issue that could benefit from additional attention. As for degradation needs, many of these examples do not necessarily require tests in fuel cells.

Water Management

Many water-management advances have been made (Figure 2). However, fundamental properties associated with two-phase flow in porous media, including capillary pressure, liquid water content and, intrinsic and relative permeabilities, are still lacking and are needed for predictive models. This situation is due to the absence of measurement methods for thin materials. Contact porosimetry [57] does not presently allow measurement of the permeabilities or during wetting (to evaluate the presence of hysteresis). Fluorescence microscopy [58] can provide additional information about liquid water flow through porous media.

Membrane density change as a result of water absorption/desorption has hardly been included in models and is partially linked to mechanical compression [59] and possibly even Schroeder's paradox [60]. The effect of mechanical compression on other membrane properties, such as water diffusion and electroosmotic drag, would be equally important and should include testing over the targeted applications range of operating conditions. The current trend to develop high temperature membranes suggests particular attention should be given to develop efficient and fast measurement methods since currently tested materials are likely going to change. The acquired knowledge should accelerate the understanding of the next generation material class.

The appearance of flow visualization methods [61, 62, 63, 64] has made possible the study of two-phase flows in flow field channels. These methods should be perfected considering the potential measurement artifacts introduced by the transparent element (change in thermal and current distribution, and flow field channel surface properties). Mathematical representations of the pressure drop in presence of two-phase flow will be needed to modify existing stack reactant flow distribution models [65].

Validation

A closer examination of the search B results with respect to the model dimension and associated validation activities led to Figure 4. Dimension here refers to the number of coordinate system axes used to derive a model or to the number of measurement method capabilities to resolve selected axial directions considered for model validation (Figure 3). Again, several major

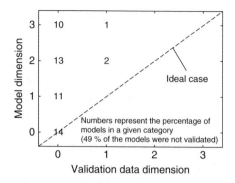

FIGURE 4. Percentage distribution of the PEMFC (including DMFC) mathematical models for the 2005 to the present period according to the dimension of both model and validation data.

trends are observed. Models were developed to all detail levels (0–3 dimensional) providing the desired level of completeness depending on the need, from scaling fuel cell system components to designing flow fields and gas diffusion electrodes. More importantly, little model validation was completed since 49% of the models were not validated to any level and another 37% were validated to a level less than ideal (validation was not completed for all coordinate axes used to derive a particular model). The number of 0 dimensional model validated using 0 dimensional data only reaches 14 % of all the models published in 2005 to the present (the only point located on the ideal case curve of Figure 4). This observation may be attributable to the relatively low investment in time and cost associated with model development (no need for a fuel cell, support systems, measurement equipment and laboratory space). However, such a situation is not ideal, negatively impacting the need to build confidence in model predictions to gain acceptance. This is especially true for 3 dimensional models, which, in the best cases, were validated with data resolving the channel length coordinate (a 2 dimensional difference from the ideal case curve).

The most common 0 dimensional validation data set is the polarization curve. The situation is quite different for methods sensitive to the along-the-channel y direction (Figure 3), the most common 1 dimensional measurement method. Although methods able to measure impedance spectra, current density, and species concentration are available [66, 67, 68, 69, 70, 71], their use have not yet significantly been adopted for model validation (Figure 4). This situation arose despite the fact that some of these methods have been available for quite some time. Cost, availability, and fuel cell design specificity of existing equipment may represent significant hurdles toward general use of these methods. Potential mitigation approaches could include measurement equipment standardization including design of a flexible fuel cell interface and intensified collaboration between research groups. None of the surveyed models was validated using x- or z-sensitive (Figure 3) measurement methods. Neutron imaging [72, 73] could eventually resolve liquid water content in the x direction. The current resolution does not appear to be sufficient for model validation and the resulting data are confounded (average over the membrane/electrode assembly thickness including liquid water in both gas diffusion electrodes). Neutron imaging can also be used as a z-sensitive method as demonstrated for membrane water content [74]. However, the method has presently insufficient resolution to contribute relevant validation data. Magnetic resonance imaging offers greater resolution than neutron imaging. However, liquid water is a complication that needs to be minimized thus significantly limiting the scope of research activities. The potential of magnetic resonance imaging to resolve the z direction was demonstrated for the membrane water content [75, 76]. Recently, a better resolution of 5 microns was achieved in the z direction (sufficient to resolve a gradient in a thin 25 μm membrane), and the x and y capability was also demonstrated [77].

Improved model validation requires the continuous development of existing measurement methods (x and z direction) and the search for new approaches. The relevance of x- and z-sensitive methods is best illustrated by reiterating some existing problems. The temperature distribution within a membrane/electrode assembly has been modeled (z direction) and is useful, for example, to understand degradation mechanisms (in particular as it relates to the membrane glass transition temperature and the formation of pinholes). However, the localization of the heat of reaction, which can influence the temperature distribution, is still debated [78]. The hydrogen oxidation reaction was previously assumed to be isothermal, exothermic or endothermic [79]. The availability of a z-sensitive temperature measurement method could potentially resolve the debate and contribute toward a better understanding of degradation mechanisms. A z-sensitive reactant concentration measurement method would be useful to discriminate between the numerous catalyst layer models. The distribution of liquid water within gas diffusion electrodes was predicted [80, 81]. Computational results have shown a preferential accumulation under the flow field landings limiting reactant transport at these locations. Confirmation of these findings by an x-sensitive measurement method would contribute toward the design of more effective flow fields and gas diffusion electrodes for high current density operation (a cost reduction approach).

Data used to construct Figure 4 were separated into the four development needs (Figure 5). Two major differences from Figure 4 are observed. Low dimensional order models were developed for degradation and freezing. Additionally, only average values were used to validate these models (y-sensitive measurement methods have apparently not been used). These observations support the conclusions reached in the previous sections. More attention should be given to develop more detailed and better validated degradation and freezing models.

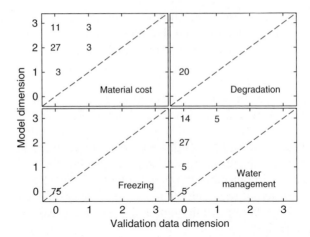

FIGURE 5. Percentage distribution of the PEMFC (including DMFC) mathematical models for the 2005 to the present period according to the dimension of both model and validation data, and topic. The dotted line represents the ideal case. Numbers represent the percentage of models in a given category. 54, 80, 25, and 45% of, respectively, material cost, degradation, freezing, and water-management models were not validated. 65 papers were examined, 3 of them having duplicate objectives, which were both recorded in the appropriate topical categories.

Conclusion

Future modeling needs, and analysis of the mathematical model literature search results on the basis of current PEMFC technology development needs and model features, have revealed the existence of several important gaps at the fundamental level. PEMFC degradation and freezing aspects are relatively and poorly addressed despite the existence of unresolved fundamental aspects, the emergence of a significant literature and the potentially low cost for many of the required experiments. Modeling needs also extend beyond electrochemical reactor engineering, which will stretch and challenge the experience of existing research groups. Further acceptance of PEMFC mathematical model activities requires a more widespread use and combination of the measurement methods available for validation, and the continuous development of additional capabilities, especially x- and z-measurement methods. Of particular relevance would be the development of in situ, structural, compositional and operational parameter-sensitive measurement methods to more directly relate cell performance to fuel cell design characteristics. Mathematical models have significantly preceded experimental fuel cell knowledge requiring the generation of extensive model parameter data sets covering the whole fuel cell applications operating condition and design ranges. A more detailed and larger scope literature search covering more relevant technical journals and mathematical model advantages would be worthwhile, potentially revealing additional clues to reconsider the present research directions.

References

[1] www.hydrogen.energy.gov/presidents_initiative.html.
[2] www.fuelcellscanada.ca.
[3] www.fuelcellmarkets.com/fuel_cell_markets/industry/2,1,1,7.html.
[4] K. Haraldsson, K. Wipke, *J. Power Sources*, **126** 88 (2004).
[5] K. Z. Yao, K. Karan, K. B. McAuley, P. Oosthuizen, B. Peppley, T. Xie, *Fuel Cells*, **4** 3 (2004).
[6] C.-Y. Wang, *Chem. Rev.*, **104** 4727 (2004).
[7] D. Cheddie, N. Munroe, *J. Power Sources*, **147** 72 (2005).
[8] E. Fontes, G. Lindbergh, C. Oloman, *Fuel Cell Modelling Handbook*, Elsevier, 2010.
[9] www.eere.energy.gov/hydrogenandfuelcells/about.html.
[10] www.hydrogen.energy.gov/pdfs/hydrogen_posture_plan_dec06.pdf.
[11] www1.eere.energy.gov/hydrogenandfuelcells/mypp/pdfs/fuel_cells.pdf.
[12] www.sciencedirect.com/science/journal/03787753.
[13] mobilemag.com/content/100/102/C3790.
[14] www.3g.co.uk/PR/Sept2005/1917.htm.
[15] A. A. Kulikovsky, *Electrochem. Commun.*, **4** 318 (2002).
[16] F. Jaouen, G. Lindbergh, G. Sundholm, , *J. Electrochem. Soc.*, **149** A437 (2002).
[17] K. T. Jeng, C. P. Kuo, S. F. Lee, *J. Power Sources*, **128** 145 (2004).
[18] Z. N. Farhat, *J. Power Sources*, **138** 68 (2004).
[19] W. Sun, B. A. Peppley, K. Karan, *Electrochim. Acta*, **50** 3359 (2005).
[20] K.-M. Yin, *J. Electrochem. Soc.*, **152** A583 (2005).
[21] D. Song, Q. Wang, Z. Liu, M. Eikerling, Z. Xie, T. Navessin, S. Holdcroft, *Electrochim. Acta*, **50** 3347 (2005).
[22] J. Mcbreen, *J. Electrochem. Soc.*, **132** 1112 (1985).
[23] W.-J. Liu, B.-L. Wu, C.-S. Cha, *J. Electroanal. Chem.*, **476** 101 (1999).
[24] A. Gamez, D. Richard, P. Gallezot, F. Gloaguen, R. Faure, R. Durand, *Electrochim. Acta*, **41** 307 (1996).
[25] D. Tsiplakides, S. G. Neophytides, O. Enea, M. Jaksic, C. G. Vayenas, *J. Electrochem. Soc.*, **144** 2072 (1997).
[26] E. Gülzow, M. Schulze, N. Wagner, T. Kaz, R. Reissner, G. Steinhilber, A. Schneider, *J. Power Sources*, **86** 352 (2000).
[27] D. A. Blom, J. R. Dunlap, T. A. Nolan, L. F. Allard, *J. Electrochem. Soc.*, **150** A414 (2003).
[28] M. V. Williams, E. Begg, L. Bonville, H. R. Kunz, J. M. Fenton, *J. Electrochem. Soc.*, **151** A1173 (2004).
[29] H.-S. Chu, C. Yeh, F. Chen, *J. Power Sources*, **123** 1 (2003).
[30] K. T. Jeng, S. F. Lee, G. F. Tsai, C. H. Wang, *J. Power Sources*, **138** 41 (2004).
[31] A. Z. Weber, J. Newman, *J. Electrochem. Soc.*, **152** A677 (2005).
[32] A. C. West, T. F. Fuller, *J. Appl. Electrochem.*, **26** 557 (1996).
[33] S. W. Cha, R. O'Hayre, Y. Saito, F. B. Prinz, *J. Power Sources*, **134** 57 (2004).
[34] A. Kumar, R. G. Reddy, *J. Power Sources*, **114** 54 (2003).

[35] S. M. Senn, D. Poulikakos, *J. Power Sources*, **130** 178 (2004).
[36] D. P. Davies, P. L. Adcock, M. Turpin, S. J. Rowen, *J. Appl. Electrochem.*, **30** 101 (2000).
[37] M. S. Wilson, F. H. Garzon, K. E. Sickafus, S. Gottesfeld, *J. Electrochem. Soc.*, **140** 2872 (1993).
[38] R. M. Darling, J. P. Meyers, *J. Electrochem. Soc.*, **150** A1523 (2003).
[39] R. M. Darling, J. P. Meyers, *J. Electrochem. Soc.*, **152** A242 (2005).
[40] C. H. Paik, T. D. Jarvi, W. E. O'Grady, *Electrochem. Solid-State Lett.*, **7** A82 (2004).
[41] J. St-Pierre, D. P. Wilkinson, S. Knights, M. L. Bos, *J. New Mater Electrochem. Syst.*, **3** 99 (2000).
[42] J. St-Pierre, N. Jia, *J. New Mat. Electrochem. Systems*, **5** 263 (2002).
[43] K. H. Kangasniemi, D. A. Condit, T. D. Jarvi, *J. Electrochem. Soc.*, **151** E125 (2004).
[44] T. E. Springer, T. Rockward, T. A. Zawodzinski, S. Gottesfeld, *J. Electrochem. Soc.*, **148** A11 (2001).
[45] K. K. Bhatia, C.-H. Wang, *Electrochim. Acta*, **49** 2333 (2004).
[46] S. Enbäck, G. Lindbergh, *J. Electrochem. Soc.*, **152** A23 (2005).
[47] G. J. M. Janssen, *J. Power Sources*, **136** 45 (2004).
[48] A. A. Kulikovsky, H. Scharmann, K. Wippermann, *Electrochem. Commun.*, **6** 75 (2004).
[49] S.-J. Lee, C.-D. Hsu, C.-H. Huang, *J. Power Sources*, **145** 353 (2005).
[50] B. Wetton, G.-S. Kim, K. Promislow, J. St-Pierre, in *FUELCELL2006 – Fourth International Conference on Fuel Cell Science, Engineering and Technology*, ASME, paper FUELCELL2006-97027.
[51] D. P. Wilkinson, J. St-Pierre, in *Handbook of Fuel Cells – Fundamentals, Technology and Applications*, Vol. 3 – Fuel Cell Technology and Applications, Part 1, Chapter 47, Edited by W. Vielstich, H. Gasteiger, A. Lamm, John Wiley and Sons, 2003, p. 611.
[52] J. St-Pierre, J. Roberts, K. Colbow, S. Campbell, A. Nelson, *J. New Mater Electrochem. Syst.*, **8** 163 (2005).
[53] M. Sundaresan, R. M. Moore, *J. Power Sources*, **145** 534 (2005).
[54] B. K. Datta, G. Velayutham, A. P. Goud, *J. Power Sources*, **106** 370 (2002).
[55] E. Cho, J.-J. Ko, H. Y. Ha, S.-A. Hong, K.-Y. Lee, T.-W. Lim, I.-H. Oh, *J. Electrochem. Soc.*, **150** A1667 (2003).
[56] R. C. McDonald, C. K. Mittelsteadt, E. L. Thompson, *Fuel Cells*, **4** 208 (2004).
[57] J. T. Gostick, M. W. Fowler, M. A. Ioannidis, M. D. Pritzker, Y. M. Volfkovich, A. Sakars *J. Power Sources*, **156** 375 (2006).
[58] S. Litster, D. Sinton, N. Djilali, *J. Power Sources*, **154** 95 (2006).
[59] A. Z. Weber, J. Newman, *AIChE J.*, **50** 3215 (2004).
[60] P. Choi, R. Datta, *J. Electrochem. Soc.*, **150** E601 (2003).
[61] K. Sugiura, M. Nakata, T. Yodo, Y. Nishiguchi, M. Yamauchi, Y. Itoh, *J. Power Sources*, **145** 526 (2003).
[62] K. Tüber, D. Pócza, C. Hebling, *J. Power Sources*, **124** 403 (2006).
[63] A. Hakenjos, H. Muenter, U. Wittstadt, C. Hebling, *J. Power Sources*, **131** 213 (2004).
[64] X. G. Yang, F. Y. Zhang, A. L. Lubawy, C. Y. Wang, *Electrochem. Solid-State Lett.*, **7** A408 (2004).
[65] P. A. C. Chang, J. St-Pierre, J. Stumper, B. Wetton, *J. Power Sources*, **162** 340 (2006).

[66] D. J. L. Brett, S. Atkins, N. P. Brandon, V. Vesovic, N. Vasileiiadis, A. Kucernak, *Electrochem. Solid-State Lett.*, **6** A63 (2003).
[67] X.-G. Yang, N. Burke, C.-Y. Wang, K. Tajiri, K. Shinohara, *J. Electrochem. Soc.*, **152** A759 (2005).
[68] Q. Dong, J. Kull, M. M. Mench, *J. Power Sources*, **139** 106 (2005).
[69] S. J. C. Cleghorn, C. R. Derouin, M. S. Wilson, S. Gottesfeld, *J. Appl. Electrochem.*, **28** 663 (1998).
[70] J. Stumper, S. A. Campbell, D. P. Wilkinson, M. C. Johnson, M. Davis, *Electrochim. Acta*, **43** 3773 (1998).
[71] A. Hakenjos, C. Hebling, *J. Power Sources*, **145** 307 (2005).
[72] R. Satija, D. L. Jacobson, M. Arif, S. A. Werner, *J. Power Sources*, **129** 238 (2004).
[73] D. Kramer, J. Zhang, R. Shimoi, E. Lehmann, A. Wokaun, K. Shinohara, G. G. Scherer, *Electrochim. Acta*, **50** 2603 (2005).
[74] R. J. Bellows, M. Y. Lin, M. Arif, A. K. Thompson, D. Jacobson, *J. Electrochem. Soc.*, **146** 1099 (1999).
[75] S. Tsushima, K. Teranishi, S. Hirai, *Electrochem. Solid-State Lett.*, **7** A269 (2004).
[76] K. Teranishi, S. Tsushima, S. Hirai, *Electrochem. Solid-State Lett.*, **8** A281 (2005).
[77] A. V. Ouriadov, R. P. MacGregor, B. J. Balcom, *J. Magn. Reson.*, **169** 174 (2004).
[78] J. Ramousse, J. Deseure, O. Lottin, S. Didierjean, D. Maillet, *J. Power Sources*, **145** 416 (2005).
[79] M. J. Lampinen, M. Fomino, *J. Electrochem. Soc.*, **140** 3537 (1993).
[80] D. Natarajan, T. V. Nguyen, *J. Electrochem. Soc.*, **148** A1324 (2001).
[81] T. Berning, N. Djilali, *J. Electrochem. Soc.*, **150** A1589 (2003).

1
Modeling of PEMFC Catalyst Layer Performance and Degradation

Jeremy P. Meyers

1.1. Introduction

The proper construction of a stable, well-dispersed, three-dimensional catalyst layer is one of the most critical determinants of performance for a PEM fuel cell. The membrane isolates the reactants from one another and provides an ionic current path from one electrode to another, and the flow fields and gas-diffusion layers distribute the reactants to the catalyst layer, but all of the relevant electrochemical reactions are carried out in the catalyst layers themselves. It is the proper construction of the so-called three-phase interface that allows the reactants and products to be brought into intimate contact and makes possible the operation of the fuel cell. Indeed, it is the tailoring of this layer by Raistrick et al. [1] in 1991 that demonstrated the practical feasibility of lowering precious metal loadings by a factor of 40 over previous designs and helped to usher in the past decade of increased activity and investment in fuel cell development.

As is common knowledge and a common lament for PEM fuel cell technologists, the kinetics of the oxygen reduction reaction (ORR) in acid conditions and at temperatures where proton-exchange membranes can operate (generally 60–100°C) are quite poor. On the anode, the search for catalysts that are resistant to poisoning by CO—a species formed either as a byproduct of the fuel-reforming process or as an intermediate in the direct oxidation of methanol and other fuels—has proven to be a challenge as well [2]. The difficulties in finding appropriate electrocatalysts for oxygen reduction in acid electrolytes and CO-tolerant anode catalysts have resulted in the fact that essentially all PEM fuel cells of practical interest are constructed with platinum or platinum alloys for catalysts.

In order to reduce the material costs of fuel cells to meet automotive cost targets, it is necessary to lower the total amount of precious metals in the system [3]. There are several ways to reduce the amount of platinum required to deliver a unit of net power from a fuel cell power plant: (1) reducing parasitic losses at the power plant level, (2) increasing the power delivered per unit active area of a fuel cell stack at fixed catalyst loadings, and (3) decreasing the amount of catalyst per unit area of electrode. While reduction of the parasitic losses at the power plant can be enabled in part by proper catalyst layer and cell design for simplified reactant and water management systems, the specific requirements of catalyst layer design are generally a function of the cell and system in which they reside; it is difficult to provide general guidance for catalyst layer operation for a more efficient system. We will therefore restrict our discussion to understanding the issues related to stable performance at high currents and to stable high specific activity.

In general, a stable, high-performance catalyst layer coupled with proper system design will allow operation at high power density (in excess of 1 A/cm^2). In order to sustain that high performance over the duration of the power plant lifetime, however, both the materials and the structure of the catalyst layer must be resistant to changes in catalytic and transport properties. Furthermore, lowering the catalyst loading renders the fuel cell more susceptible to changes in catalytic activity or electrochemically active area, because a lowered catalyst loading takes away margin from the design. A great deal of modeling work has been dedicated to understanding the requirements for high performance at the beginning of the life of a fuel cell stack, but we will direct our attention to the vulnerabilities of practical catalyst layers to degradation, and to some of the models that have been constructed to predict decay.

In this chapter, we will review first the requirements for high-performance catalyst layers, beginning with a general discussion of the means of enhancing the reaction rate by dispersion of small catalysts, and of the trade-off of ionic, electronic, and gas-phase transport that must be considered in the design of a multicomponent electrode to bring all reactants and products into intimate contact at the electrocatalyst. We will then discuss the kinetics of the principal electrochemical reactions in the fuel cell, namely, oxygen reduction and hydrogen oxidation. Translating this intrinsic activity into a practical device is discussed via the porous-electrode theory and the explicit arrangement of multiple phases into a single catalyst layer. After reviewing the requirements for a catalyst layer at the beginning of life, we will discuss at a general level some of the critical mechanisms of degradation of material and structural properties that have been identified and the models that have been employed to predict their degradation. Finally, we will examine possible means of decay mitigation that have been discussed in the open literature. In many cases, though, this chapter will pose more questions and opportunities for future research than it will provide thorough review and references, as performance degradation and durability are research areas that have only begun receiving thorough coverage quite recently, and mathematical models of those modes of failure are, as of this writing, even rarer in the literature than experimental papers on the subject.

1.1.1. Catalyst Layer Design

A proper catalyst layer design is centered around the task of enhancing the rates of charge transfer that produce the current that is delivered by the fuel cell. As alluded to in the introductory section, the kinetics for oxygen reduction are quite poor at the temperatures and pressures at which PEM fuel cells are operated, and so precious metals—generally platinum or platinum alloys—are employed to enhance the reaction rate of the reaction. The development of fuel cell catalyst layers has been a sustained effort to improve the activity per mass of precious-metal catalyst. A review of the history of fuel cell catalyst layers is provided by Kocha [4] and reveals the gradual progression of catalyst-layer structure over more than 150 years since William Grove demonstrated the first fuel cell [5].

In order to enhance the faradaic reaction rates, either the activity of the catalyst or the surface area of the catalyst must be enhanced. Following the work of Gasteiger et al. [6], if one were to construct a fuel cell with a catalyst layer approximating the surface area of a planar electrode with a Tafel slope of 70 mV/decade at 80°C, one could expect a cell potential of approximately 0.8 V for a fuel cell operating at approximately 5 mA/cm^2, neglecting gas-phase mass-transport effects and assuming the same ohmic resistance as that which has been demonstrated in conventional fuel cell stacks. Practical PEM fuel cells have been demonstrated with a performance of at least 200 mA/cm^2 at the same cell potential [7]; this 40-fold increase in power density at a fixed polarization gives an indication of the degree to which this enhancement has been achieved through careful selection of materials and tailoring of the catalyst layer structure.

It is worth noting that one cannot simply accept an arbitrarily low charge-transfer rate for a given polarization and operate at a correspondingly low current density. Power density is a critical driver for market acceptance for transportation and portable applications [3], and product cost will also be driven by power density [8]:

$$C = (c_{non-Pt} + c_{Pt}\, m_{Pt})/p + c_{Assembly}$$

Where C is cost in $ per kW at rated power, $c_{non\text{-}Pt}$ is the manufacturing cost per unit area of the cell, excluding the platinum, c_{Pt} is the cost of platinum per unit mass, m_{Pt} is the mass loading of catalyst in mass per unit area of the cell, p is power density of the cell at rated power conditions, and $c_{Assembly}$ is the cost of assembly in $ per kW. This is, admittedly, a very simple means of calculating the cost of the system, but it shows the strong effect that power density has on overall costs per unit delivered power. Unless the material and processing costs are negligible compared to the cost of the platinum, one cannot hope to achieve product cost targets by operating at lower power density. It is therefore critical to design for high power density at low catalyst loadings.

1.1.2. *Interfacial Kinetics of Supported Catalysts*

One of the primary challenges of fuel cell engineering is translating the performance of a well-characterized electrode surface to performance in a practical fuel cell system. From the standpoint of fundamental understanding, it is desirable to measure electrochemical reaction rates on simple electrode surfaces that can be fully characterized. Of course, the catalysts used in practical systems must be made inexpensively, and enhancement of active area is more important to the engineer than complete characterization of the surface. As a result, there is a disconnect between what many electrochemists study in rotating ring-disk electrode (RRDE) experimental studies and the supported catalyst particles that are used in fuel cell catalyst layers. There are many studies of the relevant electrochemical reactions on single-crystal electrodes in acid electrolytes, but the translation of single-crystal data to practical electrode performance is, at this stage of development, as much an art as a science. The details of single-crystal electrochemistry have been reviewed elsewhere by Ross [9], and the interested reader is directed to those references for a thorough treatment, but it is important to consider at least some of the critical features of interfacial electrochemistry.

1.2. Oxygen Reduction Reaction

The oxygen reduction reaction is covered in considerable detail in a review text by Kinoshita [10]. The charge-transfer reaction itself is quite complicated, and controversy still exists around the details of which of the many possible charge-transfer mechanisms determine electrode performance. The two generalized pathways that are considered are the direct four-electron reaction:

$$O_2 + 4H^+ + 4e^- \rightleftharpoons 2H_2O$$

and the peroxide pathway, a two-electron charge-transfer reaction, whereby peroxide is formed as an intermediate and then either reduced further to produce water or decomposed to form water and oxygen.

$$O_2 + 2H^+ + 2e^- \rightleftharpoons H_2O_2$$

Kinoshita's review suggests that the 4-electron pathway is predominant on noble-metal electrocatalysts such as platinum and that the 2-electron peroxide intermediate mechanism is the primary pathway on graphite and most other carbons, oxide-covered metals, transition-metal oxides, and some macrocycles. Certainly, though, peroxide *is* formed to some degree in proton-exchange membrane fuel cells and will be discussed later as an unwanted intermediate that can lead to ionomer degradation in both the catalyst layer and the separator of the fuel cell [11].

While the precise reaction mechanism is critically important for alternate catalyst discovery and for suppression of unwanted side reactions, the techniques that would allow rational catalyst design and selection—that is, the selection of novel catalysts *a priori* by modeling rather than an Edisonian design of experiments for discovery—are still being developed. Until a more reliable linkage between quantum chemical simulations and experimental data at practical scales is established, combinatorial techniques will likely be used to reduce the time and effort associated with discovery of new materials [12]. At present, the maturity of the theories that bridge the atomic and molecular scales to the macroscopic is insufficient to provide us with true predictive capability. As such, we will restrict our discussion to the characterization of known electrode materials and the task of assembling a highly active electrode from those materials. Readers who are interested in quantum chemical simulations of electrode surfaces, and the rational design of catalyst materials, are directed to a review by Koper of density functional theory and dynamic Monte Carlo simulations [13].

Given a specific reactivity for an electrode surface—generally defined by an exchange current density which might be a function of composition and temperature, and a Tafel slope [14]—one would like to enhance the surface area of the surface and therefore increase the effectiveness of the catalyst. The process of achieving high activity for a dispersed catalyst is complicated, however, by the so-called particle-size effect. Data suggest that not all surface atoms are equally reactive; Kinoshita reviewed available data on oxygen reduction in a variety of electrolytes and suggested that the data could be described by treating platinum crystallites as a cubo-octahedral structures, and by assuming that the (100) and (111) crystal faces are active sites but not the edge or corner sites; this yields a maximum in mass activity at a particle size of approximately 3 nm [15]. Kinoshita's review also cites the work of Peuckert et al. [16], who suggest that the structure sensitivity for Pt particles may be associated with the strength of adsorption of surface oxide species.

Markovic et al. [17] review the data on platinum particles and suggest that the data in dilute sulfuric acid is consistent with Kinoshita's model and further suggest that essentially all of the reactivity can be attributed to the (100) surface. They go on to suggest that this difference in reactivity between the crystal faces is due to structure sensitivity of anion adsorption that impedes the reaction. They point out that in PEM systems, where anion adsorption by the sulfonic acid groups is unlikely, there might be considerably less of a particle-size effect. Still, most PEM catalyst layers employ platinum particles on the order of 3 nm, roughly the same size as the maximum in mass activity identified by Kinoshita.

1.3. Hydrogen Oxidation

As Gasteiger et al. point out in their paper on research opportunities for alloy and nonplatinum catalysts [6], it is quite simple to envision reducing the platinum loadings on the anode side of the fuel cell, simply because the hydrogen reaction is so fast. This is true as long as the feedstock of hydrogen is sufficiently pure to avoid poisoning from unwanted contaminants. However, carbon monoxide and other impurities can severely deactivate the hydrogen electrode, and one must either purify the hydrogen external to the fuel cell stack or design a system with "air bleed"—admitting dilute quantites of oxygen to the anode—to promote CO oxidation. The kinetics of CO oxidation on supported catalysts are modeled by Springer et al. [18] Their model suggests that CO adsorption has an effect on the rate of dissociative adsorption of hydrogen at adjacent sites. By their model, one can anticipate the effect of CO or other poisons on the reactivity of hydrogen in a PEM system. While the kinetics for hydrogen oxidation are very rapid, it is critical to maintain a sufficiently clean surface to allow hydrogen oxidation to proceed.

1.4. Implications of Kinetic Limitations on Catalyst Layer Design

The interfacial charge-transfer reactions at the anode and at the cathode are the fundamental reactions that define the fuel cell. It is critical to sustain high rates of reaction at these interfaces, and if it were possible to enhance the specific reactivity of the electrode surface by material selection or design, the other challenges of fuel cell design would generally be much easier to overcome. For instance, the polarization of the anode in a system with pure hydrogen as fuel is a negligible contribution to the overall cell polarization; both the activation losses and the mass-transport losses at the anode are small enough to be negligible, whereas the losses at the cathode, with its poor kinetics and the mass-transport losses to and through the catalyst layer, dominate the shape of the polarization curve. The amount of platinum required to construct a fuel cell and the cost of fuel to deliver a given amount of energy will both be decreased, if one can coax better activity out of the catalysts.

How, then, does one go about enhancing the activity of electrocatalysts? There are essentially two approaches: improving the intrinsic activity of the catalyst and increasing the accessible surface area of the catalyst. The former task is accomplished through discovery of new materials or surface modifications that improve the charge-transfer characteristics of the catalyst; the latter is accomplished through design. In order to accomplish this task, one must assemble a catalyst layer that maximizes the activity and accessible surface area of the catalyst. Neglecting mass-transport limitations and particle-size effects, one could simply shrink the size of individual catalyst particles to enhance the surface-to-volume ratio of the crystallite and achieve a considerable increase in performance per unit superficial area of electrode:

$$i_{superficial} = i_{specific}\, 4\pi r_{particle}^2\, N\, l_{catalyst\ layer}$$

where $i_{superficial}$ is the current density per unit superficial area of electrode, $i_{specific}$ is the current density per unit interfacial area of the electrode at a given polarization, $r_{particle}$ is the particle size of the catalyst, N is the number of catalyst particles per unit volume of electrode, and $l_{catalyst\ layer}$ is the catalyst layer thickness. In this unconstrained case, one can enhance the current per unit area by simply increasing the number of particles or making the catalyst layer thicker.

One can also calculate m_{Pt}, the mass loading per unit interfacial area of the catalyst layer:

$$m_{Pt} = 4/3 \pi r_{particle}^3\, N\, l_{catalyst\ layer} \rho_{Pt}$$

where ρ_{Pt} is the density of the catalyst in g/cm^3.

It is then possible to express the maximum theoretical enhancement in performance in terms of the mass loading:

$$i_{superficial} = i_{specific} \times 3\, m_{Pt}/(\rho_{Pt} r_{particle})$$

This expression suggests that one could increase the current density per unit area of superficial area by increasing the mass loading of the catalyst or by decreasing the particle size. The former is, of course, counter to the goal of reducing the cost of catalyst in the system; the latter is constrained by the difficulty in manufacturing catalyst particles at an arbitrarily small and precise size (by way of example, in Kinoshita's analysis [15], a particle of 1.5-nm diameter consists of just 201 atoms) and further diminished by the particle-size effect, described in the previous section.

If one tries to arrange all of these catalyst particles in a single plane, it will be straightforward to access all of the catalyst surface area, but the degree of enhancement per unit superficial area that can be achieved is severely constrained. Let us consider a hexagonal close packing of catalyst particles, the most efficient geometric packing structure that one can achieve for spherical particles. For a dispersion of platinum with a radius of 1.5 nm in a hexagonal close-packed structure just one particle deep, one can only deliver a loading of $\sim 4\,\mu g/cm^2$, with a surface area enhancement of just $2\pi\sqrt{3}$, a roughness factor of approximately 3.6. In common practice, one routinely selects catalyst loadings approximately 100 times higher on a mass-per-superficial area basis than what this "perfect" single-plane hexagonal packing would allow. Even though one would like to achieve lower loadings than what is common practice today (roughly $100\,\mu g/cm^2$ for a cathode loading has been proposed as a reasonable target for automotive applications) [3], this reduced loading still exceeds what can be delivered in a single planar layer. This implies that in order to achieve the loadings and enhancements that render fuel cells practical, one must create a catalyst layer of finite thickness, not simply a planar arrangement of single particles. This is illustrated in Figure 1.1.

Once one chooses to create a catalyst layer with finite thickness, however, one must then consider the transport of reactants through the thickness of that layer. At the most fundamental level, an examination of the two defining reactions of a fuel cell reveals the engineering challenge that must be confronted in designing a catalyst layer. These reactions are, of course

$$H_2 \rightleftharpoons 2H^+ + 2e^-$$

and

$$O_2 + 4H^+ + 4e^- \rightleftharpoons 2H_2O$$

It is important to note that there are at least three different phases that must be brought into intimate contact in order for the reactions to proceed. Hydrogen and oxygen exist primarily in the gaseous phase and are only sparingly soluble in liquid water or in the ionomeric materials that conduct

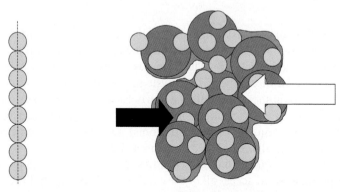

a. A single plane of catalyst particles, uniformly accessible.

b. A catalyst layer of finite thickness, with arrows to represent mass-transport limitations (ohmic limitations tend to concentrate the reaction zone near the membrane; gas- and liquid-phase transport limitations near the diffusion media).

FIGURE 1.1. Schematic diagram illustrating the enhanced surface area possible with a catalyst layer of finite thickness relative to a single plane, and the mass-transport limitations associated with that thickness.

protons in a PEM fuel cell [19]. The protons must be carried in an aqueous phase (or, more generally, in a polar solvent) [20], and the electrons must move through an electronically conducting phase. All three phases must be present, or the rate of the reaction will be severely limited by the restricted rate of transport of reactants to the catalytic surface.

It is necessary to facilitate the transport of gases, protons, and electrons through the layer and, under fully saturated conditions, which are commonly encountered in at least some regions of the fuel cell [3], the removal of product water from the location where it is produced. If one is constrained by the rate of transport of any reactant, then the rate of reaction will be constrained as well. In order to predict how the catalyst layer will perform, one needs to be able to describe how reactants will move to and through the catalyst layer to access all of the dispersed catalyst. We turn our attention to this challenge in the next section.

1.5. Porous Electrodes

Porous electrodes are used in fuel cells in order to maximize the interfacial area of the catalyst per unit superficial area. As described above, this is critical not only for lowering the total amount of catalyst needed for the fuel cell but also to increase the power that can be delivered per unit superficial area. The electrode must be designed to minimize the resistances to mass transport in the electrolytic and gas phases, and the electronic resistance in the solid phase. Porous-electrode theory is commonly used to address the intimate relationship between transport and kinetics, and to do so with a minimum of computational complexity.

While there are exceptions, most fuel cell developers have gravitated toward a common means of making a catalyst layer. Generally speaking, most catalyst layers employ a structure of platinum particles supported on carbon. The primary carbon particles are approximately 40 nm in diameter [21], and the platinum crystallites are approximately 2–3 nm in diameter [22]. The catalyst is then mixed with a solubilized form of ionomer, which is then used as an ink to print a thin catalyst layer on the order of 10 microns thick. The catalyst layer, then, is comprised of the solid, electronically conducting carbon phase, decorated with platinum, a largely contiguous network of ionomer to provide ionic transport through the thickness of the layer, and the voids that provide a pathway for gas and liquid access.

The behavior of porous electrodes is considerably more complicated than that of planar electrodes because of the intimate contact between the solid and fluid phases. Reaction rates can vary widely through the depth of the electrode due to the interplay between the ohmic drop in the solid phase, kinetic resistances, and concentration variations in the fluid phases. The number and complexity of interactions occurring make it difficult to develop analytic expressions describing the behavior of porous electrodes, except under limiting conditions.

Porous-electrode theory has been used to describe a variety of electrochemical devices including fuel cells, batteries, separation devices, and electrochemical capacitors. In many of these systems, the electrode contains a single solid phase and a single fluid phase. Newman and Tiedemann reviewed the behavior of these flooded porous electrodes [23]. Many fuel-cell electrodes, however, contain more than one fluid phase, which introduces additional complications. Typical fuel cell catalyst layers, for example, contain both an electrolytic phase and a gas phase in addition to the solid electronically conducting phase. An earlier review of gas-diffusion electrodes for fuel cells is provided by Bockris and Srinivasan [24].

1.6. Porous-Electrode Theory

In order to completely describe transport and reaction throughout the thickness of the catalyst layer, one would need detailed geometric and interfacial data about every particle, the structure of the electrolyte film providing ionic transport through the layer, and the voids that comprise the layer, as well as information about the connectedness between these elements. Completely describing an ideal catalyst layer with uniformly sized and regularly spaced particles would be computationally exhausting; describing a manufactured catalyst layer, with irregularities in shape, size, and distribution is even more difficult.

Instead, modelers frequently employ a macroscopic approach commonly referred to as porous-electrode theory for modeling electrodes, as described by Newman and Tiedemann [23]. In the macroscopic approach, the exact geometric details of the electrode are neglected. Instead, the electrode is treated as a randomly arranged porous structure that can be described by a small number of variables such as porosity and surface area per unit volume. Furthermore, transport properties within the porous structure are averaged over the electrode volume. Averaging is performed over a region that is small compared to the size of the electrode but large compared to the pore structure.

The macroscopic approach to modeling can be contrasted to models based on an assumption of a geometric description of the pore structure. Many early models of flooded porous electrodes treated the pores as straight cylinders arranged perpendicular to the external face of the electrode. Further examples of this type of approach are the flooded-agglomerate models of Giner and Hunter [25] and Iczkowski and Cutlip [26]. These models are frequently used to describe fuel cells and treat the electrode as a collection of flooded catalyst-containing agglomerates, which are small compared to the size of the electrode and are connected by hydrophobic gas pores.

Even among the models that employ porous-electrode theory, there have been differences in how the various models choose to describe the electrode. For example, consider the catalyst layer in a state-of-the-art PEM fuel cell containing a supported-platinum-on-carbon (or platinum-alloy-on-carbon) catalyst, a polymeric membrane material, and, in some cases, a void volume. Whether this void volume is considered explicitly, or whether gas- and liquid-phase transport is simply described via permeability through the ionomer is one of the key differences between the various models.

Weber et al. [27] describe four separate phases in the catalyst layer: the solid phases (catalyst and support) are described as a single phase, and the polymeric membrane is treated as another separate phase, yielding four distinct phases: the solid phase, the membrane phase, the gas phase, and a liquid phase. This last phase corresponds to liquid water infiltrating the gas pores. Weber's model allows for a saturation level that can vary with capillary pressure in the pores of the catalyst layer and adjacent gas-diffusion layer; the degree of saturation, in turn, affects the permeability of the liquid water in the pores. A very low level of saturation implies a very low permeability (and therefore, restricted liquid-phase transport); a high level of saturation implies facile liquid transport but severe gas-phase mass transport restrictions (i.e., flooding). The curves that define the relationship of saturation versus capillary pressure are wedded to the physical properties of the porous media: the composite contact angle of the pores and the pore size distribution of those pores. This model differs, conceptually, from those of Bernardi and Verbrugge [28] and Springer et al. [29] by explicitly accounting for void volume containing both gas and liquid water. In Bernardi and Verbrugge's model, oxygen and hydrogen within the catalyst layer travel as dissolved species within the ionomer in the catalyst layer. Springer et al. propose a similar picture but fit the permeability of the catalyst layer to experimental data. The basic mathematics of the different models are fundamentally similar, a strength of the macroscopic approach. Each phase is assumed to be electrically neutral.

1.7. Modeling of Decay Mechanisms

Now that we have described the attributes of a high-performance catalyst layer and the simultaneous requirements for high specific reactivity, contiguous pathways for electronic and ionic currents, and gas- and liquid-phase mass transport, one would like to understand the mechanisms for degrading that structure with time. There is a substantial amount of data detailing the degree to which performance does degrade with time in fuel cells [30], but models that predict the loss of activity (or the onset of catastrophic failure) are relatively rare in the literature. We will examine some of the degradation mechanisms that have been identified, and where available, review the models that have been proposed to describe these degradation modes quantitatively.

1.8. Loss of Catalyst Surface Area

It is perhaps not immediately obvious that the precious-metal catalysts that are employed for use in PEM fuel cells will be subject to degradation, agglomeration, and even dissolution. Most of us are familiar with platinum as an example of a noble metal, which, according to its definition, means that it resists chemical action and does not corrode. Yet there is compelling evidence that platinum can degrade under conditions experienced in the fuel cell operating environment. Within the catalyst and separator of the fuel cell, the conditions are quite acidic, and the presence of oxygen results in an environment that is extremely oxidative.

Kinoshita refers to platinum surface area loss in his review of oxygen catalysts in phosphoric acid systems and indicates that potential cycling accelerates the degradation [10]. One possible mode of platinum dissolution—in particular, one that is accelerated by potential cycling—is proposed by Darling and Meyers [31], who suggest that platinum can be oxidized to form a mobile charged species, which is then free to diffuse away from the catalyst layer to other regions of the cell, or to be redeposited in regions of lower potential. They propose that the equilibrium concentration of the mobile species increases with potential when bare platinum is exposed to solution, but that, at higher potentials, an oxide layer is formed which serves to passivate the surface. By cycling the current or the potential in a fuel cell, the cathode can be exposed to high potentials, thereby drastically increasing the rate of platinum dissolution from the surface in the period before the passivating layer is completely formed.

Figure 1.2 shows the thermodynamic equilibrium for mobile platinum versus potential for an acidic solution in equilibrium with an exposed platinum surface and a lower branch of equilibrium concentration at potentials more positive than approximately 1.1 volts relative to a hydrogen electrode, where the plotted concentration denotes the concentration of mobile species in equilibrium with a platinum oxide layer covering the surface. This shows that excursions to higher potential can rapidly increase the rate of platinum dissolution prior to passivation of the surface. Once the surface is passivated, the dissolution stops and redeposition can occur, albeit incomplete redeposition, as the platinum, once rendered mobile, is free to redeposit on larger particles or diffuse away from the catalyst layer altogether [32].

The mechanism of catalyst dissolution occurs to differing degrees, depending upon the degree to which platinum is alloyed and what other elements are included in the alloy. Work by Piela et al. [33] suggests that PtRu alloys, as employed in direct methanol fuel cell and reformate anodes to reduce sensitivity to CO, are extremely unstable and that operation leads to ruthenium

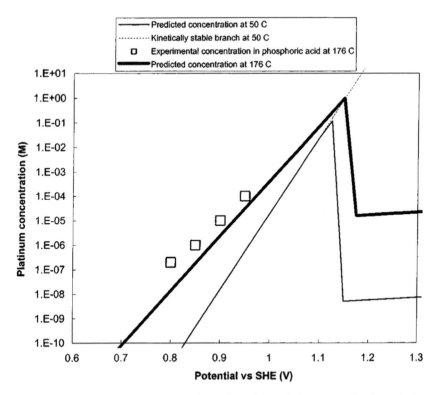

FIGURE 1.2. Equilibrium concentration of mobile platinum species in solution *versus* platinum electrode relative to standard hydrogen electrode R. M. Darling and J. P. Meyers "Mathematical Model of Platinum Movement in PEM Fuel Cells," *J. Electrochem. Soc.*, **152**, A242 (2005).. From Ref. [31]. Reproduced with permission of the Electrochemical Society.

migration across the fuel cell from the anode to the cathode, where both activity and mass-transport losses are magnified. They note that the Pourbaix diagram for ruthenium stability has a wide region of instability in the region near where the anode will operate. Other alloys, however, show improved stability with respect to dissolution and cycling. Binary and ternary alloys including cobalt (PtCo and PtIrCo) have been shown to be significantly more stable to cycling than platinum alone, even though unalloyed cobalt would be expected to dissolve and migrate in the fuel cell operating environment [34]. Whether these alloying elements change the equilibrium concentration of platinum in solution at a given potential, or change simply the kinetics, or induce passivation at lower potentials has not been established and is an area ripe for further study by both experiment and modeling.

Of course, the dissolution mechanism is one possible mechanism for platinum surface area loss but likely not the only one or even the most important one for cells that do not operate intermittently between low and high potentials. It is quite likely that agglomeration of particles and Ostwald ripening also occur. While these mechanisms have been noted in PEM fuel cells [35,36] and the general mechanism described elsewhere [37], there is little detailed modeling of their occurrence in PEM fuel cells. In general, the evolution of surface area for a particle-size distribution is complicated by the distribution of particle sizes and the differences of kinetics between large and small particles; as particles coalesce, simulations show a broadening and a growth of the particle size distribution [38]. More recent work by Campbell et al. [39] laments the lack of a kinetic model that accurately predicts long-term sintering based upon short-term measurements; their works suggests that some of the assumptions that have been used to develop simplified models of Ostwald ripening have been based upon an assumption of constant excess surface energy, which this paper suggests is an assumption that breaks down for very small particle sizes.

Another critical mechanism of electrochemical activity loss is due to carbon corrosion. Carbon corrosion has been cited as a major concern in higher temperature environments, such as phosphoric acid fuel cells [40] but is relatively benign at potentials less than 1 V RHE at the temperatures at which PEM fuel cells normally operate. Several papers have noted that in the case of gross fuel starvation, cell voltages can become negative, as the anode is elevated to very positive potentials, and the carbon is consumed instead of the absent fuel [41].

A recent paper by Reiser et al. [42] suggests, however, that transient conditions or *localized* fuel starvation can induce local potentials on the air electrode significantly higher than 1 V and thereby induce corrosion of the carbon supports that results in permanent loss of electrochemically active area. The mechanism they describe suggests that the highly conductive bipolar plates of the fuel cell allow for sufficient redistribution of current in the plane of the current collectors and that all regions of the cell experience the same potential difference.

In the regions of the cell where fuel is present on the anode, the fuel cell behaves normally. In the regions of the cell where there is no fuel present, there is no proton or electron source at lower potentials, so the electrodes must shift to significantly higher potentials to maintain the potential difference imposed by the active part of the cell while conserving current. Thus, a reverse current is established, and current is driven from the positive electrode to the negative electrode in the fuel-starved region, opposite the direction of normal current flow in the active portion of the cell. The only reactions that can sustain this current in the fuel-starved region are oxygen evolution and carbon corrosion on the positive electrode, and oxygen reduction from crossover on the negative electrode. This mechanism is shown schematically in Figure 1.3. The damage inflicted by these potential excursions is illustrated in Figure 1.4 by data taken at General Motors [43] by cycling a cell from 1 V RHE to successively higher potentials. The phenomenon has also been modeled by Meyers and Darling [44],

1. Modeling of PEMFC Catalyst Layer Performance and Degradation   33

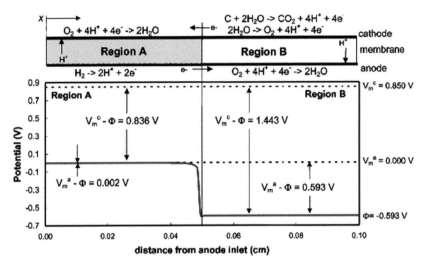

FIGURE 1.3. Potential distribution in a cell with nonuniform fuel distribution on the anode. From Ref. [42]. Reproduced with permission of the Electrochemical Society.

who simulate the evolution of composition and potential under transient conditions and examine the effect of operational variables such as rate of fuel purge and potential and current control on the location and extent of carbon corrosion.

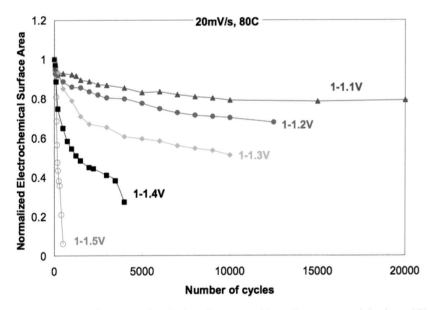

FIGURE 1.4. Loss in electrochemical surface area with cycling to potentials above 1 V relative to a standard hydrogen electrode. From Ref. [43].

1.9. Surface Property Changes

The previous sections on catalyst layer degradation have described the losses of electrochemical area due to loss of catalyst surface area, either through loss or agglomeration of the catalyst itself, or through corrosion of the carbon support which must maintain electrical contact with the particle. Changes in surface properties will also have a marked effect upon catalyst layer performance, specifically in the range where mass transport rates are important. As noted in the section on gas-phase and liquid-phase transport, the degree of saturation of the pores in the catalyst layer depend upon both the pore-size distribution and the composite contact angle of those pores. Changes either to the pore size or to the surface properties can, in turn, change the degree of saturation in the pores, rendering the fuel cell prone to either dryout or flooding. Changes in wettability have been observed for carbon supports after prolonged exposure to the fuel cell operating environment [40], but there is still no mechanistic model to predict the rate of surface property changes in the PEM fuel cell environment, or to correlate those changes to reduced performance, save for rerunning the beginning-of-life models subject to the new structural and surface properties.

1.10. Membrane Degradation

An additional failure mode that often contributes to sudden, catastrophic failure is membrane degradation. While this can be induced by puncture or elements external to the catalyst layer [7], one mode of failure that is widely accepted is that of chemical attack of the membrane by peroxide radicals, followed by gradual thinning and, ultimately, a failure of the membrane to provide an effective barrier to reactant crossover [11]. Peroxide is formed on platinum in the presence of oxygen. Membrane failure is not strictly a catalyst layer problem, but it is in the catalyst layer that this critical intermediate is formed. Understanding of the mechanisms of formation and movement of peroxide and peroxide radical across the fuel cell is critical to designing catalyst layers that can suppress these unwanted side reactions, but there has been very little in the literature to inform the design as yet. This topic has been the subject, however, of some recent papers and publications that promise to provide new insights into the problem [45].

1.11. Freeze

A topic that has recently garnered some attention in the literature is the subject of degradation associated with freeze/thaw cycles [46] and, more specifically, with drawing current from the fuel cell when the temperature of the cell itself is below freezing, a procedure that will be frequently experienced for fuel cell power plants deployed in automotive applications [47]. Given the volume change associated with freezing water, one expects that any water retained in the pores of the catalyst layer or at the catalyst layer interface with the gas-diffusion layers will expand and can create considerable stresses on the porous materials, possibly deforming them from their initial state.

The idea that freeze/thaw cycling can effect changes on pore structure is borne out by the work of Cho et al. [46], who demonstrate a marked increase in the mean pore size and width of the pore size distribution of the catalyst layer after thermal cycling. Modeling of the phase transition associated with thawing and freezing, however, as well as the coupled phenomena of water management and thermal management under partially frozen conditions, is very limited in the open literature. Oszcipok et al. [48] attempt to correlate fuel cell start-up performance with the degree of hydration of the membrane and suggest that once the membrane is fully saturated, the current density will quickly drop to zero as the catalyst layers flood. Their data are taken from isothermal cells, however, and their analysis does not fully consider the effect of waste heat on the temperature of the stack. The dynamics of water movement in porous media in the presence of a temperature gradient has been considered in other fields, notably by Rempel et al. [49] in a paper describing the dynamics of frost-heave phenomena in porous media selected to simulate the phenomena in soils. The approach has also been applied to layered porous structures with disparate pore sizes and contact angles and finds that the capillary pressure of the porous materials is a key parameter governing water redistribution among adjacent porous layers during freeze [50]. While these models have begun to describe the evolution of saturation and pressure under transient conditions at or near freezing temperatures, we still lack a model that captures simultaneously the physics of water movement, heat transfer, and the deformational stresses that can change the porous structures of the catalyst layers and predict the performance losses associated with repeated frozen starts of a fuel cell.

Summary

It is an engineering challenge to create a high-performance catalyst layer, because the nature of the catalyst layer and the triple-phase boundary demands operation at the cusp of the limitations of multiple modes of transport. Because ionic transport and gas-phase transport take place in distinct phases—and the ionic and gas-phase reactants moving through those phases necessarily move to or from opposite ends of that catalyst layer—the optimization suggests that if one of those transport processes has a higher limiting current than the others, the optimization is not complete. In increasing the volume fraction of the phase that is now "overperforming," one is decreasing the volume fraction of the competing phase, and transport through that phase will therefore be constrained relative to the other and will ultimately limit the rate of reaction. Precise balance must be maintained between these simultaneous processes. Similarly, one wants to maintain a high surface area for catalysis, which suggests a highly dispersed, highly accessible structure.

Fuel cell technologists have done an impressive job at constructing catalyst layers that meet all of these requirements, but these structures, which have been so painstakingly constructed to provide high performance at the beginning of life, must be maintained over thousands of hours and up to millions of cycles. To build that robustness into the design, the engineer must understand mechanistically the modes by which those structures can be degraded and conceive of means to reduce either the likelihood or the impact of those mechanisms.

In this chapter, we have considered some of the modes of degradation that have been identified already and reviewed the models that have been used to describe them mathematically. While detailed models of degradation are still relatively rare in the literature, a review of upcoming conferences and works in progress suggests that the fuel cell community is bringing considerable resources to bear on the problem. Once this occurs, the industry's collective understanding of fuel cell durability will be supplemented by detailed mathematical models that can be exploited in the design of fuel cell power plants.

References

[1] I. D. Raistrick, in *Proceedings of the Symposium on Diaphragms, Separators, and Ion-Exchange Membranes*, (J. W. Van Zee, R. E. White, K. Kinoshita, and H. S. Burney, eds.), p. 172, The Electrochemical Society, Inc., Pennington, NJ, (1986).

[2] H. Yano, C. Ono, H. Shiroishi, and T. Okada, "New CO Tolerant Electro-Eatalysts Exceeding Pt-Ru for the Anode of Fuel Cells," *Chemical Communications*, **9**, 1212 (2005).

[3] M. Mathias, H. Gasteiger, R. Makharia, S. Kocha, T. T. Xie, and J. Pisco, "Can Available Membranes and Catalysts Meet Automotive Polymer Electrolyte Fuel Cell Requirements," 228th National Meeting of the ACS Meeting, Philadelphia, August (2004).

[4] S. S. Kocha, "Principles of MEA preparation," in *Handbook of Fuel Cells-Fundamentals, Technology, and Applications, Volume 3: Fuel Cell Technology and Applications* (W. Vielstich, H. A. Gasteiger, and A. Lamm, eds.) John Wiley & Sons, Ltd. New York, 2003.

[5] W. R. Grove, "On a Gaseous Voltaic Battery," *Philos. Magazine J. Sci.*, **21**, 417, (1842).

[6] H. A. Gasteiger, S. S. Kocha, B. Sompalli, and F. T. Wagner, "Activity Benchmarks and Requirements for Pt, Pt-alloy, and non-Pt Oxygen Reduction Catalysts for PEMFCs," *Appl. Catal. B: Environ.*, **56**, 9 (2005).

[7] S. J. C. Cleghorn, "Developing Durable, Cost-Effective MEAs for Automotive Fuel Cells," SAE TOPTEC Symposium, April 9, 2003, Dearborn, Michigan. http://www.gore.com/MungoBlobs/fuel_cells_presentation_SAE_auto.pdf

[8] H. Tsuchiya, "Fuel Cell Cost Study by Learning Curve," Annual Meeting of the International Energy Workshop Jointly organized by EMF/IIASA, 18–20 June 2002 at Stanford University, USA. http://www.iiasa.ac.at/Research/ECS/IEW2002/docs/Paper_Tsuchiya.pdf

[9] P. N. Ross, "Oxygen reduction reaction on smooth single crystal electrodes," in *Handbook of Fuel Cells- Fundamentals, Technology and Applications, Volume 2: Electrocatalysis.* (W. Vielstich, H. A. Gasteiger, and A. Lamm, eds.) John Wiley & Sons, Ltd., New York, 2003.

[10] K. Kinoshita, *Electrochemical Oxygen Technology*, John Wiley & Sons, Ltd., New York, 1992.

[11] D. E. Curtin, R. D. Lousenberg, T. J. Henry, P. C. Tangeman M. E. Tisack, "Advanced Materials for Improved PEMFC Performance and Life," *J. Power Sources*, **131**, 41 (2004).

[12] R. Liu and E. Smotkin, "Array Membrane Electrode Assemblies for High Throughput Screening of Direct Methanol Fuel Cell Anode Catalysts," *J. Electroanal. Chem.*, **535**, 49 (2002).

[13] M. T. M. Koper, "Numerical simulations of electrocatalytic processes," in *Handbook of Fuel Cells – fundamentals, Technology, and Applications, Volume 2: Electrocatalysis* (W. Vielstich, H. A. Gasteiger, and A. Lamm, eds.), John Wiley & Sons, Ltd., New York (2004).

[14] K. Vetter, *Electrochemical Kinetics: Theoretical Aspects*, Academic Press, New York, (1967).

[15] K. Kinoshita, "Particle Size Effects for Oxygen Reduction on Highly Dispersed Platinum in Acid Electrolytes," *J. Electrochem. Soc.*, **137**, 845 (1990).

[16] M. Peuckert, T. Yoneda, R. A. Dalla Betta, and M. Boudart, "Oxygen Reduction on Small Supported Platinum Particles," *J. Electrochem. Soc.*, **133**, 944 (1986).

[17] N. Markovic, H. Gasteiger, and P. N. Ross, "Kinetics of Oxygen Reduction on Pt(hkl) Electrodes: Implications for the Crystallite Size Effect with Supported Pt Electrocatalysts," *J. Electrochem. Soc.*, **144**, 1591 (1997).

[18] T. E. Springer, T. Rockward, T. A. Zawodzinski, and S. Gottesfeld, "Model for Polymer Electrolyte Fuel Cell Operation on Reformate Feed: Effects of CO, H2 Dilution, and High Fuel Utilization," *J. Electrochem. Soc.*, **148**, A11 (2001).

[19] K. Broka and P. Ekdunge, "Oxygen and Hydrogen Permeation Properties and Water Uptake of Nafion 117 Membrane and Recast Film for PEM Fuel Cell," *J. Appl. Electrochem.*, **27** 117 (1997).

[20] B. Pivovar, "High Temperature Polymer Electrolytes Based Upon Ionic Liquids," DOE Hydrogen Program 2004 Progress Report, available at http://www.eere.energy.gov/hydrogenandfuelcells/pdfs/annual04/ivb12_pivovar.pdf

[21] M. Uchida, Y. Fukuoka, Y. Sugawara, N. Eda, and A. Ohta, "Effects of microstructure of carbon support in the catalyst layer on the performance of polymer-electrolyte fuel cells," *J. Electrochem. Soc.*, **143**, 2245 (1996).

[22] T. R. Ralph, G. A. Hards, J. E. Keating, S. A. Campbell, D. P. Wilkinson, M. Davis, J. St-Pierre, and M. C. Johnson, "Low cost electrodes for proton exchange membrane fuel cells – Performance in single cells and Ballard stacks," *J. Electrochem. Soc.,* **144**, 3845 (1997).

[23] J. Newman and W. Tiedemann, "Porous-Electrode Theory with Battery Applications," *AIChE J.*, **21**, 41 (1975).
[24] J. Bockris and S. Srinivasan, *Fuel Cells: Their Electrochemistry*, McGraw-Hill, New York (1969).
[25] J. Giner and C. Hunter, "The Mechanism of Operation of the Teflon-Bonded Gas Diffusion Electrode: A Mathematical Model," *J. Electrochem. Soc.*, **116**, 1124 (1969).
[26] R. P. Iczkowski and M. B. Cutlip, "Voltage Losses in Fuel Cell Cathodes," *J. Electrochem. Soc.*, **127**, 1433 (1980).
[27] A. Z. Weber, R. M. Darling, and J. Newman, "Modeling Two-Phase Behavior in PEFCs," *J. Electrochem. Soc.*, **151**, A1715 (2004).
[28] D. M. Bernardi and M. W. Verbrugge, "A Mathematical Model of the Solid-Polymer Electrolyte Fuel Cell," *J. Electrochem. Soc.*, **139**, 2477 (1992).
[29] T. E. Springer, T. A. Zawodzinski, and S. Gottesfeld, "Polymer Electrolyte Fuel-Cell Model," *J. Electrochem. Soc.*, **138**, 2334 (1991).
[30] J. St-Pierre, D. P. Wilkinson, S. Knights, and M. L. Bos, "Relationships between water management, contamination and lifetime degradation in PEFC," *J. New Mater. Electrochem. Systems*, **3**, 99 (2000).
[31] R. M. Darling and J. P. Meyers, "Kinetic Model of Platinum Dissolution," *J. Electrochem. Soc.*, **150**, A1523 (2003).
[32] R. M. Darling and J. P. Meyers "Mathematical Model of Platinum Movement in PEM Fuel Cells," *J. Electrochem. Soc.*, **152**, A242 (2005).
[33] P. Piela, C. Eickes, E. Brosha, F. Garzon, and P. Zelenay, "Ruthenium Crossover in Direct Methanol Fuel Cell with Pt-Ru Black Anode," *J. Electrochem. Soc.*, **151**, A2053 (2004).
[34] J. Meyers and L. Protsailo, "Development of High-Temperature Membranes and Improved Cathode Catalysts," DOE Hydrogen Program, FY 2004 Progress Report, http://www.eere.energy.gov/hydrogenandfuelcells/pdfs/annual04/iva2_meyers.pdf
[35] M. S. Wilson, F. H. Garzon, K. E. Sickafus, and S. Gottesfeld, "Surface Area Loss of Supported Platinum in Polymer Electrolyte Fuel Cells," *J. Electrochem. Soc.*, **140**, 2872 (1993).
[36] J. Xie, D. L. Wood III, K. L. More, P. Atanassov, and R. L. Borup, "Microstructural Changes of Membrane Electrode Assemblies During PEFC Durability Testing at High Humidity Conditions," *J. Electrochem. Soc.*, **12**, A1011 (2005).
[37] P. Wynblatt and N. A. Gjostein, "Particle Growth in Model Supported Metal Catalysts –I. Theory," *Metall. Acta*, **24**, 1165 (1976).
[38] B. J. McCoy, "A New Population Balance Model for Crystal Size Distributions," *Journal of Colloid and Interface Science*, **240**, 139 (2001).
[39] C. T. Campbell, S. C. Parker, and D. E. Starr, "The Effect of Size-Dependent Nanoparticle Energetics on Catalyst Sintering," *Science*, **298**, 811 (2002).
[40] K. Kinoshita, *Carbon: Electrochemical and Physicochemical Properties*, John Wiley & Sons, New York (1988).
[41] A. Taniguchi, T. Akita, K. Yasuda, and Y. Miyazaki, "Analysis of Electrocatalyst Degradation in PEMFC Caused by Cell Reversal During Fuel Starvation," *J. Power Sources*, **130**, 42 (2004).
[42] C. A. Reiser, L. Bregoli, T. W. Patterson, J. S. Yi, J. D. Yang, M. L. Perry, and T. D. Jarvi, "A Reverse-Current Decay Mechanism for Fuel Cells," *Electrochem. Solid-State Lett.*, **8**, A273 (2005).

[43] G. Skala, "Automotive PEM Stack Freeze Requirements & Suggested Fundamental Studies," presentation at 2005 DOE Freeze Workshop on Fuel Cell Operations at Subfreezeing Temperatures, February 1–2, 2005, Phoenix, AZ, http://www.eere.energy.gov/hydrogenandfuelcells/pdfs/02_skala_lanl.pdf

[44] J. P. Meyers and R. M. Darling, "Model of Carbon Corrosion in PEM Fuel Cells," *J. Electrochem. Soc.*, submitted.

[45] S. Burlatsky, N. Cipollini, D. Condit, T. Madden, and V. Atrazhev, "*Multi-scale modeling considerations for PEM fuel cell durability,*" 208th Meeting of The Electrochemical Society, Los Angeles, California, October 16-21, 2005.

[46] E. Cho, J.-J. Ko, H. Y. Ha, S.-A. Hong, K.-Y. Lee, T.-W. Lim, and I.-H. Oh, "Characteristics of the PEMFC Repetitively Brought to Temperatures Below 0°C," *J. Electrochem. Soc.*, **150**, A1667 (2003).

[47] J. P. Meyers, "Fundamental Issues in Subzero PEMFC Startup and Operation," presentation at 2005 DOE Workshop on Fuel Cell Operations at Subfreezing Temperatures, February 1-2, 2005, Phoenix, Arizona. http://www.eere.energy.gov/hydrogenandfuelcells/pdfs/03_meyers_distribution.pdf

[48] M. Oszcipok, D. Riemann, U. Kronewett, M. Kreideweis, and M. Zedda, "Statistic Analysis of Operational Influences on the Cold Start Behaviour of PEM Fuel Cells," *J. Power Sources*, **145**, 407 (2005).

[49] A. W. Rempel, J. S. Wettlaufer, and M. G. Worster, "Premelting Dynamics in a Continuum Model of Frost Heave," *J. Fluid Mech.*, **498**, 227 (2004).

[50] S. He and M. Mench, "One-dimensional transient model for frost heave in polymer electrolyte fuel cells," *J. Electrochem. Soc.*, **153**, 9, A1724 (2005).

2
Catalyst Layer Operation in PEM Fuel Cells: From Structural Pictures to Tractable Models

B. ANDREAUS AND M. EIKERLING

2.1. Catalyst Layers: The Pacemakers

While the membrane represents the heart of the fuel cell, determining the type of cell and feasible operating conditions, the two catalyst layers are its pacemakers. They fix the rates of electrochemical conversion of reactants. The anode catalyst layer (ACL) separates hydrogen or hydrocarbon fuels into protons and electrons and directs them onto distinct pathways. The cathode catalyst layer (CCL) rejoins them with oxygen to form liquid water. This spatial separation of reduction and oxidation reactions enables the electrons to do work in external electrical appliances, making the Gibbs free energy of the net reaction, $-\Delta G$, available to them.

Under conditions of electrochemical equilibrium, the chemical composition determines the electromotive force (EMF) of the complete cell. The EMF is given by the Nernst-equation, i.e. for an $H_2|O_2$ fuel cell as the simplest system,

$$E^{eq}_{cell} = -\frac{\Delta G}{2F} = E^0 + \frac{RT}{2F} \ln\left(\frac{[H_2][O_2]^{1/2}}{[H_2O]}\right) \qquad (2.1)$$

where $E^0 = 1.23\text{V}$ is the standard EMF (at 25°C, standard concentrations).

In an operating fuel cell, kinetically hindered reactions like oxygen reduction (cathode) or methanol oxidation (anode) incur major voltage losses relative to the EMF. On the anode side, hydrogen oxidation is a rapid reaction with negligible kinetic losses. Poisoning by trace amounts of carbon monoxide in the hydrogen feed or by reaction intermediates of methanol oxidation in direct methanol fuel cells may incur substantial voltage losses in the ACL. Moreover, anode processes give rise to a wealth of transient phenomena [1–4]. Most efforts in electrode theory and experiment focus, however, on the CCL, where oxygen reduction is responsible for voltage losses of about $0.3 - 0.5\text{V}$ at relevant current densities.

At operating temperatures dictated by the proton-conducting PEM ($< 90°\text{C}$ with Nafion or $\sim 150°\text{C}$ with high-T PEM), only Pt-based catalysts provide requisite reaction rates and sufficient stability in the corrosive environment of PEMFCs [5–7]. This has not changed since the first fuel cell experiments of Sir William Grove in 1839, in spite of intensive research on non-Platinum Group Metal catalysts in recent years [7,8].

The primary objective in catalyst layer development is to obtain highest possible rates of desired reactions with a minimum amount of the expensive Pt (DOE target for 2010: 0.2 g Pt per kW). This requires a huge electrochemically active catalyst area and small barriers to transport and reaction processes. At present, random multiphase composites comply best with these competing demands. Since a number of vital processes interact in a nonlinear way in these structures, they form inhospitable systems for systematic theoretical treatment. Not surprisingly then, most cell and stack models, in particular those employing computational fluid dynamics, treat catalyst layers as infinitesimally thin interfaces without structural resolution.

There are, however, several good reasons to consider catalyst layers as supreme targets for detailed structural modeling. In the first place, they play a key role for fuel cell performance. Catalyst layers control the circulation and interconversion of reactants, protons, electrons, and product water. Thereby, they largely determine efficiency, power density, and water-handling capabilities of the whole cell. Furthermore, since Pt is costly and its resources are limited, the search for inexpensive catalyst materials and advanced catalyst layer design with minimized Pt loading is a major thrust in fuel cell research and development. Finally, the complexity of catalyst layer design and operation implies a highly stimulating diversity in the relevant research topics, including molecular modeling, nanoscience, interfacial electrochemistry and surface science, statistical geometry of random composite materials, and multiphase flow in porous media.

For the most part of this chapter, we will focus on steady-state operation of the CCL. This corresponds to a nonequilibrium situation with constant but nonuniformly distributed fluxes and conversion rates. Understanding how structure, composition, and operating conditions control rates of electrocatalytic processes and their complex spatial distributions constitutes the basic task of catalyst layer modeling. This entails a complex phenomenology on multiple scales, as outlined in Figure 2.1. The optimum of the structure-performance relationship is a result of the trade-off between competing requirements. Increasing, for instance, the ionomer content automatically decreases the volume portion of the catalyst, substrate and gas pores; by overdoing it, one may loose in reactivity, gas diffusion, and in the end even electronic conductivity if the connectivity of the percolating metal/carbon network is impaired.

How are we going to disentangle this "mess"? The strategy of the modeling approaches reviewed in this contribution is to start from appropriate structural elements, identify relevant processes, and develop model descriptions that capture major aspects of catalyst layer operation. In the first instance, this program requires theoretical tools to relate structure and composition to relevant mass transport coefficients and effective reactivities. The theory of random

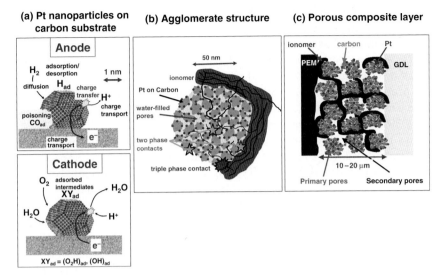

FIGURE 2.1. Structure and composition of catalyst layers at three different scales: At the nanoparticle level, anode and cathode processes are depicted, including possible anode poisoning by CO. At the agglomerate level, ionomer functions as binder and proton-conducting medium are indicated, and points with distinct electrochemical environments are shown (double- and triple-phase boundary). At the macroscopic scale, the interpenetrating percolating phases of ionomer, gas pores, and solid Pt/Carbon are shown, and the bimodal porous structure is indicated.

heterogeneous media provides a variety of approaches for obtaining such structure-property relations [9,10]. The level of detail of structural information that could and should be included depends on how precisely and reproducibly the relevant structures can be fabricated and how well they can be characterized by ex situ diagnostics. In the second part of the theory, homogenized expressions for mass transport coefficients and reaction rates are used in deterministic mathematical models of CCL operation, formulated on the basis of conservation laws for particle numbers, mass, and energy exchange.

Overall, this structure-based approach has proven its assets in electrode diagnostics and in providing guidelines for the optimized structural design of catalyst layers [11–15]. Vital performance aspects that can be rationalized are catalyst utilization, distributions of reactants, reaction rates, and electrode potential, as well as effects on the overall fuel cell power density and water balance.

2.2. Structural Picture

At the outset, it is instructive to consider the basic structural elements and some key expressions in electrode theory. The two major steps in catalyst layer development during the last 10–20 years have been the advent of highly dispersed Pt catalysts with particle sizes in the range of 1–10 nm, deposited on high surface area mesoporous carbon, as well as the impregnation with Nafion ionomer. The first step guarantees a large surface area of the catalyst. The second step, in cooperation with sufficient gas porosity, facilitates the simultaneous supply and removal of protons, electrons, reactant molecules, and product water to and from the reaction sites. Together, these modifications enabled the reduction of catalyst loading from about 4–10 mg cm^{-2} (in the 1980s) to about 0.4 mg cm^{-2} [16].

The resulting three-phase composite structure that fulfils interrelated requirements of large catalyst surface area and good mass transport is depicted in Figure 2.1c [17–19]. Typical thicknesses of such layers lie in the range $L \approx 10 - 20 \mu m$. This picture agrees with experimental data that suggest an agglomerated structure with bimodal pore size distribution (psd) in the mesoporous region [14,20–23]. Carbon particles (10 − 20 nm) aggregate and form agglomerates. Primary pores (3 − 10 nm) exist inside agglomerates. Larger, secondary pores (10 − 50 nm) build the pore spaces between agglomerates. The relative volume portions of the two types of pores depend on the type of carbon substrate and the ionomer content. As indicated in Figure 2.1c, the competition between proton-conducting ionomer and void spaces for gaseous diffusion mainly unfolds in the secondary pores. Ionomer molecules in primary pores function mainly as binder, not as proton conductor.

The macroscopic catalyst layer composition is specified in terms of the volume portions of solid Pt/C (X_{PtC}) and membrane electrolyte (X_{el}). The remaining volume is occupied by pores with total porosity $X_p = 1 - X_{PtC} - X_{el}$. For modeling purposes, it is expedient to utilize a bimodal log-normal psd, as distributions of this type are often found experimentally [24]

$$\frac{dX_p(r)}{dr} = \frac{1 - X_{PtC} - X_{el}}{\sqrt{\pi}\{\ln s_\mu + \chi_M \ln s_M\}} \frac{1}{r} \left\{ \exp\left[-\left(\frac{\ln(r/r_\mu)}{\ln s_\mu}\right)^2\right] \right.$$
$$\left. + \chi_M \exp\left[-\left(\frac{\ln(r/r_M)}{\ln s_M}\right)^2\right] \right\} \quad (2.2)$$

This psd is normalized to X_p. The parameter χ_M controls the contributions of primary and secondary pores to total porosity, r_μ, r_M determine the positions of the two peaks and s_μ, s_M their widths. Porosities due to primary and secondary pores are given by

$$X_\mu = \int_0^{r_{cut}} dr' \frac{dX_p(r')}{dr'} \text{ and } X_M = \int_{r_{cut}}^{\infty} dr' \frac{dX_p(r')}{dr'} \quad (2.3)$$

where r_{cut} ($r_\mu < r_{cut} < r_M$) separates the two domains.

The liquid water saturation S_r is obtained by integration over the distribution function

$$S_r = \frac{1}{X_p} \int_0^{r_c} dr' \frac{dX_p(r')}{dr'} \tag{2.4}$$

where r_c is the capillary radius. Pores with radius $r' \leq r_c$ are filled with liquid water and pores with $r' > r_c$ are filled with gas. In an operating fuel cell, S_r depends on details of the psd, as well as boundary conditions, stationary fluxes of species, and rates of current generation and net evaporation [25]. These conditions determine distributions of gas pressures and liquid water pressure, which control the local value of r_c as explored in Ref. [25].

Model psds with varying values of X_μ and X_M are shown in Figure 2.2. At zero current and other operating conditions as specified in the figure, capillary equilibrium exists in pores with radius $r^{eq} = 5.5$ nm.

Wetting of primary pores is essential in view of attaining high catalyst utilization and evaporation rates, while gas-filled secondary pores form the major percolation pathways for the diffusion of oxygen and water vapor. Obviously, the balance between primary and secondary pores steers the interplay of electrochemical activity, evaporation, and gaseous transport, as discussed in Section 2.6.

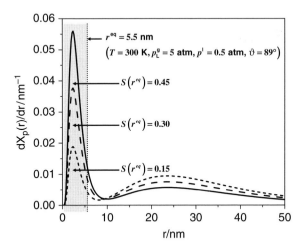

FIGURE 2.2. Model pore size distributions that represent bimodal porous structures in CCLs, calculated with Eq. (2.2) [25]. The three distributions posses the same total porosity but varying volume portions of primary and secondary pores. The equilibrium capillary radius, r^{eq}, is shown for the specified operating conditions. Values of the corresponding liquid water saturations (areas under the distribution functions within the grey box) are given.

Basically, a fuel cell electrode can, thus, be seen as a highly dispersed interface between Pt and electrolyte (ionomer or water). Due to the random composition, complex spatial distributions of electrode potential, reaction rates, and concentrations of reactants and water evolve under PEMFC operation. A subtle electrode theory has to establish the links between these distributions.

2.3. Fundamentals of Electrode Processes

2.3.1. Faradaic Current Density

Transition state theory (TST) provides a general framework to calculate rates of chemical reactions [26,27]. Since, in this chapter, we will be mainly concerned with cathode processes, we start by considering the net rate at the cathode, established as the difference of reduction (cathodic reaction) and oxidation (anodic reaction) rates [28,29]

$$v_{net} = K_{red} c_{ox}^S - K_{ox} c_{red}^S \tag{2.5}$$

where c_{ox}^S, c_{red}^S are surface concentrations of oxidized and reduced species. In a simple phenomenological treatment of electron transfer, the rate constants can be written as [27,29]

$$K_{red} = A \exp\left(-\frac{\Delta G_{red}^\dagger(E)}{RT}\right), K_{ox} = A \exp\left(-\frac{\Delta G_{ox}^\dagger(E)}{RT}\right), A = \kappa_{el} \frac{\omega_{eff}}{2\pi}, \tag{2.6}$$

where the molar Gibbs free energies of activation for reduction and oxidation, $\Delta G_{red}^\dagger(E)$ and $\Delta G_{ox}^\dagger(E)$, are functions of the electrode potential E. The latter is defined as the difference in Galvani potentials between the bulk of the metal and the bulk of the electrolyte solution, $E = \phi_{PtC} - \phi_{el}$. The pre-exponential factor A contains an effective frequency factor ω_{eff}, which incorporates contributions from degrees of freedom that reorganize during the transition, and an electronic transmission factor κ_{el} [27]. Expanding $\Delta G_{red}^\dagger(E)$ and $\Delta G_{ox}^\dagger(E)$ about the standard equilibrium potential E^0 up to terms of first order gives

$$\Delta G_{red}^\dagger(E) = \Delta G_{red}^\dagger(E^0) + \alpha F(E - E^0) \tag{2.7}$$

and

$$\Delta G_{ox}^\dagger(E) = \Delta G_{ox}^\dagger(E^0) - \beta F(E - E^0) \tag{2.8}$$

with transfer coefficients

$$\alpha = \frac{1}{F}\frac{\partial \Delta G^{\dagger}_{\text{red}}}{\partial E}\bigg|_{E^0} > 0 \text{ and } \beta = -\frac{1}{F}\frac{\partial \Delta G^{\dagger}_{\text{ox}}}{\partial E}\bigg|_{E^0} > 0 \qquad (2.9)$$

The relations

$$\Delta G^{\dagger}_{\text{ox}}(E^0) = \Delta G^{\dagger}_{\text{red}}(E^0) = \Delta G^{\dagger,0} \text{ and } \alpha + \beta = 1 \qquad (2.10)$$

lead to expressions for the rate constants, which are valid for simple "outer-sphere reactions" without specific adsorption of reactants

$$K_{\text{red}} = k^0 \exp\left\{-\frac{\alpha F(E - E^0)}{RT}\right\}, K_{\text{ox}} = k^0 \exp\left\{\frac{(1-\alpha)F(E - E^0)}{RT}\right\}, \qquad (2.11)$$

with equilibrium rate constant (units [cm s^{-1}])

$$k^0 = A \exp\left\{-\frac{\Delta G^{\dagger,0}}{RT}\right\} \qquad (2.12)$$

and cathodic transfer coefficient α. The resulting Faradaic current density for the reaction at the cathode, Red → Ox + e$^-$(Me), corresponding to the net rate in Eq. (2.5), is

$$j_F = Fk^0\left\{c^s_{\text{ox}}\exp\left(-\frac{\alpha F(E - E^0)}{RT}\right) - c^s_{\text{red}}\exp\left(\frac{(1-\alpha)F(E - E^0)}{RT}\right)\right\}, \qquad (2.13)$$

where the condition

$$E < E^0 + \frac{RT}{F}\ln\frac{c^s_{\text{ox}}}{c^s_{\text{red}}}$$

is necessary for the net reaction to proceed in cathodic direction (i.e. with $j_F > 0$). Eq. (2.13) links local values of current density, reactant concentrations, and electrode potential. It is widely known as the Butler-Volmer equation. As indicated, it can be derived from transition state theory and electron transfer theory for simple outer-sphere one-electron transfer processes. For reactions encountered in fuel cell electrocatalysis, which are typically complex multistep processes, it should be regarded in an effective way, as a phenomenological expression. Its applicability and relevant parameters (rate constants, transfer coefficients, and equilibrium potentials) have to be validated by experimental data.

In the given form, the Butler-Volmer equation is applicable rather broadly, for flat model electrodes, as well as for heterogeneous fuel cell electrodes. In the latter case, concentrations in Eq. (2.13) are local concentrations, established by mass transport and reaction in the random composite structure. At equilibrium, $j_F = 0$, concentrations are uniform. These externally controlled equilibrium concentrations serve as the reference (superscript *ref*) for defining the equilibrium electrode potential via the Nernst equation,

$$E^{eq} = E^0 + \frac{RT}{F} \ln\left(\frac{c_{ox}^{ref}}{c_{red}^{ref}}\right). \tag{2.14}$$

Defining the cathodic overpotential as the (positive) difference

$$\eta = E^{eq} - E, \text{ with } E = \phi_{PtC} - \phi_{el}, \tag{2.15}$$

we can rewrite Eq. (2.13) in the following equivalent form

$$j_F(z) = j^0 \left\{ \frac{c_{ox}(z)}{c_{ox}^{ref}} \exp\left(\frac{\alpha F \eta(z)}{RT}\right) - \frac{c_{red}(z)}{c_{red}^{ref}} \exp\left(-\frac{(1-\alpha)F\eta(z)}{RT}\right) \right\} \tag{2.16}$$

with exchange current density

$$j^0 = Fk^0 \left(c_{red}^{ref}\right)^\alpha \left(c_{ox}^{ref}\right)^{(1-\alpha)}. \tag{2.17}$$

Throughout this chapter we will neglect Ohmic losses due to electron transport in the CCL, assuming a high electronic conductivity of the Pt/C phase (>10 S cm^{-1}). For a given fuel cell current density, j_0, this phase is thus equipotential, $\phi_{PtC} = $ const. Using this condition, the total overvoltage incurred by the cathode is given by the local value of the overpotential at $z = 0$

$$\eta_0 = \eta(z = 0). \tag{2.18}$$

Eq. (2.16) can be simplified further, when all transport processes in the layer are fast and, thus, reaction rates and concentrations are distributed uniformly,

$$j_F = j^0 \left\{ \exp\left(\frac{\alpha F \eta}{RT}\right) - \exp\left(-\frac{(1-\alpha)F\eta}{RT}\right) \right\}. \tag{2.19}$$

This is the original Butler-Volmer equation. It has, however, rather limited applicability. It should be used only when electrode potential and all concentrations are uniform. Such conditions are barely encountered in fuel cell electrodes.

In distributed electrode theory, Eqs. (2.13) or (2.16) are the appropriate relations that determine local distributions of relevant functions. In most practical cases, the overpotential is large, $\eta(z) \gg RT/F$, and only the cathodic branch needs to be considered in Eq. (2.16),

$$j_F(z) \approx j^0 \frac{c_{ox}(z)}{c_{ox}^{ref}} \exp\left(\frac{\alpha F \eta(z)}{RT}\right) \qquad (2.20)$$

corresponding to the Tafel-limit.

An important detail is that the spatially varying concentrations of reduced or oxidized species appear in Eqs. (2.13) or (2.16) inside the curly brackets only, $c_{ox}(z)$ in the reduction and $c_{red}(z)$ in the oxidation term. Unfortunately, in the literature on fuel cell modeling, concentration dependencies are often written in front of the curly brackets, as if they would apply equally to forward and back reactions. This is, however, in contradiction to fundamental principles of transition state theory, as obvious from Eq. (2.15).

2.3.2. Exchange Current Density

The exchange current density is the key property of catalyst layers. It determines the value of the overpotential needed to attain the targeted fuel cell current density. This property, thus, links fundamental electrode theory with practical aspects of fuel cell performance. The following parameterization distinguishes explicitly the effects of different structural characteristics,

$$j^0 = 2 \cdot 10^3 [m_{Pt}] j^{0*} \varepsilon^{S/V} \Gamma_a g(S_r) \frac{f(X_{PtC}, X_{el})}{X_{PtC}} \qquad (2.21)$$

The real-to-apparent surface area ratio of a distributed electrode, frequently also termed heterogeneity or roughness factor, corresponds to the ratio j^0/j^{0*}. The total catalyst utilization is

$$\zeta_{Pt} = \varepsilon^{S/V} \Gamma_a g(S_r) \frac{f(X_{PtC}, X_{el})}{X_{PtC}}. \qquad (2.22)$$

The factor $2 \cdot 10^3$ on the right-hand side of Eq. (2.21) is an estimate of the total surface area per real electrode surface area that would be obtained by spreading a catalyst loading of 1mg cm^{-2} as an ideal monoatomic layer (assuming a surface atom density of $1.5 \cdot 10^{15}$ cm^{-2}). The catalyst loading $[m_{Pt}]$ has to be specified in units mg cm^{-2}. From a practical perspective, j^0 should be as large as possible to maximize the power density of the fuel cell. On the other hand, cost minimization requires a small value of $[m_{Pt}]$. Obviously, the remaining factors on the right-hand side of Eq. (2.21) should be as large as possible in order to find the best compromise of these competing optimization targets.

The generic exchange current density j^{0*} for the relevant fuel cell reactions can be studied at planar electrodes. Such studies are useful in determining the performance of different crystal facets [30–32]. In fuel cell electrocatalysts, catalyst particle sizes, particle-substrate interactions, and the surface structure of the catalyst determine the corresponding corrugated potential energy landscape at the surface. These effects play a key role in steering the balance of involved kinetic processes and in fixing the net turnover rates of desired reactions and, thus, j^{0*} [33,34]. Section 2.4 and the chapters by Marc Koper and Phil Ross in this book will address fine details of electrocatalytic processes. In Equation, an effective value of j^{0*} has to be employed, averaged over different particle sizes and surface structures.

The other factors in Eq. (2.21) account for nonideal utilization of available catalyst atoms. At the nanoparticle level, $\varepsilon^{S/V}$ is the surface-to-volume ratio of atoms in the particles. This factor depends on the detailed geometrical particle shapes (e.g. octahedral, cubo-octahedral). As an estimate, we calculate it for cubo-octahedral particles, the preferred crystalline form of supported fcc metal catalyst nanoparticles [35,36]. For this geometry, the particle radius can be defined as

$$r = \frac{a_{Me}}{2} \sqrt[3]{\frac{3}{2\pi} N_{tot}},$$

where a_{Me} is the lattice constant of the catalyst metal and N_{tot} is the total number of catalyst atoms [37]. For well-defined particle geometries, $\varepsilon^{S/V}$, as well as the size dependent fractions of specific facets and edge or kink sites are readily calculated [38]. Figure 2.3 depicts $\varepsilon^{S/V}$ as a function of r. For $r > 3.5$ nm less than 20% of the Pt atoms are on the surface, whereas

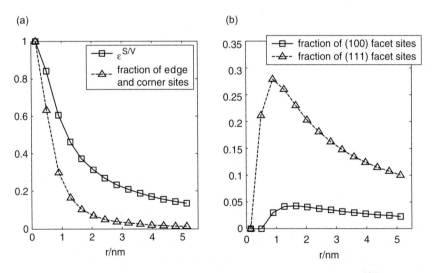

FIGURE 2.3. Calculated volume averaged fraction of total surface sites, $\varepsilon^{S/V}$, edge and corner sites, as well as different facet sites as a function of the radius of a cubo-octahedral crystallite.

$\varepsilon^{S/V} > 0.5$ for $r < 1.2$ nm. The factor Γ_a is the effectiveness of catalyst utilization in single agglomerates. It will be discussed in Section 2.5.

Next, the functions $f(X_{PtC}, X_{el})$ and $g(S_r)$ account for reduced Pt utilization due to random composition. The factorization of the effects of structure and composition into two functions f and g implies a separation of length scales. At mesoscopic scale (single agglomerate size ~ 50 – 100 nm, Figure 2.1b), catalyst utilization is determined by the internal wetted pore fraction $g(S_r)$. At this scale, rates of proton and oxygen transport are Sufficiently large, even in liquid water [39]. The wetting of pores is, however, critical in view of the accessibility of catalyst sites to protons. This requirement is expressed by $g(S_r)$. At macroscopic scale (~ 10 μm), utilization of catalyst sites demands simultaneous percolation in the three interpenetrating phases of Pt/C, ionomer, and pore network. These requirements are expressed in dependence of the statistical geometry of the composite via the function f [17–19].

Structures and compositions that give the highest values of f and g and, thus, of j^0 may not be ideal in view of the overall fuel cell performance. The full optimization has to involve all the trade-offs between rates of reactions, vaporization, and mass transport. This complex interplay will be considered in Section 2.6.

So far, we have focused on the formal description of current generation in the catalyst layer and discussed major effects of structure and composition on exchange current density and catalyst utilization. In the remainder of this chapter, we will explore in detail, how electrocatalytic activity interferes with other processes at the catalyst surface (e.g. surface diffusion) and transport in the bulk phases. The key measure of catalyst layer performance is the current density that could be extracted from a cell for a given cell potential. This links the spatially varying concentrations and reaction rates with the global performance, rated in terms of power density and fuel cell efficiency.

2.4. Nanoparticle Reactivity

2.4.1. Particle Size Effects in Electrocatalysis

The intrinsic exchange current density, j^{0*}, is not a mere materials constant, but it depends on size distributions of catalyst nanoparticles, their surface structure, as well as surface composition in the case of alloy catalysts like PtRu. In this section, we discuss modeling approaches that highlight particle size effects and the role of surface heterogeneity in fuel cell electrocatalysis.

At atomistic resolution, the rates of electrocatalytic processes may differ significantly among different surface sites. In an extreme scenario, reactions could proceed exclusively at a certain fraction of surface sites, the so-called active sites, while all other sites are electrocatalytically inactive. This effect is most obvious for alloyed catalyst when a second catalyst material is added to act as active sites through the bifunctional mechanism [40]. As we shall see, it could become important for pure metals as well.

For a given crystallite geometry, the size of the catalyst particles not only fixes the surface-to-volume ratio, $\varepsilon^{S/V}$, as depicted in Figure 2.3, but also determines the fractions of surface atoms corresponding to facets, edges and corners. This contributes to an observable particle size effect in specific activity of the catalyst. Furthermore, substrate-particle interactions may alter the electronic structure of catalyst surface atoms at the rims with the substrate [41], and the substrate may serve as a source or sink of reactants via the so-called spillover effect [42]. Finally, the electronic structure of small catalyst particles in the range of a few nanometers starts to deviate from the electronic structure of the bulk metal phase, affecting the potential energy surface experienced by molecules on the catalyst surface [41]. All of the listed phenomena contribute to an observable dependence of the electrocatalytic activity on particle size.

Let us dwell on the effect of particle sizes on surface heterogeneities and electrocatalytic activities of pure Pt catalyst particles. As pointed out by Kinoshita [37] for cubo-octahedral particles, the mass averaged distribution of (111) and (100) facets shows a maximum for a particle diameter in the range of 3–4 nm, whereas the fraction of edge and corner sites is a strongly decreasing function with increasing particle size, cf. Figure 2.3. In the case of oxygen reduction on Pt nanoparticles, a size effect in activity with a maximum activity for particles sizes around 4 nm has been observed [37,43]. As reported by Ross [30], this structure sensitivity, which follows the trends predicted by Kinoshita, is dominated by the inhibiting effects of anion adsorption on (111) and (110) facets that leave only the (100) facet sites active for oxygen reduction.

A significant size effect in activity has also been observed for CO [34] and methanol electrooxidation [44,45]. For these reactions, a significant drop in activity was observed when the particle diameter was below 2–3 nm.

2.4.2. CO Electrooxidation on Pt Nanoparticles

In the following, we present a scenario for CO electrooxidation on Pt nanoparticles to elucidate the role of particle size and surface structure. The model employs the active site concept and highlights the role of finite surface mobility of adsorbed CO.

The main steps in the reaction scheme for CO electrooxidation are (Langmuir-Hinshelwood):

$$\mathrm{CO} + n\mathrm{Pt} \underset{k_{des}}{\overset{k_{ad}}{\rightleftharpoons}} n\mathrm{Pt} - \mathrm{CO}_{ad} \qquad [\text{CO adsorption}] \qquad (2.23)$$

$$\mathrm{H_2O} + \nu\mathrm{Pt} \underset{k_b}{\overset{k_f}{\rightleftharpoons}} \nu\mathrm{Pt} - \mathrm{OH}_{ad} + \mathrm{H}^+ + \mathrm{e}^- \qquad [\text{water splitting}] \qquad (2.24)$$

$$CO_{ad} + OH_{ad} \xrightarrow{k_{ox}} CO_2 + H^+ + e^- + (v+n)Pt \quad \text{[recombination]} \quad (2.25)$$

According to Gilman [46], the recombination reaction in Eq. (2.25) involves a chemical step of formation of $COOH_{ad}$ and an electrochemical step of removal of $COOH_{ad}$. The latter is still disputed, and several pathways have been proposed [47–49].

The influence of the catalyst morphology on electrocatalytic activity becomes manifest in strong experimental [50–52] and theoretical [53,54] evidence that the recombination step, Eq. (2.25), proceeds preferentially at defect, edge, or step sites. Such sites may therefore act as active sites for the reaction. On well-defined stepped single crystal Pt surfaces, Koper and coworkers have related the active site fraction to the step site density [52,55].

When only a fraction of surface sites constitutes reaction centers, surface diffusion of reactants to these sites becomes a pivotal process. On extended surfaces, potential step experiments from Koper's group on stepped Pt(111) suggest that CO_{ad} surface mobility is very high in comparison to on-site reactivity. On small nanoparticles, a stronger CO_{ad}-catalyst interaction through trapping of CO_{ad} at low-coordination sites and stronger binding on crystal facets may lead to reduced CO_{ad} mobility and hence reduced overall activity. This is corroborated by experimentally determined activation energies of surface diffusion on CO_{ad}/Pt systems [56–58]. Indeed, NMR experiments indicate increasing activation energies of surface diffusion with decreasing particle sizes, from 0.27 eV for 10 nm particles to 0.44 eV for 1.2 nm particles [58]. In the large particle limit, CO_{ad} surface mobility approaches that of extended, planar surfaces, which is considered infinitely high for modeling purposes.

2.4.3. Active Site Concept

In general, kinetic models of electrocatalytic processes at nanoparticles should account for distributions of rates of involved processes, e.g. in Eqs. (2.23–2.25), between different sites on the heterogeneous particle surface. The minimalist's modeling approach is to use a simple two-state model with a fraction ξ_{tot} of electrocatalytically active sites as exclusive sites at which OH_{ad} can be formed, and a fraction $(1 - \xi_{tot})$ of inactive sites, at which water splitting is unlikely under feasible experimental conditions. The inclusion of finite surface mobility of adsorbed reactants to active sites leads to a complex interplay of reactant surface mobility and on-site reactivity.

A full analytical treatment of such an active site model for CO_{ad} electrooxidation is feasible only with a number of simplifying assumptions. Such a solution was presented in Ref. [34]. This model includes limited diffusivity of CO_{ad} on a circular particle toward active sites on the particle's circumference.

A comparison with potential step measurements showed that the model reproduces major experimental trends. In particular, it corroborated the idea of extremely low CO_{ad} surface mobility on small catalyst particles. The model has, however, also major restrictions: The continuum approach to surface diffusion and the simple surface geometry cannot reproduce realistic distributions of active sites. Moreover, only the product of ξ_{tot} and an effective recombination rate that includes the steps in Eqs. (2.23–2.25) could be determined.

More detailed studies of electrocatalytic processes, which incorporate heterogeneous surface geometries and finite surface mobilities of reactants, require kinetic Monte Carlo simulations. This stochastic method has been successfully applied in the field of heterogeneous catalysis on nanosized catalyst particles [59,60]. Since these simulations permit atomistic resolution, any level of structural detail may easily be incorporated. Moreover, kinetic Monte Carlo simulations proceed in real time. The simulation of current transients or cyclic voltammograms is, thus, straightforward [61].

It is instructive to consider first a mean field approximation with active sites in the limit of fast CO_{ad} surface diffusion, described in detail in Ref. [62]. Assuming infinite surface mobility of adsorbates, their individual positions are irrelevant, and the kinetics can be described in terms of surface coverages of adsorbates. This simplification permits a straightforward deterministic formulation of the kinetic equations, which serves as the basis for the Monte Carlo procedure. This model has full analytical solutions for relevant limiting cases. It generalizes previous analytical approaches. Systematic fitting procedures can be used with this model, which can provide starting values for more elaborate fitting of experimental data with the Monte Carlo approach.

2.4.4. Mean Field Model with Active Sites

The state variables of the model are the surface coverage of CO_{ad}, θ_{CO}, on inactive sites, the CO_{ad}-free fraction of active sites, θ_ξ, and the fraction of active sites covered by OH_{ad}, θ_{OH}. In the MF approach, these coverages represent local averages, normalized to the disjoint surface fractions of active and inactive sites. The ranges of variation are $0 \leq \theta_{CO} \leq 1$, i.e. normalized to $(1 - \xi_{tot})$, as well as $0 \leq \theta_{OH} \leq 1$ and $0 \leq \theta_\xi \leq 1$, each normalized to ξ_{tot}.

We consider the initial formation of free active sites with a rate

$$\nu_N = k_N(1 - \theta_\xi) \qquad (2.26)$$

corresponding to a first-order nucleation law. The full balance of OH_{ad} coverage on CO_{ad}-free active sites involves OH_{ad} formation, with rate ν_f, back reaction of OH_{ad} to H_2O with rate ν_b, and electrooxidation of CO_{ad} with OH_{ad} on a neighboring site with rate ν_{ox}. For the latter process, we distinguish between removal of CO_{ad} from an active site, ν_{ox}^{a-a}, and from an inactive site, ν_{ox}^{a-n}. The balance equations of the three variables are, thus,

$$\begin{aligned}\frac{d\theta_\xi}{dt} &= \nu_N + \nu_{ox}^{a-a} + \mu^{a\leftrightarrow n} \\ &= k_N(1-\theta_\xi) + z_{aa}k_{ox}(1-\theta_\xi)\theta_{OH} + \mu^{a\leftrightarrow n}\end{aligned} \quad (2.27)$$

$$\begin{aligned}\frac{d\theta_{CO}}{dt} &= \frac{\xi_{tot}}{1-\xi_{tot}}\left[-\nu_{ox}^{a\leftrightarrow n} + \mu^{a\leftrightarrow n}\right] \\ &= \frac{\xi_{tot}}{1-\xi_{tot}}\left[-z_{an}k_{ox}\theta_{CO}\theta_{OH} + \mu^{a\leftrightarrow n}\right]\end{aligned} \quad (2.28)$$

$$\begin{aligned}\frac{d\theta_{OH}}{dt} &= \nu_f - \nu_b - \nu_{ox}^{a-n} - \nu_{ox}^{a-a} \\ &= k_f[\theta_\xi - \theta_{OH}] - k_b\theta_{OH} - z_{an}k_{ox}\theta_{CO}\theta_{OH} - z_{aa}k_{ox}(1-\theta_\xi)\theta_{OH}\end{aligned} \quad (2.29)$$

All rate constants k_f, k_{ox}, k_N, and k_b are defined per individual process and per second. z_{an} and z_{aa} are the average nearest neighbor numbers of active/inactive and active/active sites, respectively. Introducing z_{an} and z_{aa} permits a more detailed representation of the surface structure. The term $\mu^{a\leftrightarrow n}$ allows CO_{ad} hopping between neighboring active and inactive sites.

$$\begin{aligned}\mu^{a\leftrightarrow n} &= \mu^{a\to n} - \mu^{n\to a} \\ &= z_{an}k_d^{a\to n}(1-\theta_\xi)(1-\theta_{CO}) - z_{an}k_d^{n\to a}\theta_{CO}(\theta_\xi - \theta_{OH})\end{aligned} \quad (2.30)$$

with hopping rate constants $k_d^{a\to n}$ and $k_d^{n\to a}$.

The net current density is

$$j = e_0\gamma_s\xi_{tot}\left(\nu_{ox}^{a-a} + \nu_{ox}^{a-n} + 2\nu_N + \nu_f - \nu_b\right) \quad (2.31)$$

where $e_0 = 1.602 \cdot 10^{-19}$C is the elementary charge and γ_s is the surface atom density of the catalyst. Here, it is assumed that each nucleation step involves the transfer of two electrons, one for the OH_{ad} formation and the other for the net recombination step in Eq. (2.25).

In general, the system of Eqs. (2.27–2.29) must be solved numerically. Analytical solutions can be obtained under simplifying assumptions. When active-inactive site hopping is negligible, $\mu^{a\leftrightarrow n} = \mu^{a\to n} = \mu^{n\to a} = 0$, analytical expressions can be found for the limits of fast and slow OH_{ad} formation, $k_f \gg k_{ox}\theta_{CO}, k_b$ and $k_b \gg k_{ox}\theta_{CO}, k_f$.

In the limit of fast OH_{ad} formation, the solution is

$$\theta_\xi(t) = \theta_{OH}(t) = \frac{c_\xi e^{k_N t} e^{z_{aa} k_{ox} t} - 1}{c_\xi e^{k_N t} e^{z_{aa} k_{ox} t} + \kappa}; \qquad (2.32)$$

$$\theta_{CO}(t) = \theta_{CO}^0 \left[\frac{(c_\xi + \kappa) e^{k_N t}}{c_\xi e^{k_N t} e^{z_{aa} k_{ox} t} + \kappa} \right]^{\left(\frac{\xi_{tot}}{(1-\xi_{tot})} \frac{z_{an}}{z_{aa}}\right)} \qquad (2.33)$$

with $c_\xi = \left(1 + \kappa \theta_\xi^0\right)\left(1 - \theta_\xi^0\right)$, $\kappa = z_{aa} k_{ox} k_N$. Superscripts '0' refer to the initial coverages. The transient current density is

$$j(t) = e_0 \gamma_s \xi_{tot} \left[2 k_N (1 - \theta_\xi) + 3 z_{aa} k_{ox} (1 - \theta_\xi) \theta_\xi + 2 z_{an} k_{ox} \theta_{CO} \theta_\xi \right] \qquad (2.34)$$

Note that in the homogeneous surface limit, $\xi_{tot} = 1$, Eqs. (2.32) and (2.34) reduce to well-known expressions [61,63,64],

$$j(t) = 2 e_0 \gamma_s \frac{k_{ox} \exp\{k_{ox}(t - t_{max})\}}{[1 + \exp\{k_{ox}(t - t_{max})\}]^2} \text{ with } t_{max} = -\frac{1}{k_{ox}} \ln\left\{\frac{1 - \theta_{CO}^0}{\theta_{CO}^0}\right\}, \qquad (2.35)$$

valid for fast OH_{ad} formation.

In the limit of slow OH formation, $k_b \gg k_{ox}\theta_{CO}, k_f$, the form of Eqs. (2.32) and (2.33) is preserved, however, with $\theta_{OH} = k_f/k_b \theta_\xi$, and the replacement $k_{ox} \to k_{ox} k_f/k_b$.

Chronoamperometric current transients calculated according to Eq. (2.34) are in general asymmetric, with a noticeable tailing after the current maximum. They reproduce main characteristics of experimental transients for large catalyst nanoparticles. The parameters that define the surface structure in the model, ξ_{tot}, z_{aa} and z_{an}, influence the degree of asymmetry and the position of the current maximum. A detailed parameter study can be found in Ref. [62].

The general analytical model, Eqs. (2.27–2.29), shows very good agreement with experimental current transients for Pt nanoparticles with diameter > 4 nm, as shown in Ref. [62]. The comparison with experimental data provides rate constants and Tafel parameters (or transfer coefficients) of the individual processes, as well as the equilibrium potential of OH_{ad} formation. Effects of surface structure can be rationalized. Tafel plots reveal that at low potentials the overall kinetics is limited by low OH_{ad} concentrations. The OH_{ad} formation reaction, Eq. (2.24), exhibits a Tafel slope of 120 mV/dec with equilibrium potential around 0.7 V vs. NHE. k_{ox} shows a similar potential dependence as the OH_{ad} formation, with a Tafel slope of ~105 mV/dec. When the step potentials deviate by ~50 mV from the OH_{ad} equilibrium potential ($k_b \approx k_f$), Eq. (2.34) may be used to calculate the current transients.

2.4.5. Kinetic Monte Carlo Simulation with Finite Surface Mobility

In order to fully account for finite surface mobilities and the heterogeneous surface structure, we have to employ a stochastic description of the surface processes. The kinetic Monte Carlo method enables the incorporation of structural details at an atomistic level. This method has been applied successfully in the field of heterogeneous (electro-) catalysis [59–61,65] and is further discussed in the chapter by Phil Ross in this book. In the model, hexagonal grids represent catalyst particles. The active sites are randomly distributed on the grid. Adsorbates are considered to bind to on-top sites. The first reaction method was used [66,67].

The effect of reduced CO_{ad} mobility on the overall oxidation kinetics is seen in Figure 2.4a. The current peak is reduced and a significant current tailing develops at long times. We can compare results of the kMC simulations with the MF model in the limit of high CO_{ad} mobility. In the MF approach, all points on the surface are equivalent. It neglects structural correlations between positions of active sites. In the kMC approach, such correlations due to formation of clusters of active sites are explicitly accounted for. The extent of such correlations is determined by z_{aa} and z_{an}, as well as by ξ_{tot}. In cases with significant degree of cluster formation, the MF approach gives inaccurate results. The condition for good agreement between MF and kMC approaches is $z_{aa}k_{ox} \leq k_N$, for which correlation effects are insignificant. On the other hand, systematic deviations of MF results from the kMC results are expected for surfaces with pronounced cluster formation of active site and slow nucleation rates.

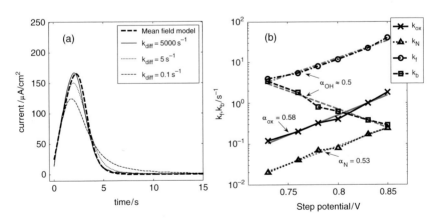

FIGURE 2.4. (a) Effect of finite CO_{ad} surface hopping rate, k_{diff}, on the shape of current transients. For high mobility, the kinetic MC simulations approach the transients of the MF model. With decreasing mobility a more pronounced current tailing arises at long times (parameters used in calculations: $\xi_{tot} = 0.3$, $k_{ox} = 1 s^{-1}$, $k_N = 0.2 \ s^{-1}$, $k_f = 50 \ s^{-1}$ and $k_{ox} = 0.5 \ s^{-1}$). (b) Potential dependence of fitted kinetic parameters for small nanoparticles (1.8 nm diameter). Transfer coefficients for nucleation, OH_{ad} formation, and recombination are indicated.

The increased current tailing at longer times along with a shift of the current peak to longer times found in kMC simulations with low CO surface mobility, *cf.* Figure 2.4a, is characteristic for experimental transients on small nanoparticles (~1.8 nm). Overall, the simulated transients capture all the essential features of experimental current transients. Analogous as for large nanoparticles, the model fits chronoamperometric current transients for various potentials and, thereby, explore effects of particle size and surface structure on rate constants, Tafel parameters (transfer coefficients), and equilibrium potentials. Due to the stochastic nature of the MC approach, systematic optimization of the fits is a much more delicate task.

The best approximations to the experimental data, obtained so far, are obtained with active site fraction of $\xi_{tot} \approx 0.05 - 0.1$. The fits reveal an approximately 10 times lower rate k_{ox} on small particles than on large particles (diameter > 4 nm). The OH_{ad} formation process exhibits only a slight positive shift in the equilibrium potential to 0.73 V in comparison with large nanoparticles, see Figure 2.4b. The same potential dependence as for large nanoparticles was found for k_{ox}. Fitting required a low surface diffusion coefficient for CO_{ad}, $D_{CO_{ad}} \approx 10^{-16} cm^2 s^{-1}$, as reported in Ref. [34]. This low value is in the range of CO_{ad}-diffusivities on nanoparticles that were reported by NMR measurements, $10^{-13} cm^2 s^{-1} > D_{CO_{ad}} > 10^{-16} cm^2 s^{-1}$ [58,68].

As conclusion to this section, we have stressed the influence of the heterogeneous surface structure on catalyst activity. Two complementary model approaches have been presented that could be used to establish the effective value of j^{0*}. Kinetic MC simulations provide the most versatile tool for exploring structure-reactivity relations. It allows incorporating detailed surface structures and finite mobilities of adsorbates. In the limit of high surface mobilities of adsorbates, an analytical MF model could replace the stochastic description. The MF model still accounts for the heterogeneous surface structure. It is, thus, more realistic for real catalyst systems in fuel cells than homogeneous surface models.

Successful modeling at this level relies heavily on well-defined experimental systems [69]. In return, the knowledge of surface textures, which either favor or impede electrocatalytic activity bears great potential for significant improvements of j^{0*} with existing catalyst materials. This understanding could be highly useful for the design of atomistically structured catalyst surfaces.

2.5. Catalyst Utilization and Performance at the Agglomerate Level

At the level of single agglomerates, *cf.* Figure 2.1b, complex distributions of reactants, reaction rates, and electrode potential emerge, mainly caused by electrostatic effects. These effects lead to a modified Tafel law equal in form to Eq. (2.20) but including a factor for the effectiveness of catalyst utilization in agglomerates, Γ_a.

The typical agglomerate size is in the range of 50–100 nm. The internal porosity of agglomerates is vital for catalyst utilization and evaporation rates, as discussed below. Structural elements resembling agglomerates have been observed in TEM pictures of catalyst layer cross-sections [14,70,71]. At present, the mechanisms of agglomerate formation are not understood. It is, however, tentatively known, how distinct procedures in catalyst layer preparation lead to distinct agglomerate compositions and structures. Preparation methods using the colloidal approach lead to bimodal pore size distributions with primary and secondary pores [22,23]. With the ink preparation method, for which the ionomer solidifies from solution during the drying process, a significant portion of the primary pores may be filled with ionomer. Resulting pore size distributions are, thus, essentially unimodal, consisting mainly of secondary pores [69].

The motives for modeling of performance at the single agglomerate level are, thus, multifold. There are few open reports on the effect of composition inside agglomerates on mass transport, electrochemical reactions, and catalyst utilization. It is poorly understood whether pores inside agglomerates should be hydrophobic or hydrophilic. Agglomerates are of vital importance in determining the true active area and the effectiveness of catalyst utilization.

Spherical and planar model geometries of agglomerates will be considered. The planar geometry leads to important conclusions about the performance of ultrathin catalyst layers.

2.5.1. Spherical Agglomerates

As illustrated in Figure 2.1b, ideal locations of Pt particles are at the true triple-phase boundary, highlighted by the big star. Catalyst particles with nonoptimal double-phase contacts are indicated by the smaller stars. Pt|gas interfaces are inactive due to the inhibited access to protons. Bulky chunks of ionomer on the agglomerate surface build the percolating network for proton conduction in secondary pores. Only individual or loosely connected ionomer molecules seem to be able to penetrate the small primary pores. It is unlikely that they could sustain notable proton conductivity. They merely act as a binder. Proton transport inside agglomerates, thus, predominantly occurs via water-filled primary pores, toward Pt|water interfaces.

The simplified model system, considered in detail in Ref. [39], is depicted in Figure 2.5. The agglomerate is spherical with radius R_a. It consists of a Pt/C matrix, with primary pores completely wetted by liquid water. For simplicity, we assume a uniform ionomer film on the agglomerate surface. Proton and oxygen concentrations (c_{H^+}, c_{O_2}) and electric potential in water (ϕ_{el}) are functions of the radial coordinate r. Uniform boundary conditions apply at $r = R_a$. Since the Pt/C phase is assumed to be equipotential, we can consider variations in overpotential, $\eta(r)$, instead of $\phi_{el}(r)$. The functions $c_{H^+}(r)$,

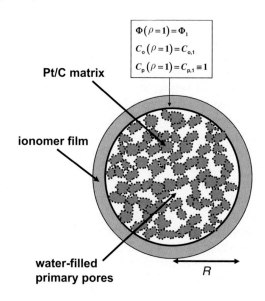

FIGURE 2.5. Model of spherical agglomerate with radius $R_a \approx 100 - 200$ nm and two-phase composition (Pt/C and water-filled primary pores), as explored in Ref. [39]. Uniform boundary conditions are assumed on the agglomerate surface.

$c_{O_2}(r)$, and $\eta(r)$ determine the Faradaic current density (in [A cm^{-3}]), given by the Tafel equation, cf. Eq. (2.20),

$$i_F(r) = i_a^0 \left(\frac{c_{O_2}(r)}{c_{O_2}^{ref}}\right)^{\gamma_{O_2}} \left(\frac{c_{H^+}(r)}{c_{H^+}^{ref}}\right)^{\gamma_{H^+}} \exp\left(\frac{\alpha F}{RT}\eta(r)\right) \qquad (2.36)$$

Here, it is reasonable to define the reference oxygen concentration $c_{O_2}^{ref}$ as the oxygen concentration at the CCL|GDL interface. The reference proton concentration $c_{H^+}^{ref}$ on the surface of the agglomerate is determined by the type of ionomer, in particular its IEC, and the volume portion of ionomer with which the CCL is impregnated. For an orientation, a typical value is $c_{H^+}^{ref} \approx 0.5$ mol l^{-1}.

Eq. (2.21) can be used to determine the exchange current density (per unit volume) at the agglomerate level, i_a^0. The factor $f(X_{PtC}, X_{el})X_{PtC}$, which accounts for macroscopic percolation effects in the three-phase composite, has to be omitted. Moreover, assuming complete flooding of primary pores, i.e. $g(S_r) \equiv 1$, we obtain

$$i_a^0 = \frac{2 \cdot 10^3 [m_{Pt}] \varepsilon^{S/V} j^{0*}}{(X_{PtC} + X_\mu)} \cdot \frac{1}{L}. \qquad (2.37)$$

According to Parthasarathy et al. [72], at high current density the one-electron transfer step $O_2 + H^+ + e^- \rightarrow (O_2H)_{ad}$ is rate determining. Therefore, the reaction orders γ_{O_2} and γ_{H^+} are taken as unity.

The steady-state flux of hydrated protons in the agglomerate is due to diffusion and migration in the internal electric field. It is dictated by the Nernst-Planck equation, with a sink term, i_F, due to electrochemical reactions at the dispersed Pt|water interfaces,

$$-FD_{H^+}^{\text{eff}} \frac{1}{r^2} \frac{d}{dr}\left[r^2\left(\frac{dc_{H^+}}{dr} + \frac{F}{RT}c_{H^+}\frac{d\eta}{dr}\right)\right] = -i_F. \tag{2.38}$$

We use Bruggemann formula to relate the effective diffusion coefficient of protons in micropores, $D_{H^+}^{\text{eff}}$, to the relative volume portion of micropores in agglomerates and the proton diffusion coefficient in water,

$$D_{H^+}^{\text{eff}} = D_{H^+}^{w}\left(\frac{X_\mu}{X_{PtC} + X_\mu}\right)^{3/2}. \tag{2.39}$$

The potential variation is obtained from the Poisson equation,

$$-\varepsilon\varepsilon_0 \frac{1}{r^2}\frac{d}{dr}(r^2 \frac{d\eta}{dr}) = -Fc_{H^+} \tag{2.40}$$

where ε is the relative dielectric permittivity of water in pores and ε_0 is the dielectric constant. Eqs. (2.38) and (2.40) form the basis of the Poisson-Nernst-Planck (PNP) theory, which is widely used in the context of ion transport through biological membranes [73–75]. Oxygen diffusion is given by Fick's law

$$-FD_{O_2}^{\text{eff}} \frac{1}{r^2}\frac{d}{dr}(r^2 \frac{dc_{O_2}}{dr}) = -\frac{i_F}{4} \tag{2.41}$$

where the effective oxygen diffusion coefficient is related by Bruggemann's relation

$$D_{O_2}^{\text{eff}} = D_{O_2}^{w}\left(\frac{X_\mu}{X_{PtC} + X_\mu}\right)^{3/2} \tag{2.42}$$

to agglomerate composition and bulk diffusion coefficient of oxygen in liquid water, $D_{O_2}^{w}$.

We rewrite Equations, (2.38), (2.40) and (2.41) in dimensionless form

$$-\frac{1}{\rho^2}\frac{d}{d\rho}\left[\rho^2\left(\frac{dC_p}{d\rho} + C_p\frac{d\Phi}{d\rho}\right)\right] = -K_1 C_o C_p \exp(\alpha\Phi) \tag{2.43}$$

$$\frac{1}{\rho^2}\frac{d}{d\rho}\left(\rho^2 \frac{d\Phi}{d\rho}\right) = -K_2 C_p \tag{2.44}$$

$$-\frac{1}{\rho^2}\frac{d}{d\rho}\left(\rho^2\frac{dC_o}{d\rho}\right) = -\delta K_1 C_o C_p \exp(\alpha\Phi) \qquad (2.45)$$

where $\rho = r/R_a$, $C_o = c_{O_2}/c_{O_2}^{\text{ref}}$, $C_p = c_{H^+}/c_{H^+}^{\text{ref}}$, $\Phi = \frac{F}{RT}\eta$ and $K_1 = \frac{i_a^0}{Fc_{H^+}^{\text{ref}} D_{H^+}^{\text{eff}}} R_a^2$, $\delta = \frac{D_{H^+}^{\text{eff}} c_{H^+}^{\text{ref}}}{4 D_{O_2}^{\text{eff}} c_{O_2}^{\text{ref}}}$, $K_2 = \frac{R_a^2}{\lambda_D^2}$ with the Debye-length $\lambda_D = \left(\frac{\varepsilon\varepsilon_0 RT}{F^2 c_{H^+}^{\text{ref}}}\right)^{1/2}$. Spherical symmetry imposes the boundary conditions $\frac{d\Phi}{d\rho}\big|_{\rho=0} = 0$, $\frac{dC_o}{d\rho}\big|_{\rho=0} = 0$, $\frac{dC_p}{d\rho}\big|_{\rho=0} = 0$. The effectiveness factor

$$\Gamma_a(R_a, \Phi_1, C_{o,1}, C_{p,1}) = \frac{3\int_0^1 \rho^2 d\rho C_o(\rho) C_p(\rho) \exp(\alpha\Phi(\rho))}{C_{o,1}\exp(\alpha\Phi_1)}, \qquad (2.46)$$

is defined as the ratio of the overall electrocatalytic turnover rate to the total rate that would be obtained if reactions were distributed uniformly. It can be used to rationalize effects of boundary values $\Phi(\rho=1) = \Phi_1$, $C_o(\rho=1) = C_{o,1}$, $C_p(\rho=1) = C_{p,1} \equiv 1$, and agglomerate structure on performance [39]. For $\Gamma_a < 1$ reaction rates in agglomerates are distributed non-uniformly, leading to a modified Tafel law that should be incorporated into models of CCL operation at the macroscopic scale in Section 2.6.

A different situation arises, when proton flux inside agglomerates is controlled by Ohm's law instead of the Nernst-Planck equation. Assuming a high intrinsic proton conductivity, proton concentration and potential are uniform, $C_p = C_{p,1} = 1$ and $\Phi = \Phi_1$. In this case, a familiar analytical solution of Eq. (2.41) is obtained [76]

$$C_o = \frac{C_{o,1}\sinh(\theta\rho)}{\rho\sinh(\theta)} \text{ and } i_F = i_a^0 \frac{C_{o,1}\exp(\alpha\Phi_1)\sinh(\theta\rho)}{\rho\sinh(\theta)} \qquad (2.47)$$

where the dimensionless parameter

$$\theta = \sqrt{\delta K_1}\exp\left(\frac{\alpha\Phi_1}{2}\right) \qquad (2.48)$$

is commonly known as the Thiele modulus. Eqs. (2.47) and (2.48) are frequently used to describe the reactivity of agglomerates that are impregnated with proton-conducting ionomer [20,21,77,78]. Due to the size relations between intra-agglomerate pores ($\sim 3 - 10$ nm) and proton-conducting pores in ionomer ($\sim 3 - 5$ nm) it seems impossible, that ionomer could form a proton-conducting phase inside agglomerates. Although the physical situation underlying Eqs. (2.47) and (2.48) seem, unrealistic they were considered in Ref. [39] for the sake of comparison.

2.5.2. Effectiveness of Catalyst Utilization and Effective Tafel Law

The highly simplified model for water-filled agglomerates allows to rationalize effects of geometrical parameters (composition, radius R_a), boundary conditions ($c_{O_2}(R_a)$, $c_{H^+}(R_a)$, $\eta(R_a)$), and reaction kinetics (transfer coefficient α) on effectiveness factors and distributions of catalyst utilization. These effects were studied in detail in Ref. [39]. It was found that oxygen concentrations are distributed rather uniformly. Oxygen diffusion is not the limiting step due to the small size of agglomerates with $R_a \approx 50 - 100$ nm. As expected, $c_{H^+}(r)$ decreases rapidly toward the center of the agglomerate, over a region determined by the Debye-length λ_D. The electrostatic field, that builds up according to Eqs. (2.43) and (2.44), counteracts the diffusive proton flux and, thus, forces protons to accumulate near the agglomerate surface, in order to minimize repulsive electrostatic interactions. The overpotential $\eta(r)$ increases, therefore, toward the center. The resulting distribution of $i_F(r)$ incorporates the opposite trends in $c_{H^+}(r)$ and $\eta(r)$. As a result, the decrease in $i_F(r)$ toward the center is much less pronounced than the decrease in $c_{H^+}(r)$. In this balance, the transfer coefficient α plays an important role, since it determines, how strongly $i_F(r)$ is affected by $\eta(r)$. The effect of α on the reaction rate distribution is shown in Figure 2.6.

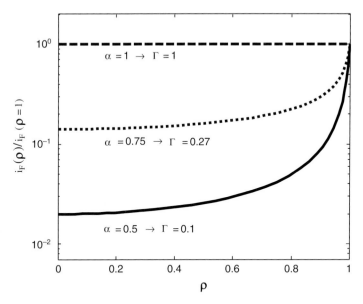

FIGURE 2.6. Effect of the cathodic transfer coefficient, α, on the reactivity distribution within an agglomerate. For $\alpha = 1$, the reactivity remains uniform throughout the agglomerate so that an effectiveness of $\Gamma_a = 1$ is obtained. For $\alpha = 0.5$, relative reactivity drops steeply toward the agglomerate center. The corresponding effectiveness factor is strongly reduced, $\Gamma_a \approx 0.095$, i.e. only $\sim$10% of the accessible catalyst is effectively used for reactions in this case.

The effect of α can be studied explicitly by considering a few simplifications in Eqs. (2.43–2.45). We assume that the oxygen concentration is uniform, $C_o(\rho) = C_{o,1}$. Moreover, due to the sluggishness of the ORR, the electrochemical reaction rate can be neglected in first approximation in Eq. (2.43), giving

$$\frac{d\ln C_p}{d\rho} = -\frac{d\Phi}{d\rho} \qquad (2.49)$$

and after integration,

$$C_p(\rho) = \exp[-\tilde{\Phi}(\rho)] \text{ with } \tilde{\Phi}(\rho) = \Phi(\rho) - \Phi_1 \qquad (2.50)$$

$\tilde{\Phi}(\rho)$ is obtained from a Poisson-Boltzmann-type equation in spherical coordinates

$$\frac{1}{\rho^2}\frac{d}{d\rho}\left(\rho^2 \frac{d\tilde{\Phi}}{d\rho}\right) = -K_2 \exp(-\tilde{\Phi}) \qquad (2.51)$$

with boundary conditions $\left.\frac{d\tilde{\Phi}}{d\rho}\right|_{\rho=0} = 0$ and $\tilde{\Phi}(\rho = 1) = 0$. The profiles of $\tilde{\Phi}(\rho)$ and $C_p(\rho)$ are solely determined by K_2 and they are independent of Φ_1. We can then calculate the effectiveness factor defined in Eq. (2.46)

$$\Gamma_a = 3 \int_0^1 \rho^2 d\rho \exp[-(1-\alpha)\tilde{\Phi}(\rho)] \qquad (2.52)$$

which depends solely on α and $K_2 \propto R_a^2$. For $\alpha = 1$, the approximate analytical solution predicts $\Gamma_a \equiv 1$, i.e. ideal catalyst utilization at the single agglomerate level. Eq. (2.52) links catalyst utilization in agglomerates and fundamental electrode kinetics, at a rather rudimentary level. As can be seen in Figure 2.6, effectiveness factors are markedly reduced for $\alpha < 1$.

An effective Tafel equation can be written for the Faradaic current density generated by a single agglomerate, which includes the effectiveness factor

$$i_F^{\text{eff}} = i_a^0 \Gamma_a \frac{c_{O_2}(\rho=1)}{c_{O_2}^{\text{ref}}} \exp\left(\frac{\alpha F \eta(\rho=1)}{RT}\right). \qquad (2.53)$$

In the case $\alpha = 1$, Equation reduces to the original Tafel law with $i_F^{\text{eff}} = i_F$ due to the uniform reaction rate distribution.

The effective law of current generation in single agglomerates, Eq. (2.53), can be incorporated into macroscopic models of CCL operation in Section 2.6, viz. as an effective Faradaic current density in REVs in Eq. (2.73). In that context, the surface values $\eta(\rho=1)$ and $c_{O_2}(\rho=1)$ have to be considered as functions of the position z within the CCL.

2.5.3. Implications for Ultrathin Planar Catalyst Layers

The length scale $R_a \sim 50 - 100$ nm determines the effectiveness of catalyst utilization for spherical agglomerates. Analogous relations apply for ultrathin planar catalyst layers with similar thickness, $L \sim 100 - 200$ nm. We consider layers that consist of Pt, water-filled pores and potentially an electronically conducting substrate. With these assumptions, we can put $f(X_{PtC}, X_{el})X_{PtC} = 1$ and $g(S_r) = 1$. The volumetric exchange current density is, thus,

$$i_{uc}^0 = 2 \cdot 10^3 [m_{Pt}] \varepsilon^{S/V} \frac{j^{0*}}{L}. \tag{2.54}$$

The main characteristics of such two-phase composite layers can be studied in full analogy to the spherical agglomerates.

With the assumptions of uniform oxygen concentration, $C_o(\xi = 1) = C_{o,1}$, and negligible reaction term in the corresponding PNP equation, a full analytical solution can be obtained [39]. The profiles of Φ and C_p are found from

$$\frac{d^2\Phi}{d\xi^2} = -K_2 \exp(\Phi_0) \exp(-\Phi) \tag{2.55}$$

and

$$C_p(\xi) = \exp[-(\Phi(\xi) - \Phi_0)] \tag{2.56}$$

with dimensionless coordinate $\xi = z/L$ and boundary conditions at the CCL|GDL interface, $\xi = 1$,

$$\left.\frac{d\Phi}{d\xi}\right|_{\xi=1} = 0, \quad \left.\frac{dC_p}{d\xi}\right|_{\xi=1} = 0, \Phi(\xi = 1) = \Phi_1 \tag{2.57}$$

and $C_p = 1$ at the PEM|CCL interface, $\xi = 0$. Note, that here C_p is normalized to the proton concentration in the PEM,

$$C_p = \frac{c_{H^+}}{c_{H^+}^{PEM}},$$

That is, $c_{H^+}^{ref} = c_{H^+}^{PEM}$. Eq. (2.55) can be solved analytically, giving

$$\Phi(\xi) = \Phi_0 + 2\ln\left\{\frac{\cos\left[\sqrt{\frac{K_2}{2}}\exp\left(\frac{\Phi_0-\Phi_1}{2}\right)(1-\xi)\right]}{\cos\left[\sqrt{\frac{K_2}{2}}\exp\left(\frac{\Phi_0-\Phi_1}{2}\right)\right]}\right\} \tag{2.58}$$

where Φ_0 and Φ_1 are related by

$$\Phi_1 = \Phi_0 - 2\ln\left\{\cos\left[\sqrt{\frac{K_2}{2}}\exp\left(\frac{\Phi_0 - \Phi_1}{2}\right)\right]\right\} \quad (2.59)$$

The effectiveness factor is

$$\Gamma_{uc} = \int_0^1 d\xi \frac{\cos^{2(1-\alpha)}\left[\sqrt{\frac{K_2}{2}}\exp\left(\frac{\Phi_0 - \Phi_1}{2}\right)\right]}{\cos^{2(1-\alpha)}\left[\sqrt{\frac{K_2}{2}}\exp\left(\frac{\Phi_0 - \Phi_1}{2}\right)(1-\xi)\right]} \quad (2.60)$$

A modified Tafel law can be written for the current-voltage performance

$$j_0 = Li_{uc}^0 \Gamma_{uc} \frac{c_{O_2}(z=L)}{c_{O_2}^{ref}} \exp\left(\frac{\alpha F \eta_0}{RT}\right). \quad (2.61)$$

Γ_{uc} accounts for the deviations form the generic Tafel characteristics. These deviations are mainly affected by L and α. For α close to 1, catalyst utilization becomes uniform, $\Gamma_{uc} \approx 1$ and the unmodified Tafel law is recovered,

$$j_0 = Li_{uc}^0 \frac{c_{O_2}(z=L)}{c_{O_2}^{ref}} \exp\left(\frac{\alpha F \eta_0}{RT}\right).$$

The nonuniform distribution of protons and potential in water-filled agglomerates and ultrathin catalyst layers is predominantly an electrostatic effect. It is determined by the Debye length, λ_D. Resulting reaction rate distributions and effectiveness factors depend on the characteristic sizes of agglomerates (R_a) or ultrathin CCLs (L) and on the transfer coefficient α.

As a major conclusion, primary pores inside agglomerates and ultrathin catalyst layers should be hydrophilic (maximum wetting). Under such conditions effectiveness of catalyst utilization can approach 100%. Moreover, the microscopic mechanism of the electrochemical reaction, represented by the transfer coefficient α, is essential for the effectiveness of catalyst utilization.

2.6. Macrohomogeneous Catalyst Layer Modeling

At the macroscopic scale, the full competition of reactant diffusion, electron and proton migration, and charge transfer kinetics unfolds. The water balance further complicates this interplay. Moreover, performance is subject to operation conditions and complex boundary conditions at interfaces to membrane and gas diffusion layer. A vast list of structural characteristics steers this interplay, including thickness, composition, pore size distributions, and wetting properties of pores.

Understanding the rules of macroscopic catalyst layer operation is crucial for optimal catalyst utilization, water management, and overall successful performance of the cell. A comprehensive catalyst layer model would have to incorporate effects on subordinate structural levels of nanoparticles and agglomerates, considered in Sections 2.4 and 2.5. These links are, however, currently not well developed. The majority of catalyst layer approaches consider the heterogeneous layer as a continuum. All processes in it are averaged over representative elementary volume elements (REVs) utilizing concepts of the theory of random composite media [9,10]. The size of these REVs is much larger than the microscopic scale, which is associated with the heterogeneities. At the same time, it is much smaller than the characteristic length of the macroscopic sample. Due to this separation of length scales, microstructural details including nanoparticles and agglomerates reside completely in effective properties, e.g. j^0, whereas at macroscopic scale distributions of physical processes can still be studied.

Due to the thickness range, $L \sim 10 - 20\mu m$, the operation of conventional catalyst layers depends decisively on the availability of sufficient gas porosity for the transport of reactants. The pertinent theory of gas diffusion electrodes dates back to the 1950s with major contributions by A.N. Frumkin and O.S. Ksenzhek [79–81]. Specifically, catalyst utilization and specific effective surface area in composite electrodes were always the focus of attention (see, e.g. [82,83] and the articles quoted therein).

In the early 1990s, T. E. Springer et al. established a one-dimensional macrohomogeneous model of electrode operation in PEFC [84]. Since then, other groups have adopted similar approaches. These models could relate global performance of CCLs to immeasurable local distributions of reactants, electrode potential and reaction rates, identify limiting processes for the overall performance, define a penetration depth of the active zone and suggest an optimum range of current density and catalyst layer thickness with minimal performance losses and highest catalyst utilization.

In Refs. [18,19], the macrohomogeneous theory was extended to include concepts of percolation theory. The resulting structure-based model correlates the performance of the CCL with the volumetric amounts of Pt, C, ionomer, and pores. A detailed review of macroscopic catalyst layer theory can be found in Ref. [17]. A further extension of this theory in Ref. [25] explores the key role of the CCL for the fuel cell water balance. This function is closely linked to the pore size distribution. Major principles of these models will be reproduced here. The details can be found in the literature cited.

2.6.1. Relations Between Structure and Effective Properties

The theory of random heterogeneous materials provides a variety of sophisticated tools to explore the relations between microstructure and the required effective properties [9,10,85]. Information on volumetric composition, pore size distribution, and wetting properties can be used to predict proton conductivity, diffusion coefficients of gases, liquid permeability, exchange current density, and net evaporation rates.

Simple parameterizations of these effective properties based on percolation in the interpenetrating networks of electron-conducting, proton-conducting, and porous phases have been utilized in Ref. [25]. In principle, more detailed models incorporating higher-order structural information, i.e. pore and particle shapes and correlations in the distributions of the distinct components, could be devised [9]. In the present situation, where only rather limited structural information is available from experiment, more detailed structure-property relations are, however, not warranted.

Proton conductivity in the layer is given by

$$\sigma_{el} = \sigma_0 \frac{(X_{el} - X_c)^2}{(1 - X_c)^2} \Theta(X_{el} - X_c) \tag{2.62}$$

where σ_0 is the bulk electrolyte conductivity. Previously, this parameterization gave reasonable results with percolation threshold $X_c \approx 0.12$ and the critical exponent 2 [85]. $\Theta(X)$ is the Heaviside step function. In Eq. (2.62), it is assumed that σ_{el} is not significantly affected by S_r. This assumption is valid as long as the catalyst is neither very dry nor extensively flooded.

The remaining effective properties depend decisively on the water distribution in the pore space. In simplest approximation, details of the pore size distribution and pore shapes are, however, neglected. Only the volumetric information about pore space filling with liquid water, determined by S_r, is incorporated and contributions due to primary and secondary pores are distinguished.

Transport of oxygen (superscript "o") and vapor (superscript "v") is given by Knudsen diffusivity with percolation-type dependence (percolation threshold X_c)

$$D^{o,v}(S_r) = D_0^{o,v} \frac{(X_P - X_\mu - X_c)^{2.4}}{(1 - X_c)^2 (X_P - X_c)^{0.4}} \left\{ \left[\frac{(1 - S_r)X_P - X_c}{X_P - X_\mu - X_c} \right]^{2.4} \Theta\left(S_r - \frac{X_\mu}{X_P}\right) + \Theta\left(\frac{X_\mu}{X_P} - S_r\right) \right\} \tag{2.63}$$

with

$$D_0^{o,v} = \sqrt{\frac{2RT}{\pi M^{o,v}}} \frac{4}{3} r_{crit},$$

where r_{crit} is a critical pore radius (bottleneck pores), e.g. obtained from critical path analysis [86], and $M^{o,v}$ is the molar mass of the diffusing gas molecules. Since the characteristic pore radii r_{crit} in CCLs are usually smaller than 100 nm, Knudsen-type diffusion prevails [86,87]. It is assumed that gas diffusion occurs only through air-filled pores. Gas diffusion through pores filled with liquid water is neglected. Moreover, contributions of primary pores are neglected since they have small pore sizes, the pathways through them are presumably highly tortuous, and they are more likely to be flooded. This parameterization, suggested in Ref. [25], expands on the composition-dependent diffusivities considered in Refs. [17–19]. Moreover, Eq. (2.63) resembles expressions considered in Refs. [88,89] for percolation in the open pore space of a partially saturated porous medium.

The relation between liquid permeability and porous structure is given by

$$K^l(S_r) = \frac{\delta}{24\tau^2}\left\{r_{el}^2\varepsilon_{el}X_{el} + r_\mu^2\left[S_rX_p\Theta\left(\frac{X_\mu}{X_p} - S_r\right) + X_\mu\Theta\left(S_r - \frac{X_\mu}{X_p}\right)\right]\right.$$
$$\left. + \tau^2 r_M^2 \frac{[S_rX_p - X_\mu - X_c]^2}{(1-X_c)^2}\Theta\left(S_r - \frac{X_\mu}{X_p}\right)\Theta(S_rX_p - X_\mu - X_c)\right\}$$

(2.64)

Here, we distinguish three different contributions due to water-filled pores in the ionomer (radius r_m, corresponding water volume portion ε_m), micropores (radius r_μ), and mesopores (radius r_M). δ is a constrictivity factor and τ is a tortuousity factor [90].

The functions $f(X_{PtC}, X_{el})$ and $g(S_r)$ in the expression for the exchange current density, Eq. (2.21), account for the nonideal utilization of the Pt surface area due to complex heterogeneous composition and partial saturation of the pore space. At macroscopic scale, only REVs with simultaneous access to Pt/C, ionomer, and pore space are utilized. This condition is expressed by [91]

$$f(X_{PtC}, X_{el}) = P(X_{el})P(X_{PtC})\left\{(1-\chi_{ec})\left[1 - [1-P(X_P)]^M\right]\right.$$
$$\left. + \chi_{ec}[1-P(X_P)]^M\right\},$$

(2.65)

with the density of the percolating cluster

$$P(X) = \frac{X}{[1+\exp(-a(X-X_c))]^b}$$

where $a = 53.7$, $b = 3.2$, and $M = 4$. The small parameter χ_{ec} allows for the effect of residual activity at nonoptimal reaction spots, cf. Figure 2.1b.

At the mesoscopic scale (size of single agglomerate, $\sim 50 - 100$ nm), the accessibility of catalyst surface sites to protons is critical, and it is determined by the water-filled pore fraction,

$$g(S_r) = \Omega \int_0^{r_c} dr' \frac{dX_P(r')}{dr'} \frac{1}{r'} \tag{2.66}$$

where Ω is a normalization factor that ensures $g \to 1$ for $S_r \to 1$. The overall effect of composition and porous structure on catalyst utilization is determined by $f(X_{PtC}, X_{el})g(S_r)/X_{PtC}$.

The liquid/vapor interfacial area ratio is the most difficult parameter to be determined [92]. It depends on porosity, pore radii, pore space connectivity, and saturation. The following form is suggested

$$\xi^{lv}(S_r) = \Upsilon L \int_0^{\infty} dr' \frac{dX_P(r')}{dr'} \frac{1}{r'} h(r', r_c) \tag{2.67}$$

The function $h(r', r_c)$ accounts for the possibility that due to hysteresis effects, the interface does not advance completely to pores with capillary radius r_c with increasing S_r; it depends on the detailed topology of the pore space and, thus, demands more detailed microstructure characterization. Υ is a factor of order 1, depending on pore geometry and wetting properties.

The above relations reveal how the composite porous structure steers effective properties. This interplay results in a complex dependence of performance on S_r. The functions $D^{\circ}(S_r)$ and $f(X_{PtC}, X_{el})g(S_r)X_{PtC}$ are plotted in Figure 2.7 for the pore size distributions in Figure 2.2. Evidently, the subdivision of the pore space into primary and secondary pores plays a key role in balancing different catalyst layer functions. A large porosity due to primary pores is beneficial for the exchange current density, since it guarantees that a large fraction of catalyst sites inside agglomerates could be reached by protons (discussion in Section 2.5). Moreover, increasing the fraction of primary pores could result in a larger liquid/vapor interfacial area and, thus, higher rates of evaporation. Secondary pores on the other hand largely contribute to the gas diffusivity in the layer.

Catalyst layer operation fails at two extremes. If the layer is dry, $S_r \to 0$, the exchange current density will go to 0 ($g(S_r = 0) = 0$). If it is completely flooded, $S_r \to 1$, gaseous transport is blocked ($D^{\circ}(S_r = 1) = 0$). The optimum value of S_r is subdued to porous structure and operating conditions.

The example depicted in Figure 2.7 and the limiting cases illustrate how electrode composition and porous structure tune the fine interplay between kinetic processes and mass transport. Understanding the rules of this interplay requires a mathematical model, discussed next

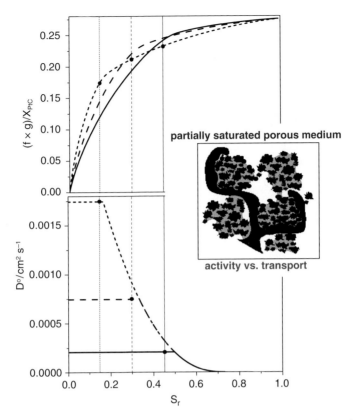

FIGURE 2.7. Catalyst utilization at macroscopic scale, i.e. the catalyst utilization factor $f(X_{PtC}, X_{el})g(S_r)/X_{PtC}$ in the exchange current density (top panel, cf. Eqs. (2.65) and (2.66)), and oxygen diffusion coefficient (bottom panel, cf. Eq. (2.63)) as functions of the liquid water saturation, plotted for the three different pore size distributions in Figure 2.2 [35]. The plots reveal the effect of porous structure on the basic competition between activity (top) and mass transport (bottom) in the CCL. The structure with a large fraction of primary pores is beneficial for catalyst utilization and detrimental for gas diffusion, and vice versa.

2.6.2. Transport and Reactions in Catalyst Layers

The main fluxes in catalyst layers and boundary conditions are depicted in Figure 2.8. Processes inside the layer are related by mass conservation laws for involved species, expressed in the form of continuity equations. The general form of the continuity equation is

$$\frac{\partial \rho(\vec{r}, t)}{\partial t} + \nabla \cdot \vec{j}(\vec{r}, t) = Q(\vec{r}, t) \tag{2.68}$$

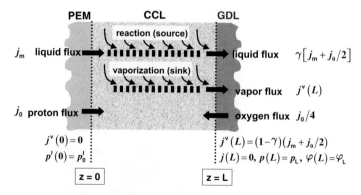

FIGURE 2.8. Fluxes of water oxygen and protons in the CCL and boundary conditions to be used in the macrohomogeneous model of CCL performance and water balance.

where $\rho(\vec{r}, t)$ is the density field of the considered species, $Q(\vec{r}, t)$ is a source/sink density, and $\vec{j}(\vec{r}, t)$ is a flux density. Here, only steady state phenomena are explored, $\partial \rho(\vec{r}, t)/\partial t = 0$.

Continuity equations including all species and processes are provided in Ref. [25]. A number of simplifications lead to the following set of equations that relate distributions and flux densities of protons (potential $\phi_{el}(z)$, proton current $j^p(z)$), oxygen (partial pressure $p(z)$, flux density $j^{ox}(z)$), water vapor (partial pressure $q(z)$, flux density $j^v(z)$), liquid water (liquid water pressure $p^l(z)$, flux density $j^l(z)$),

$$\frac{d^2\eta(z)}{dz^2} = \frac{1}{\sigma_{el}} Q^{ec}(z) \tag{2.69}$$

$$\frac{4F}{RT} D^o(S_r) \frac{dp(z)}{dz} = j_0 - j^p(z) \tag{2.70}$$

$$-\frac{F}{RT}\frac{d}{dz}\left[D^v(S_r)\frac{dq(z)}{dz}\right] = Q^{lv}(z) \tag{2.71}$$

$$\frac{F}{V_m \mu^l} K^l(S_r) \frac{dp^l(z)}{dz} = \left(n + \frac{1}{2}\right) j^p(z) + j^v(z) - j_m - \frac{j_0}{2} \tag{2.72}$$

where $nj^p(z)$ is the contribution of electroosmotic drag to the water flux inside the CCL (n denotes the electroosmotic drag coefficient). Furthermore,

$$Q^{ec}(z) = \frac{j^0(S_r)}{L}\frac{p(z)}{p^{ref}}\exp\left(\frac{\alpha F\eta(z)}{RT}\right). \tag{2.73}$$

is the sink term due to electrochemical reaction. The oxygen partial pressure is related to the oxygen concentration via the ideal gas law, $p(z) = RTc_{O_2}(z)$. Since as usual Ohmic losses due to electronic transport in the Pt/C phase are neglected, $\phi_{el}(z)$ can be replaced by $\eta(z)$ in Eq. (2.69).

$$Q^{lv}(z) = \frac{e_0\kappa^e}{L}\xi^{lv}(S_r)\{q_r^s(T) - q(z)\} \tag{2.74}$$

is the source term of water vapor due to net evaporation. κ^e denotes the intrinsic rate constant of evaporation. The saturated vapor pressure $q_r^s(T)$ is determined by the Kelvin equation,

$$q_r^s(T) = q^{s,\infty}(T)\exp\left(-\frac{2\sigma\cos(\vartheta)V_m}{RTr_c}\right), q^{s,\infty}(T) = q^0\exp\left(-\frac{E_a}{k_B T}\right) \tag{2.75}$$

with saturation pressure $q^{s,\infty}(T)$ for planar vapor|liquid interface, surface tension σ, molar volume of liquid water V_m, wetting angle ϑ, and molar gas constant R. The activation energy of evaporation is $E_a \approx 0.44$ eV (corresponding roughly to the strength of two hydrogen bonds). In addition to the conservation laws, the condition of capillary equilibrium, which is valid locally, is needed in deriving these equations. Local pore filling is accordingly controlled by the Young-Laplace equation,

$$p^c(z) = \frac{2\sigma\cos(\vartheta)}{r_c(z)} = p^g(z) - p^l(z) = p(z) + q(z) + p^r - p^l(z). \tag{2.76}$$

It relates the local distributions of liquid and gas pressures to the local capillary radius, r_c, which further determines the local liquid water saturation, S_r, via Eq. (1.4). Thereby, all transport coefficients are determined and a closed system of equations is provided. Due to complex relations $p^c \to r_c \to S_r$ and $S_r \to D^{o,v,r}(S_r), K^l(S_r), i^0(S_r), \xi^{lv}(S_r)$, the system of equations is highly nonlinear. Obtaining self-consistent solutions warrants, in general, numerical procedures.

2.7. Main Results of the Macrohomogeneous Approach

In the past, studies of the macrohomogeneous model have explored the effects of thickness and composition on performance and catalyst utilization. The relevant solutions have been discussed in detail in Refs. [17–19,84]. Here, we will briefly review the main results. Analytical relations for reaction rate distributions and relations between fuel cell current density and overvoltage losses in the CCL were discussed for limiting cases of fast oxygen diffusion and fast proton transport [18,84]. In Ref. [93], double layer charging was included into the model. With this extension into the transient domain, the complex impedance response of the CCL could be calculated. The model of impedance amplifies the diagnostic capabilities of the model, e.g. providing the proton conductance of the CCL from the linear branch of impedance spectra (in Cole-Cole representation) in the high frequency limit.

These studies did not, however, explore effects of liquid water formation, vaporization and two-phase transport in the porous structure and its impact on the fuel cell water balance. Recent findings on these latter performance issues will be discussed in more detail in a subsequent section.

2.7.1. Optimum Thickness and Performance

The structural approach relates local distributions of reactant concentrations, electrode potential, and reaction rates in the catalyst layer to the global performance of the layer. Characteristic length scales, which represent the interplay between mass transport and kinetic processes, could be defined. They help distinguishing active and inactive zones in the catalyst layer.

Three different regimes of current density could be identified in electrode polarization curves, as indicated in Figure 2.9: (1) a kinetic regime at small

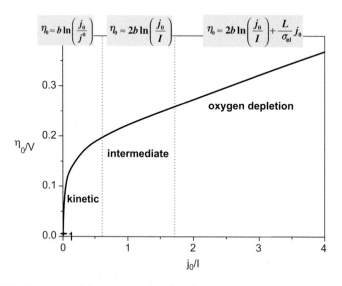

FIGURE 2.9. Overpotential vs. current density plot of the CCL. Three major regions of performance are indicated. The corresponding approximate relations $\eta_0(j_0)$ are also shown. For parameters used in calculations, cf. Ref. [17].

current densities, $j_0 \ll I$ with simple Tafel dependence $\eta_0 \approx b\ln(j_0/j^0)$ and $b = \frac{RT}{\alpha F}$, (2) an intermediate regime for $j_0 \sim I$ with dominating double Tafel slope behavior

$$\eta_0 \approx 2b\ln\left(\frac{j_0}{I}\right), \text{ with } I = \frac{4FD^\circ p_L}{RTL}, \qquad (2.77)$$

and (3) an oxygen depletion regime for $j_0 \gg I$ in which all oxygen is consumed in a sublayer of thickness δ_{eff}, defined below.

For a given composition, the thickness and the target current density of fuel cell operation should be adjusted in order to operate the catalyst layer in the intermediate regime, since it represents the best compromise between transport losses and kinetic losses. Although reaction rates distributions are slightly nonuniform in this regime, all parts of the layer are used for reactions. There are, thus, no inactive parts. As long as the CCL is operated in the intermediate regime, overvoltage losses are almost independent of the thickness. In Ref. [17], these findings were displayed in the form of a phase diagram.

The existence of a maximum thickness beyond which the performance degrades is due to the concerted impact of oxygen and proton transport limitations. Considered separately, each of these limitations would only serve to define a minimum thickness, below which performance worsens due to insufficient electroactive surface. The characteristic thickness of the effective layer, in which the major fraction of the current density is generated, is

$$\delta_{\text{eff}} = \frac{I}{j_0} L \qquad (2.78)$$

In the oxygen depletion regime, $j_0 \gg I$, only a thin sublayer with thickness $\delta_{\text{eff}} \ll L$, adjacent to the GDL, is active. The remaining sublayer with thickness $(L - \delta_{\text{eff}}) \approx L$, adjacent to the membrane, is not used for reactions, due to the starvation in oxygen. For this situation with rather nonuniform reaction rate distribution, catalyst is used very ineffectively. The inactive part causes overpotential losses due to proton transport in the polymer electrolyte, which could cause limiting current behavior, if the proton conductivity is low.

In summary, $\delta_{\text{eff}} \sim L$ should be warranted. This corresponds to operation of the CCL in the intermediate regime, which ensures high performance (i.e. minimal overpotential losses) and most effective catalyst utilization.

2.7.2. Composition Effects

The fact which transport limitations prevails in the CCL depends on the composition. If it has insufficient porosity, but a well-developed network of polymer electrolyte, it will have severe gas transport limitations but good proton transport and the other way around. The macrohomogeneous model variations can be used to determine an ideal state of saturation and the corresponding ideal capillary radius r^{eq}. Under equilibrium conditions, the adjustment of the operating conditions (partial pressures of gaseous components and liquid pressure) forces liquid water to retreat to pores of radius r^{eq}, according to Eq. (2.76).

Under stationary operation with $j_0 > 0$ and nonzero fluxes of oxygen and water, indicated in Figure 2.8, local states of saturation deviate from the ideal state. These deviations can be quantified by displaying the values of the local capillary radius, $r_c(z)$. With increasing j_0, $r_c(z)$ will increase relative to r^{eq}, corresponding to increasing S_r.

In Ref. [25], a simplified δ-distribution was used for the psd,

$$\frac{dX_p(r)}{dr} = X_\mu \delta(r - r_\mu) + X_M \delta(r - r_M) \qquad (2.79)$$

in order to study the water fluxes in the CCL and critical effects on fuel cell operation due to nonideal local saturation. This simplification permits a full analytical solution and readily reveals major principles of catalyst layer operation. It still captures essential physical processes, critical phenomena, operation conditions, and major structural features such as typical pore sizes (r_μ, r_M) and distinct contributions to porosity from micropores (X_μ) and mesopores (X_M).

With the psd in Eq. (2.79), the continuum of possible wetting states collapses into three distinct states in which the CCL could exist locally, shown in Figure 2.10: a dry state with $S_r = 0$, an optimum wetting state with $S_r = X_\mu / X_p$, and a fully flooded state with $S_r = 1$. In the dry and the fully flooded states, the CCL operates poorly due to impaired exchange current density or impaired reactant diffusion, respectively. In the optimum wetting state, local capillary equilibrium between liquid and gas phase is established in primary pores ($\sim 3 - 10$ nm radius), favoring large evaporation rates. At the same time, secondary pores ($10 - 50$ nm) are open for gaseous transport of reactants and products.

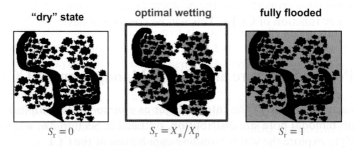

FIGURE 2.10. The three wetting states of the CCL that are permitted if the pore size distribution is represented by the bimodal δ-distribution in Eq. (2.79).

Using the simple model psd in Eq. (2.79) and the assumption of fast proton transport, a full solution of the system of conservation equations in Section 2.6 could be found, providing the pressure distributions $p(z)$, $q(z)$, $p^l(z)$, the capillary pressure $p^c(z) = p(z) + q(z) + p^r - p^l(z)$, the current density distribution $j(z)$, and the overall relation between overpotential and total current density, $\eta_0(j_0)$. This solution highlights several vital functions of CCLs for the fuel cell water balance.

2.8.1. Liquid-to-Vapor Conversion Capability

The first performance aspect focuses on the stationary water fluxes through the CCL and on the conversion between liquid and vapor fluxes. The liquid-to-vapor conversion capability of the CCL determines how much water will exit toward GDL as liquid or as vapor. Complete liquid-to-vapor conversion is possible for current densities satisfying the condition

$$j_0 < j_{\text{crit}}^{\text{lv}} = \frac{2q_2^s f D_2^v}{\lambda_v} \tanh\left(\frac{L}{\lambda_v}\right) - 2j_m \tag{2.80}$$

where j_m is the net water flux from the membrane side and

$$\lambda_v = \left(\frac{f D_2^v L}{e_0 \kappa^e \xi_2^{\text{lv}}}\right)^{1/2} \tag{2.81}$$

is an "evaporation-penetration depth." The subscript '2' in D_2^v and ξ_2^{lv} refer to the case when the whole CCL is in the optimal wetting state, cf. Figure 2.10. For $j_0 > j_{\text{crit}}^{\text{lv}}$, the minimum liquid water flux at $z = L$ is finite,

$$j^{l,\min}(L) = \frac{j_0}{2} + j_m - \frac{q_2^s f D_2^v}{\lambda_v} \tanh\left(\frac{L}{\lambda_v}\right). \tag{2.82}$$

The parameter $j_{\text{crit}}^{\text{lv}}$ is the maximum current density for which complete conversion of liquid water into vapor is possible. It is an important characteristic for evaluating successful CCL design. High rates of evaporation and vapor diffusion facilitate liquid-to-vapor conversion. The minimum liquid water flux relative to the total water flux at $z = L$ is given by

$$\gamma_{\min} = \begin{cases} 0 & \text{for } j_0 \leq j_{\text{crit}}^{\text{lv}} \\ \frac{j_0 - j_{\text{crit}}^{\text{lv}}}{j_0 + 2j_m} & \text{for } j_0 > j_{\text{crit}}^{\text{lv}} \end{cases} \tag{2.83}$$

It is shown in Figure 2.11 as a function of j_0 for the different psds in Figure 2.2.

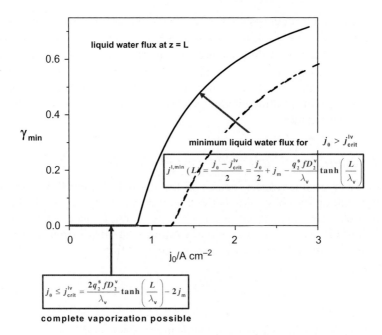

FIGURE 2.11. Liquid water fluxes at the CCL|GDL boundary, calculated in the model of water management for the CCL with different fractions of primary and secondary pores as depicted in Figure 2.2 [25].

2.8.2. Critical Liquid Water Formation in the CCL

Next, the second important aspect of CCL water balance will be considered: At which current density does local liquid water formation lead to excessive flooding of the layer, corresponding to the fully flooded state, and where in the CCL does this happen first? Figure 2.12 shows the capillary radius $r_c(z)$ of pores as a function of z for various j_0. For $r_\mu < r_c(z) < r_M$, the CCL is operated in the optimal wetting state. At $j_0 = 0$, the capillary radius is equal to the equilibrium radius everywhere, $r_c(z) = r^{eq}$. As j_0 grows, $r_c(z)$ increases relative r^{eq}. At a critical current density, $j_0 = j_{crit}^{fl}$, $r_c(z)$ reaches r_M, indicating the blocking of pathways for gaseous transport in the corresponding parts of the layer. It can be seen that excessive flooding arises first in the interior, close to the CCL|GDL boundary and not at the PEM|CCL boundary.

Figure 2.13 depicts the effect of porous structure on the relations of η_0 vs. j_0, using different fractions of primary and secondary pores that correspond to psds in Figure 2.2. The two characteristic current densities, j_{crit}^{lv} and j_{crit}^{fl}, are shown for the different graphs. Effects of various parameters on j_{crit}^{lv} and j_{crit}^{fl} have been discussed in Ref. [25]. The model reveals sensitive dependencies of CCL operation on porous structure, thickness, wetting angle, total cathodic

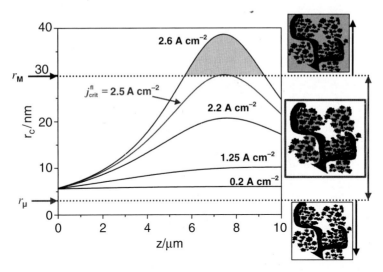

FIGURE 2.12. Capillary pressure as a function of position in the CCL for different fuel cell current densities as indicated on the graphs. For small j_0, $r_c(z) \simeq r^{eq}$, indicating operation in the optimal wetting state. For $j_0 > j_{crit}^{fl}$, parts of the CCL are completely flooded, i.e. $r_c(z) > r_M$, and the fuel cell performance is critically impaired [25].

gas pressure, and net liquid water flux. With rather favorable parameters (10 μm thickness, 5 atm cathodic gas pressure, 89° wetting angle), j_{crit}^{fl} is found in the range 2 – 3 A cm^{-2}. At twice the thickness of the reference case, i.e. for $L = 20$ μm, j_{crit}^{lv} is slightly improved due to enhanced evaporation, but the critical current density j_{crit}^{fl} decreases by about a factor two, $j_{crit}^{fl} \approx 1.3$ A cm^{-2}. The maximum current density of fuel cell operation is, thus, strongly reduced. Likewise strong effects are found when the pores are considered slightly more hydrophilic ($j_{crit}^{fl} \approx 1.3$ A cm^{-2} for $\vartheta = 87°$) or when the total gas pressure on the cathode side is reduced from 5 atm to 2.5 atm, giving $j_{crit}^{fl} \approx 0.73$ A cm^{-2}.

The effect of allowing a net flux of liquid water, $j_m = 0.3 \cdot j_0$ (at fixed electroosmotic coefficient $n = 1$), is astounding. Permitting a larger j_m suppresses liquid water formation in CCLs, as indicated by the considerably larger value of $j_{crit}^{fl} = 4.9$ Acm^{-2}. Indeed, a larger net water flux from membrane toward the cathode outlet may be unhealthy for overall fuel cell operation, causing problems with flooding in GDLs and flow field plates. But it is beneficial for the local capillary equilibrium that controls pore filling in CCLs. The increased value of j_m means that less hydraulic flux toward the membrane has to be generated in the CCL, requiring smaller liquid pressure gradients. Therefore, capillary pressures remain larger, and local capillary equilibrium persists in smaller pores.

In a nutshell, the CCL acts like a watershed in the fuel cell, regulating the balance of opposite water fluxes toward membrane and cathode outlet. Due to a benign porous structure, the CCL represents the prime component in

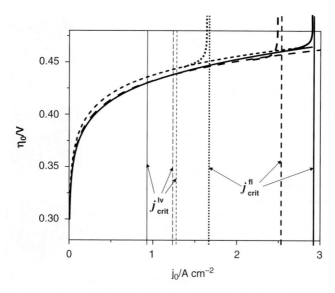

FIGURE 2.13. Relations between overpotential and current density, $\eta_0(j_0)$, for the porous structures with different fractions of primary pores and secondary pores, corresponding to Figure 2.2. The two critical current densities, representing the liquid-to-vapor conversion capability, j_{crit}^{lv}, and the resistance to liquid water formation of the CCL, j_{crit}^{fl}, are indicated for the different graphs [25].

PEFCs for the conversion of liquid to vapor fluxes. Our results also strongly suggest the CCL as a critical fuel cell component in view of excessive flooding that could give rise to limiting current behavior in $\eta_0(j_0)$ relations.

2.9. Structure-Function Optimization: Goals and Strategies

The theoretical tools discussed in this contribution address various optimization tasks in PEMFC research: (i) highest system efficiencies and fuel cell power densities and, thus, minimum overvoltage losses in CCLs; (ii) optimum catalyst utilization and, thus, minimal Pt loading (and minimal cost), and (iii) water-handling capabilities of CCLs and their impact on the water balance of the complete fuel cell. Structural parameters, as well as operating and boundary conditions that control the complex interplay of processes enter at three major levels of the theory.

Electrical current is generated at highly dispersed catalyst nanoparticles. Atomistic surface structures of catalyst particles determine their specific activities. Detailed models in nanoparticle electrocatalysis have to distinguish the contributions of distinct surface sites to the relevant electrocatalytic processes, as discussed in Section 2.4. Such models, if validated in experiment, could be highly useful in determining relevant kinetic parameters and in designing the catalyst surface at the subnanometer scale. Moreover, particle sizes and their geometrical shapes determine the atomistic surface-to-volume ratio, $\varepsilon^{S/V}$, a factor that affects catalyst utilization and exchange current density. Understanding these factors could help identify particle sizes and surface textures for highest activity and best catalyst utilization.

At the agglomerate level, electrostatic effects determine the effectiveness of catalyst utilization. The approach considered in Section 2.5 uses an effective exchange current density, averaged over distributions of catalyst particle sizes, surface structures, and different kinetic steps. Nonuniform distributions of protons and overpotential inside agglomerates lead to a modified, effective Tafel law, in which reaction rates are averaged over the agglomerate volume. This Tafel law should then be incorporated into macroscopic models of CCL operation to represent the current generation in REVs. The closer the original Tafel parameter α is to 1, the better will be the effectiveness of catalyst utilization and the smaller will be the deviations of the effective Tafel law from the original one.

It should be noted that at present the existence of agglomerates and the mechanisms of agglomerate formation are poorly corroborated in experiment. Their *a priori* assumption seems to be justified only if there is a spontaneous tendency of carbon particles to aggregate and if pores in these aggregates could not be penetrated by ionomer. Otherwise, for uniform penetration of all pores by ionomer, an entirely homogeneous approach would be appropriate.

At the macroscopic scale, all processes and operating conditions have to be included and the full competition between the different optimization tasks unfolds. Effects of composition and thickness on performance and catalyst utilization have been considered in Section 2.7. Constraints due to the statistical geometry of the three-phase composite alone give an upper limit of $\sim 20\%$ for the catalyst utilization. The porous structure, determined by the type of substrate and the composition of the electrode, steers the interplay of two major functions, viz. vaporization and gaseous transport. In the optimum wetting state, local capillary equilibrium between liquid and gas phase exists in primary pores, favoring large evaporation rates; at the same time, secondary pores are open for gaseous transport of reactants and products. As discussed in Section 2.8, the cathode catalyst layer plays a major role in the overall PEMFC water balance, acting like a watershed for the cell.

In spite of the complexity of catalyst layers, the presented theoretical tools contribute to the fundamental understanding of structure-function relations, and they provide guidelines for upgraded diagnostics and design. Parasitic voltage losses that are caused by reduced activity of the catalyst, impaired mass transport, and insufficient water management could be minimized.

The best strategy to approach the optimization of catalyst layers would be a concerted experimental-theoretical effort. Since the theory inevitably has to invoke quite a number of simplifying assumptions, often of uncontrollable nature, offering a pure theoretically driven optimization would be irresponsible. Ex situ diagnostics is needed to characterize structural details and explore their relations to effective properties. The availability of such experimental data defines the level of detail of structure-property relationships that the theory could employ. In situ experimental studies, exploring the performance and comparing it with the theoretical predictions, provide the essential benchmark for the optimization. The theory, corroborated by these systematic experimental procedures, could then be used to (i) identify salient features of good or bad catalyst layer performance, (ii) explain causes of catalyst layer failure, (iii) suggest catalyst particle sizes, thickness, composition, porous structure, and operating conditions for attaining highest fuel cell efficiencies and power densities.

Naturally, the success of this strategy depends decisively on the ability to fabricate catalyst layers in a reproducible and stable way. Moreover, processes in catalyst layers are coupled strongly and nonlinearly to the operation of other components in the fuel cell. As emphasized in Ref. [97], fuel cells have to be designed as a whole, not as a collection of stand-alone parts. A full optimization has to account for these cooperative phenomena at all scales and in all components.

List of Symbols

Acronyms

ACL	anode catalyst layer
CCL	cathode catalyst layer
EMF	electromotive force
PEM	polymer electrolyte membrane
PEMFC	polymer electrolyte membrane fuel cell
GDL	gas diffusion layer
ORR	oxygen reduction reaction
REV	representative elementary volume

Constants

Elementary charge	$e_0 = 1.6022 \cdot 10^{-19} \, \mathrm{C}$
Permittivity in vacuum	$\varepsilon_0 = 8.8542 \cdot 10^{-12} \, \mathrm{C}^2 \mathrm{J}^{-1} \mathrm{m}^{-1}$
Faraday constant	$F = 96485 \, \mathrm{C\,mol}^{-1}$
Gas constant	$R = 8.3144 \, \mathrm{J\,mol}^{-1} \mathrm{K}^{-1}$
Molar volume of liquid water	$V_m = 18.2 \cdot 10^{-5} \, \mathrm{m}^3 \mathrm{mol}^{-1}$

2. Catalyst Layer Operation in PEM Fuel Cells 85

Roman Letters

c_{H^+}, c_{O_2}	Proton and oxygen concentrations	[mol cm^{-3}]
$D_{H^+}^{eff}, D_{O_2}^{eff}$	Effective proton and oxygen diffusivities in water-filled agglomerate	[cm^2 s^{-1}]
D^o, D^v	Effective oxygen and water diffusivities in CCL	[cm^2 s^{-1}]
E^0, E^{eq}, E	Standard, equilibrium, and local electrode potential	[V]
$\frac{f(X_{PtC}, X_{el})}{X_{PtC}}$	Probability of having simultaneous access to Pt/C, ionomer, and pore space in REV	
$g(S_r)$	Internal wetted pore fraction (nomalized)	
i_F	Volumetric Faradaic current density	[A cm^{-3}]
i_F^{eff}	Effective Volumetric Faradaic current density including effectiveness factor of agglomerates	[A cm^{-3}]
i_a^0, i_{uc}^0	Volumetric exchange current density of agglomerate or ultrathin CCL	[A cm^{-3}]
I	Characteristic current density of oxygen diffusion	[A cm^{-2}]
j_F	Faradaic current density	[A cm^{-2}]
j^0, j^{0*}	Effective and generic exchange current density	[A cm^{-2}]
j^p, j^v, j^l	Local proton, water vapor and liquid water flux densities in CCL	[A cm^{-2}]
j_m, j_0	Liquid water and proton flux density at PEM|CCL boundary	[A cm^{-2}]
j_{crit}^{lv}	Critical current density for complete liquid-to-vapor conversion in the CCL	[A cm^{-2}]
j_{crit}^{fl}	Critical current density for flooding within the CCL	[A cm^{-2}]
K^l	Effective liquid permeability in CCL	[cm^2]
k_{ox}	$CO_{ad} + OH_{ad}$ recombination rate constant	[s^{-1}]
k_N	Nucleation rate constant	[s^{-1}]
k_f, k_b	OH_{ad} formation and backreaction rate constants	[s^{-1}]
L	Thickness of CCL	[cm]
$[m_{Pt}]$	Catalyst content of CCL in mg cm^{-2} (geometric area)	

p, p^l, p^g	Local pressures of oxygen, liquid water, and total gas pressure	[atm]
p^c	Local capillary pressure	[atm]
q, q_r^s	Local and saturated water vapor pressure	[atm]
r_c, r^{eq}	Local and equilibrium capillary radius	[cm]
R_a	Agglomerate radius	[cm]
S_r	Local liquid water saturation	
T	Temperature	[K]
X_{PtC}, X_{el}, X_p	Volume portions of Pt/C, electrolyte, and pore space	
X_μ, X_M	Volume portions of primary and secondary pores	
X_c	Percolation threshold	
z_{an}, z_{aa}	Average number of nearest active/inactive and active/active sites.	

Greek Letters

α	Cathode transfer coefficient	
γ	Relative liquid water flux at the CCL|GDL interface	
Γ_a, Γ_{uc}	Effectiveness factors of agglomerates and ultrathin catalyst layers	
δ_{eff}	Thickness of the effective layer in the CCL	[cm]
ε	Relative permittivity (of water $\sim$78)	
$\varepsilon^{S/V}$	Surface to volume fraction of catalyst atoms	
ζ_{Pt}	Total catalyst utilization	
η	Local overpotential	[V]
η_0	Total overpotential of CCL	[V]
ϑ	Wetting angle of water on CCL pore walls	[°]
θ_{CO}	Surface coverage of CO_{ad} on inactive sites	
θ_ξ, θ_{OH}	CO_{ad}-free and OH_{ad}-covered fraction of active sites	
κ^e	Intrinsic rate constant of evaporation	[atm^{-1} cm^{-2} s^{-1}]
λ_v	Evaporation-penetration depth	[cm]
μ^l	Liquid water viscosity	[N s cm^{-2}]
ξ^{lv}	Ratio of the distributed liquid|vapor interfacial area to the apparent electrode surface area	
ξ_{tot}	Active site fraction	
σ_{el}	effective proton conductivity in CCL	[S cm^{-1}]

References

[1] N.M. Marković and P.N. Ross, *Surf. Sci. Rep.*, **45**, 117–229 (2002).
[2] K. Krischer, in: *Modern Aspects of Electrochemistry*, **32**, 1–142, B.E. Conway, J. o'M. Bockris, and R.E. White, Editors, Plenum Press, New York (1999).
[3] K. Krischer and H. Varela, in: *Handbook of Fuel Cells*, **2**, 679–701, W. Vielstich, A. Lamm, and H. Gasteiger, Editors, John Wiley & Sons, Ltd, New York (2003).
[4] F. Hajbolouri, B. Andreaus, G.G. Scherer, and A. Wokaun, *Fuel Cells*, **4**, 160–168 (2004).
[5] J.K. Nørskov, J. Rossmeisl, A. Logadottir, L. Lindqvist, J.R. Kitchin, T. Bligaard, and J. Jónsson, *J. Phys. Chem. B*, **108**, 17886–17892 (2004).
[6] B. Hammer, Y. Morikawa, and J.K. Nørskov, *Phys. Rev. Lett.*, **76**, 2141–2144 (1996).
[7] T.R. Ralph and M.P. Hogarth, *Platinum Metals Rev.*, **46**, 3–14 (2002).
[8] H. Liu, C. Song, L. Zhang, J. Zhang, H. Wang, and D.P. Wilkinson, *J. Power Sources*, **155**, 95–110 (2006).
[9] S. Torquato, *Random Heterogeneous Materials*, Springer, New York (2002).
[10] G.W. Milton, *The Theory of Composites*, Cambridge University Press, New York (2002).
[11] A. Havránek and K. Wippermann, *J. Electroanal. Chem.*, **567**, 305–315 (2004).
[12] T. Navessin, M. Eikerling, Q. Wang, D. Song, Zh. Liu, J. Horsfall, and K. Lovell, *J. Electrochem. Soc.*, **152**, A796–A805 (2005).
[13] Q. Wang, M. Eikerling, D. Song, S. Liu, T. Navessin, Z. Xie, and S. Holdcroft, *J. Electrochem. Soc.*, **151**, A950–A957 (2004).
[14] Z. Xie, T. Navessin, K. Shi, R. Chow, Q. Wang, D. Song, B. Andreaus, M. Eikerling, Z. Liu, and S. Holdcroft, *J. Electrochem. Soc.*, **152**, A1171–A1179 (2005).
[15] D. Song, Q. Wang, Z. Liu, M. Eikerling, Z. Xie, T. Navessin, and S. Holdcroft, *Electrochim. Acta*, **50**, 3347–3358 (2005).
[16] M.S. Wilson and S. Gottesfeld, *J. Electrochem. Soc.*, **139**, L28–L30 (1992).
[17] M. Eikerling, A.A. Kornyshev, and A.A. Kulikovsky, *Physical Modeling of Cell Components, Cells and Stacks*, in: *Encyclopedia of Electrochemistry*, Volume 5: Electrochemical Engineering, Editors by Digby D. Macdonald and P. Schmuki, chapter 8.2, p. 447–543, VCH-Wiley, Weinheim, (2007).
[18] M. Eikerling and A.A. Kornyshev, *J. Electroanal. Chem.*, **453**, 89–106 (1998).
[19] M. Eikerling, A.A. Kornyshev, and A.S. Ioselevich, *Fuel Cells*, **4**, 131–140 (2004).
[20] F. Jaouen, G. Lindbergh, and G. Sundholm, *J. Electrochem. Soc.*, **149**, A437–A447 (2002).
[21] J. Ihonen, F. Jaouen, G. Lindbergh, A. Lundblad, and G. Sundholm, *J. Electrochem. Soc.*, **149**, A448–A454 (2002).
[22] M. Uchida, Y. Aoyama, N. Eda, and A. Ohta, *J. Electrochem. Soc.*, **142**, 463–468 (1995).
[23] M. Uchida, Y. Aoyama, N. Eda, and A. Ohta, *J. Electrochem. Soc.*, **142**, 4143–4149 (1995).
[24] R.E. Baltus, *J. Membrane Sci.*, **123**, 165–184 (1997).
[25] M. Eikerling, *J. Electrochem. Soc.*, **153**, E58–E70 (2006).
[26] R. Zwanzig, *Nonequilibrium Statistical Mechanics*, Oxford University Press, Oxford (2001).

[27] A.M. Kuznetsov and J. Ulstrup, *Electron Transfer in Chemistry and Biology*, Wiley & Sons, Chichester (1999).
[28] A.J. Bard and L.R. Faulkner, *Electrochemical Methods: Fundamentals and Applications*, 2nd Edition, Wiley-VCH, Weinheim (2001).
[29] W. Schmickler, *Interfacial Electrochemistry*, Oxford University Press, Oxford (1996).
[30] P.N. Ross, in: *Handbook of Fuel Cells*, **2**, 465–480, W. Vielstich, A. Lamm, and H. Gasteiger, Editors, John Wiley & Sons, Ltd, New York (2003).
[31] S. Mukerjee and S. Srinivasan, in: *Handbook of Fuel Cells*, **2**, 465–480, W. Vielstich, A. Lamm, and H. Gasteiger, Editors, John Wiley & Sons, Ltd, New York (2003).
[32] P. Waszczuk, A. Crown, S. Mitrovski and A. Wieckowski, in: *Handbook of Fuel Cells*, **2**, 465–480, W. Vielstich, A. Lamm, and H. Gasteiger, Editors, John Wiley & Sons, Ltd, New York (2003).
[33] A. Wieckowski, E.R. Savinova, and C.G. Vayennas, Editors, *Catalysis and Electrocatalysis at Nanoparticle Surfaces*, Marcel Dekker, New York (2003).
[34] F. Maillard, M. Eikerling, O.V. Cherstiouk, S. Schreier, E. Savinova, and U. Stimming, *Faraday Discuss.*, **125**, 357–377 (2004).
[35] W. Romanowski, *Surf. Sci.*, **18**, 373–388 (1969).
[36] U.A. Paulus, A. Wokaun, G.G. Scherer, T.J. Schmidt, V. Stamenkovic, V. Radmilović, N.M. Marković, and P.N. Ross, *J. Phys. Chem. B*, **106**, 4181–4191 (2002).
[37] K. Kinoshita, *J. Electrochem. Soc.*, **137**, 845–848 (1990).
[38] R. van Hardeveld and F. Hartog, *Surf. Sci.*, **15**, 189–230 (1969).
[39] Q. Wang, M. Eikerling, D. Song, and S. Liu, *J. Electroanal. Chem.*, **573**, 61–69 (2004).
[40] M. Watanabe and S. Motoo, *J. Electroanal. Chem.*, **60**, 275–283 (1975).
[41] S. Mukerjee, in: *Catalysis and Electrocatalysis at Nanoparticle Surfaces*, 501–530, A. Wieckowski, E.R. Savinova, and C.G. Vayenas, Editors, Marcel Dekker, New York (2003).
[42] M. Eikerling, J. Meier, and U. Stimming, *Z. Phys. Chem.*, **217**, 395–414 (2003).
[43] M. Peuckert, T. Yoneda, and R.A. Dalla Betta, *J. Electrochem. Soc.*, **133**, 944–947 (1986).
[44] A. Kabbabi, F. Gloaguen, F. Andolfatto, and R. Durand, *J. Electroanal. Chem.*, **373**, 251–254 (1994).
[45] T. Frelink, W. Visscher, and J.A.R. Van Veen, *J. Electroanal. Chem.*, **382**, 65–72 (1995).
[46] S. Gilman, *J. Phys. Chem.*, **68**, 70–80 (1964).
[47] J. Narayanasamy and A.B. Anderson, *J. Electroanal. Chem.*, **554–555**, 35–40 (2003).
[48] C. Saravanan, B.D. Dunietz, N.M. Marković, G.A. Somorjai, P.N. Ross, and M. Head-Gordon, *J. Electroanal. Chem.*, **554–555**, 459–465 (2003).
[49] T.E. Shubina, Ch. Hartnig, and M.T.M. Koper, *Phys. Chem. Chem. Phys.*, **6**, 4215–4221 (2004).
[50] J.S. Luo, R.G. Tobin, D.K. Lambert, G.B. Fisher, and C.L. DiMaggio, *Surf. Sci.*, **274**, 53–62 (1992).
[51] N.M. Marković, B.N. Grgur, C.A. Lucas, and P.N. Ross, *J. Phys. Chem. B*, **103**, 487–495 (1999).

[52] N.P. Lebedeva, A. Rodes, J.M. Feliu, M.T.M. Koper, and R.A. van Santen, *J. Phys. Chem. B*, **106**, 9863–9872 (2002).
[53] B. Hammer, O.H. Nielsen, and J.K. Nørskov, *Catal. Lett.*, **46**, 31–35 (1997).
[54] B. Hammer and J.K. Nørskov, *Adv. Catal.*, **45**, 71 (2000).
[55] N.P. Lebedeva, M.T.M. Koper, J.M. Feliu, and R.A. van Santen, *J. Phys. Chem. B*, **106**, 12938–12947 (2002).
[56] B. Poelsema, L.K. Verheij, and G. Comsa, *Phys. Rev. Lett.*, **49**, 1731–1735 (1982).
[57] J.E. Reutt-Robey, D.J. Doren, Y.J. Chabal, and S.B. Christman, *J. Chem. Phys.*, **93**, 9113–9129 (1990).
[58] L.R. Becerra, C.A. Klug, C.P. Slichter, and J.H. Sinfelt, *J. Phys. Chem.*, **97**, 12014–12019 (1993).
[59] V.P. Zhdanov and B. Kasemo, *Surf. Sci. Rep.*, **39**, 25–104 (2000).
[60] V.P. Zhdanov and B. Kasemo, *Surf. Sci.*, **545**, 109–121 (2003).
[61] M.T.M. Koper, A.P.J. Jansen, R.A. van Santen, J.J. Lukkien, and P.A.J. Hilbers, *J. Chem. Phys.*, **109**, 6051–6062 (1998).
[62] B. Andreaus, F. Maillard, J. Kocylo, E. Savinova, and M. Eikerling, *J. Phys. Chem. B*, **110**, 21028–21040 (2006).
[63] A.V. Petukhov, *Chem. Phys. Lett.*, **277**, 539–544 (1997).
[64] M. Bergelin, E. Herrero, J.M. Feliu, and M. Wasberg, *J. Electroanal. Chem.*, **467**, 74–84 (1999).
[65] C. Saravanan, N.M. Markovic, M. Head-Gordon, and P.N. Ross, *J. Chem. Phys.*, **114**, 6404–6412 (2001).
[66] D.T. Gillespie, *J. Comp. Phys.*, **22**, 403–434 (1976).
[67] K. Binder, in: *Monte Carlo Methods in Statistical Physics*, Topic in Current Physics, **7**, K. Binder, Editor, Springer, Berlin (1986).
[68] J.P. Ansermet, PH.D. Thesis, University of Illinois (1985).
[69] R. Schuster and G. Ertl, in: *Catalysis and Electrocatalysis at Nanoparticle Surfaces*, 211–238, A. Wieckowski, E.R. Savinova, and C.G. Vayenas, Editors, Marcel Dekker, New York (2003).
[70] N.P. Siegel, M.W. Ellis, D.J. Nelson, and M.R. Spakovsky, *J. Power Sources*, **115**, 81–89 (2003).
[71] T. Navessin, X. Zhong, *private communication* (2005).
[72] A. Parthasarathy, S. Srinivasan, A.J. Appleby, and C.R. Martin, *J. Electrochem. Soc.*, **139**, 2530–2537 (1992).
[73] R.S. Eisenberg, (Review), *J. Membrane Biol.*, **171**, 1–24 (1999).
[74] A.E. Cárdenas, R.D. Coalson, and M.G. Kurnikova, *Biophys. J.*, **79**, 80–93 (2000).
[75] B. Corry, S. Kuyucak, and S.-H. Chung, *Biophys. J.*, **84**, 3594–3606 (2003).
[76] H. Scott, *Elements of Chemical Reaction Engineering*, 2nd Edition, 615, Prentice-Hall, New Jersey (1992).
[77] R.P. Iczkowski and M.P. Cutlip, *J. Electrochem. Soc.*, 127, 1433–1440 (1980).
[78] M.L. Perry, J. Newman, and E.J. Cairns, *J. Electrochem. Soc.*, **145**, 5–15 (1998).
[79] A.N. Frumkin, *Zh. Fiz. Khim.*, **23**, 1477 (1949).
[80] O.S. Ksenzhek and V.V. Stender, *Dokl. A NSSSR*, **107**, 280 (1956).
[81] Yu.A. Chizmadzev and Yu.G. Chirkov, *Electrodics: Transport*, in: *Comprehensive Treatise of Electrochemistry*, 6(5), 317–391, J. o'M. Bockris, Yu.A. Chizmadzhev, B.E. Conway, S.U.M. Khan, S. Sarangapani, S. Srinivasan, R.E. White, and E. Yeager, Editors, Plenum Press, New York, (1983).

[82] I. Rousar, K. Micka, and A. Kimla, *Electrochemical Engineering II, Part F*, Elsevier, Amsterdam (1986).
[83] K. Mund and F.V. Sturm, *Electrochim. Acta*, **20**, 463–467 (1975).
[84] T.E. Springer, M.S. Wilson, and S. Gottesfeld, *J. Electrochem. Soc.*, **140**, 3513–3526 (1993).
[85] D. Stauffer and A. Aharony, *Introduction to Percolation Theory*, 2nd Edition, Taylor & Francis, London (1994).
[86] V.N. Ambegaokar, B.I. Halperin, and J.S. Langer, *Phys. Rev. B*, **4**, 2612–2620 (1971).
[87] W. Kast and C.R. Hohenthanner, *Int. J. Heat Mass Transfer*, **43**, 807–823 (2000).
[88] A.G. Hunt and R.P. Ewing, *Soil. Sci. Soc. Am. J.*, **67**, 1701–1702 (2003).
[89] P. Moldrup, T. Olesen, T. Komatsu, P. Schjonning, and D.E. Rolston, *Soil. Sci. Soc. Am. J.*, **65**, 613–623 (2001).
[90] F.A.L. Dullien, *Porous Media*, Academic Press, New York (1979).
[91] A.S. Ioselevich, A.A. Kornyshev, and W. Lehnert, *Solid State Ionics*, **124**, 221–237 (1999).
[92] P.C. Reeves and M.A. Celia, *Water Resour. Res.*, **32**, 2345–2358 (1996).
[93] M. Eikerling and A.A. Kornyshev, *J. Electroanal. Chem.*, **475**, 107–123 (1999).
[94] S.J. Lee, S. Mukerjee, J. McBreen, Y.W. Rho, Y.T. Kho, and T.H. Lee, *Electrochim. Acta*, **43**, 3693–3701 (1998).
[95] E. Passalacqua, F. Lufrano, G. Squadrito, A. Patti, and L. Giorgi, *Electrochim. Acta*, **46**, 799–805 (2001).
[96] M.K. Debe, in: *Handbook of Fuel Cells: Fundamentals, Technology, and Applications*, **3**, 576–589, W. Vielstich, A. Lamm, and H. Gasteiger, Editors, John Wiley & Sons, Ltd, New York (2003).
[97] M. Eikerling, A.A. Kornyshev, and A.A. Kulikovsky, *The Fuel Cell Review*, Dec. 2004/Jan. 2005, 15–25.

3
Reactor Dynamics of PEM Fuel Cells

Jay Benziger

3.1. Introduction

Polymer electrolyte membrane (PEM) fuel cells are complex multiphase chemical reactors, whose principal products are water and an electric current. The basic operation of fuel cells has been reviewed previously in this volume. Hydrogen and oxygen are fed on opposite sides of an ion-conducting polymer. Hydrogen is oxidized to protons at a catalytic anode and the protons are conducted across the membrane where they react with oxygen and electrons to make water at a catalytic cathode. The proton current is driven by the chemical potential difference of hydrogen between the anode and cathode. When an external load is connected across the anode and cathode, an electron current passes through the external load, matched by a proton current through the ion-conducting membrane. The current is limited by both the external load impedance and the internal resistance of the ion-conducting membrane.

The internal resistance of the polymer electrolyte membrane depends on the water content of the membrane. The water ionizes acid moieties providing mobile protons, like protons in water [1–3]. Absorbed water also swells the membrane, which may affect the interface between the polymer electrolyte and the electrodes. Nafion, a Teflon/perfluorosulfonic acid copolymer, is the most popular polymer electrolyte because it is chemically robust to oxidation and strongly acidic. The electrodes are commonly Pt nanoparticles supported on a nanoporous carbon support and coated onto a microporous carbon cloth or paper. These structures provide high three-phase interface between the electrolyte/catalyst/reactant gas at both the anode and cathode.

There are multiple transport and reaction steps in a fuel cell. Many of these reaction and transport processes are discussed in other chapters. PEM fuel cell designs have been heuristically derived to achieve high power output. Many proprietary methods of membrane-electrode assemblies have been developed, as well as complex structures of the flow fields, to provide the reactants to the fuel cell. Fuel cell models vary in complexity from relatively simple single phase one-dimensional models to complex models that attempt to account for multiphase flow and spatial variations in the water content, current density and reactant concentrations [3–8]. Unfortunately most fuel cell data are limited to steady state, integrated current and conversion data. The integral response of fuel cells is not amenable for model discrimination. Most models have a number of adjustable parameters, and the fit to an integral current-voltage curve is not unique.

We describe here an experimental approach to design PEM fuel cell reactors where the complexities of macroscopic design parameters are chosen to obtain data in the least ambiguous form, not to optimize the overall power output. The fuel cell designs presented here can be used with virtually any membrane-electrode assembly; these are fuel cell test stations. The fuel cell reactors described here can be thought of as building blocks for more complex fuel cells designs.

Almost all PEM fuel cell test stations employ serpentine flow channels to distribute the gas flow across the active area of the gas diffusion layer [8–11]. These flow channel designs mimic those found in large-scale commercial fuel cells. The flow channels typically have a small cross-sectional area resulting in a high gas velocity that pushes liquid water drops through the flow channels to avoid flooding. As water is a product of the fuel cell reaction, the water concentration increases along the length of the flow channels from the inlet to the outlet. The partial pressures of hydrogen at the anode and oxygen at the cathode decrease from the inlet to the outlet of the flow channels. The variable water and reactant concentrations result in a variable current density throughout the fuel cell. The dynamics to changes in feed flow rates, integrated current density and temperature are complex and difficult to interpret with the serpentine flow channels.

To facilitate PEM fuel cell modeling efforts, fuel cell reactors with simplified characteristics have been developed. In this chapter, we will describe two model fuel cell reactors – the differential fuel cell reactor and the segmented anode parallel channel fuel cell reactor. These reactors have simplified geometries that mimic idealized one-dimensional and two-dimensional fuel cells. We will show how these model fuel cells are ideal for elucidating fuel cell dynamics. The one-dimensional reactor demonstrates the existence of steady state multiplicity and the associated critical water activity in the polymer membrane for fuel cell ignition and extinction. Complex dynamics, including autonomous oscillations, are shown to arise from changes in the PEM with hydration and dehydration. The two-dimensional reactor demonstrates how ignition fronts propagate along the flow channels in PEM fuel cells. Flow configurations are shown to impact the location of current ignition in the flow channel and the direction of propagation. We show how these simplified reactor designs provide improved physical understanding of the operation and design of PEM fuel cells.

3.2. The Differential PEM Fuel Cell

The differential PEM fuel cell reactor is motivated by considering a small element in the serpentine flow channel fuel cell as shown in Figure 3.1 [12]. Mathematical models of fuel cells use differential mass, momentum and energy balances around the differential element as the defining equations for modeling larger and more complex flow fields [12]. In the differential element the only compositional variations are transverse to the membrane. The key element of a differential fuel cell is that the compositions in the gas phases in the flow channels at the anode and cathode are uniform.

We constructed a differential PEM fuel cell by replacing the flow channels with open gas plenums at the anode and cathode, as shown schematically in Figure 3.2. Compositional uniformity in the gas plenums is assured by creating sufficient volume (V_g) and choosing the reactant feed rates (F_A and F_C at the anode and cathode, respectively) so that the residence time of gases in the plenums ($\tau_R = V_g/F$) is long compared to the gas phase diffusion time in the plenums ($\tau_D = V_g^{2/3}/D_{gas}$). When the ratio $\tau_R/\tau_D > 1$, diffusive mixing dominates over convective flow, and there will be homogeneity in the gas phase composition. We typically operate our differential PEM fuel cells with $\tau_R/\tau_D \sim 1$–10. By comparison, a GlobeTech fuel cell test station with serpentine flow channels is typically operated with $\tau_R/\tau_D < 0.02$. The model fuel cell shown in Figure 3.2 is one-dimensional, with the only gradients transverse to the membrane. Because the gas compositions are uniform by virtue of diffusional mixing, it satisfies the condition for a stirred tank reactor (STR), and we refer to this model fuel cell as the Stirred Tank Reactor PEM fuel cell (STR PEMFC).

It is possible to operate the serpentine flow channel test station in a differential mode by limiting the reactant conversion, so the concentration

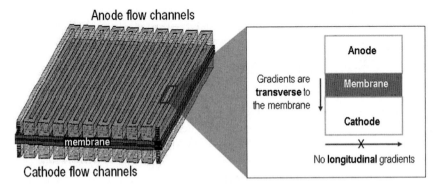

FIGURE 3.1. The differential element in a serpentine flow PEM fuel cell. In a small volume along the flow channels there are no longitudinal gradients, and the only compositional gradients are transverse to the membrane.

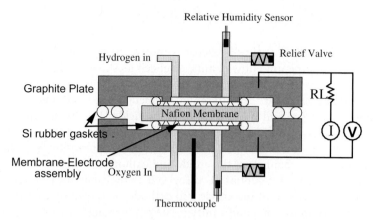

FIGURE 3.2. Schematic of the STR PEM fuel cell. The exposed electrode area is ~1 cm² on each side, with gas volumes above the anode and cathode of ~0.2 cm³. The MEA employed ETEK electrodes and a Nafion 115 membrane. The original versions of the fuel cell did not permit liquid water to freely drain; they could only be operated with water activity less than one (no liquid water). Newer versions have the cell rotated by 90° and the effluents designed to permit liquid water to drain without accumulating in the gas plenums.

gradients *along the flow channel* are small. Keeping the fractional conversion of the reactants <5% will give nearly homogeneous compositions at the anode and cathode. However, the current density is limited to <50 mA/cm² for a serpentine flow channel test station to be approximately differential. The STR PEM fuel cell has no limit on the reactant conversion; we have operated with current densities as high as 1.5 A/cm².

The greatest utility of the STR PEM fuel cell reactor is to study the dynamics of PEM fuel cell operation. Specific questions that have been explored are as follows:

1. How long does it take for a PEM fuel cell to start up from various initial conditions? How do the system parameters affect start-up of the fuel cell?
2. How does the PEM fuel cell respond dynamically to changes in load? Temperature? Gas flow rate?
3. How should the system parameters be controlled under conditions of variable load, such as those encountered in automotive applications?

To assist the analysis of the STR PEM fuel cell data, we choose the simplest reactor engineering model that retains the essential physics. The fuel cell can be thought of as a set of reactors connected through a set of flow regulators, as shown in Figure 3.3. Hydrogen molecules are oxidized to protons and electrons at the anode. The resistances in the membrane and external load regulate the current in the fuel cell (i.e., the flow of protons and electrons). The protons and electrons meet up at the cathode along with the oxygen to produce water. The external load resistance is analogous to a valve that

3. Reactor Dynamics of PEM Fuel Cells   95

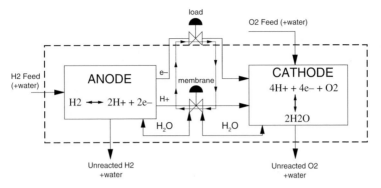

FIGURE 3.3A. Conceptual reactor coupling in a fuel cell. The heavy dashed line represents the physical boundary of the fuel cell. The feed flow and composition at the anode and cathode are system parameters shown as inputs. The effluents leaving the anode and cathode are system variables. The membrane and the external load resistance for the fuel cell are analogous to valves that regulate flow of the intermediate products from the anode to the cathode. The dashed line going through the valves indicates that the resistance to flow of those two regulating valves is in series. The load resistance is a system parameter and is drawn external to the fuel cell boundary. Water is shown moving between the cathode and anode through the membrane. The water flux depends on the concentration gradients and the current (via electro-osmotic drag).

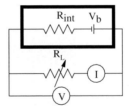

FIGURE 3.3B. Simplified equivalent circuit for the fuel cell reactor model. The battery voltage, V_b, is the result of the chemical potential difference for hydrogen between the anode and cathode. The internal resistance, R_{int}, is primarily the resistance of the membrane. R_L is the external load impedance. V is the voltage across the external load and I is the current through the load.

regulates the flow of product out of the anode reactor to the cathode reactor. The system parameters (independent variables under direct control by the operator) are the feed flow rates, composition, the heat input and the external load resistance. These independent system parameters have been drawn outside the heavy dashed line in Figure 3.3A to indicate that they can be directly controlled. Inside the heavy dashed line, the system variables (current, voltage, membrane water content, etc.) are the quantities whose evolution results from the inputs to the system.

We present data with the external load resistance as the independent parameter. This is different from the traditional electrochemistry approach where PEM fuel cells are operated under galvanostatic or potentiostatic control (*constant current or constant voltage*). When the chemical reaction is driven by the imposition of an external electrical driving force, such as with electrolysis of water, the current or voltage is a system parameter that can be independently manipulated. However, in a fuel cell, the chemical reaction drives the current through the external load, and the load resistance is the system parameter that can be manipulated. Constant current or voltage requires a feedback controller that adjusts the external resistance to maintain the current or voltage. We want to first understand the operation of the fuel cell operating autonomously, and then we can impose control and separate the system dynamics from the controller dynamics.

3.2.1. The STR PEM Design

Photographs of one of our STR PEM fuel cells are shown in Figure 3.4. The membrane-electrode-assembly (MEA) was pressed between two machined graphite plates press fit into Teflon blocks and sealed with a silicon rubber gasket. Gas plenums of volume $V_g \sim 0.2\,\mathrm{cm}^3$ were machined in graphite plates above a membrane area of $\sim 1\,\mathrm{cm}^2$. There were several pillars matched between the two plates to apply uniform pressure to the MEA. Hydrogen

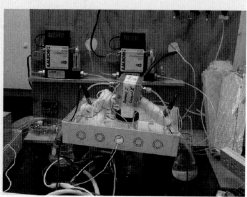

FIGURE 3.4. Photographs of the STR PEM fuel cell. The anode and cathode gas plenums with the pillars are easily seen. There are no horizontal surfaces for water to accumulate in the gas plenums, so the water drains and does not hinder gas diffusion to the electrode/electrolyte interface. The photograph to the right shows the overall fuel cell including the temperature and mass flow controllers. Everything is connected to a computer that continuously records the cell temperature, the gas flows to the anode and cathode, the relative humidities in the anode and cathode effluents, the current through the circuit and the voltage across the external load.

and oxygen were supplied from commercial cylinders through mass flow controllers at flow rates $F \sim 1$–$20 \, cm^3/min$ (mL/min). The residence times of the reactants in the gas plenums ($V_g/F \sim 1.2$–$12 \, s$) were greater than the characteristic diffusion time ($V_g^{2/3}/D \sim 0.3$–$1 \, s$), ensuring uniformity of the gas compositions. The cell temperature was controlled by placing the graphite plates between aluminum plates fitted with cartridge heaters connected to a temperature controller. The entire fuel cell assembly was mounted inside an aluminum box to maintain better temperature uniformity.

Gas pressure was maintained in the cell by placing spring loaded pressure relief valves at the outlets. The effluents were bled into 10 mL graduated cylinders to collect the water effluent. Tees were placed in the outlet lines with relative humidity sensors in the dead legs of the tees. The water content of the outlet streams was measured with humidity sensors (Sensirion SHT7X).

Any suitable membrane-electrode-assembly (MEA) can be tested in the STR PEM fuel cell. We report here results using MEAs we fabricated, consisting of a Nafion™ 115 membrane pressed between 2 E-TEK electrodes (these consist of a carbon cloth coated on one side with a Pt/C catalyst). The catalyst weight loading was $0.4 \, mg\text{-}Pt/cm^2$. The electrodes were brushed with solubilized Nafion solution to a loading of $\sim 0.4 \, mg\text{-}Nafion/cm^2$ before placing the membrane between them [13]. The assembly was hot pressed at 130°C and 10 MPa. Copper foils were pressed against the graphite plates and copper wires were attached to connect to the external load resistor.

The current and voltage across the load resistor were measured as the load resistance was varied. A 10-turn 0–20 Ω, 6 W potentiometer was connected in series with a 10-turn 0–500 Ω, 6 W potentiometer. The load resistance was varied from 0 to 20 Ω to obtain a polarization curve (IV). To examine the low current range of the polarization curve the resistance had to be increased over the range of 0–500 Ω. The voltage across the load resistor was read directly by a DAQ board. The current through the load resistor was passed through a 0.1 Ω sensing resistor and the differential voltage across the sensing resistor was read by the DAQ board. An IV curve was typically collected and stored in $\sim 100 \, s$. Some tests were also performed with the fuel cell hooked up to an Arbin Instruments MSTAT4+ test station. The Arbin system was employed to obtain data under non-autonomous conditions, i.e. under galvanostatic or potentiostatic control.

3.3. The Segmented Anode Parallel Channel PEM Fuel Cell

Most fuel cell designs employ flow channels to distribute the reactants across the MEA area and to assist pushing liquid water out of the fuel cell; the serpentine flow channel represented in Figure 3.1 is the most popular flow channel design. Because reactants are consumed and water is produced in the fuel cell reaction, there is a decrease in the reactant concentrations between the flow channel inlet and outlet. There is also an increase in the water concentration from the inlet to the outlet. Decreases in reactant composition result in a decrease in the chemical potential across the membrane. Increases in the water concentration will decrease the membrane resistance for proton transport and increase the mass transport resistance across the gas diffusion layer.

The current/voltage response from a fuel cell with flow channels is an integrated average across the membrane area. To examine the current density variations resulting from compositional variations in a PEM fuel cell, we constructed a model two-dimensional fuel cell [14, 15]. The model two-dimensional fuel cell had a segmented anode and parallel channels at the anode and cathode; it is referred to as the segmented anode parallel channel (SAPC) fuel cell. The anode flow can be co-current or counter-current to the cathode flow. Local current densities were measured by segmenting the anode with insulators placed between the anode segments, while the cathode was made as a single unit. The MEA was made as a single assembly. The reactor was designed with the lateral separation between anode segments being $\sim$10 times the transverse separation between the anode and cathode, thus assuring that the transverse current is large compared to the lateral currents.

3.3.1. The SAPC Fuel Cell Design

The cathode was a machined block of graphite with three parallel flow channels 2 mm × 2 mm × 30 mm long. The entire graphite block was press fit into a larger piece of Teflon for electrical isolation. The flow channels initiated and terminated in common manifolds in the Teflon block at either end of the graphite flow channels. The anode had the same channel structure, except that it was made of six graphite pieces separated by Teflon spacers inserted into a Teflon block, as shown in Figure 3.5A. An MEA was placed between the cathode and anode and sealed in the same way as for the differential fuel cell.

FIGURE 3.5A. Segmented anode fuel cell. The flow channels and divisions of the anode are shown. The basic setup for temperature control, reactant feed, and relative humidity and water flow measurements for the segmented anode fuel cell was the same as the one employed with the differential PEM fuel cell.

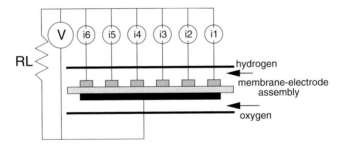

FIGURE 3.5B. Equivalent circuit for the segmented anode fuel cell. The current through each segment, as well as the voltage drop across the load resistance, were monitored. The voltage across the load resistance was recorded, as well as the currents through the six segments of the anode. The total current was determined by summing the individual currents through the six segments of the anode.

Each of the graphite segments was connected individually to a lead wire and run through a $0.1\,\Omega$ sensing resistor. The six leads from the anode were connected together, and the common lead was connected through a $0\text{–}20\,\Omega$ 10-turn potentiometer to the cathode. The entire assembly was mounted between two aluminum blocks that were temperature controlled with cartridge heaters. A computer DAQ board read the voltage drop across the potentiometer (that served as the load resistance) and the currents through each of the six segments of the anode. The electrical equivalent of the segmented anode fuel cell is shown in Figure 3.5B.

Hydrogen and oxygen were supplied from commercial cylinders through mass flow controllers at flow rates, $F \sim 1\text{–}20\,\text{cm}^3/\text{min}$ (mL/min). The effluents were bled into 10 mL graduated cylinders with a small hydrostatic head ($\sim 2\,\text{cm}\,H_2O$) so that the cell pressure was effectively 1 bar. The flows at the anode and cathode could be either co-current or counter-current. The fuel cell could be oriented with the flow channels either vertical or horizontal. The data we present here are with the flow channels oriented vertically and the cathode flow downward. We found that the orientation of the flow channels has a dramatic effect on the fuel cell operation under conditions where liquid water forms. The data presented here were collected under conditions where gravity causes the liquid water to drain from the flow channels, so the results are not plagued by liquid pools hindering mass transport from the flow channels to the electrode/electrolyte interface.

3.4. Dynamics of Start-Up of a Pem Fuel Cell

A polymer electrolyte must contain sufficient water for the fuel cell to function; but how much water is sufficient? Water is a product of the fuel cell reaction. Can the fuel cell make enough water to keep the fuel cell functioning? If the feed streams are humidified, how much water needs to be introduced? The membrane is a reservoir for water, and the resistance of the membrane changes as the water inventory changes. The electrical resistance is coupled to the water content resulting in positive feedback. We examined the operation of both the one-dimensional and two-dimensional PEM fuel cells under autohumidification conditions to elucidate the dynamics of fuel cell start-up. Autohumidified fuel cell operation employs dry feeds; the water to humidify the electrolyte membrane is provided by the fuel cell reaction. However, the membrane must initially have sufficient water content to permit the current to become self-sustaining.

3.4.1. Ignition in an STR PEM Fuel Cell

To control the initial water content in the polymer electrolyte membrane, we modified the fuel cell to permit direct injection of water. Tees were placed in the gas inlets to the fuel cell with a septum on the run and the gas flow coming in through the branch. A 0.5 µL syringe was used to directly inject water into the fuel cell.

Prior to start-up of the STR PEM fuel cell, the initial water content in the membrane and the load resistance were fixed. The polymer membrane was dried by flowing dry oxygen through the cathode chamber at $\sim$100 mL/min and dry hydrogen through the anode chamber at $\sim$100 mL/min for $\sim$12 h at 80°C with the fuel cell at open circuit. With a dry membrane a finite resistance of 0.2–20 Ω could be connected between the anode and cathode, and the fuel cell and the current through the circuit is less than 1 mA. After drying out the membrane the cell temperature was reduced to 60°C, the flow rates were reduced to 10 mL/min at both the anode and cathode, and the fuel cell was permitted to equilibrate for $\sim$2 h at open circuit. After the temperature was stable at 60°C, the flow to both the anode and cathode was shut off, and an aliquot of 0.5–2.5 µL of water was injected to the anode and permitted to equilibrate for $\sim$5 min. Hydrogen flow at 8 mL/min to the anode and oxygen flow at 4 mL/min to the cathode were initiated, and the current through the load resistor (set at 2 Ω) was measured as a function of time. The current response for start-up of the autohumidified STR PEM fuel cell is shown in Figure 3.6. For initial membrane water concentrations of ≤ 0.5 mg/cm^2, the fuel cell current decayed with time to near zero (the fuel cell current was "extinguished"). When the initial water concentration in the membrane was $\sim$1.0 mg/cm^2, the fuel cell current "ignited," rising from an initial value of $\sim$85 mA/cm^2 to a final steady value of $\sim$340 mA/cm^2. When the initial membrane water content is greater than the critical concentration of 1.0 mg/cm^2 at this temperature, flow rate, and external load, the water production was sufficient to sustain the water content in the membrane. At lesser initial water content the resistance to proton current is too great, and evaporation of water from the membrane exceeds water production, thereby dehydrating the membrane and extinguishing the current.

3. Reactor Dynamics of PEM Fuel Cells   101

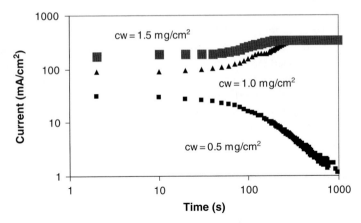

FIGURE 3.6A. STR PEM fuel cell start-up from different initial membrane water contents. Aliquots of water were directly injected into a 1 cm^2 dry fuel cell to set the initial water content. The cell temperature was 60°C, with a flow of 8 mL/min of H_2 to the anode and 4 mL/min O_2 the cathode. The load resistance was 2 Ω.

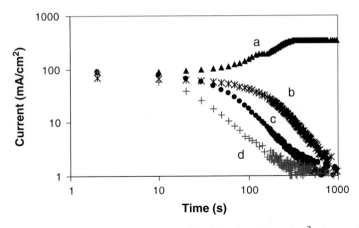

FIGURE 3.6B. STR PEM fuel cell start-up after injecting 1.0 mg/cm^2 of water into a 1 cm^2 dry fuel cell. (a) Cell temperature – 60°C, feed flows: anode – 8 mL/min of H_2 cathode – 4 mL/min O_2, load resistance – 2 Ω. (b) Cell temperature – 60°C, feed flows: anode – 8 mL/min of H_2 cathode – 4 mL/min O_2, load resistance – 5 Ω. (c) Cell temperature – 60°C, feed flows: anode – 12 mL/min of H_2 cathode – 6 mL/min O_2, load resistance – 2 Ω. (d) Cell temperature – 70°C, feed flows: anode – 8 mL/min of H_2 cathode – 4 mL/min O_2, load resistance – 2 Ω.

Figure 3.6B compares three experiments where the initial water loading in the membrane was the same (1.0 mg/cm^2), the difference being a change in either the feed flow rate, the temperature, or the external load resistance. Increasing any one of those parameters extinguished the fuel cell current.

The ignition phenomenon results from a positive feedback between the membrane water activity and the reaction rate [16]. Increased membrane water activity decreases the membrane resistance, increasing the fuel cell current. The increased current produces more water that will further increase the water activity in the membrane. The current increase is self-limiting. As the reactants are consumed, the transport of hydrogen or oxygen to the electrode/electrolyte interface will limit the cell current. This corresponds to a shift in the rate-limiting step of the fuel cell reaction. When the water activity is low, proton transport across the membrane is rate limiting; when the water activity is high, reactant transport from the gas flow channel to the cathode catalyst surface becomes rate limiting.

The ignition phenomenon reported here shows a direct analogy to thermal ignition for exothermic reactions in stirred tank reactors. In autothermal stirred tank reactors, the energy balance between heat generation by reaction and heat removal by convection results in steady state multiplicity [17–20]. In the PEM fuel cell, there is an analogous balance between water produced by reaction and water removed by convection. The steady state mass balance between water produced by reaction (1/2 the proton current) and the water removed by convection is given by Eq. (3.1).

$$\text{water generated} = \frac{i_{H^+}}{2} = \text{water removed by convection}$$

$$i_{H^+} = \frac{(V_{oc} - V_{op})}{R_{int} + R_L} = \frac{V_b}{R_{int} + R_L} = \frac{F_A^{out} P_w^A + F_C^{out} P_w^C}{2RT} \mathcal{F} \quad (3.1)$$

The Fs are volume flow rates, Ps are the water partial pressures, F is Faraday's constant, and i_{H+} is the current. The fuel cell current equals the effective cell voltage divided by the sum of the load resistance and membrane resistance. The effective cell voltage, V_b, is the thermodynamic potential (V_{oc}) reduced by the overpotential (V_{op}) associated with the oxidation/reduction reactions at the anode and cathode.

Proton conduction in Nafion requires water to ionize the sulfonic acid groups and establish percolation paths through the membrane [2]. The water uptake by the PEM, λ = number of water molecules per sulfonic acid residue, and the membrane resistivity, R, are functions of the water activity, a_w; they have negligible dependence on temperature [21, 22].

$$R \approx 10^5 * \exp(-14(a_w)^{0.2})\text{-cm}^2 \quad (3.2)$$

$$\lambda = 1.75 \frac{15 a_w (1 - 10 a_w^9 + 9 a_w^{10})}{(1 - a_w)(1 + 14 a_w - 15 a_w^{10})} \quad H_2O/SO_3 \quad (3.3)$$

More detailed analyses of the relationship between proton conductivity and water activity are found elsewhere in this volume.

Figure 3.7 shows the water balance for the STR PEM fuel cell based on Eq. (3.1). Water production is equal to half the current density as given by the left-hand side of Eq. (3.1). Water removal is by convection of water exiting the fuel cell, which is the right-hand side of Eq. (3.1). Because the membrane resistance decreases exponentially with water activity, the proton current is a sigmoidal function of membrane water activity. The water removal plotted as a function of water activity is a straight line, whose slope increases with temperature and reactant flow rate. Steady state is represented by the intersections of the water production and the water removal curves.

Fixing the load resistance, the cell temperature and reactant flow rates results in either one or three intersections of the water production and water removal curves that correspond to steady states. At a high load resistance and high temperature, there is a single low current or "extinguished" steady state. At moderate load resistances and low temperature three steady states exist. In addition to the extinguished state, there is a high current or "ignited" steady state, as well as an intermediate steady state.

Of the three steady states only two are stable. Both the ignited and extinguished steady states are stable to fluctuations. For example, a positive fluctuation from the ignited state increases the water content in the membrane; at that condition, the water removal is greater than the water generation so the

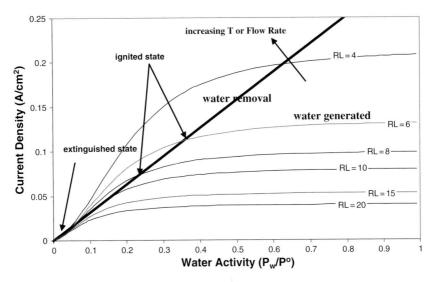

FIGURE 3.7. Water production (fuel cell current) and water removal rates as functions of the membrane water activity for an autohumidification PEM fuel cell. The rates are expressed in terms of the current through the external load resistor. A set of curves represents the water production rates for different external load resistances. The water production is based on Eq. (3.1), substituting the membrane resistance as a function of membrane water activity given by Eq. (3.2). Water removal is linearly dependent on the vapor pressure.

system will return to steady state. The middle steady state is unstable, positive fluctuations in the water content from that steady state will cause the system to generate more water than is removed, and the fuel cell will evolve toward the ignited steady state; negative fluctuations in water content from the intermediate state will drive the system toward the extinguished state.

The critical water content shown in Figure 3.6A corresponds to the water content of the unstable steady state in Figure 3.7. Membrane resistance is not very sensitive to temperature and does not depend on reactant flow rates, so the water generation curves in Figure 3.7 only change with changes in the load resistance. In contrast, the water removal is sensitive to the temperature and the flow rates, but does not depend on the load resistance; the slope of the water removal line increases with temperature and reactant flow rates. From Figure 3.7 it is easy to rationalize the results in Figure 3.6B. Increasing load resistance, increasing temperature and increasing flow rate will all shift the intersection of the water production and water removal curves to lower water activity, until they no longer intersect and only the extinguished steady state can exist.

3.4.2. Ignition Front Migration in the SAPC PEM Fuel Cell

The dynamics of ignition show different features in a fuel cell with flow channels. Water is convected along the flow channels so the membrane water content and current density change. The current density profiles depend on the flow rates, temperature, and load resistance. Ignition will be initiated locally in the flow channel at a location where the water content is highest, and then convection and diffusion of water can result in an ignition front motion. We built the segmented anode parallel channel (SAPC) PEM fuel cell to explicitly measure the current density along the flow channel, searching for ignition and ignition front propagation. At high reactant flow rates, high temperature, and high load resistance, current ignition is inhibited. By reducing all three of these system parameters it is possible to achieve conditions where current ignition is feasible.

Ignition and the motion of ignition fronts in the SAPC PEM fuel cell is illustrated in Figure 3.8 [15]. Local current through each segment of the anode is plotted as a function of time using an intensity scale. The fuel cell was extinguished by open circuit operation at 80°C and dry feeds of 15 mL/min at both the anode and cathode. The temperature was decreased to 25°C, the flow rates reduced to 3.5 mL/min at both the anode and cathode and the load resistance decreased to 0.25 Ω. Starting with a "dry" membrane, the fuel cell ignited and the current distribution evolved over time. There was an induction period of hours before any significant current was recorded in any element of the fuel cell circuit. With co-current flow (Figure 3.8A) ignition first occurred at the outlet of the fuel cell; the current in anode element 6, nearest the outlet, rose from <1 mA to $\sim$ 100 mA over a period of 5 min. An ignition front then propagated toward the entrance of the flow channels over

3. Reactor Dynamics of PEM Fuel Cells   105

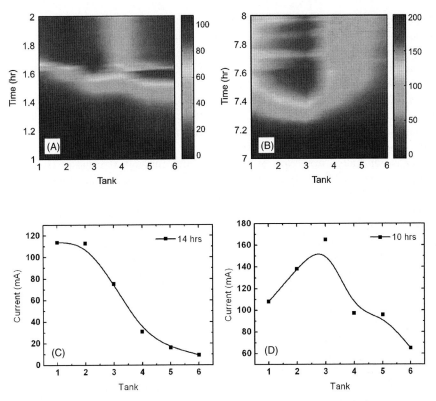

FIGURE 3.8. Ignition in segmented anode PEM fuel cell with co-current (left) and counter-current (right) flow. The color intensity represents the local current at the anode. In co-current flow (A) the current ignites after ~1.5 h at the exit to the flow channels and propagates upstream to the flow channel entrance. The steady state current profile (C) is stable after 6 h, with the current decreasing from the inlet to the exit. In counter-current flow (B) ignition first occurs at the center of the flow channels after an induction period of ~ 7 h. The current density fans out over time. After 10 h, the steady state current profile (D) shows the highest current at the center with the current declining at either end of the fuel cell.

a period of 15–20 min. As the ignition front moved toward the flow channel inlets, the current at the exits of the flow channels dropped. The steady state current distribution for co-current flow is shown in Figure 3.8C. There is a high current in element 1, the highest current is seen with element 2, and the current then decreased toward the outlet.

Counter-current flow of hydrogen and oxygen produces a much different ignition pattern (Figure 3.8B). With the same pre-treatment, temperature and flow rates, the induction time with counter-current flow was substantially longer than with co-current flow. Ignition first occurred at element 3, at the center of the flow channel. The ignition front then fanned outward, but the highest current always occurred in the center of the flow channel. The steady state current was almost twice as great for counter-current flow as compared to co-current flow.

Ignition of the current in the PEM fuel cell requires sufficient water buildup in the membrane to sustain the current to produce more water. When the reactants are fed co-currently the small amount of water produced is continually transported down-stream toward the exits of the flow channels. The water slowly accumulates near the exit until a critical level of water is achieved that results in ignition. After ignition near the exit of the flow channels, the water is conducted upstream by diffusion both through the gas phase and through the membrane. The diffusion of water upstream leads to the propagation of the current ignition toward the inlet of the flow channels.

Diffusion of water along the membrane results in the distinctly different ignition fronts with counter-current flow. The water made at the cathode of the fuel cell is partitioned between the membrane and the cathode gas flow channel. The water in the membrane diffuses to the anode where the water activity is lower, and then it can enter the anode gas flow channel.

In co-current flow, the water in both the anode and cathode flow channels is convected toward the respective exits. With counter-current flow the water in the cathode stream will humidify the membrane toward the exit of the cathode flow channel, which is the entrance to the anode flow channel; and water in the anode stream will humidify the membrane at the exit of the anode flow channel, which is the entrance to the cathode flow channel. The counter-current flow tends to homogenize the water distribution, creating the highest water concentration toward the middle of the flow channel. Ignition occurs at the point of highest water concentration. The fronts fan out from there as the water is dispersed by both convection and diffusion.

3.4.3. Parameters that Affect Ignition

One can understand qualitatively why flow rate, temperature, and load resistance all affect ignition in the PEM fuel cell.

(i) Flow rate – increasing the flow rate dilutes the concentration of water in the gas streams, reducing the total amount of water that is absorbed into the membrane. Only at low flow rates, when the gas streams are sufficiently humidified by the water formed upstream, will the fuel cell ignite.
(ii) Temperature – increasing the temperature increases the vapor pressure of water. For the same amount of water formed, less water is retained in the membrane at higher temperature; this dries out the membrane and extinguishes the fuel cell.
(iii) Load resistance – increasing the load resistance reduces the current through the fuel cell circuit and hence decreases the water production. With less water formed, the water activity decreases, and hence the fuel cell extinguishes.

The parameter space for ignition and extinction is different between the STR fuel cell and the SAPC fuel cell because of different rates of convection and diffusion. The relative rate of convection to gas phase diffusion is characterized by the dispersion number, $N_{dispersion} = (F/A_{channel})/(D/l_{channel})$, where $A_{channel}$ is the cross-sectional area for flow through the channel, and $l_{channel}$ is the distance for flow along the channel. The ratio of dispersion numbers for the STR and SAPC fuel cells we designed is $N_{SAPC}/N_{STR} = 5$; convection is more important in the SAPC fuel cell. Greater diffusion in the STR fuel cell results in "back-mixing," where the water formed mixes with the feed to give uniform composition. Back-mixing helps maintain the membrane water content sufficiently high to keep the fuel cell ignited. Convection is more important with long flow channels. Convection will tend to sweep out the water formed in the fuel cell and extinguish the fuel cell if the feeds are dry. Flow rates need to be much less in the SAPC fuel cell to observe ignition.

The dynamics of ignition and front propagation in PEM fuel cells can be predicted using a straightforward extension of the 1-D model that predicted ignition in the differential PEM fuel cell. The key elements in the model that account for ignition and front propagation are (1) an exponential dependence of proton conductivity in the polymer electrolyte with membrane water content, (2) the dynamics of water absorption into the polymer electrolyte membrane, (3) the membrane capacity as a reservoir for water, and (4) transport of water upstream either in the membrane or in the flow channel. There is a close relationship between current density along the fuel cell flow channels and the local water content in the polymer membrane, as model simulations reveal. Simulations based on the model presented for the STR PEM fuel cell also predict the correct trends for the effects of reactant flow rates, cell temperature, and load resistance on ignition. Our current STR PEM model is only semi-quantitative because the electrode/electrolyte interface evolves dynamically changing the interfacial resistance and gas transport to the flow channels. Improved STR PEM fuel cell design and better instrumentation should provide better data to elucidate these phenomena.

3.5. Dynamics of Changes in Load

When used for automotive applications, fuel cells must respond to changes in the load. Changing the load alters the water production, changing the balance between water produced and water removed, resulting in a change of the membrane water content. The effect of the load resistance on the water activity can be seen in the polarization curves for the STR PEM fuel cell shown in Figure 3.9A [23]. The STR PEM fuel cell was operated in the autohumidification mode. The STR PEM fuel cell was equilibrated at 80°C for 12 h with a fixed load resistance (either 0.2 Ω or 20 Ω). After equilibration, the polarization curve was obtained by sweeping the load resistance between 0.2 and 20 Ω in 100 s. The relative humidity in the anode and cathode streams

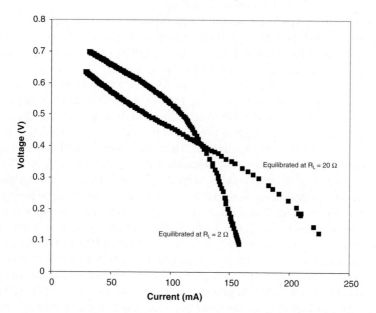

FIGURE 3.9A. "Instantaneous" polarization curves for STR PEM fuel cell equilibrated with a fixed load resistance at 80°C for 12 h. The IV curves were recorded by sweeping the load resistance from 0 to 20 Ω in a period of 100 s.

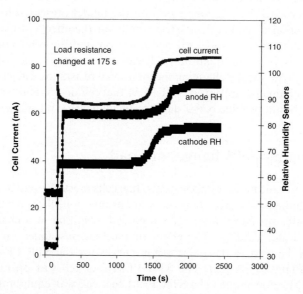

FIGURE 3.9B. Dynamic response of the STR PEM fuel cell for switching the load resistance from 20 Ω to 7 Ω at 80°C. The flow rates were 5 mL/min H_2 and 10 mL/min O_2. The resistance was switched at 175 s.

changed by <2% while obtaining these polarization curves; these polarization curves can be considered as "constant" membrane water activity.

Figure 3.9A illustrates that the "constant water content" polarization curve does not represent a unique characterization of the PEM fuel cell. Operation with different load resistances for extended periods of time resulted in different membrane water activities. *The membrane water activity is critical in defining the polarization curve.* The striking feature about Figure 3.9A is that the two polarization curves cross. Extended operation with a low load resistance produced an MEA with "high" water content, while extended operation with a high load resistance produces an MEA with "low" water content. The MEA with the high water content shows a higher voltage at low currents, indicating a lower activation polarization. At high currents, the "high" water content of the MEA shows a lower voltage, suggesting that the water is limiting mass transport of oxygen to the cathode. The "low" water content MEA has greater activation polarization, but a lower mass transport resistance.

The dynamic response of the STR PEM fuel cell to a change in resistive load shows an unusual multi-step process. Figure 3.9B shows an immediate step response of the current to the change in load, followed by decay to a plateau value. There was a subsequent jump in the fuel cell current after 1500 s. The time constant for the rise to the initial plateau was $\sim$1 s. The change in water activity at the anode was a delayed by $\sim$100 s relative to the changes in current and water activity at the cathode. The jump in current after 1500 s occurred with no changes to any external parameter and was completely unexpected. The cathode relative humidity response tracked the current response; the anode relative humidity response tracked the current but was delayed by $\sim$100 s.

The 100 s time constant for water transport through the membrane is evident in the delay of the response of the water activity in the anode effluent compared to the rise in current. The 1500 s time constant for the second jump in the current shown in Figure 3.9B is still not well understood. Benziger and co-workers have hypothesized that water uptake swells the polymer membrane, altering the electrode/electrolyte three-phase interface; the delayed jump in current is attributed to the polymer swelling into the pores of the electrode altering the three-phase contact [23]. Nazarov and Promislow suggested that the long time transients result from slow-moving hydration fronts moving laterally in the membrane [24]. Such fronts could arise from water diffusion under the pillars in the STR PEM fuel cell. A third possible explanation is that liquid water builds up at the interface between the cathode and the membrane until the hydrostatic pressure can overcome the surface forces and push the water through the hydrophobic gas diffusion layer [25]. All three explanations require coupling of mechanical and chemical processes. We have measured the time constants for the viscoelastic creep and stress relaxation in the Nafion and found them to be 10^3–10^5 s [26]. The mechanical relaxation time constants for Nafion are the same magnitude as the time constants in the PEM fuel cell, suggesting that PEM fuel cell dynamics are controlled by mechanical/chemical coupling.

The unusual dynamic performance with a long time delay was only observed over limited regions parameter space. Two stable ignited states were observed with a 1 cm^2 autohumidified STR PEM fuel cell, flow rates of 3–10 cm^3/min at both the anode and cathode, temperatures 70–95°C, and load resistances of 4–12 Ω. Figure 3.10 shows the "steady state" I-V or polarization curve for the STR PEM fuel cell that was operated in the autohumidification mode at 95°C with an anode feed of 10 mL/min H$_2$ and a cathode feed rate of 10 mL/min O$_2$. Three days were required to obtain a steady state polarization curve, because the membrane water content needed 3–4 h to stabilize after each change to the load resistance. After permitting the fuel cell to come to steady state with a load resistance of ∼0.2 Ω for 24 h, the load resistance was increased stepwise by ∼1 Ω every 4 h, after which the current and voltage were recorded. The water activity at the cathode and anode are not constant along the steady state polarization curve. The curvature of the polarization curve is the result of the water activity at the cathode increasing with the decreasing load resistance.

This "steady state" polarization curve showed that the current decreased smoothly and the voltage increased smoothly as the load resistance was increased from 0.2 to 11 Ω. When the load resistance was increased from 11 to 12 Ω both the steady state current and voltage decreased abruptly, but the fuel cell did not extinguish. The resistance was then increased stepwise from

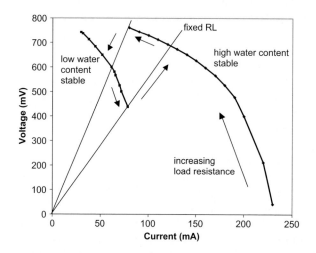

FIGURE 3.10. Steady state polarization curves for the autohumidification PEM fuel cell at 95°C with H$_2$ flow and O$_2$ flow of 10 mL/min. The high water content curve was taken after establishing steady state with a load resistance of 2 Ω. The low water content curve was taken after establishing steady state with a load resistance of 20 Ω. Each point on the polarization curves was taken after a 4-h equilibration time.

12 to 20 Ω; the current decreased and voltage increased smoothly but along a different path. Starting with a load resistance of 20 Ω, the process was reversed, and the load resistance was decreased stepwise by ~1 Ω every 4 h. The current and voltage followed the same path as seen for the high load resistance polarization curve, but there was no abrupt change in the current and voltage at 11 Ω. Instead the I-V data varied smoothly until the resistance was decreased below 5 Ω, when a jump was observed that matched the polarization curve taken with decreasing load resistance.

The polarization curves in Figure 3.10 show a hysteresis loop, where the steady state current and voltage depend on the direction of approach. It was possible to go around the hysteresis loop shown in Figure 3.10 reproducibly many times. When the load resistance was between 5 and 11 Ω, the steady state current and voltage were dependent on the direction of approach. These multi-valued steady states were stable; the current and voltage at any point on the steady state polarization curve was steady for periods of >24 h.

The steady state polarization curve shown in Figure 3.10 identified 3 *stable* steady states (which imply a total of 5 steady states). The three states are the dry state (extinguished current), low water content state (intermediate current), and high water content state (high current). The uniform composition in the STR PEM fuel cell provides a unique correlation between composition and current density, which is not available for cells with long flow channels where the concentration varies spatially.

3.6. Characteristic Time Constants for PEM Fuel Cells

The ignition/extinction results and responses to changes in load provide information about the time scales for the response of the fuel cell. The time constant for transitioning to steady state during startup is ~ 100 s. Five of the key time constants associated with PEM fuel cells are listed in Table 3.1. They include the characteristic reaction time of the PEM fuel cell (τ_1), the time for gas phase transport across the diffusion layer to the membrane electrode interface (τ_2), the characteristic time for water to diffuse across the membrane from the cathode to the anode (τ_3), the characteristic time for water produced to be absorbed by the membrane (τ_4), and the characteristic time for water vapor to be convected out of the fuel cell (τ_5). Approximate values for the physical parameters have been used to obtain order of magnitude estimates of these time constants.

The initial fuel cell response to changes in load show is rapid with a time constant ~ 1 s; this corresponds to the convective flow into the fuel cell and the diffusion across the gas diffusion layer, τ_1 and τ_2. Diffusion across the polymer membrane from the cathode/electrolyte interface to the anode, τ_3, is evident in Figure 3.9B, where the relative humidity change at the anode lags the change at the cathode by ~100 s.

TABLE 3.1 Characteristics Times for PEM Fuel Cells.

	Physical Significance			Approximate Value
τ_1	Characteristic time for reaction rate relative to reactor volume	$\tau_1 = \dfrac{V_R}{i}$	$\sim \dfrac{(0.1\ cm^3/cm^2)}{(1\ A/cm^2)}$	0.1–1 s
τ_2	Characteristic diffusion time across gas diffusion layer	$\tau_2 = \dfrac{(\ell_{diffusion\ layer})^2}{(D_{gas}^{eff})}$	$\sim \dfrac{(0.03\ cm)^2}{(0.01\ cm^2/s)}$	0.1 s
τ_3	Characteristic diffusion time for water across membrane from cathode to anode	$\tau_3 = \dfrac{(\ell_{membrane})^2}{(D_{water}^{membrane})}$	$\sim \dfrac{(0.01\ cm)^2}{(10^{-6}\ cm^2/s)}$	10–100 s
τ_4	Characteristic time for water production relative to sulfonic acid density	$\tau_4 = \dfrac{\lambda N_{SO3}}{i}$	$\sim \dfrac{5(2.3 \times 10^{-5}\ mol/cm^2)}{(1\ A/cm^2)}$	100–1000 s
τ_5	Characteristic time for water removal from the fuel cell	$\tau_5 = \dfrac{\lambda N_{SO3}}{P_w^o(F_A + F_C)/RT}$	$\sim \dfrac{5(2.3 \times 10^{-5}\ mol/cm^2)(1\ cm^2)}{(3 \times 10^{-6}\ mol/s)}$	100–1000 s

Response times of ~ 100 s are associated with water uptake and transport through the membrane; these are the processes that govern the dynamics of fuel cell ignition. The water content in the membrane (and hence the membrane resistance) must equilibrate to changes in the external load resistance, reactant flow rates, and temperature that alter the balance between water production and water removal. *The polymer membrane is a reservoir for water and the filling and draining of that reservoir dominate PEM fuel cell dynamics.* The times for ignition and extinction are not the same; ignition depends on filling the membrane with water, which occurs with time constant

τ_4. In contrast, extinction depends on water removal with time constant τ_5. The extinction time becomes much longer at lower temperatures where the vapor pressure of water is less.

The long delayed responses of the fuel cell to changes in load have been attributed to mechanical property changes in the polymer. We have initiated measurements of polymer stress relaxation. The stress relaxation and viscoelastic creep of Nafion is both temperature and water concentration dependent. Response times vary from ~ 1 s to $\sim 10^6$ s, which can give a wide range of characteristic response times for PEM fuel cells.

To capture the dynamics of PEM fuel cell startup it is essential to include the dynamic mass balance for the water content in the membrane, N_w^m. If water accumulation in the hydrophobic gas diffusion layer is neglected, then the water accumulation in the membrane is the difference between the water produced and the net water removed by convection from the anode and cathode, as given by Eq. (3.4). We measure all the quantities on the right-hand side of Eq. (3.4), so we can determine the accumulation rate of water.

$$\frac{\partial N_w^m}{\partial t} = \left(\frac{F_A^{in} P_w^{A,in}}{RT} - \frac{F_A^{out} P_w^{A,out}}{RT} \right) + \left(\frac{F_C^{in} P_w^{C,in}}{RT} - \frac{F_C^{out} P_w^{C,out}}{RT} \right) + \frac{i}{2F} \quad (3.4)$$

Water is produced at the cathode/membrane interface, and it must be transported to the anode and cathode flow channels to be removed. At present, we do not have a direct measurement of the water activity in the membrane (the partial pressure of water in the membrane, $P_w^{membrane}$), but we do know the water content in the effluent streams. The experimental data may be integrated from the known initial water content (after the water injection) to the steady state current and partial pressures of water. Integration of Eq. (3.4) gives the steady state membrane water content for known water partial pressure at the anode and cathode. Effective mass transfer coefficients for water from the cathode/membrane interface to the cathode gas flow channel ($k_w^{cathode}$) and from the cathode/membrane interface to the anode gas flow channel (k_w^{anode}) can also be estimated from the data. At steady state water transport from the cathode/membrane interface must equal the convection of water from the fuel cell as given by Eqs. (3.5 and 3.6).

$$\frac{F_C^{out} P_w^{C,out}}{RT} = k_w^{cathode} \left(P_w^{membrane} - P_w^{C,out} \right) \quad (3.5)$$

$$\frac{F_A^{out} P_w^{A,out}}{RT} = k_w^{anode} \left(P_w^{membrane} - P_w^{A,out} \right) \quad (3.6)$$

Estimates for mass transfer coefficients at 80°C were obtained from the STR PEM fuel cell startup, $k_w^{cathode} \sim 10^{-5} mol/cm^2 - s, k_w^{anode} \sim k_w^{cathode}/2$. Direct measurements of transport coefficients in PEM fuel cells are rare and the STR PEM fuel cell is one of the few ways to get in-situ values.

3.7. Autonomous Oscillations in STR PEM Fuel Cells

After extended operation of an STR PEM fuel cell with the same membrane electrode assembly (> 2500 h), autonomous oscillations were observed under conditions where the STR PEM fuel cell exhibited 5 steady states [23]. An example of the oscillations is shown in Figure 3.11. These oscillations have periods of 10^3–10^5 s and show characteristics of a capacitively coupled switch. The oscillations transition very rapidly (<10 s) between high and low states with an overshoot on the rise and undershoot and recovery on the fall. The period, magnitude and on/off times for these oscillations varied with temperature, and load resistance. Benziger and co-workers have suggested that these unusual dynamics are associated with mechanical relaxations of the polymer membrane driven by changing water content, but the detailed physical processes causing these unusual dynamics are not yet understood.

Oscillations have been reported in PEM fuel cells with carbon monoxide impurities in the hydrogen feed [27]. Those oscillations result from poisoning and periodic oxidation of islands of adsorbed CO on the anode and have relatively short periods of ~10 s. There are no reports of highly regular oscillations with pure hydrogen feed in serpentine flow channel fuel cells.

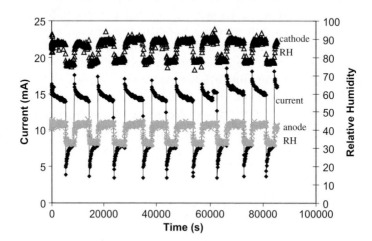

FIGURE 3.11. Characteristic autonomous oscillations observed with the STR PEM fuel cell during 5000–12,000 h of operation. These are stationary states. The feeds were dry with flow rates 10 mL/min H_2, 10 mL/min O_2. The cell temperature is 80°C, with a load resistance of 10 Ω.

However, there are reports of irregular current fluctuations in fuel cells operated with low feed humidification [28]. The one-dimensional STR PEM fuel cell provides dynamic data that is more amenable to analysis than the chaotic data from the serpentine flow channel fuel cell; coupled oscillations in different parts of the flow channel can give rise to complicated data. The data from the STR PEM fuel cell provide important validation criteria for fuel cell models and can be employed to identify conditions where fuel cell operation becomes hard to control and should be avoided.

3.8. Fuel Cell Response to Process Control

The electrical elements in a fuel cell circuit can respond quickly, but the overall system response is dictated by the dynamics of physical transport in the fuel cell. There are time delays for fuel fed to the fuel cell and gases transported across the gas diffusion layer to the electrode/electrolyte interface. A simple example of the process control dynamics in a PEM fuel cell arise when maintaining the power output by controlling the fuel feed [29]. An STR PEM fuel cell was set up with feedback control on the anode and cathode feeds to follow a set point change in the current through a fixed load resistance. This is analogous to changing the gas flow to a car motor when pressing on the gas pedal. PID control on the mass flow of hydrogen and oxygen followed steps up and down in the current through a 1 Ω load. The hydrogen flow to the anode was set to give 100% fuel utilization. The oxygen flow to the cathode was set from 100 to 150% stoichiometric. Results for 100% hydrogen utilization and oxygen stoichiometry of 130% are shown in Figure 3.12. The process time constant for the step changes were ~8 s; this time corresponds to the residence time of gas in the gas plenums of the fuel cell. The residence time in the gas flow channels will dominate the system response of fuel cells. It is possible to reduce the flow channel volume smaller and shorten the residence time; however, this would hinder liquid water flow through the flow channels and create large pressure drops.

The results in Figure 3.12 show very stable currents, while the hydrogen flow fluctuates quite a bit. The volume in the flow channels acts as a capacitor and damps out the fluctuations. Successful control is critically dependent on feed stoichiometry. Running at 100% hydrogen utilization it was only possible to maintain system stability if the oxygen stoichiometry was >130%. At oxygen stoichiometry of 120% and hydrogen stoichiometry of 100% it was impossible to stably operate the fuel cell; the system oscillated, as shown in Figure 3.13. These oscillations occur when the reactant concentrations at the both the anode/electrolyte and cathode/electrolyte interfaces are both transport limited. There was a significant delay of ~1000 s between when the cathode flow was changed and the moment the system began to deviate from the set point. This delay is attributed to the time for water concentrations to equilibrate at the cathode.

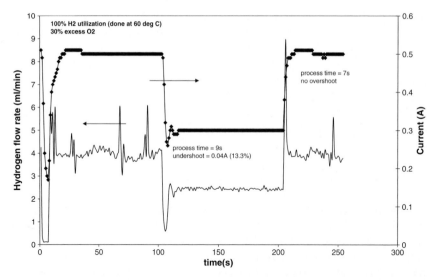

FIGURE 3.12. Control of a PEM Fuel Cell. The power (or current) delivered to a fixed load resistance was controlled by controlling the reactant flows to the fuel cell. The cell was operated at 60°C with dry feeds and a 1 Ω load. The hydrogen fuel utilization was 100%. The cathode feed was 130% stoichiometric. PID control based on current measurement was used. Set point changes in the current from 0.5 A to 0.3 A at 100 s and from 0.3 to 0.5 at 200 s are shown.

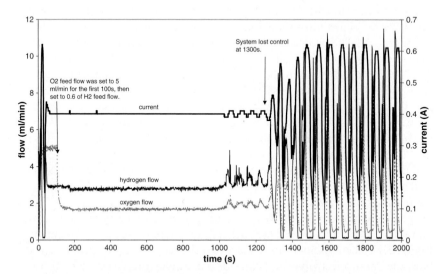

FIGURE 3.13. Control of PEM Fuel Cell. The power (or current) delivered to a fixed load resistance was controlled by controlling the reactant flows to the fuel cell. The cell was operated at 60°C with dry feeds and a 1 Ω load. The hydrogen fuel utilization was 100%. The cathode feed was changed from 5 cm^3/min to 120% stoichiometric at 100 s. The feed flow controls were PID control as above.

3.9. Other Applications of the STR PEM Fuel Cell

We have stressed the utility of model fuel cells to elucidate the dynamics of cell operation. The model fuel cells can also help provide detailed information about steady state performance in a well-defined system, and thus to test the detailed transport and reaction models presented in other chapters in this book. Because the STR PEM fuel cell is a one-dimensional system one does not need to deal with spatial gradients along the channels, complicating the analysis of the current/voltage/load data. We have employed the STR PEM fuel cell to obtain data about water transport and gas transport in PEM fuel cells.

Figure 3.14 presents data of the water activity in the effluents from an autohumidified STR PEM fuel cell [23]. The effluent compositions are the same as the internal gas compositions at the anode and cathode in the STR PEM fuel cell. This makes it simple to correlate cell performance with composition, since we can directly measure the effluent composition. The surprising result illustrated in Figure 3.14 is that the water activity in the anode effluent is nearly the same as the water activity in the cathode effluent.

We have also used the STR PEM fuel cell to better understand the mass transport controlled regime of fuel cell operation. It is well known that the current or power output from a PEM fuel cell is reduced when air is used at the cathode instead of oxygen. Two effects can contribute to diminished fuel cell performance; either the thermodynamic potential is reduced or the oxygen transport across the gas diffusion layer is reduced because of the presence of an inert gas. Figure 3.15 shows a series of I-V curves taken in the STR PEM

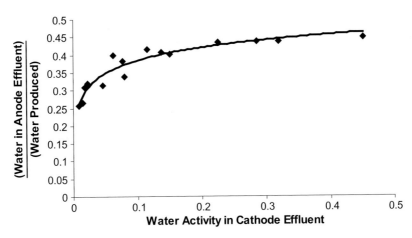

FIGURE 3.14. The fraction of water produced in the fuel cell that exited in the anode effluent as a function of the water activity in the cathode effluent. These data were obtained in the autohumidification STR PEM fuel cell. The water activity in the cathode effluent was varied by changing the external load resistance, which changed the amount of water produced in the fuel cell. Data from operation at temperatures from 35 to 105°C are shown and the water partitioning does not appear to be temperature dependent.

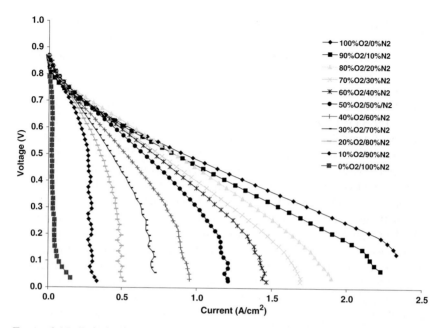

FIGURE 3.15. Polarization curves for a series of dilutions of oxygen in the cathode feed stream in an autohumidified STR PEM fuel cell. These data were taken at 60°C and 1 bar total pressure. The feed was 8 cm^3/min at both the anode and cathode. The anode feed was pure hydrogen and the cathode feed was an oxygen/nitrogen mixture. The cell was operated for 2 h with a cathode feed of pure O$_2$ feed and a fixed current density of 0.5 A/cm^2. The cathode feed was then switched to the specified composition allowed to equilibrate for 1 h at a current density of 0.5 A/cm^2. (At 0–20% oxygen composition the current density was the maximum current – limited by availability of oxygen). A polarization curve was obtained by sweeping the voltage from 0.9 V to 0.025 V at 0.02 V/s; two polarization curves were obtained 1 h apart.

fuel cell by systematically changing the partial pressure of oxygen in the cathode feed from 1 bar to 0 bar [30]. The data show a continuous change in the onset of the mass transport limited polarization regime, with little or no change in the activation polarization regime.

Because there are no spatial compositional variations complicating the data, it is easy to conclude that the inert gas reduces mass transport across the gas diffusion layer, moving the onset of the mass transport polarization regime to lower current densities.

3.10. Conclusions

Two model fuel cell reactors, the stirred tank reactor- polymer electrolyte membrane fuel cell and the segmented anode parallel channel fuel cell, have been shown to be effective in the study of PEM fuel cell dynamics. Simplified geometry and flow patterns provide idealized one-dimensional and two-dimensional fuel cells where the steady state and dynamic behavior are easily analyzed for parameter estimation and model discrimination.

A one-dimensional PEM fuel exhibited steady state multiplicity, resulting from positive feedback between proton conduction in the membrane and water production from the fuel cell reaction. The critical membrane water activity necessary for "ignition" of the fuel cell current was identified. The time for current ignition is $\sim$ 100–1000 s, resulting from the titration of the sulfonic acid residues in the polymer membrane from the water formed by the fuel cell reaction.

Local current density along the flow channels was measured with a split anode parallel flow channel PEM fuel cell. In co-current flow water convected downstream accumulated at the outlet where the fuel cell ignites; in contrast, ignition occurs in the interior of the flow channels in counter-current flow fuel cells. Ignition fronts propagate along the flow channels when diffusion is comparable to convective flow. In co-current flow the ignition front propagates from the exit toward the entrance of the flow channels. In counter-current flow the ignition waves spread out from the center toward both ends of the fuel cell.

Model fuel cells reveal complex non-linear dynamics of PEM operation, including autonomous oscillations and loss of controllability with high reactant utilization. The one-dimensional STR PEM fuel cell provides well-defined data that can help elucidate the underlying physics underlying the non-linear dynamics.

Model fuel cells are also well suited to obtain data for model discrimination and verification. For example, mass transfer limitations from oxygen dilution at the cathode are very well defined in the one-dimensional STR PEM fuel cell; the mass transfer effects limit the current output from a fuel cell.

Using a well-established approach from chemical reactor engineering we have introduced simplified model reactors as the starting point to effectively test complex reactor models. The model reactors have demonstrated that the dynamics of PEM fuel cells are dominated by the balance between water production and water removal. Because of the strong non-linear relationship between membrane water content and proton conductivity, PEM fuel cells exhibit interesting and complex behavior such as ignition, front propagation and autonomous oscillations. It is critical to understand the underlying physics behind these phenomena to optimize operation and avoid fuel cell failure.

Acknowledgments

We thank the National Science Foundation (CTS -0354279 and DMR-0213707) for support of this work. Andy Bocarsly and Supramaniam Srinivasan were instrumental in introducing me to PEM fuel cells. I want to thank all the undergraduate students (J.F. Moxley, C. Teuscher, E. Karnas,

C. Woo, R. Mejia-Ariza) and graduate students (E.-S. J.Chia, W. Hogarth, B. Satterfield) for their contributions in the lab. I especially want to thank my collaborator Ioannis Kevrekidis for encouragement to pursue this work, and developing mathematical models for the complex system dynamics.

Nomenclature

$A_{channel}$	cross-sectional area for flow channel
a_w	water activity = P_w/P_w^o
D	Gas phase diffusivity
F	volumetric flow rate of reactant feeds
I	current
i_{H+}	proton current
k_w	mass transfer coefficient for water vapor
$l_{channel}$	length of flow channel
N_{SO_3}	number of sulfonic acid residues in membrane
N_w^m	water content in membrane
P_w	partial pressure of water
P_w^o	vapor pressure of water
R	gas constant
R_{int}	internal resistance for fuel cell membrane electrode assembly
R_L	external load resistance
T	temperature
V	voltage drop across the load
V_g	gas flow volume at fuel cell electrodes
V_b	battery voltage for fuel cell
V_{oc}	open circuit voltage of fuel cell
V_{op}	activation polarization overpotential
λ	number of water molecules per sulfonic acid residue
τ_R	residence time of fuel cell
τ_D	characteristic diffusion time
τ_i	characteristic time constants
F	Faraday's constant

References

[1] Thampan, T., et al., *PEM fuel cell as a membrane reactor*. Catalysis Today, 2001. **67**(1–3): pp. 15–32.
[2] Gierke, T.D., G.E. Munn, and F.C. Wilson, *Morphology of perfluorosulfonated membrane products – wide-angle and small-angle X-Ray studies*. Acs Symposium Series, 1982. **180**: pp. 195–216.
[3] Weber, A.Z. and J. Newman, *Modeling transport in polymer-electrolyte fuel cells*. Chemical Reviews, 2004. **104**(10): pp. 4679–4726.
[4] Wang, C.Y., *Fundamental models for fuel cell engineering*. Chemical Reviews, 2004. **104**(10): pp. 4727–4765.

[5] Bernardi, D.M. and M.W. Verbrugge, *A Mathematical-model of the solid-polymer-electrolyte fuel-cell*. Journal of the Electrochemical Society, 1992. **139**(9): pp. 2477–2491.

[6] Bernardi, D.M. and M.W. Verbrugge, *Mathematical-model of a gas-diffusion electrode bonded to a polymer electrolyte*. Aiche Journal, 1991. **37**(8): pp. 1151–1163.

[7] Natarajan, D. and T. Van Nguyen, *A two-dimensional, two-phase, multicomponent, transient model for the cathode of a proton exchange membrane fuel cell using conventional gas distributors*. Journal of the Electrochemical Society, 2001. **148**(12): pp. A1324–A1335.

[8] Su, A., Y.C. Chiu, and F.B. Weng, *The impact of flow field pattern on concentration and performance in PEMFC*. International Journal of Energy Research, 2005. **29**(5): pp. 409–425.

[9] Cha, S.W., et al., *Geometric scale effect of flow channels on performance of fuel cells*. Journal of the Electrochemical Society, 2004. **151**(11): pp. A1856–A1864.

[10] Cha, S.W., et al., *The scaling behavior of flow patterns: a model investigation*. Journal of Power Sources, 2004. **134**(1): pp. 57–71.

[11] Yan, W.M., et al., *Effects of flow distributor geometry and diffusion layer porosity on reactant gas transport and performance of proton exchange membrane fuel cells*. Journal of Power Sources, 2004. **125**(1): pp. 27–39.

[12] Benziger, J.B., et al., *A differential reactor polymer electrolyte membrane fuel cell*. AIChE Journal, 2004. **50**(8): pp. 1889–1900.

[13] Raistrick, I.D., *Electrode assembly for use in a solid polymer electrolyte fuel cell*. 1989, US: U.S. Department of Energy.

[14] Benziger, J., Chia, J.E.-S., E. Kimball, and I.G. Kevrekidis, *Reaction Dynamics in a Parallel Flow Channel PEM Fuel Cell*. Journal of the Electrochemical Society, 2007. **154**(8): pp. B835–B844.

[15] Benziger, J.B., Chia, J.E.-S., Y. DeDecker, and I.G. Kevrekidis, *Ignition Front Propagation in Polymer Electrolyte Membrane Fuel Cells*. Journal of Physical Chemistry C, 2007. **111**: 2330–2334.

[16] Moxley, J.F., S. Tulyani, and J. Benziger, *Steady-state multiplicity in polymer electrolyte membrane fuel cells*. Chemical Engineering Science, 2003. **58**: pp. 4705–4708.

[17] Froment, G.F. and K.B. Bischoff, *Chemical Reactor Analysis and Design*. Second ed. 1979, New York: John Wiley & Sons. 765.

[18] Folger, H.S., *Elements of chemical reaction engineering*. Third ed. 1999, Upper Saddle River, NJ: Prentice Hall. 967.

[19] Liljenroth, F.G., *Starting and stability phenomena of ammonia-oxidation and similar reactions*. Chemical and Metallurgical Engineering, 1918. **19**: pp. 287–293.

[20] van Heerden, C., *Autothermic processes: properties and reactor design*. Industrial and Engineering Chemistry, 1953. **45**(6): pp. 1242–1247.

[21] Thampan, T., et al., *Modeling of conductive transport in proton-exchange membranes for fuel cells*. Journal of the Electrochemical Society, 2000. **147**(9): pp. 3242–3250.

[22] Yang, C.R., *Performance of Nafion/Zirconium Phosphate Composite Membranes in PEM Fuel Cells*, in *Department of Mechanical Engineering*. 2003, Princeton NJ: Princeton University.

[23] Benziger, J., et al., *The dynamic response of PEM fuel cells to changes in load*. Chemical Engineering Science, 2005. **60**: pp. 1743–1759.

[24] Nazarov, I. and K. Promislow, *Ignition waves in a stirred PEM fuel cell*. Chemical Engineering Science, 2006. **61**(10): pp. 3198–3209.

[25] Benziger, J., Nehlsen, J., Blackwell, D., T. Brennan, and J. Itescu, *Water flow in the gas diffusion layer of PEM fuel cells*, Journal Of Membrane Science, 2005. **261**(1–2): 98–106.
[26] Satterfield, M.B. and J.B. Benziger. *Investigation of PEM Fuel Cell Behavior: Swelling and Viscoelastic Properties*. in *ACS National Meeting*. 2005. Washington DC: American Chemical Society.
[27] Zhang, J.X. and R. Datta, *Sustained potential oscillations in proton exchange membrane fuel cells with PtRu as anode catalyst*. Journal of the Electrochemical Society, 2002. **149**(11): pp. A1423–A1431.
[28] Lee, W.K., Van Zee, J.W., S. Shimpalee, and S. Dutta, *Effect of Hunidition on PEM Fuel Cell Performance, Part I: Experiments*. Proceedings of the ASME IMECE, Nashville, TN, HTD 364–1, pp. 359–366.
[29] Woo, C.H.-K. and J.B. Benziger, *PEM Fuel Cell Current Regulation by Fuel Feed Control*, Chemical Engineering Science, 2007. **62**: pp 957–968.
[30] Mejia-Ariza, R., Effect of gas compositon on mass transfer to the cathode/membrane interface, 2005. http://www.princeton.edu/~pccm/outreach/REU2005/REU2005Presentations/mejiaariza.pdf.

4
Coupled Proton and Water Transport in Polymer Electrolyte Membranes

J. Fimrite, B. Carnes, H. Struchtrup and N. Djilali

4.1. Introduction

Solid polymer electrolytes, typically perfluorosulfonic acid (PFSA) membranes, are at the core of Polymer electrolyte membrane fuel cells (PEMFCs). These membranes electrically and mechanically isolate the anode and cathode while, when appropriately humidified, allowing for effective ion migration. Nafion, manufactured by DuPont, is one of the most thoroughly used and studied membranes in the PFSA family. Another family of membranes that holds some promise for use in PEMFCs is the group of sulfonated polyaromatic membranes, typically sulfonated polyetherketones. While research is being performed on other types of membranes, as well as hybrid membranes that might have been better-suited properties, information on these is scarce [1–10].

The functionality of polymer electrolyte membranes depends on an array of coupled transport phenomena that determine water content and conductivity. This Chapter synthesizes understanding of the salient phenomena, provides a critical examination of classical and recently proposed macroscopic models, and describes a general theoretical framework that allows improved modelling of transport in membranes, particularly in the context of ongoing efforts to develop more comprehensive computational fuel cell models [11–15] that allow analysis and optimization of fuel cells in a design and development environment. Kreuer et al. [16] recently presented a comprehensive review of both microscopic and macroscopic modelling aspects of transport phenomena in PEMs. Microscopic modelling work for PEMs, including molecular dynamics simulations [17] and statistical mechanics modelling [18–21], has focused primarily on Nafion membranes and has provided insight into some of the fundamental transport mechanisms. In the context of multi-dimensional fuel cell modelling, practical considerations dictate the use of macroscopic models.

In this chapter, we first provide some brief background and a summary of key experimental observations related to membrane conductivity, membrane hydration and sorption isotherms. We then examine the coupled transport mechanisms occurring within the "bulk" solvent. Of particular interest are the coupling and how the introduction of interactions with the membrane alters the transport mechanisms. In order to elucidate some of the outstanding formulation issues, we present an analysis of the binary friction and dusty fluid model, that shows that the binary friction model provides a general and rational framework for modelling transport phenomena in polymer electrolyte membranes. Finally, we use this framework to develop a new binary friction membrane transport model (BFM2) which

- relies on rationally derived transport equations based on the physics of multi-component transport in the membrane,
- removes the redundant viscous terms,
- is not restricted by the assumption of equimolar counter diffusion [22],
- accounts for the effect of temperature on the sorption isotherm.

Following the derivation of the BFM2, we show how this model's unknown transport parameters can be determined by considering the limit of a uniformly hydrated membrane to match the conditions of AC impedance conductivity measurements. Using empirically fitted transport parameters, the predictive ability of the model is then assessed for Nafion 1100 equivalent weight (EW) membranes. The material presented in this chapter is largely based on recent developments presented in Refs. [23, 24], but also includes new results illustrating the predictive abilities of the BFM2 in a complete fuel cell model.

4.2. Background

Transport of protons and water are the two phenomena of prime interest. Prior to examining the mechanisms that govern their transport, it is useful to review some of the background briefly that informs model formulation, including relevant aspects of membrane morphology, hydration behaviour and sorption isotherms.

4.2.1. Membrane Families

Sulfonated fluoropolymer membranes (also referred to as perfluorinated ion exchange membranes) or perfluorosulfonic acid membranes (PFSAs) such as Nafion, are currently the membranes of choice in low-temperature fuel cells as they exhibit high conductivity (when adequately hydrated), good stability (both mechanical and chemical) within the operating environment of the fuel cell, and high permselectivity for non-ionised molecules to limit crossover of reactants [25]. Sulfonated fluoropolymer membranes are based on a polytetrafluoroethylene (PTFE) backbone that is sulfonated by adding a side chain ending in a sulfonic acid group ($-SO_3H$) to the PTFE backbone. The resulting macromolecule contains both hydrophobic and hydrophilic regions. Altering the length of the chains, and location of the side chain on the backbone, alters the equivalent weights of sulfonated fluoropolymer membranes. The equivalent weight (EW) and its inverse, the ion exchange capacity (IEC), are defined as

$$\text{EW} = \frac{1}{\text{IEC}} = \frac{\text{Weight of dry polymer sample in grams}}{\text{Number of moles of acid groups}} \quad (4.1)$$

There is now general consensus that a hydrated PFSA membrane forms a two-phase system consisting of a water-ion phase distributed throughout a partially crystallized perfluorinated matrix phase [25, 26, 27]. The crystallized portion of the membrane cross-links the polymer chains, preventing complete dissolution of the polymer at temperatures below the glass transition temperature of the polymer [25] (~405 K for Nafion [26]). For a detailed review and discussion of membrane morphology, readers are referred to Weber and Newman [27], and to Kreuer et al. [16]

Based on earlier work, Weber and Newman [27] postulate the formation of approximately spherical clusters in regions with a high density of sulfonate heads, and an interfacial region that under vapour-equilibrated conditions consists of collapsed channels (Figure 4.1a) that can fill with water to form a liquid channel when the membrane is equilibrated with liquid water (Figure 4.1b). In their collapsed form, the channels allow for conductivity, since sorbed waters can dissociate from the sulfonate heads, but the amount of water sorbed is not sufficient to form a continuous liquid pathway [27].

In addition to Nafion, the family of sulfonated fluoropolymers includes Dow chemical membranes and Membrane C. Weber and Newman predict that the clusters formed within Dow membranes are smaller than in Nafion due to the higher elastic deformation energy [27]. For sulfonated polyetherketone membranes, which are under investigation due to their potential in lowering costs, separation into hydrophobic and hydrophilic domains is not as well defined as in Nafion [28]. As a result, their structure consists of narrower channels and clusters that are not as well connected as in Nafion [27, 28].

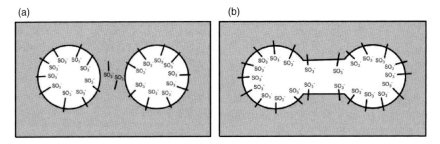

FIGURE 4.1. Schematics of (a) vapour-equilibrated membrane showing the collapsed interconnecting channel, (b) liquid-equilibrated membrane showing interconnecting channel swollen (after Ref. [27]).

4.2.2. Membrane Hydration

The protonic conductivity is strongly dependent on the membrane water content. In order to understand the water transport and swelling behaviour of PFSA membranes, we first examine the processes that take place as the membrane sorbs water molecules, focusing on Nafion, for which observations are more readily available. Due to similarities in morphology, other PFSA membranes are expected to exhibit similar behaviour.

Water sorption behaviour of PEMs is commonly considered in terms of λ, the number of sorbed waters per sulfonate head. The anhydrous form ($\lambda = 0$) of the membrane is not common, since complete removal of water requires raising the temperature to a point where decomposition of the membrane begins to occur. Approximately one and a half waters per sulfonate head are considered to remain in a membrane that is not in contact with any vapour or liquid water [26].

The first waters sorbed cause the sulfonate heads to dissociate, resulting in the formation of hydronium ions [26]. The water that hydrates the membrane forms counter-ion clusters localized on sulfonate sites with the sulfonate heads acting as nucleation sites [26]. Given the hydrophobic nature of the backbone, and the hydrophilic nature of the sulfonate heads, it is reasonable to consider that all water molecules sorbed by the membrane at this low water content are associated with the sulfonate heads. Moreover, the hydronium ions will be localized on the sulfonate heads, and, because the amount of water sorbed is insufficient for the formation of a continuous water phase, the conductivity will be extremely low. Figure 4.2a is a schematic

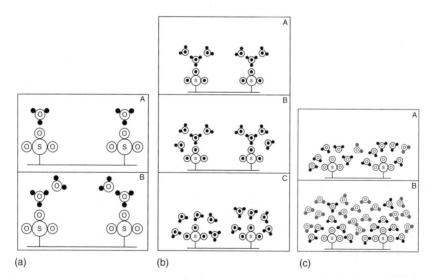

FIGURE 4.2. Schematic hydration diagram for Nafion: (a) for $\lambda = 1$ and $\lambda = 2$, (b) for water contents of $\lambda = 3-5$, and (c) for $\lambda = 6$ and $\lambda = 14$. Free waters are shown in grey.

of the state of a membrane for λ in the range [1,2]. Note that sulfonate heads might cluster together, thus some transport is possible even at lower water contents (λ ~ 2).

For λ in the range [1,2], the hydrogen bonds have approximately 80% of the strength of those in pure water, but as more water is added to the counter-ion clusters, the hydrogen bonds become weaker since the cluster shape does not allow for the formation of stronger bonds [26]. In the range λ = [3–5], the counter-ion clusters continue to grow while the excess charge (proton) is mobile over the entire cluster [26]. For λ greater than 2, the membrane will conduct some protons as the excess protons are mobilized on the counter-ion clusters and some pathways may be formed through the membrane to allow for conductivity. Figure 4.2 b illustrates the hydration state for λ = [3,5]. The number of water molecules forming the primary hydration shell for Nafion is expected to lie in the range [4,6] [29]. Molecular dynamics simulations indicate that the primary hydration shell for the sulfonate head grows to a maximum of five waters, and any additional waters are not as strongly bound and thus form a free phase [30, 31].

For λ ≥ 6, counter-ion clusters coalesce to form larger clusters, and eventually a continuous phase is formed with properties that approach those of bulk water [26, 28, 32, 28]. The free water phase is screened (or shielded) from the sulfonate heads by the strongly bound water molecules of the primary hydration shell [28, 29]. Figure 4.2 c is a schematic representation of the hydration states for λ = 6 (near the conductivity threshold) and 14 (saturated vapour equilibrated).

Figure 4.3 shows conductivity measurements, which show that the membrane exhibits low conductivity for λ less than five (Figure 4.3 a); as λ approaches five, the membrane becomes more conductive as some counter-ion clusters may connect, but there is still insufficient water for all clusters to coalesce [26]. In the range λ= [2,5], corresponding to roughly RH =

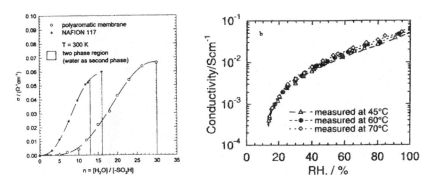

FIGURE 4.3. Conductivity of Nafion and a sulfonated polyaromatic membrane as a function of water content [33] at room temperature (left); and conductivity dependence on temperature and relative humidity for the E-form of Nafion [34] (right).

[13–60%], the conductivity changes sharply by nearly two orders of magnitude, as shown in Figure 4.3b. On the other hand, for λ changing from 5 and 14 (RH = 60–100%), the increase in conductivity is of less than one order of magnitude. This highlights how significant the transition to a continuous phase is.

Variations on this hydration scheme are expected for other PFSA membranes, as, among other factors, the number of waters in the primary hydration shell will vary according to the strength of the charge on the acid group, and the distance between sulfonate heads will affect the conductivity threshold, which will vary with the amount of water needed to connect the clusters.

Often, as in Figure 4.3b, the conductivity is measured as a function of the activity of the solvent with which the membrane is equilibrated. In order to relate these measurements to the actual water content, one can use experimentally determined sorption isotherms as shown in Figure 4.4 for Nafion and a sulfonated polyaromatic membrane. The sorption isotherms will be revisited in more detail to discuss their critical role in membrane transport models [21].

Both membranes in Figure 4.4 exhibit the so-called Schroeder's paradox, an observed difference in the amount of water sorbed by a liquid-equilibrated membrane and a saturated vapour-equilibrated membrane, with both reservoirs at the same temperature and pressure [27, 35, 36]. This difference leads to the jump in lambda when the membrane is water equilibrated (activity = 1), as shown in Figure 4.4. The underlying mechanisms for this behaviour are not completely resolved, but Choi and Datta [29] proposed a good explanation, arguing that an additional capillary pressure causes the vapour-equilibrated membrane to sorb less water than the liquid-equilibrated membrane from an external solvent with the same activity.

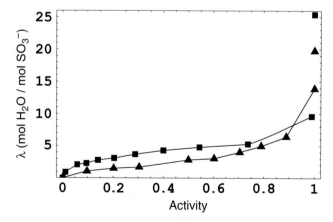

FIGURE 4.4. Water sorption isotherm for Nafion 117 (*triangles*) and a sulfonated polyaromatic membrane (*squares*) at 300 K [33].

4.2.3. Transport Mechanisms

We now turn our attention to conductivity, a key performance parameter in fuel cells. The high mobility of protons is afforded by the fact that the excess protons within the hydrogen bonded water network become indistinguishable from the "sea" of protons already present [38]. Kreuer et al. [16] recently provided a state-of-the-art review of proton transport mechanisms, and we will only briefly summarize relevant aspects. An excess proton in bulk water is typically found as a member of one of two structures, the first being a hydronium (H_3O^+), that is a proton donor to three other strongly bound waters [37]. The three strongly bound waters form the primary hydration shell of a hydronium, and the result is an "Eigen" ion [32, 37, 38] $(H_9O_4)^+$. The excess proton may also reside between two water molecules forming a "Zundel" ion [32, 37, 38] $(H_5O_2)^+$. The Zundel and Eigen ions are part of a fluctuating complex [37], with the structure fluctuating between the Zundel and Eigen ions on a time scale of the order of 10^{-13} seconds [32].

Proton diffusion can occur via two mechanisms, structural diffusion and vehicle diffusion [37]. It is the combination of these two diffusion mechanisms that confers protonic defects exceptional conductivity in liquid water. The conductivity of protons in aqueous systems of "bulk" water can be viewed as the limiting case for conductivity in PFSA membranes. When aqueous systems interact with the environment, such as in an acidic polymer membrane, the interaction reduces the conductivity of protons compared to that in bulk water [37]. In addition to the mechanisms described above, transport properties and conductivity of the aqueous phase of an acidic polymer membrane will also be effected by interactions with the sulfonate heads, and by restriction of the size of the aqueous phase that forms within acidic polymer membranes [32]. The effects of the introduction of the membrane can be considered on the molecular scale and on a longer-range scale, see Refs. [16, 32]. Of particular relevance to macroscopic models are the diffusion coefficients. As the amount of water sorbed by the membrane increases and the molecular scale effects are reduced, the properties approach those of bulk water on the molecular scale [32].

Another phenomenon linked to membrane conductivity is electro-osmotic drag—the process whereby water molecules are dragged by protons as these flow through the liquid phase of the membrane. Zawodzinski et al. [35] found that for a vapour-equilibrated membrane the number of water molecules dragged per proton (the electro-osmotic drag coefficient) has a value of approximately one over a wide range of water vapour activities. At lower water contents, all the waters within the collapsed channels (Figure 4.1a) are strongly bound to the sulfonate heads, while the lower concentration of sulfonate heads means that this portion of the membrane is more hydrophobic than areas where clusters form. Thus, there is no free water phase present in the collapsed channels. Consequently, we cannot expect large hydrated structures to diffuse through the membrane, as in bulk water. Instead, we expect the hydronium ions delocalized on the water molecules hydrating the sulfonate heads within the collapsed channels to allow for conductivity between clusters. Therefore, we have hydronium ions diffusing through the membrane liquid phase, which corresponds to an electro-osmotic drag coefficient of one, as is expected for a vapour-equilibrated membrane.

Under liquid-equilibrated conditions (Figure 4.1 b), the membrane liquid phase is well interconnected, and the effect of the sulfonate heads on the free water is reduced due to shielding; larger structures, such as Eigen and Zundel ions, can diffuse through the membrane liquid phase and, thus, more waters are dragged through the membrane per proton. Approximately 2.5 water molecules accompany each proton through the membrane for a liquid-equilibrated membrane [27, 39].

To simplify our modelling efforts we will assume that one water is carried through the membrane per proton over a wide range of water vapour activities, which is commonly done [40, 22], and approximately 2.5 water are carried through per proton when liquid-equilibrated.

4.3. Membrane Transport Models

Membrane modelling has been considered from both the nano/microscopic and the macroscopic viewpoints, but little has been done to bridge these two limits. The breadth of microscopic modelling work for PEMs encompasses molecular dynamics simulations [17] and statistical mechanics modelling [18–21]. Most applications have focused on Nafion, and interestingly, some models even apply macroscopic transport relations to the microscopic transport within a pore of a membrane [41]. While the focus of this Chapter is on macroscopic models required for computational simulations of complete fuel cells [12, 13, 15], the proposed modelling framework is based on fundamental relations describing molecular transport phenomena.

Macroscopic models can be classified into two broad categories: (i) membrane conductivity models and (ii) mechanistic models, typically for fuel cell water management purposes. The latter usually require the use of a conductivity model, a fit to empirical data, or the assumption of constant conductivity (e.g. fully hydrated membrane at all times), and can be further classified into hydraulic models, in which a water transport is driven by a pressure gradient, and diffusion models, in which transport is driven by a gradient in water content.

4.3.1. Hydraulic Models

One of the earliest hydraulic models is that of Bernardi and Verbrugge [42, 43] and is based on the Nernst-Planck equation for the transport of species within the fluid phase, and on the Schloegl equation to describe fluid transport,

4. Coupled Proton and Water Transport in Polymer Electrolyte Membranes

$$N_i^{N-P} = -z_i \frac{F}{RT} D_i c_i \nabla \Phi - D_i \nabla c_i + c_i v_s, \quad v_s = \frac{k_\phi}{\eta} z_f c_f F \nabla \Phi - \frac{k_p}{\eta} \nabla p \quad (4.2)$$

In the formulation of Bernardi and Verbrugge, the membrane is assumed fully hydrated, and the gases are taken to be dissolved in the pore fluid [42]. A more general variant of this hydraulic model was proposed by Eikerling et al. [44] and allows water content variation, and dependence of conductivity, permeability, and electro-osmotic drag coefficient on the local water content.

One issue with hydraulic models is that in membranes with lower water contents interactions between the sulfonate heads and the backbone are significant, and the water molecules are localized on the sulfonate heads (see Figures 4.2 a and 2 b). The water is less "bulk–like" and the clusters are no longer well connected. Conceptually, the concentration gradient seems to be a more appropriate driving force than the pressure gradient [27].

4.3.2. Diffusion Models

The distinguishing feature of the classical diffusion model of Springer et al. [39] (hereafter SZG) is the consideration of variable conductivity. SZG relied on their own experimental data to determine model parameters, such as water sorption isotherms and membrane conductivity as a function of the water content. Alternative approaches include the use of concentrated solution theory to describe transport in the membrane [45], and invoking simplifying assumptions such as thin membrane with uniform hydration [46].

SZG's ground-breaking model has been particularly valuable in determining membrane resistance in computational fuel cell models at higher/intermediate water content conditions. This model has, however, several limitations. The equations used are not based on the physics of conductivity, but are essentially a curve fit and, thus, the model constants have no physical significance and the model is restricted to 1100 *EW* Nafion. Even with parameter adjustments, SZG's model is not expected to be useful in predicting or correlating the behaviour of other types of membranes.

The model reads [39]

$$\sigma_{springer} = \exp\left[1268\left(\frac{1}{303} - \frac{1}{273 + T_{cell}}\right)\right]\sigma_{30}, \quad (4.3)$$

with $\sigma_{30} = 0.005139\lambda - 0.00326(\lambda > 1)$

where T_{cell} is the cell temperature in degrees centigrade and σ_{30} is the conductivity (with units of S cm^{-1}) at 30°C that is measured to be a linear function of λ.

One of the problems with the model of SGZ and other diffusion models, is that under conditions close to full hydration (Figure 4.2 c), there is essentially no water concentration gradient, and diffusion models are unable to produce a water concentration profile. In such regimes, a hydraulic model is more appropriate. Hence, diffusion models represent correctly the behaviour at low water contents, while hydraulic models represent better the behaviour in saturated membranes [27].

An approach that is conceptually simpler and does not require the prescription of transport to hydraulic or diffusion mechanisms was proposed by Janssen [47], and Thampan et al. [22] (hereafter TMT) based on the use of chemical potential gradients in the membrane. More recently, Weber and Newman [27] developed a novel model where the driving force for vapour-equilibrated membranes is the chemical potential gradient, and for liquid-equilibrated membranes it is the hydraulic pressure gradient. A continuous transition is assumed between vapour- and liquid-equilibrated regimes with corresponding transition from 1 to 2.5 for the electro-osmotic drag coefficient.

4.4. Membrane Conductivity Models

The model of TMT is one of the few models solely targeted at predicting conductivity behaviour of a membrane, and in contrast to the model of SZG, is based on physical rather than purely empirical considerations [22]. It is in this vein that they invoke the dusty fluid model (DFM) to model transport in the membrane. Before considering the model of TMT we examine the background of the DFM and the binary friction model (BFM).

4.4.1. The Binary Friction Model

The BFM is developed in Ref. [49] by considering transport within a pore structure and applying the Stefan-Maxwell equations to the fluid mixture [48]. In Ref. [24] we showed that it can be written as

$$-\frac{1}{RT}\nabla_T \mu_i^e = \sum_{j=1}^{n} \frac{X_j}{D_{ij}^e}\left(\frac{\mathbf{N}_i}{c_i} - \frac{\mathbf{N}_j}{c_j}\right) + \frac{1}{D_{iM}^e}\left(\frac{\mathbf{N}_i}{c_i}\right), i = 1,\ldots,n, \quad (4.4)$$

where μ_i^e is the electrochemical potential of species i, $X_i = c_i/c_t$ with $c_t = \sum c_i$ as the total mole density of the fluid (refer to Ref. [49] for details), D_{ij}^e is the effective Stefan-Maxwell interaction coefficient between species i and j, and D_{iM}^e is the effective interaction coefficient between species i and the porous medium.

The terms on the right-hand side describe the interaction among species i and j, and between species i and the membrane M, respectively. D_{ij}^{S-M} and D_{iM}^e should not be interpreted as diffusion coefficients but rather as interaction terms equivalent to an inverse friction coefficient between species. Physically, the D_{ij}^e relate changes in *relative* species fluxes to

4. Coupled Proton and Water Transport in Polymer Electrolyte Membranes

gradients in the electrochemical composition of the mixture arising from species-species interactions, while the D^e_{iM} relate *absolute* species fluxes to gradients in individual electrochemical potential gradients, arising from species-medium interactions.

In the original BFM, as well as in the DFM and dusty gas model (DGM) discussed in the next section, the structure of the porous media is considered independent of the transport equations. The transport equations are first written with the *pore*-averaged fluxes N'_i, and are cast per unit of *pore* surface area. The flux is then corrected to a flux per unit of *membrane* cross sectional area by multiplying the flux by a correction factor that includes the porosity λ and tortuosity factor [48, 49, 50] τ

$$N_i = \frac{\varepsilon}{\tau} N'_i \qquad (4.5)$$

The porosity and tortuosity factor can be brought into the diffusion coefficients [51] by scaling the standard binary diffusion coefficients using the porosity τ and tortuosity factor λ according to

$$D^e_{ij} = \frac{\varepsilon}{\tau} D_{ij} \qquad (4.6)$$

The effective diffusion coefficients D^e_{iM} between the species and porous medium are assumed to follow the same scaling for appropriate reference values D_{iM}. An alternative to the above correction that avoids the use of the tortuosity is the Bruggeman correction [22]

$$D^e_{ij} = (\varepsilon - \varepsilon_0)^q D_{ij} \qquad (4.7)$$

where ε_0 is the threshold porosity, which is the minimum fraction of the volume that must be occupied by the fluid to allow transport. The Bruggeman exponent q is either used as a fitted parameter or is given the value of 1.5 Note that the Bruggeman correction is equivalent to setting $\tau = \varepsilon/(\varepsilon-\varepsilon_0)^q$.

Another model referred to in the literature as a "diffusion" model [50] is similar in nature to the BFM, but is derived by assuming the membrane can be modelled as a dust component (at rest) present in the fluid mixture. The equations governing species transport are developed from the Stefan-Maxwell equations with the membrane as one of the mixture species. The resulting equation for species i is identical to Eq. (4.4) [50], thus the BFM and this diffusion model are equivalent.

4.4.2. The Dusty Fluid Model

The dusty fluid model (DFM) shares some similarities with the BFM. It was derived based on the dusty gas model (DGM), which describes gas flow through porous media [50]. Space does not allow a detailed discussion of the DFM [24], but a key distinction of this model compared to the BFM is the presence of additional viscous terms. Proponents of the DFM have argued that the BFM does not account for viscous transport, and that an additional convective velocity, calculated using the Schloegl equation, must be added to the diffusive velocity.

This interpretation of velocities and the resulting additional terms are in fact erroneous as we have shown in Ref. [24], and amount to double accounting. The BFM was on the other hand shown to implicitly contain the viscous terms, i.e. the Schloegl equation [24]. The new membrane model developments presented in this Chapter are therefore based on the correct and rational BFM framework.

4.4.3. Conductivity Model Based on the DFM

Having examined the governing equations used to model transport in porous structures, we now turn our focus to an important stepping stone in theoretical modelling of membrane conductivity, the model of Thampan et al. (TMT) [22], which is based on the DFM equations. TMT use the Bruggeman correction, Eq. (4.7), with a percolation threshold below which conductivity is zero. This then satisfies the requirement of a minimum amount of water sorbed to represent the connectivity threshold in the water phase allowing charge conduction through the membrane. Although hydronium ion formation is the first step in the reaction of water molecules with a sulfonate head [26], the hydronium is not necessarily free to move, but for lower water contents will instead be localized on the sulfonate head. Referring to Figure 4.1b we note that at such low water contents the liquid phase within the membrane is poorly connected.

The use of the Brunauer-Emmett-Teller (BET) sorption model by TMT is problematic due to the fact that the BET is fit to sorption data at 30°C and does not consider the temperature dependence of sorption behaviour. One way the model of TMT could be improved is by using more recent models for sorption isotherms, e.g. that of Choi and Datta [29], or by using conductivity data measured as a function of water content.

In developing the transport equations, TMT make several assumptions that need to be critically re-examined. Though it is probably reasonable to assume that hydronium ions are the charge carriers for vapour-equilibrated membranes, this is not valid for liquid-equilibrated membranes, where the transport number is found to be around 2.5 [52]. For more realistic predictions in the liquid-equilibrated regime considered by TMT, this assumption needs to be modified.

TMT assume equimolar counter diffusion (closed conductivity cell), i.e. equal and opposite water and hydronium fluxes [22]. For a more general transport model, it is desirable to develop a second flux expression for water,

4. Coupled Proton and Water Transport in Polymer Electrolyte Membranes

which considers the influence of forces (i.e. gradient in water molar density) that drive water through the membrane.

TMT also assume that $D_{IM}^e \approx D_{2M}^e$ [22]. This need to be reconsidered since, due to the differences between the hydronium ions and water, we expect interaction forces with the membrane to be different for the different species. This assumption, coupled with the assumption of a closed conductivity cell, forces convection to be zero [22], thus causing fortuitously the additional viscous terms of the DFM to drop out anyway.

The resulting conductivity expression reads [22]

$$\sigma_{Thampan} = (\varepsilon - \varepsilon_0)^q \left(\frac{\lambda_1^0}{1+\delta} \right) c_{HA,0} \alpha, \tag{4.8}$$

where α is the degree of dissociation, $c_{HA,0}$ is the acid group concentration in the pore fluid, δ is the ratio D_{12}/D_{1M} and λ_1^0 is the equivalent conductance of hydronium at infinite dilution. Note also the presence of the Bruggeman correction.

On initial inspection, the model provides a good fit to the experimental data of Sone et al. [34] for vapour-equilibrated membranes, see Figure 4.5. In fact there are discrepancies masked by the log scale used in the plot. Nonetheless, TMT's model is significant in that it provides a theoretical framework based on the structure of the membrane and the physics of the transport phenomena. There are several avenues for improvement:

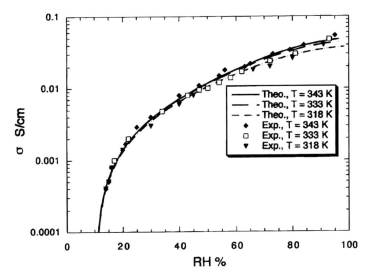

FIGURE 4.5. Conductivity of Nafion 117 equilibrated in water vapour vs. water vapour activity at different temperatures. Experimental results of Sone et al. [34]; theoretical predictions of TMT [22].

- The BFM should be used instead of the DFM, thus removing formulation inconsistencies due to the additional viscous terms. Interestingly, TMT's assumption that $D_{IM}^e \approx D_{2M}^e$ actually causes the extra viscous terms from the DFM to drop out anyway.
- The restriction to equimolar counter diffusion should be removed.
- The effect of temperature on the sorption isotherm should be accounted for.
- Comparison should be made in a format where any differences between the model and experimental data can be more readily identified and estimated.

4.5. Binary Friction Membrane Model

4.5.1. Specialization of BFM to a PEM

We now proceed to reduce the general binary friction model, Eq. (4.4), to the binary friction membrane model (BFM2) by means of scaling arguments.

We consider transport within the family of perfluorosulfonic acid (PFSA) membranes, such as Nafion. The membrane is taken as the porous medium, and only two species are assumed present in the pore fluid, proton carriers (species i = 1) and water (i = 2). Furthermore we assume that the dominant proton carrier is hydronium.

The gradient in electrochemical potential (at constant temperature T) can be expressed as

$$\nabla_T \mu_i^e = RT(\nabla \ln X_i + \nabla \ln \gamma_i) + V_{M,i} \nabla_p + z_i F \nabla \Phi \quad (4.9)$$

where the terms represent the effects of composition, activity, Gibb's free energy, and electrical potential. X_i is the mole fraction, γ_i is the activity, $V_{M,i}$ is the specific molar volume, z_i is the charge, and Φ is the ionic potential.

Introducing the above expression for the gradient of the electrochemical potential into the BFM and multiplying both sides by the mole fraction X_i, we obtain

$$-(\hat{\nabla} X_i + X_i \hat{\nabla} \ln \gamma_i + \beta X_i \hat{v}_i \hat{\nabla} \hat{p} + \Theta X_i z_i \hat{\nabla} \hat{\Phi})$$
$$= \sum_{j=1}^{2} \left\{ \frac{X_j \hat{N}_i}{\hat{c}_t \hat{D}_{ij}} - \frac{X_i \hat{N}_j}{\hat{c}_t \hat{D}_{ij}} \right\} + \frac{\hat{N}_i}{\hat{c}_t \hat{D}_{iM}}, i = 1, 2 \quad (4.10)$$

In order to identify terms that may be neglected to simplify the numerical solution and to improve physical insight, the system in Eq. (4.10) is written in dimensionless form using the parameters and variables found in Table 4.1. The reference molar densities are chosen to be the inverse of the partial molar volume of water, since this does not vary significantly with temperature or

4. Coupled Proton and Water Transport in Polymer Electrolyte Membranes

TABLE 4.1 Non-dimensional quantities and associated reference values.

Non-Dimensional Quantities		Reference Values	
Length	$\hat{x} = x/L_M$	L_M = membrane thickness	
Total mole density	$\hat{c}_t = c_t/c_{ref}$	$c_{ref} = 1/V_{M,2} = 55.6 \times 10^{-3}$ mol m^{-3}	
Mole density of species i	$\hat{c}_i = c_i/c_{ref}$		
Mole flux	$\hat{N}'_i = N'_i/N_{ref}$	$N_{ref} = c_{ref} v_{ref}$, $v_{ref} = D_{ref}/L_M$	
Molar volume	$\hat{v}_1 \approx \hat{v}_2 = \dfrac{V_{M,2}}{V_{M,ref}}$	$V_{M,ref} = \dfrac{1}{c_{ref}} = V_{M,2}$	
Diffusion coefficients	$\hat{D}_{ij} = \dfrac{D_{ij}^{S-M}}{D_{ref}}$, $\hat{D}^e_{ij} = \dfrac{D^e_{ij}}{D_{ref}}$	$D_{ref} = D_{ij}^{S-M}$, or D^e_{iM}	
Pressure gradients	$\hat{\nabla}\hat{p} = \nabla p / \left(\dfrac{\Delta p_{ref}}{L_M}\right)$	$\Delta p_{ref} = 5 \times 10^5$ Nm^{-2}	
Potential gradients	$\hat{\nabla}\hat{\Phi} = \nabla\Phi / \left(\dfrac{\Delta\Phi_{ref}}{L_M}\right)$	$\Delta\Phi_{ref} = 0.3$ V	
Gradient operator	$\hat{\nabla} = L_M \nabla$	L_M = membrane thickness	
Additional coefficients	$\beta = \dfrac{\Delta p_{ref}}{RT c_{ref}}$	$\Delta p_{ref} = 5 \times 10^5$ Nm^{-2}	
	$\Theta = \dfrac{F\Delta\Phi_{ref}}{RT}$	$\Delta\Phi_{ref} = 0.3$ V	

pressure. We assume the molar volumes for water and hydronium to be the same (see Section 4.5.4).

4.5.2. Magnitude of the Coefficients for the Driving Force Terms

To perform an order of magnitude analysis of the driving force terms, we assume, as a limiting case scenario, the pressure drop across the membrane to be $\Delta p_{ref} = 5 \times 10^5$ Nm^{-2}, while the maximum potential drop across the membrane is approximately $\Delta\Phi_{ref} = 0.3$ V.

Following previous studies [22], we assume the gradients in composition are small and, thus, gradients in the activity coefficients are negligible, i.e. $(X_i \hat{\nabla} \ln \gamma_i \cong 0)$. Using the values in Table 4.1, the coefficients for the driving force terms are estimated in Table 4.2.

TABLE 4.2 Comparing the relative magnitude of the driving forces in the transport equations.

Gradient of Interest	Coefficient for Gradient Term	Coefficient Value	Approximate Order of Magnitude
$\hat{\nabla} X_1, \hat{\nabla} X_2$	1	1	~1
$\hat{\nabla}\hat{p}$	$X_1 \hat{v}_1 \beta$	$X_1(3.15 \times 10^{-3})$	~10^{-3}
	$X_2 \hat{v}_2 \beta$	$X_2(3.15 \times 10^{-3})$	~10^{-3}
$\hat{\nabla}\hat{\Phi}$	$X_1 \Theta$	$X_1(10.1)$	1–10

As can be seen from the table, compared to the potential and mole fraction gradient terms, the pressure terms are of a significantly lower order and can be neglected. Also, the potential term is the dominant term when $z_i \neq 0$, while the gradient in mole fraction term is dominant when $z_i = 0$.

Thus, Eq. (4.8) can be reduced to

$$-(\hat{\nabla} X_i + \Theta X_i z_i \hat{\nabla}\hat{\Phi}) = \sum_{j=1}^{2}\left\{\frac{X_j \hat{N}_i}{\hat{c}\hat{D}_{ij}} - \frac{X_i \hat{N}_j}{\hat{c}\hat{D}_{ij}}\right\} + \frac{\hat{N}_i}{\hat{c}\hat{D}_{iM}}, i = 1, 2 \quad (4.11)$$

Expanding this for both species ($i = 1, 2$), using that $\hat{D}_{ij} = \hat{D}_{ji}$ [49], casting the above into matrix form, and inverting the matrix to obtain an expression for the fluxes in terms of the driving forces yields

$$\begin{pmatrix}\hat{N}_1 \\ \hat{N}_2\end{pmatrix} = -\frac{\hat{c}(\varepsilon - \varepsilon_0)^q}{\frac{X_1}{\hat{D}_{12}\hat{D}_{1M}} + \frac{X_2}{\hat{D}_{12}\hat{D}_{2M}} + \frac{1}{\hat{D}^e_{1M}\hat{D}_{2M}}} \times \begin{bmatrix} \frac{X_1}{\hat{D}_{12}} + \frac{1}{\hat{D}_{2M}} & \frac{X_1}{\hat{D}_{12}} \\ \frac{X_2}{\hat{D}_{12}} & \frac{X_2}{\hat{D}_{12}} + \frac{1}{\hat{D}_{1M}} \end{bmatrix}\begin{pmatrix}\hat{\nabla}X_1 + \Theta X_1 \hat{\nabla}\hat{\Phi} \\ \hat{\nabla}X_2\end{pmatrix} \quad (4.12)$$

Here, we have also introduced the Bruggeman correction, Eq. (4.7).

Note that X_i and ε depend on the water content λ, and the D_{ij} are functions of λ and temperature T. The next two sections will consider these dependencies in further detail.

4.5.3. Mole Numbers, Volumes, Porosities Etc.

In the original presentation of the BFM2 model [23], special attention was paid to computing the degree of dissociation of the sulfonate heads as a function of water content, $\alpha(\lambda)$. However, since i) almost all sulfonate heads dissociate as soon as the water content λ exceeds 2; ii) the conductivity vanishes at small values for λ; and iii) conditioned membranes in fuel cells maintain water contents above $\lambda = 2$, it is therefore reasonable and expedient to assume $\alpha = 1$. The model is thus presented here with this simplification.

Due to electroneutrality, and since we assume that all sufonate heads are dissociated, the number of dissociated protons, n_1, equals the number of sulfonate heads, $n_1 = n_{sh}$. Due to the definition of λ, the total number of water molecules is given by $n_w = \lambda n_{sh}$, which implies that[*]

---

[*] Obviously, this restricts the presented equations to values of $\lambda > 1$, since for $\lambda < 1$ one would, for the case of complete dissociation, expect $n_1 = n_w$.

$$n_1 = n_{sh} = \frac{n_w}{\lambda} \quad (4.13)$$

We also have to account for the water molecules that are associated with protonated complexes. For a protonated complex containing ω_{pw} water molecules, then $\omega_{pw} n_1$ water molecules take part in the proton transport, while the remaining $n_2 = n_w - \omega_{pw} n_1$ water molecules form the second species. For generality, the parameter ω_{pw} could be retained in the model. However, as we are considering transport within vapour-equilibrated Nafion, it is reasonable to assume that hydronium is the protonated complex that is formed. Thus, we will assume $\omega_{pw} = 1$ throughout, so that

$$n_2 = n_w - n_1 = n_w \left(1 - \frac{1}{\lambda}\right) \quad (4.14)$$

Next we determine the molar densities, which involve partial molar volumes. We could not find data on the exact partial molar volume of hydronium or other protonated complexes under the given conditions, i.e. within a hydrated membrane. In order to be able to progress, we make the reasonable assumption that the molar volume of water and hydronium is approximately the same [23, 53], so that $V_{M,1} = V_{M,2}$.

The volumes occupied by the protonated complex, the free waters, and the membrane are

$$V_1 = V_{M,2} n_1 = V_{M,2} n_w \frac{\alpha}{\lambda}, \quad V_2 = V_{M,2} n_2 = V_{M,2} n_w \left(1 - \frac{1}{\lambda}\right),$$

$$V_M = V_{M,M} n_{sh} = V_{M,M} n_w \frac{1}{\lambda} \quad (4.15)$$

where $V_{M,M}$ is the volume of membrane per mole of acid heads, $V_{M,M} = EW/\rho_{dry}$; ρ_{dry} is the dry density of the polymer membrane.

The total volume of pore fluid is, from Eq. (4.15),

$$V_p = V_1 + V_2 = V_{M,2} n_w \quad (4.16)$$

The mole densities of protonated complexes, free water within the pore fluid, and the total mole density of the pore fluid, are accordingly given by

$$c_1 = \frac{n_1}{V_p} = \frac{1}{V_{M,2}} \frac{\alpha}{\lambda}, \quad c_2 = \frac{n_2}{V_p} = \frac{1}{V_{M,2}} \left(\frac{\lambda - \alpha}{\lambda}\right), \quad c_t = c_1 + c_2 = \frac{1}{V_{M,2}}. \quad (4.17)$$

This gives the mole fraction for the protonated complexes and the free waters as

$$X_1 = \frac{c_1}{c_t} = \frac{1}{\lambda}, \quad X_2 = \frac{c_2}{c_t} = 1 - \frac{1}{\lambda}. \quad (4.18)$$

The porosity is defined as the volume of the pore fluid divided by the total volume, and, since the total volume is the sum of all volumes that make up the system, the porosity can be written as

$$\varepsilon = \frac{V_p}{V_t} = \frac{V_1 + V_2}{V_1 + V_2 + V_M} = \frac{\lambda}{\lambda + V_{M,M}/V_{M,2}}. \quad (4.19)$$

The threshold porosity ε_0 in the Bruggeman correction is defined by specifying a minimum water content λ_{min} with $\varepsilon_0 = \varepsilon(\lambda_{min})$. Here λ_{min} is the minimum amount of water that must be sorbed by the membrane for the pore liquid phase to be sufficiently well connected to allow for transport through the membrane. Examining the conductivity data in Figure 4.1, it is clear that λ_{min} should lie somewhere between 1.5 and 2, since this is the approximate range where the conductivity bounds intersect the x-axis.

4.5.4. Diffusion Coefficients

The determination of the coefficients D_{12}, D_{1M} and D_{2M}, is required next. These coefficients depend on the PFSA material, and the state of the material, in particular on temperature T and water content λ.

One way to specify these coefficients is to fit their values to detailed measurements. This would require a painstaking effort involving systematic measurements covering all possible states of the membrane. Such measurements would be required anew for a new material.

Determination of the coefficients based on understanding of the membrane microstructure and modelling of the interaction between the membrane and the two transported species, i.e. hydronium and water, would be better. Most desirable would be a proper mathematical transition from an exact microscopic description of the interaction of membrane, hydronium and water, towards a macroscopic model. Such information and description being currently unavailable, we have to rely on guidance from knowledge on the membrane morphology to devise assumptions on the functional dependence of the coefficients on temperature and water content.

We assume the interaction between water and hydronium ions, as described by D_{12}, does not depend on water content. This seems reasonable based on studies of the Stefan-Maxwell coefficients for systems of non-ideal fluids [48]. Furthermore, since D_{12} represents binary diffusion within the fluid in the membrane, it should not depend on the water content λ. Thus, we can use the typical literature value, which at a temperature $T_0 = 303$ K is denoted as D_{12}^0.

In the absence of specific knowledge on species-membrane interaction, it is reasonable to assume the interaction, D_{1M} and D_{2M}, depends on the water content due to changes in the geometry of the membrane, and the proximity of the species to the membrane. In order to determine the empirical fit of our

4. Coupled Proton and Water Transport in Polymer Electrolyte Membranes

model parameters, we assume a simple power law dependence, so that both D_{1M} and D_{2M} are increasing functions of λ. We introduce three dimensionless constants A_1, A_2, s and write

$$D_{1M} = D_{12}A_1\lambda^s \text{ and } D_{2M} = D_{12}A_2\lambda^s. \tag{4.20}$$

The use of the same exponent, s, for both coefficients is a simplification; this could be refined by considering two separate exponents, s_1 and s_2.

Finally, we consider the temperature dependence. The coefficients D_{ij} describe the interaction between two different species, and it is physically plausible to assume that the interaction becomes weaker with temperature. We will also adopt an Arrhenius law behaviour as is typical of thermally activated interaction processes. As a first approximation, we assume that all three interaction coefficients vary in the same way, so that their temperature dependence is given through the temperature dependence of the coefficient D_{12}, which we write as

$$D_{12} = D_{12}(T) = D_{12}^0 \exp\left[\frac{E_a}{R}\left(\frac{1}{T_0} - \frac{1}{T}\right)\right], \tag{4.21}$$

where E_a is the activation energy. Again, a more refined theory could consider different Arrhenius laws for each of the three coefficients.

4.5.5. BFM2 Model

The results of the last three sections can now be combined to yield the Binary Friction Membrane Transport Model (BFM2) which, after reinserting the dimensions, can be written as

$$\begin{pmatrix} N_1 \\ N_2 \end{pmatrix} = \frac{-c_t(\varepsilon - \varepsilon_0)^q}{\lambda\left(\frac{1}{D_{12}D_{1M}} + \frac{\lambda-1}{D_{12}D_{2M}} + \frac{\lambda}{D_{1M}D_{2M}}\right)} \begin{bmatrix} \frac{1}{D_{12}} + \frac{\lambda}{D_{2M}} & \frac{-1}{D_{2M}} \\ \frac{\lambda-1}{D_{12}} & \frac{1}{D_{1M}} \end{bmatrix} \begin{pmatrix} \frac{F}{RT}\nabla\Phi \\ \nabla\lambda \end{pmatrix} \tag{4.22}$$

or, when we make all coefficients explicit, as

$$\begin{pmatrix} N_1 \\ N_2 \end{pmatrix} = \frac{-c_t\left(\frac{\lambda}{\lambda+V_{M,M}/V_{M,2}} - \varepsilon_0\right)^q D_{12}^0 \exp\left[\frac{E_a}{R}\left(\frac{1}{T_0}-\frac{1}{T}\right)\right]}{\lambda\left(\frac{1}{A_1\lambda^s} + \frac{\lambda-1}{A_2\lambda^s} + \frac{1}{A_1A_2\lambda^{2s-1}}\right)}$$
$$\times \begin{bmatrix} 1 + \frac{\lambda^{1-s}}{A_2} & \frac{-1}{A_2\lambda^s} \\ \lambda - 1 & \frac{1}{A_1\lambda^s} \end{bmatrix} \begin{pmatrix} \frac{F}{RT}\nabla\Phi \\ \nabla\lambda \end{pmatrix} \tag{4.23}$$

The model contains six unknown constants, namely ε_0 (or λ_{min}), E_a, A_1, A_2, s, and q. These must be obtained by fitting to experimental data. While Eqs.

(4.22, 4.23) describe coupled transport of water and hydronium through a PFSA membrane, it is possible to determine the coefficients from conductivity measurements, in the absence of net water transport. This is possible because one of the characteristics of the BFM2 is that the constants that appear in the general BFM2 model are *identical* to those and in the simplified conductivity form of the model as will be shown next. Once determined from conductivity experiments, the constants can be reintroduced into the BFM2, and the model applied to compute coupled water and ionic transport.

4.6. Binary Friction Conductivity Model

Conductivity is directly related to the transport of hydronium ions through the membrane, and is the best-documented transport property. In this section we reduce the BFM2 model, Eq. (4.22)) to a conductivity model. It should be emphasized that this conductivity model is derived here primarily as a *tool* to gain insight into the behaviour of the unknown transport coefficients and to specify the model constants.

4.6.1. The Conductivity Model

Conductivity measurements are commonly performed on membranes using AC impedance measurement techniques [34]. In such experiments, the membrane is generally considered to be uniformly equilibrated with water vapour and thus that the gradient in water content is zero. It follows from Eq. (4.22) that the protonic flux is then proportional to the potential gradient in the membrane as

$$N_1 = -c_t(\varepsilon - \varepsilon_0)^q \frac{D_{1M}(D_{2M} + \lambda D_{12})}{\lambda(D_{2M} + D_{12}\lambda + D_{1M}(\lambda - 1))} \frac{F}{RT} \nabla \Phi \quad (4.24)$$

The ionic current density is related to the ionic fluxes by Faraday's law

$$i = F \sum_{i=1}^{2} z_i N_i = F N_1 \quad (4.25)$$

and conductivity is defined as

$$\sigma \equiv \frac{-i}{\nabla \Phi} \quad (4.26)$$

Using Eqs. (4.25) and (4.26), we obtain the Binary Friction Conductivity Model (BFCM)

$$\sigma = c_t(\varepsilon - \varepsilon_0)^q \frac{F^2}{RT} \frac{D_{1M}(D_{2M} + \lambda D_{12})}{\lambda(D_{2M} + D_{12}\lambda + D_{1M}(\lambda - 1))}, \quad (4.27)$$

which, when the coefficients are specified according to the Section 4.5, can be written in explicit form as

$$\sigma = c_t \left(\frac{\lambda}{\lambda + V_{M,M}/V_{M,2}} - \varepsilon_0 \right)^q \frac{D_{12}^0 F^2}{RT} \exp\left[\frac{E_a}{R} \left(\frac{1}{T_0} - \frac{1}{T} \right) \right] \\ \times \frac{A_1(\lambda + A_2 \lambda^s)}{\lambda(A_2 + A_1(\lambda - 1) + \lambda^{1-s})} \quad (4.28)$$

Again we emphasize that Eq. (4.27) is a reduction of (4.22) derived in the limit of negligible water content gradient in the membrane (or in the limit of a negligible upper off-diagonal term $(-1)/D_{2M}$). Similarly to other established membrane models, Eq. (4.27) does not account for the impact of water flux on protonic current. Its most useful feature in the context of the developments presented here is that it contains the *same* transport coefficients as the general BFM2, and thus any appropriate specification of parameters for the BFCM automatically and conveniently defines a set of parameters for the more general BFM2, which can then be used to predict coupled protonic current and water fluxes as will be illustrated in Section 4.7.

4.6.2. Experimental Data

Sone et al. [34] reported conductivity data for Nafion 117 in the E-form (no heat treatment), measured using a four-electrode AC impedance method. Membranes used in fuel cells are typically heated during the manufacture of the membrane electrode assembly, and it may thus be more appropriate to fit to the data for a membrane in the N or S form; however, fitting to the E-form data allows direct comparison with Thampan et al. [22] fit to their model.

Sone et al's data is measured in the plane of the membrane, but is expected to provide a reasonable measure of the normal direction conductivity since Nafion presents no apparent ordering of the macromolecules in any preferential direction, and its properties are expected to be reasonably isotropic. We also assume that the conductivity data measured by Sone and co-workers represents the conductivity of 1100 *EW* Nafion membranes of all thicknesses.

In Ref. [34] the conductivity data (in S cm^{-1}) collected was fitted to a third degree polynomial, i.e.

$$\sigma_{Sone} = a_{Sone} + b_{Sone}x + c_{Sone}x^2 + d_{Sone}x^3 \quad (4.29)$$

where x is the relative humidity ($x = 100\, p/p_{sat}(T) = 100a$), and the coefficients (a_{Sone}, b_{Sone}, c_{Sone} and d_{Sone}) are given for various temperatures. We will use the data for the E-form of Nafion for 30°C and 70°C.

An added complication is that the conductivity data of Sone et al. [34] is given as a function of the water vapour activity outside of the membrane. However, in the BFM2 and the BFCM, as well as in the models of Springer et al. [39] and Thampan et al. [22], conductivity is defined as a function of water contents λ. Casting conductivity as a function of λ, rather than activity, is more practical in applying the membrane model since λ is taken in general as varying throughout the membrane.

To convert the data of Sone et al. [34] from a function of relative humidity x to a function of water content, we require the activity a as a function of water content λ. This is accomplished using a least squares fit of a third degree polynomial to fit the available sorption isotherm data from Sone et al. [34] with λ as the independent variable and activity as the dependent variable. The resulting expression at 30°C is

$$a_{30C} = -0.246 + 0.232\lambda - 0.0147\lambda^2 + 3.149 \times 10^{-4}\lambda^3 \quad (4.30)$$

The largest standard error was found to be SE = 0.03843, and was added or subtracted from the least squares fit to estimate the error in our curve fit, $a|_{error} = a_{30C} \pm SE$.

Assuming the data in Ref. [54] for 80°C is for a membrane with no heat-treatment (E-Form), and applying a similar procedure on Nafion 117 data at 80°C, we can fit a fourth order polynomial

$$a_{80C} = -0.00562 + 0.0146\lambda + 0.0685\lambda^2 - 0.0115\lambda^3 + 5.60 \times 10^{-4}\lambda^4 \quad (4.31)$$

The largest standard error in this case was found to be SE = 0.0312, with an error estimate for the fitting of $a|_{error} = a_{80C} \pm SE$.

We now have the data available to plot the conductivity as a function of the water content at 30 and 80°C. We will use the fitted sorption data at 80°C in our fitting of the BFCM below at 70°C.

4.6.3. Fitting Conductivity

The conductivity fit of Sone et al., Eq. (4.29), was modified to plot the conductivity as a function of water content by using Eqs. (4.30, 4.31) at 30 and 70°C, respectively. The relative error was used as a guide to maintain the error within reasonable bounds while ensuring a fit within the range of conductivity values defined by the standard error in the sorption isotherms, and the resulting parameter values are

$$\lambda_{min} = 1.65, D_{12} = 6.5 \times 10^{-9} m^2 s^{-1}, s = 0.83, q = 1.5,$$
$$A_1 = 0.084, A_2 = 0.5, E_a/R = 1800 \text{ K}. \quad (4.32)$$

The minimum water content λ_{min} was chosen such that the model conductivity threshold closely matches the data of Sone et al. A reasonable order of magnitude for D_{12} was chosen from a literature survey, and A_1, A_2, s, and D_{12} were then varied such that the BFCM results lay within the error bounds for all λ.

4. Coupled Proton and Water Transport in Polymer Electrolyte Membranes

The coefficients A_1 and A_2 reflect the relative size of D_{1M} and D_{2M} respectively. Since the fitted value of A_1 is much smaller than A_2, this implies that the resistance of the membrane to hydronium diffusion is much larger than to water. Given that hydronium ions have a net charge while water does not, this is reasonable since we expect the interaction forces between the membrane (with charged sulfonate heads) and the hydronium ions to be stronger than the forces between membrane and water. Sensitivity analysis shows [23] that regardless of the order of magnitude of D_{12} the *ratio* of the values of D_{1M} and D_{2M} remains of the same order of magnitude, which is encouraging in terms of the generality of the model.

Figure 4.6 shows the fit of the BFCM to the experimental data, using the value of $s = 0.8$. The plot fits within the anticipated range of conductivity values over almost the entire range of water contents, with the fit falling slightly outside the error bounds at a water content of approximately 5. At lower water contents (λ of about 2), the relative error is amplified by the small conductivity values, but the absolute error is in fact small.

The conductivity models of Refs. [22] and [39], Eqs. (4.3, 4.8), are compared to the BFCM in Figure 4.7, which also shows the expected range of the data. Springer's model[†] falls within the upper and lower bounds for high water contents λ, but outside for low values of λ. Thampan's model falls within the upper and lower bounds for low water contents, but deviates significantly from the experimental results for higher water contents.

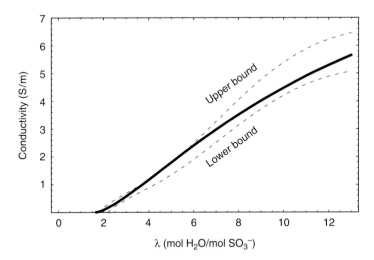

FIGURE 4.6. BFCM and anticipated upper and lower bounds on conductivity resulting from expected error in fit to sorption isotherm data at 30°C.

[†] Springer et al. omit to report whether the membrane considered was in the E-form or some heat-treated form. Here we assume that it is for the E-form for comparison purposes.

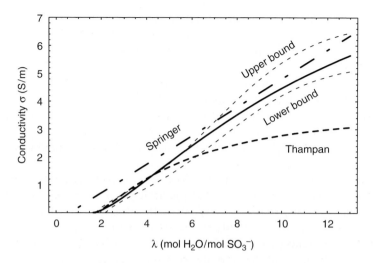

FIGURE 4.7. Comparison of calculated conductivity using BFCM (solid) and models of Springer et al. [39] and Thampan et al. [22] against experimental data of Sone et al. [34] for E-form Nafion 117 at 30°C.

For brevity, a direct comparison of the BFCM to the results using Thampan and Springer models at 70°C is not presented here, but we note that Springer et al.'s model predictions are markedly inferior to those at 30°C. The minimum relative error is approximately 20% and the computed values lie outside the upper bound of conductivity range at all water contents. Possible causes for the large discrepancy in the Springer model predictions are discussed in Ref. [23]. In any case, for this temperature, the BFCM exhibits again lower overall error for almost all water contents.

With the fitted parameters available at two temperatures, predictions at intermediate temperatures can be obtained to gauge the ability of the BFCM to correctly predict the temperature dependence of conductivity. Comparisons are made with Sone's data at 45°C. One problem in attempting this is lack of reliable sorption isotherm data at temperatures near 45°C. This was remedied by implementing the chemical model of Weber and Newman developed for determining λ for a vapour-equilibrated membrane [55] and using it to provide sorption isotherm data. A standard error of ±0.038 was used on the activity (the same standard error as was used at 30°C) to provide error bars within which conductivity is expected to lie.

As a reliability check, Weber and Newman's chemical model was used to translate experimental conductivity data from a function of activity to a function of water content at 30, 70 and 80°C. For the case of 30°C, Weber and Newman's conversion and the conversion using the fit-to-data overlap in the mid-to-high water content range. At lower water contents, the differences between the sorption isotherm model and the fit to the data for low water contents become more significant. In the conversion process for 70 and 80°C, a significant overlap was noted for both temperatures over the entire range of water contents. The effect of a 10°C temperature variation on the sorption isotherms is small, and it was deemed that using the sorption data at 80°C to convert the conductivity data at 70°C is acceptable [23]. Weber and Newman's chemical model appears to provide a reasonable enough translation of the conductivity data, and should provide a useful basis to validate the temperature dependence behaviour of our model.

Figure 4.8 compares the model predictions to experiments at 45°C. Springer's model falls outside the error across the range, while Thampan's provides the best fit at very low water contents. Although the BCFM falls marginally outside the error bounds at high and low water contents, it provides a better overall fit. The overall agreement of the BCFM predictions with experimental data over a broad range of water contents suggests that the temperature dependence of conductivity is well captured.

In closing the discussion on conductivity, we note that small deviations between the sorption isotherm model and experimental data, which would normally be perfectly acceptable, can introduce potentially large errors in the conductivity predictions. Consider that a standard error of ± 0.038 at an activity near 1 (± 3.8% variation) can cause the conductivity to range between approximately 5.2 and 4.4 S m^{-1} (± 8% variation in conductivity approximately) at a water content of around 11. This illustrates the critical

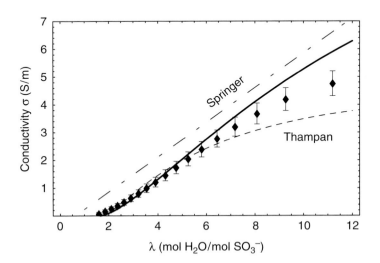

FIGURE 4.8. Prediction of membrane conductivity at 45°C with BFCM and models of Springer et al. [39] and Thampan et al. [22].

impact of sorption isotherm models (or data) on the determination of conductivity as a function of λ. A more definitive verdict on the errors will await experimental data at 45°C.

4.7. Implementation of BFM2 in a Fuel Cell Model

Having determined the transport parameters from the binary membrane conductivity model, we are now in a position to implement the full BFM2 into a complete fuel cell model to solve for *coupled* proton and water transport. In this Section we describe briefly the implementation in a fuel cell model and present sample simulations to illustrate features of the BFM2 predictions.

4.7.1. PEM Fuel Cell Model

In planar fuel cells, the membrane is part of a layered sandwich structure (the membrane-electrode assembly) consisting of a thin catalyst layer and a porous electrode (gas diffusion layer) on either side of the membrane. The oxygen reduction and hydrogen oxidation reactions take place at the cathode and anode catalyst layers, and the reactants and products are transported through the porous electrodes. A fuel cell model thus requires appropriate coupling of the membrane sub-model to the adjacent transport and electrochemical reactions. Detailed strategies for implementing complete fuel cell models have been discussed elsewhere [11–15].

For the purpose of illustrating the impact of membrane transport models on fuel cell performance predictions, we present results obtained using a 1D fuel cell model using both Springer et al's model and the BFM2. A full description of the model, which is beyond the scope of this Chapter, is given in Ref. [56]. The model resolves the catalyst layer in conjunction with Butler-Volmer kinetics and is solved numerically using a finite element method. One of the key issues in the implementation of the BFM2 in the fuel cell model is the coupling of the equation for water content (λ) to that for water mole fraction (X_w) in the porous electrode. This is dealt with using an iterative algorithm that enforces water conservation and allows the enforcement of the sorption isotherm at the membrane-catalyst interface [56].

4.7.2. Effect of Membrane Transport on Fuel Cell Performance

The direct effect of water content λ on predicted membrane conductivity can be assessed by setting the water transport to produce a linear profile

4. Coupled Proton and Water Transport in Polymer Electrolyte Membranes   149

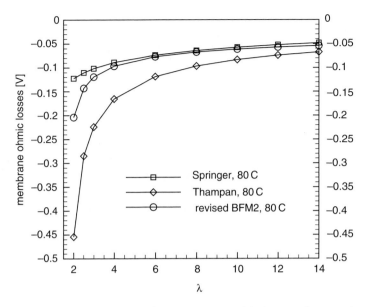

FIGURE 4.9. Comparison of membrane overpotential for varying anode water contents. Membrane models: Springer et al. [39], Thampan et al. [22] and BFM2 [23].

of λ through the membrane, effectively producing a spatially varying membrane conductivity $\sigma_m(\lambda)$. The anode water content is varied, while the cathode side is kept at $\lambda = 14$, and the total ohmic loss through the membrane computed using each model are plotted in Figure 4.9 for a constant current density of $i = 1$ A/cm^2. The membrane overpotential varies significantly between the three models, particularly at low anode water content. The BFM2 and Springer's model are in close agreement for higher values of λ. At lower values, the conductivity in the BFM2 is lower than the Springer model, but not as low as that obtained from Thampan et al's model.

A clearer comparison of the two membrane models is presented in Figure 4.10, showing isocurrent lines obtained from simulations with constant voltage drop through the membrane, with varying anode and cathode water content at the boundaries. The voltage drop is specified as $dV = 0.1$ V. This corresponds to a current density of approximately 1.4 A/cm^2 for a fully humidified membrane. As in Figure 4.9, the models are similar at higher membrane humidification levels, but differ significantly at lower anode and/or anode water contents.

The overall impact of different humidification regimes is shown in Figure 4.11 in terms of the polarization curves obtained using the two models under two different humidification regimes: (i) dry cathode operation with 75% relative humidity (RH) on the anode and 25% RH on the cathode

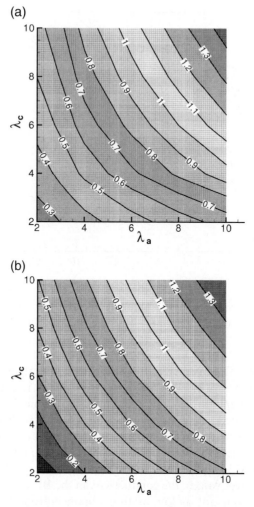

FIGURE 4.10. Current density contours computed using (a) model of Springer et al. and (b) BFM2. Computations performed for fixed membrane voltage drop of $0.1V$.

(labelled RH = 75/25); and (ii) dry anode with reversed conditions (labelled RH = 25/75). The Springer model predicts similar polarization curves for both regimes, while the BFM2 predicts a substantially larger polarization loss for the dry anode regime. The sensitivity of the BFM2 to the humidification regimes is physically realistic and should provide enhanced modelling reliability not only for low humidification conditions, such as those encountered in ambient air breathing fuel cells [57, 58] and those relying on passive humidification schemes [59], but also under typical fuel cell operating conditions in which *local* membrane drying is frequently encountered [60].

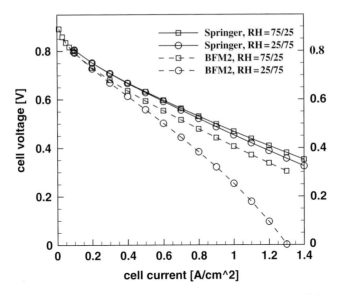

FIGURE 4.11. Polarization curves obtained using Springer et al. model and BFM2 under dry cathode (RH = 75/25) and dry anode (RH = 25/75) conditions.

4.8. Closing Remarks

We have presented a review of experimental and macroscopic modelling aspects of transport phenomena in polymer electrolyte membranes. This included examination of the connection between the hydration scheme and the behaviour of the membrane, a discussion of the so-called Schroeder's paradox, and the influence of the membrane phase on transport mechanisms. We also provided a critical examination of various approaches to modelling transport phenomena in membranes, and established that binary friction model provide a correct and rational framework for modelling membrane transport.

Based on this framework, a Binary Friction Membrane Model (BFM2) was developed to account for coupled transport of water and hydronium ions in polymer electrolyte membranes. The BFM2 was cast in a general form to allow for broad applicability to the PFSA family of membranes. As a tool to determine the model parameters, a simplified Binary Friction Conductivity Model (BFCM) was derived to represent conditions found in AC impedance conductivity measurements.

In order to validate the BFCM, we applied the model to 1100 EW Nafion. We used available conductivity and sorption data at 30 and 70°C to determine the parameters of the conductivity model (E_a, λ_{min}, D_{12}, s, q, A_1, A_2). We then used this to predict the conductivity at 45°C. The overall agreement of the BCFM predictions with experimental data over a broad range of water contents suggests that the temperature dependence of conductivity is well captured. A more rigorous analysis of the temperature dependence would, however, require additional experimental data. Ideally sorption isotherm data and conductivity data, allowing for the determination of conductivity as a function of water content, should be obtained for a wider range of temperatures to allow for a more systematic analysis. Without detracting from the features of the BFCM, it should be noted that the predictive abilities of the classical models of Springer et al. and of Thampan et al. might also benefit from accounting for temperature dependence in the sorption isotherms.

One of the inherent advantages of the simplified BFCM is the ability to gain insight into the necessary transport parameters. A key feature is that the subset of transport coefficients determined from AC impedance measurements for the BFCM are the *same* as those appearing in the general BFM2. With additional data on water transport parameters (water diffusion coefficient and electro-osmotic drag), the fully specified BFM2 with all its required parameters was applied in a fuel cell model to account for coupled proton and water transport, and we presented results illustrating the enhanced predictive ability of the BFM2 compared to existing models, particularly in capturing the physics of lower humidity operation.

While the focus of the model performance assessment was on Nafion 1100 equivalent weight (EW), the binary friction membrane transport model is quite general and should be applicable to other PFSA membranes. This is supported by preliminary results in applying the model to Dow membranes and membrane C, whereby rational changes in a single model parameter based on physical considerations of structural differences from Nafion yield conductivity predictions that are in good agreement with experimental measurements [61].

In order to apply the BFCM and the more general BFM2 to other membranes and to more systematically assess its performance for Nafion, detailed information and data are required, either from experiments or fundamental simulations. In addition to specification of the EW and the dry density of the membrane, or the molar volume (required for the porosity portion of the model), the model also requires specification, either from experiments or from a complementary dissociation model, of the fraction of dissociated acidic heads forming the charge carrying species.

The specification of the BFCM parameters (i.e. λ_{min}, D_{12}, s, A_1, and A_2) requires, at a minimum, conductivity data as a function of water content for a range of temperatures. Other relevant empirical data useful for implementing and assessing the general BFM2 are water diffusion coefficient through the membrane, and the electro-osmotic drag coefficient. As an alternative or supplement to empirical determination of the model parameters, it might be possible to obtain insight into these parameters from fundamental experiments or molecular dynamics simulations, and to elucidate, for example, the D_{ij} interaction terms.

Acknowledgements

This work was funded in part by grants to ND from the Natural Sciences and Engineering Research Council of Canada and the MITACS Network of Centres of Excellence.

References

[1] C. Chuy, V. I. Basura, E. Simon, S. Holdcroft, J. Horsfall and K. V. Lovell, *J. Electrochem. Soc.*, **147**, 4453 (2000).
[2] J. Ding, C. Chuy and S. Holdcroft, *Macromolecules*, **35**, 1348 (2002).
[3] Kerres, J, Ullrich, A., Haring, T., Baldauf, M., Gebhardt, U. and W. Preidel, *J. New Mat. Electr. Sys.*, **3**, 229 (2000).
[4] J. Kerres, W. Cui, R. Disson and W. Neubrand, *J. Membrane Sci.*, **139**, 211 (1998).
[5] C. Manea and M. Mulder, *J. Membrane Sci.* **206**, 443 (2002).
[6] M. K. Song, Y. T. Kim, J. M. Fenton, H. R. Kunz and H. W. Rhee. *J. Power Sources*, **117**, 14 (2003).
[7] V. I. Basura, C. Chuy, P. D. Beattie and S. Holdcroft, *J. Electroanal. Chem.*, **501**, 77 (2001).
[8] O. Savadogo, *J. New Mat. Electr. Sys.*, **1**, 47 (1998).
[9] J. J. Sumner, S. E. Creager, J. J. Ma and D. D. DesMarteau, *J. Electrochem. Soc.*, **145,** 107 (1998).
[10] J. S. Wainright, J. T. Wang, D. Weng, R. F. Savinell and M. Litt, *J. Electrochem. Soc.*, **142**, L121 (1995).
[11] T. Berning and N. Djilali, *J. Electrochem. Soc.,* **150**, A1589 (2003).
[12] S. Um and C. Y. Wang, *J. Power Sources*, **125** (1), 40–51 (2004).
[13] P. T. Nguyen, T. Berning and N. Djilali, *J. Power Sources*, **130** (1–2), 149 (2004).
[14] A. Z. Weber and J. Newman, *Chem. Rev.*, **104**, 4679–4726 (2004).
[15] B. Sivertsen and N. Djilali, *J. Power Sources*, **141** (1), 65–78 (2005).
[16] K. D. Kreuer, S. Paddison, E. Spohr and M. Schuster, *Chem. Rev.*, **104**, 4637 (2004).
[17] X. D. Din and E. E. Michaelides, *AIChE J.*, **44**, 35 (1998).
[18] S. J. Paddison, R. Paul and T. A. Zawodzinski, *J. Chem. Phys.*, **115**, 7753 (2001).
[19] R. Paul and S. J. Paddison, *J. Chem. Phys.*, **115**, 7762 (2001).
[20] S. J. Paddison, R. Paul and T. A. Zawodzinski, *J. Electrochem. Soc.*, **147**, 617 (2000).
[21] M. Eikerling, A. A. Kornyshev and U. Stimming, *J. Phys. Chem. B*, **101**, 10807 (1997).
[22] T. Thampan, S. Malhotra, H. Tang and R. Datta, *J. Electrochem. Soc.*, **147**, 3242 (2000).
[23] J. Fimrite, B. Carnes, H. Struchtrup and N. Djilali, *J. Electrochem. Soc.,* **152**, A1815 (2005).
[24] J. Fimrite, H. Struchtrup and N. Djilali, *J. Electrochem. Soc.,* **152**, A1804 (2005).
[25] G. Inzelt, M. Pineri, J. W. Schultze and M. A. Vorotyntsev, *Electrochim. Acta*, **45**, 2403 (2000).

[26] M. Laporta, M. Pegoraro and I. Zanderighi, *Phys. Chem. Chem. Phys.*, **1**, 4619 (1999).
[27] A. Z. Weber and J. Newman, *J. Electrochem. Soc.*, **150**, A1008 (2003).
[28] K. D. Kreuer *J. Membrane Sci.*, **185**, 29 (2000).
[29] P. Choi and R. Datta, *J. Electrochem. Soc.*, **150**, E601 (2003).
[30] J. Elliot, S. Hanna, A. M. S. Elliot, and G. E. Cooley, *Phys. Chem. Chem. Phys.*, **1**, 4855 (1999).
[31] A. Vishnyakov and A. V. Neimark, *J. Phys. Chem. B*, **104**, 4471 (2000).
[32] K. D. Kreuer, *Solid State Ionics*, **136**, 149 (2000).
[33] K. D. Kreuer, *Solid State Ionics*, **97**, 1 (1997).
[34] Y. Sone, P. Ekdunge and D. Simonsson, *J. Electrochem. Soc.*, **143**, 1254 (1996).
[35] T. A. Zawodzinski, T. E. Springer, J. Davey, R. Jestel, C. Lopez, J. Valerio and S. Gottesfeld, *J. Electrochem. Soc.*, **140**, 1981 (1993).
[36] T. A. Zawodzinski, C. Derouin, S. Radzinski, R. J. Sherman, V. T. Smith, T. E. Springer and S. Gottesfeld, *J. Electrochem. Soc.*, **140**, 1041 (1993).
[37] K. D. Kreuer, *Chem. Mater.*, **8**, 610 (1996).
[38] M. Eikerling, A. A. Kornyshev, A. M. Kuznetsov, J. Ulstrop and S. Walbran, *J. Phys. Chem. B*, **105**, 3646 (2001).
[39] T. E. Springer, T. A. Zawodzinski and S. Gottesfeld, *J. Electrochem. Soc.*, **138**, 2334 (1991).
[40] P. Berg, K. Promislow, J. St-Pierre, J. Stumper and B. Wetton, *J. Electrochem. Soc.*, **151**, A341 (2004).
[41] B. R. Breslau and I. F. Miller, *Ind. Eng. Chem. Fund.*, **10**, 554 (1971).
[42] M. Verbrugge and D. Bernardi, *AIChE J.*, **37**, 1151 (1991).
[43] M. Verbrugge and D. Bernardi, *J. Electrochem. Soc.*, **139**, 2477 (1992).
[44] M. Eikerling, Y. I. Kharkats, A. A. Kornyshev and Y. M. Volfkovich, *J. Electrochem. Soc.*, **145**, 2684 (1998).
[45] T. Fuller and J. Newman, *J. Electrochem. Soc.*, **140**, 1218 (1993).
[46] D. M. Bernardi, *J. Electrochem. Soc.*, **137**, 3344 (1990).
[47] G. J. M. Janssen, *J. Electrochem. Soc.*, **148**, A1313 (2001).
[48] R. Taylor and R. Krishna, *Multicomponent Mass Transfer*, John Wiley & Sons, Toronto (1993).
[49] P. J. A. M. Kerkhof, *Chem. Eng. J.*, **64**, 319 (1996).
[50] E. A. Mason and A. P. Malinauskas, *Gas Transport in Porous Media; The Dusty Gas Model*, Elsevier, Amsterdam (1983).
[51] R. Krishna and J. A. Wesselingh, *Chem. Eng. Sci.*, **52**, 861 (1997).
[52] T. A. Zawodzinski, J. Davey, J. Valerio and S. Gottesfeld, *Electrochim. Acta*, **40**, 297 (1995).
[53] B. E. Conway, *Ionic Hydration In Chemistry and Biophysics*, Elsevier Scientific Publishing Company, Netherlands (1981).
[54] D. R. Morris and X. Sun, *J. Appl. Polym. Sci.*, **50**, 1445 (1993).
[55] J. T. Hinatsu, M. Mizuhata and H. Takenaka, *J. Electrochem. Soc.*, **141**, 1493 (1994).
[56] B. Carnes and N. Djilali, *Modelling and Simulation of Conduction of Protons and Liquid Water in a Polymer Electrolyte Membrane Using the Binary Friction Membrane Model*, IESVic Report, University of Victoria (2005).
[57] W. Ying, Y. J. Sohn, W. Y Lee, J. Ke and C. S. Kim, *J. Power Sources*, **145**, 563–571 (2005).

[58] S. Litster, J. G. Pharoah, G. McLean and N. Djilali, *J. Power Sources,* **156**, 334 (2006).
[59] Y. Wang and C. Y. Wang, *J. Power Sources,* **147**, 148 (2005).
[60] J. Stumper, M. Löhr and S. Hamada, *Proc. Int. Symp. Fuel Cell & Hydrogen Technologies,* MetSoc, Calgary, Canada, 531 (2005).
[61] J. Fimrite, *Transport Phenomena in Polymer Electrolyte Membranes,* MASc Thesis, University of Victoria (2004).

5
A Combination Model for Macroscopic Transport in Polymer-Electrolyte Membranes

ADAM Z. WEBER AND JOHN NEWMAN

5.1. Introduction

The membrane is the heart of the fuel-cell sandwich and hence the entire fuel cell. It is this electrolyte that makes polymer-electrolyte fuel cells (PEFCs) unique and, correspondingly, the electrolyte must have very specific properties. Thus, it needs to conduct protons but not electrons as well as inhibit gas transport in the separator but allow it in the catalyst layers. Furthermore, the membrane is one of the most important items in dealing with water management. It is for these reasons as well as for others that modeling and experiments of the membrane have been pursued more than any other layer [1].

Although there have been various membranes used, none is more researched or seen as the standard than the Nafion® family by E. I. du Pont de Nemours and Company. Like the other membranes used, the general structure of Nafion is a copolymer between polytetrafluoroethylene and polysulfonyl fluoride vinyl ether. These perfluorinated sulfonic acid (PFSA) ionomers exhibit many interesting properties such as a high conductivity, prodigious water uptake, and high anion exclusion to name a few. Nafion® is the main membrane studied in this chapter.

This chapter can be broken down into various parts. First, a short general background including previous models and experimental evidence is discussed. Second, a physically based, qualitative model of the structure of the membrane is presented. Next, a mathematical description is given to the physical model in terms of general governing equations. The following sections describe the two transport modes and the transition between them. These discussions include the elucidation from literature data of the water content and various transport parameters. Finally, there is some discussion about the validation of the model and its attributes, with a focus on fuel-cell water management.

5.2. Background

A model of the membrane must contain certain key elements. Foremost among these is that it must be based on and agree with the physical reality and phenomena that have been observed with these membranes. Furthermore, expressions for the various properties of the membrane should have the relevant dependences such as on temperature and water content. The water content should also be allowed to vary in a systematic and continuous fashion. The above statements mean that although the model is basically phenomenological a serious attempt must be made at using all relevant data and observations and imparting a physical significance to everything. Finally, the model should describe the observed fluxes in the membrane.

There are three main fluxes through the membrane. A proton flux that goes from anode to cathode, a water electro-osmotic flux that develops along with the proton flux, and a water-gradient flux. This last flux is sometimes known as the water-back flux or back-diffusion flux; it is due to a difference in the chemical potential of water at the two sides of the membrane and may be in either direction although the direction is typically from cathode to anode due to water production at the cathode. In addition to the above three fluxes, there are also fluxes due to crossover of oxygen and hydrogen, which are described in Section 5.9.

It is also well documented [2–7] that fuel-cell membranes and ionomers in general typically show a phenomenon known as Schröder's paradox [8]. Schröder's paradox is the difference in water uptake (and therefore other properties) due to the type of reservoir in contact with the membrane. As seen in Figure 5.1, the water content of the membrane, λ or moles of water per mole of sulfonic acid sites, for a saturated-vapor reservoir is different from that for a liquid-water reservoir even though the chemical potential is identical. This is seen in practice and the size of the difference depends on the membrane state and history. This effect is an important issue since fuel cells are often operated with humidified gases resulting in situations where there is liquid water on the cathodic side of the membrane but only water vapor on the anodic side.

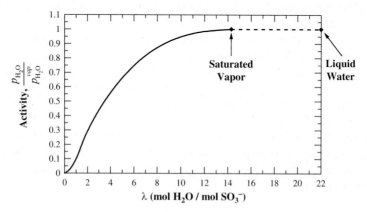

FIGURE 5.1. Water-uptake isotherm at 25°C showing the effect of Schröder's paradox.

5.2.1. Modeling Approaches

In terms of both quantitative and qualitative modeling, PEMs have been modeled within two extremes, the macroscopic and the microscopic, as discussed in various chapters in this book and in recent review articles [1, 9, 10]. The microscopic models provide the fundamental understanding of processes like diffusion and conduction in the membrane on a single-pore level. They allow for the evaluation of how small perturbations like heterogeneity of pores and electric fields affect transport, as well as the incorporation of small-scale effects.

On the other side of the scale are the macroscopic models, which are in line with the macrohomogeneous approach taken in this chapter. There are two main schools of thought regarding membrane models, those that assume the membrane is a single phase and those that assume it is two phases. The former usually leads to a diffusion model [11, 12] and the latter to a hydraulic one [13, 14]. Both models can be made to agree with experimental data, but neither describes the full range of data nor all of the observed effects, like Schröder's paradox.

The diffusion-type models assume that the membrane is a single homogeneous phase in which water dissolves and moves due to a concentration gradient. This model was popularized by Springer et al. [12] and Fuller [11] and has been modified for chemical potential gradients by Janssen [15] and Thampan et al. [16] and made more thermodynamically rigorous by Meyers and Newman [5]. A problem arises though when the membrane is saturated where it does not make sense to have only a diffusion-type flow since the concentration of water in the membrane is uniform; there is no concentration gradient. Hence, a two-phase or hydraulic model should be used in this case.

The hydraulic models treat the membrane as a heterogeneous porous medium in which water moves by convection due to a hydraulic-pressure gradient. The most recognized type of this model was done by Bernardi and Verbrugge [14, 17] and it has been modified to include effects of saturation, pore-size distribution, and hydrophobicity by Eikerling et al. [13, 18] A problem with these models arises for the case of a membrane in a low relative humidity reservoir. In such a system, there is not a continuous liquid pathway across the medium and the membrane matrix interacts significantly with the water due to the binding and solvating of the sulfonic acid sites. Thus, a concentration and not a hydraulic pressure of liquid water, which might not even be defined, seems to be the more appropriate driving force; a one-phase model should be used.

The two macroscopic models mentioned above have both advantages and limitations. They describe part of the transport that is occurring and this is why they are both still used, but the correct model is some kind of superposition between them as has been mentioned before [19, 20]. Each type of model operates properly at one limit of water concentration; they should somehow be averaged between those limits. The physical model in this chapter describes this in a coherent and physically consistent manner. The need for a combination model is also apparent in that the above models do not describe or present a physical picture of Schröder's paradox whereas a combination model could be designed to describe, or at least account for, the paradox.

5.2.2. Experimental Evidence of Membrane Structure

The general structure of Nafion® in particular, and ionomers in general, as a function of water content has been the source of many studies as recently reviewed by Mauritz and Moore [21] and Kreuer et al. [10] For the most part, the experimental data have shown that a hydrated membrane phase separates into ionic and matrix or nonionic phases. The ionic phase is associated with the hydrated sulfonic acid groups and the matrix phase with the polymer backbone. Thus, water is associated with the hydrophilic ionic phase and not the hydrophobic matrix phase. The actual way in which the phases segregate within the polymer depends on the water content.

There are different ways into which the two phases can separate. The consensus is that, at lower hydration levels, the phase separation results in ionic domains that have been idealized to be ionic clusters with hydrophilic interiors and hydrophobic exteriors. This is known as the cluster-network model first proposed by Hsu and Gierke [22]. In this idealized picture, the water is contained in a spherical domain about 4 nm in diameter into which the polymer side-chains infiltrate. As the membrane initially takes on water, the sulfonic acid groups are hydrated and dissociate creating charged groups that participate in coulombic repulsions. These interactions are opposed by the work required to deform the polymer matrix. Hence, there is a balance between the surface or electrostatic energy and the elastic or deformation energy. The clusters are connected by interfacial regions or bridging-site pathways about 1 nm in diameter. These were determined by Hsu and Gierke to be transient connections with a stability on the order of ambient thermal fluctuations in agreement with molecular-dynamics simulations [10, 23–25]. While the cluster-network model is an idealized picture, it provides a useful visualization of the polymer phase-separated microstructure.

As the membrane becomes more hydrated, the sulfonic acid sites become associated with more water allowing for a less bound and more bulk-like water to form. This is why there is a flattening out of the slope above $\lambda = 6$ in the uptake isotherm, Figure 5.1. The extreme case is when the membrane is placed in a liquid-water reservoir, where the ionic domains swell and a bulk-like liquid-water phase comes into existence throughout the membrane. The way in which water does this is unknown but is probably due to the interfacial properties of the membrane, such as the fluorocarbon-rich skin on the surface of Nafion [26, 27] or the removal of a liquid-vapor meniscus at the membrane surface [28]. In essence, the reorganization results in a porous structure with an average pore size between 1 and 2 nm [29].

5.3. Membrane Structure as a Function of Water Content

The physical model draws on the various sources of evidence in the literature and thereby derives a qualitative description of the structure of and transport in the membrane. A main focus of the physical model is how the membrane structure changes as a function of water content λ (moles of water per mol of sulfonic acid sites), which can be measured by examining the weight gain of an equilibrated membrane. The physical model of the polymer-electrolyte membrane that we use is relatively simple. It is based on the above experimental evidence and is in line with modern thoughts of transport within the membrane. The model is cast in the framework of capillary phenomena (especially for the liquid-equilibrated membrane) and thus the terms hydrophobic and hydrophilic are used, and it is proposed that these take into account all of the relevant interaction and surface energies and forces. This supposition proves very useful in describing the observed data and in the development of a macrohomogeneous mathematical model from the physical one.

The key to the physical model is how the structure of the membrane changes with water content. The structure of the hydrated membrane is affected by many factors including pretreatment procedures, operating temperature, water content, length of side chains, and equivalent weight, to name a few. However, there are still some global changes and structural elements common to all of the PFSA ionomers as a function of water content. In keeping with the choice of using the capillary framework, there are two separate regions in our model. The sulfonic acid sites are seen as being hydrophilic since they want to be hydrated with water and are in principle triflic acid; hence the ionic phase or cluster is hydrophilic. Conversely, the polymer matrix is basically Teflon®; hence it is hydrophobic. Finally, based on various experimental data, all species transported through the membrane, including water, protons, and gases, move by way of the clusters and channels and not through the fluorocarbon matrix [12, 22, 30]. Thus, the fluorocarbon matrix can be taken as inert. The matrix roles are to add mechanical strength, prevent dissolution, and add hydrophobicity.

In terms of the structure within the membrane, the idealized Hsu and Gierke cluster-network model is used as a picture where the pathways between the clusters are interfacial regions. These pathways are termed collapsed channels since they can be expanded by liquid water to form a liquid-filled channel. In essence, the collapsed channels are sulfonic acid sites surrounded by the polymer matrix having a low enough concentration such that the overall pathway between two clusters remains hydrophobic. In other words, they are composed of bridging ionic sites [31] and the electrostatic energy density is too low compared to the polymer elasticity to allow for a bulk-like water phase to form and expand the channels. In all, for a vapor-equilibrated membrane the structure is that of ionic domains that are hydrophilic and contain some bulk-like water. These clusters are connected by pathways with a fluctuating tortuosity at ambient conditions due to the movement of the bridging sulfonic acid sites and polymer in the matrix between clusters.

When the membrane is placed in liquid water, rearrangement and a phase-transition occur. This could be due to surface rearrangements wherein the fluorocarbon-rich skin of the membrane is repelled from the interface between the water and membrane. What this means is that in order to minimize the energy of the system the side chains and backbone of the polymer reorient so that the chains are now arranged at the membrane/water interface. This hypothesis agrees with the data that show that the water contact angle on the membrane surface becomes more hydrophilic after the membrane is placed in liquid water [32]. The presence of liquid water also results in the removal of a vapor-liquid meniscus, which could also aid in the above rearrangements [28].

Inside the membrane, the internal structure and existence of bulk-like water causes cluster agglomeration and higher concentrations of sulfonic-acid sites in the interfacial regions between clusters (i.e., the collapsed channels). In the next step in the physical model, it is proposed that the liquid water has enough pressure (or energy) to enter or form a separate phase in these more highly concentrated channels and expand them through a balance between pressure and electrostatic and mechanical energies. This is helped by the fact that water plasticizes the membrane thereby decreasing the required energy to expand a channel [33]. By expanding the channels, the water also helps to stabilize them and this process of expansion then continues in all of the channels for which the water pressure is sufficient. The expanded channels remain more hydrophobic than the clusters due to their lower concentration of hydrophilic sites (less columbic repulsions) and without the liquid water present, will collapse. This assumption also allows one to account for Schröder's paradox. In all, the resulting structure for the liquid-equilibrated membrane is that of a porous medium resembling strings of pearls. As mentioned earlier, the average pore size of this structure is around 1–2 nm, which agrees with the average size between an expanded channel and a shrunken cluster in our model.

In summary, Figure 5.2 is a schematic of how the membrane structure changes as a function of water content. In the first panel, the dry membrane absorbs water in order to solvate the acid groups. This initial water is associated strongly with the sites. Additional water causes the water to become less bound and ionic domains form in the polymer matrix, the second panel. With more water uptake, these domains grow and form interconnections with each other. The connections, or collapsed channels, are transitory and have low concentrations of sulfonic acid sites. The cluster-channel network forms based on a percolation-type phenomenon of the clusters; therefore, to form a transport pathway, the clusters must grow and be close enough together to be linked by the collapsed channels. From conductivity data, this percolation threshold is shown practically to occur around $\lambda = 2$ [34].

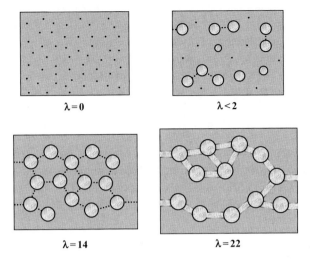

FIGURE 5.2. Evolution of the membrane structure as a function of water content, λ (moles of water per mole of sulfonic acid sites). The pictures are cross-sectional representations of the membrane where the gray area is the fluorocarbon matrix, the black is the polymer side chain, the light gray is the water, and the dotted line is a collapsed channel. (The schematic is adapted from Ref. [70] with permission of The Electrochemical Society, Inc.)

The lower left panel corresponds to a membrane that is in contact with saturated water vapor where a complete cluster-channel network has formed. When there is liquid water at the boundary of the membrane, structural reorganization and a phase transition occur allowing for bulk-like liquid water to exist in the channels resulting in a pore-like structure, the final panel in Figure 5.2. Because the channels are now filled with liquid, the uptake of the membrane has increased without a change in the chemical potential of the water (i.e., Schröder's paradox).

5.4. General Governing Equations and Modeling Approach

The physical model bridges the gap between the two types of mathematical models in the literature. Furthermore, it does so with a physically based description of the structure of the membrane. However, to put it to use in simulations a mathematical model and approach is required that describes the governing phenomena discussed above. In this section, the general governing equations based on the physical model are developed using concentrated-solution theory and the approach of having two transport modes is introduced.

5.4.1. Governing Equations

To account for all the relevant interactions between species in both transport modes concentrated-solution theory [11, 35] is used. Its use allows for the natural incorporation of coupled transport phenomena. The derivation starts with the equation of multicomponent transport

$$\mathbf{d}_i = \sum_{j \neq i} K_{i,j}(\mathbf{v}_j - \mathbf{v}_i) \qquad (5.1)$$

where $\mathbf{d}_i$ is the driving force per unit volume acting on species i, $K_{i,j}$ are the frictional interaction parameters between species i and j, and $\mathbf{v}_i$ is the velocity of i relative to a reference velocity. $K_{i,i}$ is undefined and by the Onsager reciprocal relations

$$K_{i,j} = K_{j,i} \qquad (5.2)$$

The Gibbs-Duhem equation with the assumption of isothermal operation must be satisfied,

$$\sum_i c_i \nabla \mu_i = \nabla p \qquad (5.3)$$

where μ_i and c_i are the electrochemical potential and concentration of species i, respectively, and p is the thermodynamic pressure. The use of the Gibbs-Duhem equation allows for harmony between the left side and the right side (using Eq. 5.2) of the summation of Eq. (5.1) over all species. It demonstrates that for an N species system, there are only $N-1$ independent equations of the form of Eq. (5.1). Therefore, for the three species membrane system (membrane, protons, and water), only two independent equations can be written (we neglect the membrane equation).

The left side of Eq. (5.1) can be replaced with the general isothermal driving force developed by Hirschfelder et al. [36]

$$\mathbf{d}_i = c_i \left[\nabla \mu_i - \frac{M_i}{\rho} \nabla p - \mathbf{X}_i + \frac{M_i}{\rho} \sum_j \mathbf{X}_j c_j \right] \qquad (5.4)$$

where M_i is the molar mass of species i, ρ is the density of the solution, and the $\mathbf{X}_i$ terms refer to body forces per mole acting on species i. For a membrane, Bennion [37] requires that the stresses acting on the membrane constitute the external body force $\mathbf{X}_i$, a force that is absent for the other species. This treatment is valid because any applied pressure is similar to compression by a mesh screen and thus the membrane remains stationary and water is still in equilibrium with the reservoir. If there is only a single body force acting, a mechanical force balance on each volume element of the membrane yields, at constant temperature,

$$c_m \mathbf{X}_m = \nabla p \tag{5.5}$$

The above relationship allows the simplification of the driving forces in the membrane system. Using the above formulation, substituting the resultant driving forces into the generalized transport equation, Eq. (5.1), and inverting the resultant equations yields the two independent transport equations

$$\mathbf{N}_+ = c_+ \mathbf{v}_+ = -L_{+,+} c_+^2 \nabla \mu_+ - L_{+,w} c_+ c_w \nabla \mu_w \tag{5.6}$$

and

$$\mathbf{N}_w = c_w \mathbf{v}_w = -L_{w,+} c_+ c_w \nabla \mu_+ - L_{w,w} c_w^2 \nabla \mu_w \tag{5.7}$$

where the subscripts +, w, and m signify protons, water, and membrane, respectively, $\mathbf{N}_i$ is the superficial flux density of species i, and the $L_{i,j}$'s ($= L_{j,i}$) are related directly to the $K_{i,j}$'s [11]. In determining the above equations, the stationary membrane velocity is taken as the reference.

Before proceeding, some comments should be made about the pressure body force. Meyers showed that for a one-phase system a force balance on the system requires that it cannot support a pressure gradient if it is to remain stationary [38]. This is at odds with Eq. (5.5) when used for the vapor-equilibrated membrane. However, for a two-phase system (i.e., the liquid-equilibrated membrane), there can still be a pressure gradient while the system remains stationary because at each point within the system there is a separate stress (pressure) equilibrium relation, hence a mesh screen analogy. Also, the use of two phases gives an additional degree of freedom such that the actual equilibrium relation need not be known. To avoid the above discrepancy for the vapor-equilibrated membrane, two points are made. First, although the vapor-equilibrated membrane is treated as a one-phase system, it is actually a two-phase system with water in the micelles and those clusters interact with the matrix to counterbalance the pressure gradient; the membrane remains stationary. Second, the arrangement of the driving forces in the above equations bypasses the problem and the need for an additional stress relation for the one-phase system because chemical potential is used as the driving force, which incorporates both pressure and activity; this point is discussed in more detail in a later section.

To use Eqs. (5.6) and (5.7), the frictional coefficients must be related to measurable quantities. Generally, there are $\frac{1}{2}N(N-1)$ independent transport properties needed to characterize the system. Because there are three species in the membrane, three independent transport properties are required. The $L_{i,j}$'s can be related to experimentally measured transport properties using a set of three orthogonal experiments [39]. Doing this results in the two governing transport equations for the membrane

$$\mathbf{i} = -\kappa\nabla\Phi - \frac{\kappa\xi}{F}\nabla\mu_w \tag{5.8}$$

and

$$\mathbf{N}_w = -\frac{\kappa\xi}{F}\nabla\Phi - \left(\alpha + \frac{\kappa\xi^2}{F^2}\right)\nabla\mu_w \tag{5.9}$$

where α is called the transport coefficient, κ is the ionic conductivity, F is Faraday's constant, Φ is the electrical potential defined as that measured by a hydrogen reference electrode located next to the reaction site but exposed to the reference conditions (i.e., it carries its own extraneous phases of hydrogen and platinum with it), $\mathbf{i}$ is the ionic current density, and ξ is the electro-osmotic coefficient.

In Eqs. (5.8) and (5.9), there are four unknowns: Φ, $\mathbf{i}$, $\mathbf{N}_w$, and μ_w because the temperature is specified. Hence, two more equations are necessary to solve for the unknowns. To do this, conservation of charge (electroneutrality) and mass are used for the steady-state system

$$\nabla \cdot \mathbf{i} = 0 \tag{5.10}$$

and

$$\nabla \cdot \mathbf{N}_w = 0 \tag{5.11}$$

Equations (5.8–5.11) constitute a closed set of independent equations that completely describe the transport within the membrane for a concentrated-solution system composed of water, proton, and membrane.

The boundary conditions used in conjunction with the above equations can vary and are to some degree simulation dependent. Normally, the current density, water flux, reference potential, and water chemical potential are specified; but two water chemical potentials or the potential drop in the membrane can also be used. If modeling more regions than just a membrane, additional mass balances and internal boundary conditions must be specified. In addition, for modeling the membrane in the catalyst layers, rate equations are required for kinetics and water transfer among its various phases [40]. The above equations are also valid only for the steady-state case (the time-dependent terms have been ignored).

5.4.2. Modeling Approach

In the above derivation, nothing was stated about the mode of transport and thus the equations above are general. According to the physical model, there are two separate structures for the membrane depending on what phase of water is located at the boundary and each is expected to have its own transport mechanism or mode. The two different modes can be considered independent of each other except for properties that affect the entire membrane/water system as a whole, like density and pressure, since the model is macrohomogeneous. If there is no liquid water in contact with the membrane surface, there is only the vapor-equilibrated transport mode and if there is a liquid-water boundary then there also is the liquid-equilibrated transport mode. Schematically, this is shown in Figure 5.3, where the left side of the membrane is in contact with vapor and the right side with liquid, and the transition region is shown thicker than normal to emphasize the approach.

In the figure, it is clear that there are three regions: the liquid-equilibrated mode, the vapor-equilibrated mode, and the combined region where both modes occur. Unlike other models, which would have the discontinuity in terms of water content at either the left or the right end or ignore it, the physical and mathematical models yield a smooth, continuous transition depicted by the black curve. The curve itself is a representation of the fraction of expanded channels as discussed in Section 5.7.1. Conditions were chosen such that the transition region in Figure 5.3 is broader than for typical fuel-cell conditions and Nafion® properties. The transition is normally very sharp as is shown later (for example, see Figure 5.10).

The two limiting states in the figure, $\lambda < 14$ and $\lambda = 22$, correspond to the diffusion and hydraulic models, respectively, in agreement with the physical model. In the middle region, both modes occur, and the average water

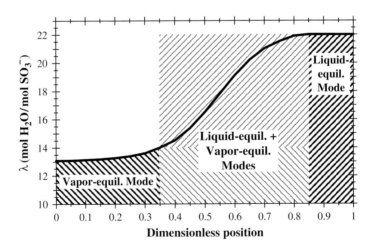

FIGURE 5.3. Water content as a function of membrane position (black curve) using parameters and conditions to broaden and emphasize the transition region. The hatched boxes show where the equations for the two different transport modes apply. (The schematic is adapted from Ref. [39] with permission of The Electrochemical Society, Inc.)

content is calculated from a linear average between the two extremes weighted by the fraction of channels that are expanded. This kind of treatment allows access to all water contents which can occur during fuel-cell operation.

When the membrane is neither fully liquid- nor vapor-equilibrated, the transport mode is assumed to be a superposition between the two; they can be treated as separate transport modes occurring in parallel. For this case, the transport-property values should be taken to be mode dependent and averaged depending on the fraction of channels that are expanded. By and their respective resistances to transport. By proposing this treatment, the model bridges the gap between the two transport modes and in essence the two types of models in the literature. In addition, the physical model goes further in that a structural parameter, namely the fraction of expanded channels is used to do this averaging and thus a continuous function and transition between the two modes is obtained that is also physically based.

Before proceeding to discuss the three cases in more detail in the following sections, it should be noted that of the four governing equations, Eqs. (5.10) and (5.11) remain the same, regardless of the transport mode, because they represent macroscopic balances and are not affected explicitly by transport. However, Eqs. (5.8) and (5.9) may have to be modified.

5.5. Vapor-Equilibrated Transport Mode

For a vapor-equilibrated membrane (i.e., one that is in contact with water vapor only), the physical model proposes that there is water in the ionic domains but none in the collapsed channels except for the bound water hydrating the few sulfonic acid sites present. Furthermore, the sulfonic acid sites that make up the collapsed channels are always fluctuating, but the clusters are close enough together to form a transport pathway after the percolation threshold has been reached. Due to the nature of the collapsed channels, the membrane is treated as a homogenous single-phase system. In this sense, the water vapor does not penetrate into the cluster-network, but instead dissolves into the membrane. Thus, the vapor-equilibrated membrane transport mechanism is similar to the single-phase transport models mentioned previously.

For the transport properties, the conductivity is explained based on the percolation concept where the percolation threshold occurs around $\lambda = 2$ and is correlated with enough clusters being hydrated and connected by hydrated sulfonic acid sites to form a complete conductive pathway across the membrane [34, 41]. With the addition of more water into the system, more clusters and channels form and the pathways become less tortuous; the conductivity increases. The electro-osmotic coefficient depends on the type of complexes that water forms with a proton and in the absence of a contiguous liquid-pore network, the complex that forms and is transported is basically a hydronium ion. Thus, the conductivity and the electro-osmotic flow are mainly due to a hydronium ion that hops from acid site to acid site through the ionic domains and collapsed channels [31]. Finally, the non-electro-osmotic water movement should be caused by the chemical-potential gradient of water.

In terms of equations, the transport mode has the same basis as the diffusion models. However, the original driving force of chemical potential is used instead of separating it into activity and pressure as many models do [1]. This separation is somewhat arbitrary and not rigorous, requires additional assumptions, and can lead to errors when there is a pressure difference across the membrane [39]. Unfortunately, our approach does not allow for the separate determination of activity and pressure in the membrane. The proposed model herein goes beyond that of other models that use the chemical-potential driving force by using membrane properties and expressions that are derived from experimental data and the physical model as well as allowing the calculation of the water content, λ.

In summary, the transport mode of a vapor-equilibrated membrane is that of a single membrane phase in which protons and water are dissolved. The chemical-potential gradient is used directly since it precludes the necessity of separating it into pressure and activity terms. Thus, Eqs. (5.8–5.11) are used directly without any modifications. Although it makes sense to use the chemical-potential driving force, most of the experimental data are a function of water content or λ. Thus, a way is needed to relate λ to the chemical potential.

5.5.1. Chemical Model for Determining Water Content, λ

Chemical potential and water content, λ, can be related through an uptake isotherm. Uptake isotherms of λ as a function of water–vapor activity or relative humidity, such as that given in Figure 5.1, are prevalent in the literature [4, 6, 42, 43]. They have been used in almost every model that deals with vapor-equilibrated membranes and treats the membrane as a single phase [1]. As discussed in the proposed physical model, the water uptake is described by the hydration of the sulfonic acid sites in the membrane clusters and a balance between osmotic, elastic, and electrostatic forces. The approach taken here is to calculate the isotherms using the chemical model of Meyers and Newman [5] with some modifications [39].

In the chemical model, equilibrium is assumed between protons and water with a hydronium ion. This equilibrium considers the tightly bound water in the membrane [13, 19, 44] and agrees with the vapor-equilibrated transport picture of a hydronium ion being the dominant proton-transfer species in the membrane. The equilibrium relates the electrochemical potentials of the species, and at the boundary the water in the membrane is in equilibrium with that in the vapor phase. Expressing the electrochemical potentials according to the thermodynamic basis of Meyers [38] and Meyers and Newman [5] for multicomponent transport in a polymer-electrolyte membrane, along with electroneutrality and a mass balance on water in the membrane, leads to two equations that must be solved simultaneously

$$\frac{\lambda_{H_3O^+}}{\left(1-\lambda_{H_3O^+}\right)\left(\lambda-\lambda_{H_3O^+}\right)} \exp\left[\phi_1 \lambda_{H_3O^+}\right] \exp[\phi_2 \lambda] = K_1$$

$$a_w = K_2\left(\lambda - \lambda_{H_3O^+}\right) \exp\left[\phi_2 \lambda_{H_3O^+}\right] \exp[\phi_3 \lambda]$$

(5.12)

where $\lambda_{H_3O^+}$ is the ratio of moles of hydronium ions to moles of sulfonic acid sites, and

$$K_1 = \frac{\lambda^*_{H_3O^+} E_W}{\lambda^*_w \lambda^*_{H^+}}$$

$$K_2 = \frac{\lambda^*_{H^+}}{E_W} \exp\left[-\frac{\mu_w^{\text{ref}}}{RT}\right]$$

$$\phi_1 = \frac{2}{E_W}\left(E^*_{w,w} - 2E^*_{H_3O^+,H^+} - 2E^*_{w,H_3O^+}\right)$$

(5.13)

$$\phi_2 = \frac{2}{E_W}\left(E^*_{w,H_3O^+} - E^*_{w,w}\right)$$

$$\phi_3 = \frac{2E^*_{w,w}}{E_W}$$

where E_W is the equivalent weight of the membrane, λ^*_i is the secondary-reference-state quantity for species i, and $E^*_{i,j}$ is a binary interaction parameter between species i and j [45]. A value of K_1 is assumed and the two equations and associated parameters above are then fit to representative data from the literature [6, 39, 43]. From Thampan et al. [16] the degree of dissociation represented by the equilibrium coefficient, K_1, is relatively high. Hence, the values of Meyers and Newman corresponding to $K_1 = 100$ are used as given in Table 5.1.

TABLE 5.1 Fit parameters for the chemical model of water uptake from the vapor phase [5].

Parameter	Value
K_1	100
K_2	0.217
$E^*_{w,w}$	−41.7 g/mol
E^*_{w,H_3O^+}	−52.0 g/mol
$E^*_{H_3O^+,H^+}$	−3721.6 g/mol

The values in Table 5.1 are valid for water uptake at 30°C. Other data have shown that water uptake decreases with increasing temperature [4, 16, 42, 46, 47]. According to these various data, the water uptake of the membrane exposed to a saturated-vapor reservoir decreases from a value of $\lambda = 14$ to a value of around $\lambda = 9$ or 10 at 60°C with perhaps a slight upswing afterwards. To account for the temperature dependence rigorously, the temperature dependence of all the parameters listed in Table 5.1 must be known. This leads to a very complicated fitting procedure with more unknowns than data points. Furthermore, the effects may counteract each other. To simplify the analysis, all the parameters are assumed independent of temperature except for K_2, where the temperature appears explicitly as seen in Eq. (5.13). Fitting the model equations to data yields an expression for K_2

$$K_2 = 0.217 \exp\left[\frac{1000}{R}\left(\frac{1}{T_{\text{ref}}} - \frac{1}{T}\right)\right] \tag{5.14}$$

where T_{ref} is the reference temperature (30°C or 303.15 K) and R equals 8.3143 J/mol·K.

The chemical model fails to predict accurately the uptake curve at very low water activity. This is because the binding energy at low λ is much higher because water is primarily hydrating the sulfonic acid sites in their first shells and is more strongly bound than water at higher λ values [10]. Thus, the chemical model must be altered. The most rigorous way to do this is to refit the chemical model parameters for this different binding energy (i.e., refit the data using K_1 as a function of λ) or perhaps add secondary terms to the original Gibbs equation presented by Edwards [45]. However, since there are not many data points to fit the new curve, and for simplicity, a slight correction to the calculation of λ is employed that affects only low water-uptake values and fits the experimental data in that regime. Although this correction is empirical, it has the same effect as increasing K_1 with decreasing λ [5]. The correction that is used has the form

$$\lambda = \lambda_o[1 + \exp(0.3 - \lambda_o)] \tag{5.15}$$

where λ_o is the value of λ calculated from the chemical model. Using the above expressions and the parameter values in Table 5.1, a family of isotherm curves can be generated as shown in Figure 5.4.

The figure demonstrates that the assumed simple temperature dependence correctly describes that λ at unit activity decreases in smaller intervals as the temperature increases and that this effect is more pronounced at higher activities. Because a simple temperature dependence is used, a minimum in the value of λ with respect to temperature is not observed. However, since the upswing is relatively small, neglecting the minimum should not be a significant source of error. The above way of treating temperature effects is consistent with the overall approach of our membrane model and is more

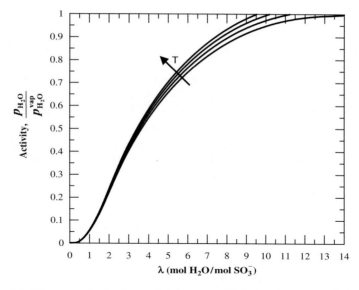

FIGURE 5.4. Water uptake isotherms from the modified chemical model for a membrane in contact with water vapor; the temperatures are 25, 45, 65, and 85°C. (The figure is reproduced from Ref. [39] with permission of The Electrochemical Society, Inc.)

rigorous than other methods which might only examine the change in vapor pressure [46]. The overall results of the modified chemical model agree with experimental data and the isotherms used and generated are from a physical basis and take into account temperature and membrane-equivalent-weight effects. A final comment should be made that the fully-vapor-equilibrated values depend also on membrane state and history, and if pretreated correctly, approach those of a liquid-equilibrated membrane.

5.6. Liquid-Equilibrated Transport Mode

In this mode, the model proposes that the structure of the membrane is more akin to a noninteracting porous medium that is filled with liquid water (i.e., a two-phase system) rather than the interacting description given above for the vapor-equilibrated membrane. This is because of the existence of bulk-like water in the middle of the ionic domains and channels. Such a structure enables a continuous percolated water pathway from domain to domain and from membrane side to side. The conductivity can be explained by the Grotthuss and vehicle mechanisms. The electro-osmotic flow is described by the vehicle mechanism and proton hydration effects, including the formation of the Zundel, $H_5O_2^+$, and Eigen, $H_9O_4^+$ ions [44, 48, 49].

Due to the presence of the bulk-like or loosely bound water network and the assumption of a two-phase system, the chemical-potential driving force becomes one of pressure alone [39]

$$\nabla \mu_{\text{w}} = \bar{V}_{\text{w}} \nabla p_{\text{L}} \tag{5.16}$$

where the subscript L denotes the liquid-phase property.

The governing equations are easily obtained by substitution of Eq. 5.16 into Eqs. (5.8) and (5.9).

One comment should be made regarding the form of the transport equations. In the literature, two-phase flow has often been modeled using Schlögl's equation [50, 51]. This equation is similar in form to Eq. (5.9), but it is empirical and ignores the Onsager cross coefficients. Equations (5.8) and (5.9) stem from concentrated-solution theory and take into account all the relevant interactions. Furthermore, the equations for the liquid-equilibrated transport mode are almost identical to those for the vapor-equilibrated transport mode making it easier to compare the two with a single set of properties (i.e., it is not necessary to introduce another parameter, the electrokinetic permeability).

As in the case for the vapor-equilibrated transport mode, the properties of the liquid-equilibrated transport mode depend on the water content and temperature of the membrane. For a fully liquid-equilibrated membrane, the properties are uniform at the given temperature. This is because the water content remains constant for the liquid-equilibrated mode unlike in the vapor-equilibrate one. From experimental data, the value of λ for liquid-equilibrated Nafion® is around 22, assuming the membrane has been pretreated correctly [6, 7, 52]. In agreement with the physical model, the water content is only a very weak function of temperature for extended (E)-form membranes (as assumed in our analysis) and can be ignored [6]. For other membrane forms, this dependence is much stronger and cannot be ignored, as discussed in the Section 5.10.1.

5.7. Occurrence of Both Modes: The Transition Region

When the membrane is in contact with liquid water on one side and vapor on the other (i.e., it is neither fully liquid nor vapor equilibrated), as can often occur during fuel-cell operation, both the liquid- and vapor-equilibrated transport modes will occur. This results in a transition between modes that exists in the membrane. As discussed in the physical model, a continuous transition between the two transport modes is assumed. Thus, transport in the transition region is a superposition between the two transport modes; they are treated as separate transport mechanisms occurring in parallel (i.e., the middle region in Figure 5.3). In this section, an approach to modeling the transition region is introduced followed by a discussion of its limitations, other approaches, and points to consider.

5.7.1. Approach and Governing Equations

For the transition region, the transport-property values are taken to be mechanism dependent, and the overall water flux and current are distributed between the two modes based on the fraction of channels that are expanded or filled with bulk-like water. Hence, the governing transport equations become

$$\mathbf{i} = S\left(-\kappa_L \nabla\Phi - \frac{\kappa_L \xi_L}{F} \bar{V}_w \nabla p_L\right) + (1-S)\left(-\kappa_V \nabla\Phi - \frac{\kappa_V \xi_V}{F} \nabla\mu_w\right) \quad (5.17)$$

and

$$\mathbf{N}_w = S\left[-\frac{\kappa_L \xi_L}{F} \nabla\Phi - \left(\alpha_L + \frac{\kappa_L \xi_L^2}{F^2}\right) \bar{V}_w \nabla p_L\right] \\ + (1-S)\left[-\frac{\kappa_V \xi_V}{F} \nabla\Phi - \left(\alpha_V + \frac{\kappa_V \xi_V^2}{F^2}\right) \nabla\mu_w\right] \quad (5.18)$$

where S is the fraction of expanded channels and is defined by

$$S = \frac{\lambda - \lambda_V|_{a_w=1}}{\lambda_L - \lambda_V|_{a_w=1}} \quad (5.19)$$

where $\lambda_V|_{a_w=1}$ and λ_L are the values of λ for the membrane in equilibrium with saturated vapor and liquid at the operating temperature, respectively. From the previous sections, $\lambda_V|_{a_w=1}$ is a function of temperature while λ_L is not (i.e., it always equals 22). Equation (5.19) also allows for the determination of λ in the transition region, assuming that S is known or can be calculated. From the relation, it is easy to see that λ will vary in a linear fashion from $\lambda_V|_{a_w=1}$ (e.g., 14 at ambient temperature) to 22 as S goes from 0 to 1. The calculation of S is presented in the following section.

In terms of a capillary framework, the fraction of expanded channels is similar to a saturation. Although averaging the two equations by this fraction is not necessarily rigorous, it has a physical basis. Furthermore, it has the correct limiting behavior (i.e., all vapor-equilibrated when there are no expanded channels (i.e., no bulk-like water), $S = 0$, and all liquid-equilibrated when all the channels are expanded (i.e., bulk-like water throughout), $S = 1$) and a relatively sharp transition, as expected for a phase transition.

When both transport modes occur, the governing equations above do not form a closed set of equations because both μ_w and p_L appear in Eqs. (5.17) and (5.18) as separate driving forces; another relationship is needed. If local equilibrium between the vapor- and liquid-equilibrated parts of the membrane is assumed, then the necessary additional equation becomes

$$\nabla \mu_w|_V = \bar{V}_w \nabla p_L|_L \tag{5.20}$$

where the fact that two different transport modes are being equated is emphasized. Thus, for the set of equations, there are two overall driving forces (ionic potential and water chemical potential) and fluxes (current density and water flux). Using this assumption, Eqs. (5.17) and (5.18) can be rewritten to be of the same form as Eqs. (5.8) and (5.9), except where effective properties are used. For example, the effective conductivity would be given by

$$\kappa^{\text{eff}} = S\kappa_L + (1-S)\kappa_V \tag{5.21}$$

In summary, when both the liquid- and vapor-equilibrated transport modes occur in the membrane they are assumed to occur in parallel. In other words, there are two separate contiguous pathways through the membrane, one with liquid-filled channels and another that is a one-phase-type region with collapsed channels. To determine how much of the overall water flux is distributed between the two transport modes, the fraction of expanded channels is used. As a final note, at the limits of $S = 1$ and $S = 0$, Eqs. (5.17) and (5.18) or their effective property analogs collapse to the respective equations for the single transport mode, as expected.

5.7.2. Calculating the Fraction of Expanded Channels, S

A channel expansion depends on three main factors, which are the coulombic or electrostatic repulsions of the hydrated sulfonic acid sites (this will include the concentration or density of sites), the deformational or elastic energy to push against the polymer, and the osmotic or liquid pressure [53, 54]. Using a capillary framework, this equilibrium between these energies can be represented by an equation of the type [55]

$$p_L - p_M = -\frac{2\gamma \cos \theta}{r_c} \tag{5.22}$$

where r_c is the expanded channel radius, γ is the surface tension, and θ is the contact angle of water with the membrane. However, this equation contains many unknown parameters, and it is easier to put it in the form of

$$r_c = \frac{\Gamma}{p_L} \tag{5.23}$$

where Γ is the only unknown and accounts for the various energetic interactions between water and the channel. We assume that Γ is a function of temperature in the same way as the elastic modulus is assumed to be (i.e., inverse relation) [56] since this is assumed to have the largest temperature dependence of the various energies for the liquid-equilibrated membrane. In line with the physical model and the capillary framework, r_c is also the incipient expanded channel radius, such that smaller radii are not expanded and larger ones are. Physically, this is because smaller radii have a lower concentration of sulfonic acid sites and are more stiff, such that a higher liquid pressure is required to expand them and form bulk-like water.

To determine the fraction of expanded channels, Γ and the channel-size distribution must be known. The channel-size distribution gives the fully expanded channel radii and is taken to be the same for different operating conditions and the same as the distribution measured for a liquid-equilibrated membrane. The reasons that this distribution is assumed to be constant are that it should not vary significantly with pressure or temperature under typical fuel-cell operating conditions and is used only when there is a separate liquid-water phase. This assumption has been used and proved valid within error tolerances [13, 18, 57]. The pore-size distribution for Nafion has been measured by the method of standard contact porosimetry [29, 58, 59]. In those studies, the distribution included both the channels and the clusters. Since only the channel-size distribution is of interest, only that regime of data is fit using the log-normal distribution [39]. The average channel radius is around 1.5 nm as expected from the physical model and other studies [23, 60, 61].

To use Eq. (5.23), Γ is still required. The value of Γ is assumed to be constant and the value that is used is an average value for the entire channel including the more hydrophilic entrance and exit regions and the less hydrophilic middle region. There are two main pieces of experimental data that allow for the deduction of Γ. The first is from Meyers's dissertation [38]: the water content in a Nafion membrane suspended in pure water vapor at 60°C is about $\lambda = 20$ ($S = 0.9$). This finding, along with the observation of water droplets on the membrane surface, hints that liquid water penetrates and expands some channels. The exact reason is currently unknown, but it may be due to the lack of inert gases, or perhaps the change of the membrane chemical potential at the lower operating pressure, or because the membrane had to be weighted in order for it not to curl up on itself (i.e., structural changes). If one assumes that the membrane channels are given by the above channel-size distribution and that the liquid-water pressure is equal to the vapor pressure of water at 60°C (pressure equilibrium), then a value of Γ can be determined. The second piece of data comes from freezing-point-depression experiments of the liquid-phase water in Nafion. Cappadonia et al. [62] correlated the freezing-point depression as a function of pore radius. To use their data, the temperature data are changed into vapor-pressure data using a handbook [63] and then the Kelvin equation is used to help calculate Γ. In accordance with both findings, a value of $\Gamma = 4 \times 10^{-5}$ N/m is used.

Now that the contact angle and channel-size distribution are known, the fraction of expanded channels can be calculated as a function of the hydraulic pressure. In accordance with the physical model, the expanded channel-cluster network is treated as a bundle of capillaries. To calculate S, a critical radius, r_c, is determined by Eq. (5.23) and the channel-size distribution is integrated from infinity to the critical radius [64] resulting in

$$S = \frac{1}{2}\left[1 - \text{erf}\left(\frac{\ln r_c - \ln(1.25)}{0.3\sqrt{2}}\right)\right] \quad (5.24)$$

Using the above equations, isotherms of the fraction of expanded channels versus liquid pressure can be generated as shown in Figure 5.5. From the curves, the temperature dependence of the saturation is not strong since the transition still occurs over a small liquid-pressure range. All of the curves show that, at a liquid pressure of 1 bar, the channels are completely expanded and filled with liquid in agreement with experimental observations. If the liquid pressure falls below about 0.15 bar, then the liquid water phase ceases to exist at all temperatures and the transport of water is solely by the vapor-equilibrated transport mode, which also agrees with the physical model. If the liquid pressure is above around 0.6 bar, then λ is 22 (only the liquid-equilibrated transport mode).

The sharpness of the transition region in the curves is determined by the channel-size distribution. The overall shape of the curves is due to treating the surface energies and interactions in terms of capillary phenomena. This shape makes physical sense in that the transition between the two types of transport modes should be relatively sharp for this kind of phase-transition

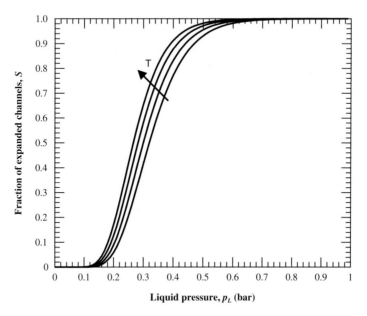

FIGURE 5.5. Fraction of expanded channels as a function of the liquid pressure using the parameters and equations given in the text; the temperatures are 25, 45, 65, and 85°C.

phenomenon. Furthermore, the connected network structure within the membrane has a propensity toward the limiting states as any porous network does (i.e., it is not energetically favored to have two similar and continuous porous networks, only one of which is filled).

5.7.3. Discussion and Limitations of the Above Approach

Before proceeding further, some comments should be made regarding the transition region. As stated above, a single driving force is used for transport in the transition region (i.e., Eq. (5.20) is valid). The repercussion of this assumption of local equilibrium is that there is a gradient in chemical potential across the transition region. This is in violation of Schröder's paradox. In essence, we have given a slight slope to the dashed line in Figure 5.1. However, as seen in Figure 5.5 and later in Figure 5.10, the transition region itself is very small and thus so is the slope of the line. In fact, from Figure 5.5 one can calculate that the chemical potential varies by only about 1 J/mol through the transition region. If desired, one can shrink the transition region such that it is infinitesimally thick and thus have a zero difference in chemical potential across the transition in harmony with Schröder's paradox. To do this in the equations, the channel-size distribution would become a delta function at its average value. Physically, this is akin to treating the paradox as a phase transition.

However, there is no evidence that the transition should occur at a single chemical potential in the membrane. For example, a liquid-to-vapor phase transition can occur in a porous medium at different chemical potentials if the medium has varying pore-size or contact-angle distributions. We have inherently assumed something similar in the physical model of the membrane and something like a Kelvin equation may be required to relate μ and λ. Also, the exact location of a single transition has to be determined. This could be done through assuming that it occurs at the same chemical potential as bulk liquid to vapor water, or by taking the average channel size, or perhaps some other means. Basically, further experimental research is required to quantify the existence and location of the transition region and to correlate it with operating conditions such as pressure, temperature, and membrane stress.

One possible away around the above situation is to use the model but keep the driving forces separate through the transition region. Such a situation would involve using Eqs. (5.17) and (5.18) without Eq. (5.20). This approach is probably the worst to use from a physical standpoint, but it allows for a basically flat chemical-potential gradient in the vapor-equilibrated transition region agreeing with Schröder's paradox. Finally, some thought should go into the actual existence of the paradox. It seems more than likely that it is basically a phase transition based on both experimental structural evidence and the fact that it occurs in many polymeric systems and at the same conditions as vapor-liquid equilibrium. However, there is a need to understand where the transition occurs in the membrane and the nature of the paradox. It is also probably related to the polymer phase segregation, ionic and nonionic moieties, membrane elastic properties, and operating conditions. For example, Schröder's paradox may narrow and cease to exist as the system pressure approaches the vapor pressure of water since there is no longer any inert gases present, resulting in the disappearance of any vapor-liquid menisci. This also would result in a loss of a degree of freedom for the system meaning that at only a single pressure can the paradox occur given a temperature of the system. Finally, the paradox is probably not a true thermodynamic inconsistency, and instead the lower fully-vapor-equilibrated values result from metastable or kinetically hindered states that depend on membrane history and state and evolve ith the long-time relaxation of the polymer. Research is currently going on to resolve the above issues.

5.8. Physical and Transport Properties

Each transport mode has its own set of three transport parameters (conductivity and electro-osmotic and transport coefficients), as well as several structural or membrane properties. Obviously, these properties are dependent on the membrane as well as the pretreatment conditions. The expressions for the properties that we use in the model come from experimental data from various sources and all are given in terms of two parameters: the temperature, T, and the water content, λ. The discussion below is abbreviated, and only notes are made as to how the property expressions agree with the physical model.

5.8.1. Thickness

It is assumed that the membrane can swell freely within the PEFC, that is, the membrane pushes against the other PEFC sandwich layers and expands. In reality, the membrane is probably constrained from swelling freely due to the applied compression upon assembly and operation. However, using a simple stress analysis, we predict that the membrane will typically swell to a large extent due to its prodigious water uptake and the large force required to compress it [56]. If desired, a fractional or degree-of-constraint term can be added to the equations below in determining the membrane thickness but such a treatment is beyond the scope of this discussion and the reader is referred to another paper if interested [56].

Since the membrane can swell freely, a function is needed to find the thickness of the membrane after swelling. This thickness is important in simulating transport in membranes because it directly affects many of the gradient driving forces. Due to a slight anisotropy in the membrane, the thickness can be related to the water content by [39]

$$l = l_o \left(1 + 0.36 \frac{\hat{\lambda} \bar{V}_w}{\bar{V}_m}\right) \qquad (5.25)$$

where l_o is the dry membrane thickness and $\hat{\lambda}$ is the average value of λ in the membrane

$$\hat{\lambda} = \frac{1}{l}\int_0^{z=l} \lambda(z)\mathrm{d}z = \int_0^1 \lambda(\varsigma)\mathrm{d}\varsigma \qquad (5.26)$$

where a dimensionless position variable has been substituted into the second equality. Treating swelling in this manner ensures conservation of the membrane mass as discussed by Meyers and Newman [65].

$\hat{\lambda}$ and l are solved for simultaneously with the rest of the equations. To do this, it is necessary to add two equations to the set of governing equations being solved. The first equation is Eq. (5.26), and the second arises from the length being a scalar quantity that is uniform

$$\frac{\mathrm{d}l}{\mathrm{d}\varsigma} = 0 \qquad (5.27)$$

These two equations are solved with the free-expansion boundary condition

$$l = l_o\left(1 + 0.36\frac{\hat{\lambda}\bar{V}_w}{\bar{V}_m}\right) \text{ at } \varsigma = 1 \qquad (5.28)$$

5.8.2. Transport Properties

Table 5.2 lists the expressions used for the transport properties in the model. Both the liquid- and vapor-equilibrated properties are given along with some general constitutive relations. The discussion below about the expressions focuses only on how they relate to the physical model; for an in-depth discussion and derivation of the expressions, the reader is referred to our paper on the subject [39].

As discussed in the physical model, the conductivity is caused by both the vehicle and Grotthuss mechanisms. The main difference between the conductivity of the vapor and liquid-equilibrated membranes stems from their respective water contents. A percolation-type equation is used for the conductivity although, at higher water contents, the conductivity should level out and ideally approach that of liquid water at infinite dilution (i.e., the polymer is dissolved in an infinite amount of water). The temperature dependence of the conductivity is due to the change in the equilibrium constant for the dissociation of the sulfonic acid sites and the activation energies for the Grotthuss mechanism and vehicle mechanisms [39]. Finally, the expressions given in Table 5.2 are valid for pure water and the

TABLE 5.2 Expressions for membrane physical and transport properties [39].

General Definitions

Water partial molar volume	$\bar{V}_w = \frac{M_w}{\rho_w}$
Membrane partial molar volume	$\bar{V}_m = \frac{EW}{\rho_{m,o}}$
Total system volume	$V = \bar{V}_m + \lambda \bar{V}_w$
Water volume fraction	$f = \frac{\lambda \bar{V}_w}{V}$
Water concentration	$c_w = \frac{\lambda}{V}$

Transport Properties

	Vapor-Equilibrated	Liquid-Equilibrated	
Conductivity	$\kappa_V = 0.5(f_V - 0.06)^{1.5} \exp\left[\frac{15000}{R}\left(\frac{1}{T_{ref}} - \frac{1}{T}\right)\right]$	$\kappa_L = 0.5(f_L - 0.06)^{1.5} \exp\left[\frac{15000}{R}\left(\frac{1}{T_{ref}} - \frac{1}{T}\right)\right]$	for $f \leq 0.45$
		$\kappa_L = \kappa_L(0.45, T)$	for $f \geq 0.45$
Electro-osmotic coefficient	$\xi_V = \lambda$ for $\lambda < 1$ $\xi_V = 1$ for $\lambda \geq 1$	$\xi_L = 2.55 \exp\left[\frac{4000}{R}\left(\frac{1}{T_{ref}} - \frac{1}{T}\right)\right]$	
Transport coefficient	$\alpha_V = \frac{c_0 v D_{\mu 0}}{RT(1-x_0 Y)}$	$\alpha_L = \frac{k_{sat}}{\mu \bar{V}_w^2}$	
	$D_{\mu 0} = 1.8 \times 10^{-5} f_V \exp\left[\frac{20000}{R}\left(\frac{1}{T_{ref}} - \frac{1}{T}\right)\right]$		

acid form of the membrane. Other forms and the presence of anions will affect the conductivity [66–68].

The electro-osmotic coefficient, sometimes termed the electro-osmotic drag coefficient, is the transport number of water in the membrane. It is a measure of the number of water molecules that are carried with each proton in the absence of a concentration gradient. As discussed in the physical model, the electro-osmotic coefficient arises mainly due to the complexes that solvate the protons and the vehicle mechanism and depends solely on temperature and water content, at least macroscopically. For the vapor-equilibrated case, it has a value at or close to unity since basically only a hydronium ion is moving through the membrane. For the liquid-equilibrated case, an Arrhenius dependence is used with an activation energy that describes the data and stems from the amount of energy needed to break a hydrogen bond in water in Nafion® [39], which is the limiting step in the Grotthuss mechanism [25, 44, 48].

The transport coefficient, α, is in essence a modified diffusion coefficient or effective permeability. It relates the water flux in the absence of current to its gradient of chemical potential. For a vapor-equilibrated membrane, the membrane is treated as a single phase, and α_V relates the flux to a chemical-potential driving force. As seen in Table 5.2, α_V is comprised of the water mole fraction and concentration in the membrane and the water diffusion coefficient. The subscript V is added to the expressions since they are for transport in the vapor-equilibrated part of the membrane only. The value of the diffusion coefficient compared to that of pure water agrees with the physical reasoning that the water in the membrane is more strongly held than in bulk water. For a liquid-equilibrated membrane, the membrane is treated as a two-phase system in which bulk-like liquid water exists in the channels between the ionic domains (see Section 5.6). The chemical-potential driving force can be rewritten as a driving force in terms of pressure alone, and Darcy's law is used yielding the expression in Table 5.2 for α_L.

As a point of comparison, the values of α_L and α_V are plotted in Figure 5.6 as a function of temperature. α_V is a function of temperature both through the way that λ_V and $\bar{V}_w$ are functions of temperature and through the Arrhenius dependence of the diffusion coefficient. The temperature dependence of α_L is mainly due to how viscosity changes with temperature. As seen in Figure 5.6, and as expected by physical reasoning, the value of α_L is much larger than that of α_V indicating that it is easier for water to travel through liquid-filled expanded channels rather than through the collapsed channels. Finally, just as with conductivity, the decrease in λ_V with temperature causes a minimum in the α_V curve around a temperature of 35°C. This decrease has the effect of having an apparent activation energy for α_V that is lower than the expected 20 kJ/mol and is close to the activation energy of α_L (12 kJ/mol).

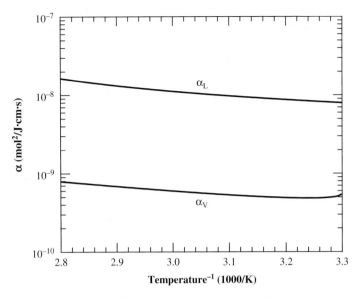

FIGURE 5.6. Arrhenius plot of the liquid- and vapor-equilibrated transport coefficients, α_L and α_V, respectively, as functions of temperature. α_V is evaluated at unit activity $\left(\lambda = \lambda_V|_{a_w=1}\right)$. (The figure is modified from Ref. [39] with permission of The Electrochemical Society, Inc.)

5.9. Gas Crossover

In a PEFC, oxygen and hydrogen crossover is important because of the obvious performance loss, the development of a mixed potential, and even durability issues due to hydrogen peroxide generation platinum migration, and possible carbon corrosion [69]. Furthermore, crossover becomes increasingly important as the membranes used become thinner. Presented in this section are the parameters and governing equations to model this phenomenon.

Because the gases in the membrane are in low concentration and do not interact significantly with each other, a dilute-solution transport equation can be used

$$\mathbf{N}_i = -\psi_i \nabla p_i \qquad (5.29)$$

where ψ_i and p_i are the permeation coefficient and partial pressure of species i, respectively. The permeation coefficient is the product of the solubility and diffusivity. It is used because it allows for a single variable to describe the transport, instead of two, each with their dependences on temperature, etc., which may even offset each other. Furthermore, using the above equation allows for a simple boundary condition of continuous partial pressure of the gas at the membrane interface. In addition, although the equation uses a partial-pressure driving force, this is due to the use of permeation coefficients and the real driving force is chemical potential and the proposed physical model is not violated; there is no separate gas phase in the membrane. Finally, one may want to add a convection term to the right side of Eq. (5.29) to account for hydrogen moving with the water flux; however, a back-of-the-envelope calculation shows that such a convective flow is at least an order of magnitude less than the permeation one and it can be disregarded to a first approximation.

The permeation coefficients, like the other transport properties, are expected to depend on water content, temperature, and the state of the membrane (i.e., collapsed or expanded channels). Fitting the experimental data [39] yields the following expressions for vapor- and liquid-equilibrated membranes respectively

$$\psi_{H_2,V} = (2.2 \times 10^{-11} f_V + 2.9 \times 10^{-12}) \exp\left[\frac{21000}{R}\left(\frac{1}{T_{ref}} - \frac{1}{T}\right)\right] \quad (5.30)$$

$$\psi_{H_2,L} = 1.8 \times 10^{-11} \exp\left[\frac{18000}{R}\left(\frac{1}{T_{ref}} - \frac{1}{T}\right)\right] \quad (5.31)$$

and

$$\psi_{O_2,V} = (1.9 \times 10^{-11} f_V + 1.1 \times 10^{-12}) \exp\left[\frac{22000}{R}\left(\frac{1}{T_{ref}} - \frac{1}{T}\right)\right] \quad (5.32)$$

$$\psi_{O_2,L} = 1.2 \times 10^{-11} \exp\left[\frac{20000}{R}\left(\frac{1}{T_{ref}} - \frac{1}{T}\right)\right] \quad (5.33)$$

for hydrogen and oxygen, respectively.

The permeation coefficients are plotted in Figure 5.7. In the top graph, Figure 5.7(a), the values of the permeation coefficients for hydrogen and oxygen as a function of water content are shown where the points at $\lambda = 22$ are the liquid-equilibrated values. As discussed in Section 5.7.1, for the transition region the vapor- and liquid-equilibrated values can be connected with a straight line (i.e., the dotted lines in the figure) when a single driving force is used (see Eq. 5.21). In Figure 5.7(b), the permeation coefficients for oxygen for the liquid-equilibrated, dry, and saturated-vapor-equilibrated cases as a function of temperature are given. Also included in that graph are the values for oxygen permeation in water and Teflon. The above expressions and Figure 5.7(a) show that the permeation coefficient of hydrogen is about 1.5 times the value of the oxygen coefficient and with a slightly lower activation energy. Thus, a plot of the hydrogen coefficients appears similar to

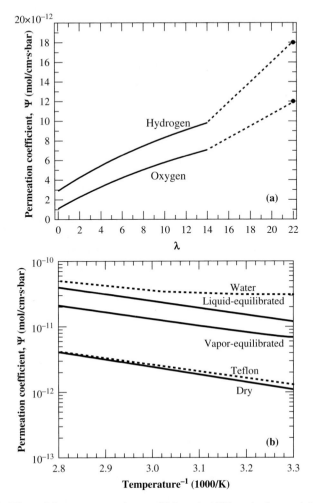

FIGURE 5.7. Plots of the gas permeation coefficients in 1100 equivalent weight Nafion®. (a) Hydrogen and oxygen permeation coefficients at 30°C as a function of water content. The dotted lines signify the transition-region values. (b) Arrhenius plot of the oxygen permeation coefficient as a function of temperature for a liquid-equilibrated membrane, a vapor-equilibrated membrane, and a dry membrane. Also plotted are the oxygen permeation coefficients in water [79] and Teflon [80, 81]. (Figure b is reproduced from Ref. [39] with permission of The Electrochemical Society, Inc.)

Figure 5.7(b) except that all the values are increased by a factor of about 1.5. As can be seen in the figure, the permeation-coefficient values are basically bounded with the liquid-equilibrated values higher than the vapor-equilibrated ones. The values are bounded because at higher water contents the gases mainly move through the bulk-like liquid water, and under dry conditions the membrane is very similar to Teflon® (see Figure 5.2).

5.10. Model Aspects

The above sections describe the membrane model both from its physical underpinnings and its mathematical treatment. In this section, some of the aspects of the model in terms of simulation results and descriptions are highlighted.

5.10.1. Validation

Although the physical and mathematical models as well as the various property-value expressions are taken from experimental results and basic physics, there is still a need to validate the approach taken. This has been done both qualitatively [70] and quantitatively [71]. The qualitative validation involved comparing trends in terms of properties, and the quantitative validation involved using the membrane model in a simple pseudo 2-D fuel-cell model to explain and agree with experimental water-balance data. The model has also been validated in a more complicated fuel-cell model, such as the ones used to examine the effects of microporous layers in fuel cells [40]. However, such comparisons serve to validate the entire model and not necessarily just the one for the membrane.

The physical model can be used to describe trends seen in experimental data. For example, the interconnectivity of the cluster network is predicted to have a profound effect on a membrane's transport properties. The percolation threshold for conductivity should increase when the clusters become smaller, which could be due to a stiffer and/or more crystalline polymer matrix. These smaller clusters would also mean that the membrane would exhibit lower electro-osmotic coefficients, larger liquid water uptakes, and a greater dependence of the various properties on water content than in Nafion®. In fact, these predictions are what is seen in such systems as sulfonated polyetherketones [19, 72] and Dow membranes [73, 74] or when the equivalent weight [22] or drying temperature [4, 6] of Nafion® is increased.

The above example verifications of the physical model show that it could be used to aid in membrane development. For example, high-temperature membranes need to remain conductive at low relative humidities. According to the physical model, one approach would be to have very small, highly interconnected, and hydrophilic ionic domains in a hydrophobic matrix. Such a structure would allow for water to remain in the ionic domains at elevated temperatures and still have a relatively short conduction pathway across the cell.

Along with the physical-model validation, the mathematical model was shown to agree quantitatively with water-balance results [71]. These results include examining how various operating conditions, flow arrangements, and membrane structures affect the net water flux through the membrane. For example, temperature has a strong effect on the water balance of the fuel

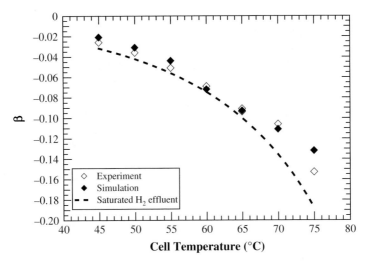

FIGURE 5.8. Net flux of water per proton flux, β, as a function of cell temperature. The open and solid symbols are data [75] and simulation [71], respectively. The conditions are $T = 50°C$, $i = 0.2$ A/cm^2, $p_a = 1.01$ bar, $p_c = 1.72$ bar, air and hydrogen stoichiometries of 1.7 and 1.6, respectively. The dashed line is the value of β that yields a saturated hydrogen stream effluent (RH = 100 %). (The figure is reproduced from Ref. [71] with permission of The Electrochemical Society, Inc.)

cell. This is shown in Figure 5.8 for the case of dry feeds where the simulation results agree well with those of the experiments. In the figure, the net water flux through the membrane normalized by the proton flux, β, decreases with increasing temperature. This is because the liquid-equilibrated transport mode ceases to be dominant over the vapor-equilibrated one, which is also related to the saturation of the anode gas. This occurs even though the value of λ decreases because the electro-osmotic flow remains about constant whereas the transport coefficient increases (see Section 5.8). The higher vapor pressures also mean that there is not enough liquid produced in the fuel cell to saturate either feed stream, and a nonuniform current-density distribution will exist (along the gas channel). This problem is also why operation and simulation at high temperatures with dry gases is difficult [75–78]. Furthermore, due to the assumption of a uniform current-density distribution, the simulations show worse agreement with the data at the higher temperatures. Also, the operating window between flooding and membrane dehydration should become narrower at high temperatures thus causing less stable performance.

5.10.2. Describing Water Management

The model can be used to match and explain experimental data, as well as examine results which cannot be obtained experimentally. This type of analysis leads to increased understanding, and eventually to predictive performance and optimization that can help guide fuel-cell developers. An example of this is operation with dry hydrogen and air similar to that described above. However, while the overall water-balance data correlate with the experimental values [71] the model allows for a more in-depth analysis.

Figure 5.9 shows the net water flux per proton flux and partial pressures of water at four positions in the fuel-cell sandwich as functions of gas-channel position. The flow is countercurrent with both air and hydrogen fed dry, and the other conditions are given in the caption. The profiles clearly demonstrate that, although both gases are fed dry, the anode humidification is much more important since it actually starts to dry out the cathode gases. This is also witnessed by the large negative value of the net water flux. The reason for the dominance of the anode is that water vapor in hydrogen diffuses much faster than in oxygen, and hence the partial pressure at the anode gas-diffusion layer (GDL) / membrane interface causes water to move mainly from the cathode to the anode. Due to this negative value of β, it may be possible to operate PEFCs with completely dry feeds; liquid product water is basically recirculated keeping the membrane hydrated. A figure such as Figure 5.9 can be used

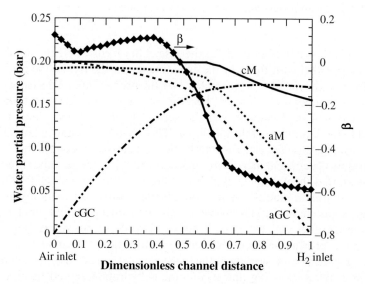

FIGURE 5.9. Water partial pressures and β as a function of position in the gas channel. The feed is countercurrent with dry hydrogen and air with the conditions: $T = 80°C$, $i = 0.6$ A/cm², $p_a = p_c = 1.5$ bar, hydrogen and air stoichiometries of 4 and 2, respectively. The partial pressures are given for the anode and cathode gas channels (aGC and cGC) and GDL / membrane interfaces (aM and cM). The water vapor pressure at this temperature is about 0.2 bar. (The figure is reproduced from Ref. [71] with permission of The Electrochemical Society, Inc.)

for deciding the optimum catalyst placement and other design issues. Thus, simulations can be used to design PEFCs for certain operating conditions. For example, dispersing the catalyst so that it has a maximum concentration near the zero-water-balance point and near the cathode inlet should decrease the overall ohmic losses and flooding problems.

5.10.2.1. Water Profiles

To examine the dry-feed case in more detail, the membrane water content is plotted as a function of both gas-channel and membrane positions in Figure 5.10, something that is not experimentally obtainable. The graph demonstrates a maximum penetration depth of liquid water into the membrane near the air inlet. This maximum is caused by the higher liquid-pressure at the membrane/cathode GDL interface, which is due to a balance between the hydrogen and air relative humidity. When there is no liquid water in the membrane, λ gradually decreases with both positions in the membrane. The transition between transport modes and the modeling of Schröder's paradox as a continuous change in water content is also clearly shown, and, as mentioned previously, the transition region is very sharp.

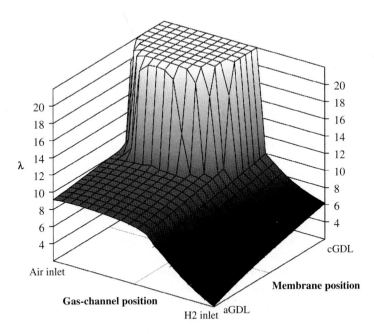

FIGURE 5.10. Membrane water content as a function of position both along the gas channel and through the thickness of the membrane for the case of countercurrent operation. The conditions are the same as in Figure 9. (The figure is reproduced from Ref. [71] with permission of The Electrochemical Society, Inc.)

There are several key parameters that affect the location of the transition region. Primarily, it depends on the current density and thickness of the membrane and secondarily on the pressure in the gas channel. As the current density increases, the membrane begins to dry out, and the transition moves closer to the cathode. The thicker the membrane, the closer to the cathode the transition occurs. For example, Nafion® 112 membranes remain fully hydrated until current densities that are at least double those for Nafion® 117. Although the gas-channel pressure greatly affects the local current-density profiles, it has only a minimal effect on the position of the transition at high current densities. It affects the profiles by creating a higher liquid pressure at the membrane/GDL interface. At high current densities, this has a minimal impact due to the high electro-osmotic flow. However, at low current densities, the higher pressure in the gas channels means a higher pressure in the membrane, which in turn means a deeper penetration depth of liquid water.

In addition, parameters such as the effective permeabilities of the GDLs and the membrane will also affect the membrane water profile. A higher value of the GDL permeability results in a lower liquid pressure at the membrane/cathode GDL interface, which approaches the value of the pressure in the gas channel. A higher value of membrane permeability results in a smaller pressure gradient inside the membrane, and hence any transition between modes occurs closer to the anode. The above can be explained by water flowing down the path of least resistance, which is determined by the various layers' permeabilities. Such an analysis also agrees with why microporous layers can show considerable gains in the limiting current [40].

Knowing where liquid water exists and how far it penetrates helps to guide design efforts like the prediction of hot spots and GDL loadings, such that the transition front is moved to maximize the conductivity of the membrane while still minimizing gas crossover and flooding. Furthermore, as expected, ohmic losses near the anode inlet are going to be large due to the very low λ values seen in Figure 5.10. This also results in a much more nonuniform current-density distribution and due to the much more resistive membrane, ohmic heating and related nonisothermal effects could become important. In all, the simulations show that the anode side of the membrane has a greater influence on the water content and properties of the membrane than the cathode side and should be fed at least partially humidified.

To examine the transport-mode-transition region in more detail, simulations were run at different current densities [71]. The resultant membrane water profiles are shown in Figure 5.11 where a vapor-equilibrated membrane at unit activity has a water content of $\lambda = 8.8$ as calculated by the modified chemical model (see Section 5.5.1). The profiles in the figure demonstrate that the higher the current density the sharper the transition from the liquid-equilibrated to the vapor-equilibrated mode as well as the lower the value of the water content at the anode GDL/membrane interface. The reason why the transition occurs at the same point in the membrane is that the electro-osmotic flow and the water-gradient flow are both proportional to the current

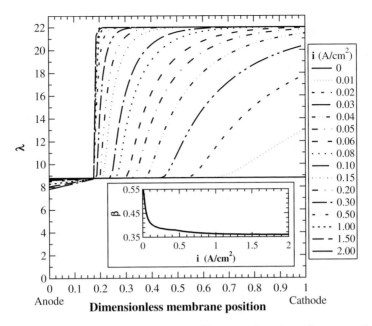

FIGURE 5.11. Membrane water-content profiles as a function of current density from 0 to 2 A/cm². (inset) Value of β as a function of current density. (The figure is reproduced from Ref. [71] with permission of The Electrochemical Society, Inc.)

density. Furthermore, the transition sharpens due to the higher liquid pressures at the higher current densities, which increase the value of λ in the liquid-equilibrated-transport-mode regions.

The inset graph in Figure 5.11 shows the value of β as a function of current density. As seen in the water profiles, β depends strongly on the transport mode in the membrane. At very low current densities, β is high because the transport is mainly through the vapor-equilibrated mode. β then drops rapidly as soon as the liquid-equilibrated mode comes into effect and then gradually approaches a constant value as the transition sharpens. Even though the value of λ at the anode GDL/membrane interface decreases slightly with current density, the average value of λ actually increases because the interface is becoming sharper.

5.11. Summary and Future Directions

In this chapter, a combination model for a polymer-electrolyte or proton-exchange membrane in a fuel cell was reviewed. The model describes the macroscopic transport phenomena that occur inside the membrane.

It strikes a balance between being complex enough to handle all of the important governing phenomena and yet remaining simple enough to run in a realistic amount of time. The membrane model has a physical basis and accounts for many effects experimentally observed but not modeled as well as provides for a modeling methodology of Schröder's paradox. In addition to the physical and mathematical descriptions of the model, some simulation results were analyzed in terms of model validation and the water profiles in the membrane during operation. This allowed for a closer look and discussion of the transition region between transport modes. The region is sharp and depends both on the membrane's properties (structural, chemical, and transport) as well as operating conditions such as current density and membrane state. Overall, the presented model bridges the various gaps of models currently in the literature and allows one to understand water management and its associated membrane effects.

There are various improvements that can be made to the presented model, some improvements could be accomplished. Foremost among these possible future-work directions is the inclusion of nonisothermal effects. Such effects as ohmic heating could be very important, especially with resistive membranes or under low-humidity conditions. Also, as mentioned, a consensus needs to be reached as to how to model in detail Schröder's paradox and the mode transition region; experiments are currently underway to examine this effect. Further detail is also required for understanding the membrane in relation to its properties and role in the catalyst layers. This includes water transport into and out of the membrane, as well as water production and electrochemical reaction. The membrane model can also be adapted to multiple dimensions for use in full 2-D and 3-D models. Finally, the membrane model can be altered to allow for the study of membrane degradation, such as pinhole formation and related failure mechanisms due to membrane mechanical effects, as well as chemical attack due to peroxide formation and gas crossover.

Acknowledgements

We would first like to thank the funding sources for this work, UTC Fuel Cells, LLC, EPA through a STAR graduate fellowship (91601301-0) and the Assistant Secretary for Energy Efficiency and Renewable Energy, Office of Hydrogen, Fuel Cell, and Infrastructure Technologies, of the U.S. Department of Energy under contract number DE-AC02-05CH11231. We would also like to thank Stephen Paddison and Keith Promislow for inviting us to write this chapter.

Notation

Roman

a_i	=	activity of species i in phase α
c_i	=	interstitial concentration of species i, mol/cm^3
$\mathbf{d}_i$	=	driving force per unit volume acting on species i in phase k, J/cm^4
$D_{i,j}$	=	diffusion coefficient of i in j, cm^2/s
D_{μ_w}	=	diffusion coefficient of water related to a chemical potential driving force, cm^2/s
$E_{i,j}^*$	=	binary interaction parameter between i and j used in the chemical model, g/mol
E_W	=	membrane equivalent weight, g/equiv
f	=	water volume fraction in the membrane
F	=	Faraday's constant, 96487 C/equiv
i	=	superficial current density, A/cm^2
k	=	effective permeability, cm^2
k_{sat}	=	saturated or absolute permeability, cm^2
$K_{i,j}$	=	frictional coefficient of interaction between species i and j, J·s/cm^5
$L_{i,j}$	=	inverted frictional coefficient of interaction between species i and j, cm^5/J·s
m_i	=	molality of species i in the membrane, mol/g
M_i	=	molecular weight of species i, g/mol
$\mathbf{N}_i$	=	superficial flux density of species i, mol/cm^2·s
p	=	total thermodynamic pressure, bar
p_i	=	partial pressure of species i, bar
p_k	=	total pressure of phase k, bar
p_w^{vap}	=	vapor pressure of water, bar
r_c	=	critical pore radius, μm
R	=	ideal-gas constant, 8.3143 J/mol·K
S	=	fraction of expanded channels
T	=	absolute temperature, K
$\mathbf{v}_i$	=	superficial velocity of species i, cm/s
V	=	cell potential, V
V	=	volume, cm^3
$\bar{V}_i$	=	(partial) molar volume of species i, cm^3/mol
x_i	=	mole fraction of species i
$\mathbf{X}_i$	=	extensive term referring to body forces per mole acting on species i

Greek

α	=	transport coefficient, mol^2/J·cm·s
β	=	net water flux per proton flux through the membrane
γ	=	surface tension, N/cm
Γ	=	energetic interaction parameter of the membrane with bulk-like water, N/cm
ζ	=	dimensionless membrane thickness
ξ	=	electro-osmotic coefficient
ρ	=	solution density, g/cm^3
θ	=	contact angle of water with the channel, degrees
κ	=	ionic conductivity, S/cm
λ	=	moles of water per mole of sulfonic acid sites

α	= transport coefficient, mol^2/J·cm·s
$\lambda_v\|_{aw} = 1$	= maximum value of λ for a membrane in contact with water vapor at unit activity and the operating temperature
$\hat{\lambda}$	= average membrane water content
$\lambda_{H_3O^+}$	= moles of hydronium ions per mole of sulfonic acid sites
λ_i^*	= secondary reference state quantity for species i, g/mol
μ	= viscosity, Pa·s
μ_i	= electrochemical potential of species i, J/mol
μ_w^*	= combination of reference states and constants for water in the membrane
ϕ_l	= chemical model parameters defined in Eq. (5.13)
Φ	= electrical potential, V
ψ_i	= permeation coefficient of species i, mol/bar·cm·s

Subscripts/Superscripts

+	= proton
eff	= effective transport property
L	= liquid phase or liquid-equilibrated membrane
m	= membrane as a species
o	= initial or reference value
ref	= parameter evaluated at the reference conditions
V	= vapor-equilibrated membrane
w	= water

References

[1] A. Z. Weber and J. Newman, "Modeling Transport in Polymer-Electrolyte Fuel Cells," *Chemical Reviews*, **104**, 4679 (2004).

[2] N. Cornet, G. Gebel, and A. de Geyer, "Existence of the Schroeder Paradox with a Nafion Membrane? Small-Angle X-Ray Scattering Analysis," *Journal De Physique Iv*, **8**, 63 (1998).

[3] F. Meier and G. Eigenberger, "Transport Parameters for the Modelling of Water Transport in Ionomer Membranes for Pem-Fuel Cells," *Electrochimica Acta*, **49**, 1731 (2004).

[4] J. T. Hinatsu, M. Mizuhata, and H. Takenaka, "Water-Uptake of Perfluorosulfonic Acid Membranes from Liquid Water and Water-Vapor," *Journal of the Electrochemical Society*, **141**, 1493 (1994).

[5] J. P. Meyers and J. Newman, "Simulation of the Direct Methanol Fuel Cell - I. Thermodynamic Framework for a Multicomponent Membrane," *Journal of the Electrochemical Society*, **149**, A710 (2002).

[6] T. A. Zawodzinski, C. Derouin, S. Radzinski, R. J. Sherman, V. T. Smith, T. E. Springer, and S. Gottesfeld, "Water Uptake by and Transport through Nafion(R) 117 Membranes," *Journal of the Electrochemical Society*, **140**, 1041 (1993).

[7] T. A. Zawodzinski, T. E. Springer, J. Davey, R. Jestel, C. Lopez, J. Valerio, and S. Gottesfeld, "A Comparative-Study of Water-Uptake by and Transport through Ionomeric Fuel-Cell Membranes," *Journal of the Electrochemical Society*, **140**, 1981 (1993).

[8] P. Schröder, "Über Erstarrungs- Und Quellungserscheinungen Von Gelatine," *Zeitschrift für physikalische Chemie*, **45**, 75 (1903).

[9] C. Y. Wang, "Fundamental Models for Fuel Cell Engineering," *Chemical Reviews*, **104**, 4727 (2004).

[10] K. D. Kreuer, S. J. Paddison, E. Spohr, and M. Schuster, "Transport in Proton Conductors for Fuel-Cell Applications: Simulations, Elementary Reactions, and Phenomenology," *Chemical Reviews*, **104**, 4637 (2004).

[11] T. F. Fuller, *Solid-Polymer-Electrolyte Fuel Cells*, Ph.D. Dissertation, University of California, Berkeley, CA (1992).

[12] T. E. Springer, T. A. Zawodzinski, and S. Gottesfeld, "Polymer Electrolyte Fuel Cell Model," *Journal of the Electrochemical Society*, **138**, 2334 (1991).

[13] M. Eikerling, Y. I. Kharkats, A. A. Kornyshev, and Y. M. Volfkovich, "Phenomenological Theory of Electro-Osmotic Effect and Water Management in Polymer Electrolyte Proton-Conducting Membranes," *Journal of the Electrochemical Society*, **145**, 2684 (1998).

[14] D. M. Bernardi and M. W. Verbrugge, "A Mathematical Model of the Solid-Polymer-Electrolyte Fuel Cell," *Journal of the Electrochemical Society*, **139**, 2477 (1992).

[15] G. J. M. Janssen, "A Phenomenological Model of Water Transport in a Proton Exchange Membrane Fuel Cell," *Journal of the Electrochemical Society*, **148**, A1313 (2001).

[16] T. Thampan, S. Malhotra, H. Tang, and R. Datta, "Modeling of Conductive Transport in Proton-Exchange Membranes for Fuel Cells," *Journal of the Electrochemical Society*, **147**, 3242 (2000).

[17] D. M. Bernardi and M. W. Verbrugge, "Mathematical Model of a Gas-Diffusion Electrode Bonded to a Polymer Electrolyte," *AIChE Journal*, **37**, 1151 (1991).

[18] M. Eikerling, A. A. Kornyshev, and U. Stimming, "Electrophysical Properties of Polymer Electrolyte Membranes: A Random Network Model," *Journal of Physical Chemistry B*, **101**, 10807 (1997).

[19] K. D. Kreuer, "On the Development of Proton Conducting Materials for Technological Applications," *Solid State Ionics*, **97**, 1 (1997).

[20] A. Z. Weber and J. Newman, "Physical Model of Transport in Polymer-Electrolyte Membranes," in *Proton Conducting Membrane Fuel Cells Iii*, J. W. Van Zee, T. F. Fuller, S. Gottesfeld, and M. Murthy, Editors, The Electrochemical Society Proceeding Series, Pennington, NJ (2002).

[21] K. A. Mauritz and R. B. Moore, "State of Understanding of Nafion," *Chemical Reviews*, **104**, 4535 (2004).

[22] W. Y. Hsu and T. D. Gierke, "Ion-Transport and Clustering in Nafion Perfluorinated Membranes," *Journal of Membrane Science*, **13**, 307 (1983).

[23] A. Vishnyakov and A. V. Neimark, "Molecular Dynamics Simulation of Microstructure and Molecular Mobilities in Swollen Nafion Membranes," *Journal of Physical Chemistry B*, **105**, 9586 (2001).

[24] A. Vishnyakov and A. V. Neimark, "Molecular Simulation Study of Nafion Membrane Solvation in Water and Methanol," *Journal of Physical Chemistry B*, **104**, 4471 (2000).

[25] E. Spohr, P. Commer, and A. A. Kornyshev, "Enhancing Proton Mobility in Polymer Electrolyte Membranes: Lessons from Molecular Dynamics Simulations," *Journal of Physical Chemistry B*, **106**, 10560 (2002).

[26] G. Gebel, P. Aldebert, and M. Pineri, "Swelling Study of Perfluorosulphonated Ionomer Membranes," *Polymer*, **34**, 333 (1993).

[27] R. S. McLean, M. Doyle, and B. B. Sauer, "High-Resolution Imaging of Ionic Domains and Crystal Morphology in Ionomers Using Afm Techniques," *Macromolecules*, **33**, 6541 (2000).

[28] P. Choi and R. Datta, "Sorption in Proton-Exchange Membranes. An Explanation of Schroeder's Paradox," *Journal of the Electrochemical Society*, **150**, E601 (2003).

[29] J. Divisek, M. Eikerling, V. Mazin, H. Schmitz, U. Stimming, and Y. M. Volfkovich, "A Study of Capillary Porous Structure and Sorption Properties of Nafion Proton-Exchange Membranes Swollen in Water," *Journal of the Electrochemical Society*, **145**, 2677 (1998).

[30] M. Falk, "An Infrared Study of Water in Perflurosulfonate (Nafion) Membranes," *Canadian Journal of Chemistry*, **58**, 1495 (1980).

[31] F. P. Orfino and S. Holdcroft, "The Morphology of Nafion: Are Ion Clusters Bridged by Channels or Single Ionic Sites?," *Journal of New Materials for Electrochemical Systems*, **3**, 285 (2000).

[32] T. A. Zawodzinski Jr., S. Gottesfeld, S. Shoichet, and T. J. McCarthy, "The Contact Angle between Water and the Surface of Perfluorosulphonic Acid Membranes," *Journal of Applied Electrochemistry*, **23**, 86 (1993).

[33] W. Y. Hsu and T. D. Gierke, "Elastic Theory for Ionic Clustering in Perfluorinated Ionomers," *Macromolecules*, **15**, 101 (1982).

[34] W. Y. Hsu, J. R. Barkley, and P. Meakin, "Ion Percolation and Insulator-to-Conductor Transition in Nafion Perfluorosulfonic Acid Membranes," *Macromolecules*, **13**, 198 (1980).

[35] J. Newman and K. E. Thomas-Alyea, *Electrochemical Systems* 3rd ed., John Wiley & Sons, New York (2004).

[36] J. O. Hirschfelder, C. F. Curtiss, and R. B. Bird, *Molecular Theory of Gases and Liquids*, John Wiley & Sons, Inc., New York (1954).

[37] D. N. Bennion, *Mass Transport of Binary Electrolyte Solutions in Membranes, Water Resources Center Desalination Report No. 4*, Tech. Rep. 66-17, Department of Engineering, University of California, Los Angeles, CA (1966).

[38] J. P. Meyers, *Simulation and Analysis of the Direct Methanol Fuel Cell*, Ph.D. Dissertation, University of California, Berkeley (1998).

[39] A. Z. Weber and J. Newman, "Transport in Polymer-Electrolyte Membranes. Ii. Mathematical Model," *Journal of the Electrochemical Society*, **151**, A311 (2004).

[40] A. Z. Weber and J. Newman, "Effects of Microporous Layers in Polymer Electrolyte Fuel Cells," *Journal of the Electrochemical Society*, **152**, A677 (2005).

[41] W. Y. Hsu, M. R. Giri, and R. M. Ikeda, "Percolation Transition and Elastic Properties of Block Copolymers," *Macromolecules*, **15**, 1210 (1982).

[42] P. C. Rieke and N. E. Vanderborgh, "Temperature-Dependence of Water-Content and Proton Conductivity in Polyperfluorosulfonic Acid Membranes," *Journal of Membrane Science*, **32**, 313 (1987).

[43] D. R. Morris and X. D. Sun, "Water-Sorption and Transport-Properties of Nafion-117-H," *Journal of Applied Polymer Science*, **50**, 1445 (1993).

[44] M. Eikerling, A. A. Kornyshev, A. M. Kuznetsov, J. Ulstrup, and S. Walbran, "Mechanisms of Proton Conductance in Polymer Electrolyte Membranes," *Journal of Physical Chemistry B*, **105**, 3646 (2001).

[45] T. J. Edwards, *Thermodynamics of Aqueous Solutions Containing One or More Volatile Weak Electrolytes*, M.S. Thesis, University of California, Berkeley (1974).
[46] C. M. Gates and J. Newman, "Equilibrium and Diffusion of Methanol and Water in a Nafion 117 Membrane," *AIChE Journal*, **46**, 2076 (2000).
[47] K. Broka and P. Ekdunge, "Oxygen and Hydrogen Permeation Properties and Water Uptake of Nafion(R) 117 Membrane and Recast Film for Pem Fuel Cell," *Journal of Applied Electrochemistry*, **27**, 117 (1997).
[48] N. Agmon, "The Grotthuss Mechanism," *Chemical Physics Letters*, **244**, 456 (1995).
[49] K. D. Kreuer, "On the Complexity of Proton Conduction Phenomena," *Solid State Ionics*, **136**, 149 (2000).
[50] R. Schlögl, "Zur Theorie Der Anomalen Osmose," *Zeitschrift für physikalische Chemie, Neue Folge*, **3**, 73 (1955).
[51] M. W. Verbrugge and R. F. Hill, "Transport Phenomena in Perfluorosulfonic Acid Membranes During the Passage of Current," *Journal of the Electrochemical Society*, **137**, 1131 (1990).
[52] R. F. Silva, A. De Francesco, and A. Pozio, "Tangential and Normal Conductivities of Nafion((R)) Membranes Used in Polymer Electrolyte Fuel Cells," *Journal of Power Sources*, **134**, 18 (2004).
[53] K. A. Mauritz and C. E. Rogers, "A Water Sorption Isotherm Model for Ionomer Membranes with Cluster Morphologies," *Macromolecules*, **18**, 483 (1985).
[54] B. Dreyfus, "Thermodynamic Properties of a Small Droplet of Water around an Ion in a Compressible Matrix," *Journal of Polymer Science Part B-Polymer Physics*, **21**, 2337 (1983).
[55] F. A. L. Dullien, *Porous Media: Fluid Transport and Pore Structure* 2nd ed., Academic Press, Inc., New York (1992).
[56] A. Z. Weber and J. Newman, "A Theoretical Study of Membrane Constraint in Polymer-Electrolyte Fuel Cells," *AIChE Journal*, **50**, 3215 (2004).
[57] E. H. Cwirko and R. G. Carbonell, "Interpretation of Transport-Coefficients in Nafion Using a Parallel Pore Model," *Journal of Membrane Science*, **67**, 227 (1992).
[58] Y. M. Volfkovich, V. S. Bagotzky, V. E. Sosenkin, and I. A. Blinov, "The Standard Contact Porosimetry," *Colloids and Surfaces a-Physicochemical and Engineering Aspects*, **187**, 349 (2001).
[59] Y. M. Volfkovich, N. A. Dreiman, O. N. Belyaeva, and I. A. Blinov, "Standard-Porosimetry Study of Perfluorinated Cation-Exchange Membranes," *Soviet Electrochemistry*, **24**, 324 (1988).
[60] M. C. Tucker, M. Odgaard, S. Yde-Anderson, and J. O. Thomas, "Abstract 1235," *203rd Meeting of the Electrochemical Society*, Paris (2003).
[61] S. Koter, "The Equivalent Pore Radius of Charged Membranes from Electroosmotic Flow," *Journal of Membrane Science*, **166**, 127 (2000).
[62] M. Cappadonia, J. W. Erning, and U. Stimming, "Proton Conduction of Nafion((R))-117 Membrane between 140 K and Room-Temperature," *Journal of Electroanalytical Chemistry*, **376**, 189 (1994).
[63] *CRC Handbook of Chemistry and Physics* 59th ed., R. C. Weast, Editor, CRC Press, Boca Raton, FL (1979).
[64] A. Z. Weber, R. M. Darling, and J. Newman, "Modeling Two-Phase Behavior in Pefcs," *Journal of the Electrochemical Society*, **151**, A1715 (2004).
[65] J. P. Meyers and J. Newman, "Simulation of the Direct Methanol Fuel Cell - Ii. Modeling and Data Analysis of Transport and Kinetic Phenomena," *Journal of the Electrochemical Society*, **149**, A718 (2002).

[66] T. Okada, G. Xie, O. Gorseth, S. Kjelstrup, N. Nakamura, and T. Arimura, "Ion and Water Transport Characteristics of Nafion Membranes as Electrolytes," *Electrochimica Acta*, **43**, 3741 (1998).

[67] C. Gavach, G. Pamboutzoglou, M. Nedyalkov, and G. Pourcelly, "Ac Impedance Investigation of the Kinetics of Ion-Transport in Nafion Perfluorosulfonic Membranes," *Journal of Membrane Science*, **45**, 37 (1989).

[68] H. L. Yeager, "Transport Properties of Perflurosulfonate Polymer Membranes," in *Perfluorinated Ionomer Membranes*, A. Eisenberg and H. L. Yeager, Editors, American Chemical Society Symposium Series, Number 180 (1982).

[69] C. A. Reiser, L. Bregoli, T. W. Patterson, J. S. Yi, J. D. Yang, M. L. Perry, and T. D. Jarvi, "A Reverse-Current Decay Mechanism for Fuel Cells," *Electrochemical and Solid State Letters*, **8**, A273 (2005).

[70] A. Z. Weber and J. Newman, "Transport in Polymer-Electrolyte Membranes. I. Physical Model," *Journal of the Electrochemical Society*, **150**, A1008 (2003).

[71] A. Z. Weber and J. Newman, "Transport in Polymer-Electrolyte Membranes. Iii. Model Validation in a Simple Fuel-Cell Model," *Journal of the Electrochemical Society*, **151**, A326 (2004).

[72] M. Ise, K. D. Kreuer, and J. Maier, "Electroosmotic Drag in Polymer Electrolyte Membranes: An Electrophoretic Nmr Study," *Solid State Ionics*, **125**, 213 (1999).

[73] R. B. Moore and C. R. Martin, "Morphology and Chemical-Properties of the Dow Perfluorosulfonate Ionomers," *Macromolecules*, **22**, 3594 (1989).

[74] Y. M. Tsou, M. C. Kimble, and R. E. White, "Hydrogen Diffusion, Solubility, and Water-Uptake in Dows Short-Side-Chain Perfluorocarbon Membranes," *Journal of the Electrochemical Society*, **139**, 1913 (1992).

[75] F. N. Büchi and S. Srinivasan, "Operating Proton Exchange Membrane Fuel Cells without External Humidification of the Reactant Gases – Fundamental Aspects," *Journal of the Electrochemical Society*, **144**, 2767 (1997).

[76] G. J. M. Janssen and M. L. J. Overvelde, "Water Transport in the Proton-Exchange-Membrane Fuel Cell: Measurements of the Effective Drag Coefficient," *Journal of Power Sources*, **101**, 117 (2001).

[77] N. Rajalakshmi, T. T. Jayanth, R. Thangamuthu, G. Sasikumar, P. Sridhar, and K. S. Dhathathreyan, "Water Transport Characteristics of Polymer Electrolyte Membrane Fuel Cell," *International Journal of Hydrogen Energy*, **29**, 1009 (2004).

[78] M. V. Williams, H. R. Kunz, and J. M. Fenton, "Operation of Nafion(R)-Based Pem Fuel Cells with No External Humidification: Influence of Operating Conditions with Gas Diffusion Layers," *Journal of Power Sources*, **135**, 122 (2004).

[79] R. H. Perry and D. W. Green, *Perry's Chemical Engineers' Handbook* 7th ed., J. O. Maloney, Editor, McGraw-Hill, New York (1997).

[80] T. Sakai, H. Takenaka, N. Wakabayashi, Y. Kawami, and E. Torikai, "Gas Permeation Properties of Solid Polymer Electrolyte (Spe) Membranes," *Journal of the Electrochemical Society*, **132**, 1328 (1985).

[81] R. A. Pasternak, M. V. Christensen, and J. Heller, "Diffusion and Permeation of Oxygen, Nitrogen, Carbon Dioxide, and Nitrogen Dioxide through Polytetrafluoroethylene," *Macromolecules*, **3**, 366 (1970).

6
Analytical Models of a Polymer Electrolyte Fuel Cell

A. A. KULIKOVSKY

6.1. Introduction

6.1.1. General Remarks

A typical cross section of a polymer electrolyte fuel cell (PEFC) is sketched in Figure 6.1. The membrane electrode assembly (MEA) is clamped between two metal or graphite plates with the channels for feed gases supply, called the "flow field". The MEA usually consists of two gas-diffusion layers (GDLs) and two catalyst layers, separated by proton-conducting membrane.

The generation of current induces fluxes of gases, liquid water, heat and charged particles in a cell. The distribution of the respective parameters (concentrations, fields etc.) is usually very non-uniform. Furthermore, the characteristic scale of parameters variation ranges from several micrometres (the thickness of the catalyst layer) to several metres (the length of the channel). In general, the problem of fuel cell modeling is multi-scale and multi-dimensional.

There are two different approaches to modelling of such a system. The first pursues simulation of real cells, taking into account as many details as possible. The number and complexity of the processes in MEA, together with sophisticated geometries of the flow fields, inevitably lead to 3D numerical models (for review of these models please see the respective chapters in this book). The models of that type are often based on commercial CFD solvers.

Ironically, the drawbacks of the CFD approach stem from its great versatility: it is difficult to resist temptation to take into account all imaginable processes, and CFD model usually includes several tens of kinetic, transport, operational and design parameters. In real systems, however, many of these parameters are poorly known. In this situation, the numerical model gives a qualitative rather than an exact picture. While they are valuable tools for understanding 3D effects, CFD models still fail to predict performance and local characteristics of cells in a wide range of operating conditions.

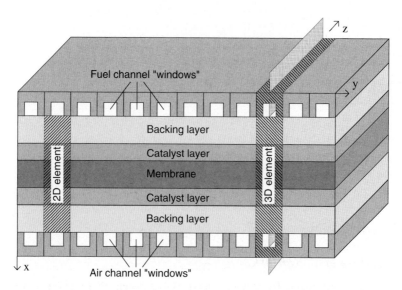

FIGURE 6.1. Sketch of the cell cross section and the system of coordinates. Local 2D models simulate single 2D element. 3D models usually deal with a single 3D element; however, the situation with 3D models is rapidly changing and nowadays these models are able to simulate the whole cell. Along-the-channel models solve equations in the x-z plane.

Some of the key parameters may be estimated through extensive comparison of experimental and numerical curves. However, CFD models are time consuming, which makes this task difficult and computationally expensive. Last but not least, cell optimization in a space of several tens of parameters using CFD simulations is hardly possible.

The approach utilized in this chapter aims at understanding the effects and trends in cell functioning rather than at simulation of real systems. This approach utilizes another philosophy: the models should involve a minimal set of parameters and should be solvable. Models of this type deliberately ignore many secondary details and focus on the principal effects in cell operation. The primary goal of this modelling is qualitative understanding rather than quantitative simulation. Several models of this type are presented below.

6.1.2. Voltage Losses in a PEFC

At zero current, the fuel cell electrodes provide the thermodynamic open circuit voltage: $V_{cell} = V_{oc}$. Connection of a load induces current I in the system and reduces V_{cell} by $\delta V(I)$. The current drawn from the fuel cell thus costs some potential; the thermodynamic voltage V_{oc} is the "capital" at our disposal. The value of V_{oc} is given by Nernst equation [1]; in this chapter, V_{oc} is assumed to be constant, and we will focus on $\delta V(I)$.

It is convenient to eliminate the cell active area A introducing a mean current density $J = I/A$. The quantities V_{cell} and δV are then functions of J via $V_{cell}(J) = V_{oc} - \delta V(J)$. A central question is then "Which processes contribute to $\delta V(J)$ and how large is each contribution?"

Electrochemical reactions on both sides of the cell occur in the high field of the double layer at the catalyst particle/electrolyte interface. In the active layers, tiny catalyst particles are mixed with polymer electrolyte. This mixture partially fills the voids of the matrix of carbon threads, which provide electronic contact between catalyst particles. The driving force for charged particles in this environment is modelled by two continuous potentials, the electrolyte potential φ_m and the carbon phase potential φ. φ_m drives protons in the electrolyte phase and φ induces electron current in the carbon phase. In the following, φ will be furnished with the subscripts "a" and "c" to distinguish the anode and the cathode side, respectively. The *polarization voltages* (overpotentials) $\eta_a = \varphi_a - \varphi_m$ and $\eta_c = \varphi_m - \varphi_c$ determine the rate of the respective electrochemical reaction.[1] Figure 6.2 sketches the distribution of potentials across the catalyst layers and membrane. Note that Figure 6.2 depicts schematic of voltage losses in a cell. Real measurable potential of a carbon phase on the anode side is $\varphi'_a = V_{oc} - \varphi_a$.

Since electronic conductivity of the carbon phase σ is usually much larger than the conductivity of electrolyte (membrane) phase, the variation of φ across the active layer can be neglected (Figure 6.2).[2] Due to excellent kinetics of hydrogen ionization (oxidation) and high hydrogen diffusivity, the

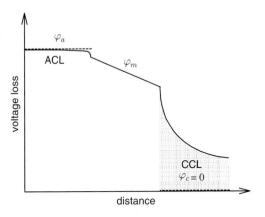

FIGURE 6.2. Sketch of voltage losses in a PEFC. ACL and CCL abbreviate anode and cathode catalyst layers, respectively. Solid line - potential of the membrane phase, dashed lines - potential of the carbon phase. Dotted areas display local polarization voltage η.

---

[1] In electrochemical studies, η is usually called overpotential. "Polarization voltage" is a more general term, which includes voltage loss due to a transport of reactants to the catalyst sites. In the following, we will use both terms as synonyms.

[2] Low σ may lead to the shift of active zone in the catalyst layer along the y-axis (Figure 6.1) [2]. This effect, however, is out of the scope of models described below.

contribution of the anode side to δV is marginal (Figure 6.2). In the following, we will focus on the cathode side and membrane.

The protonic conductivity of the membrane increases linearly with water content. Hence, the transport of water through the membrane determines its local resistivity. The structure of the membrane and the physics of water and proton transport through the polymer electrolyte is a subject of meso- and nano-scale studies. In this chapter, the membrane is treated as a continuum with given (experimental) transport coefficients. Transport of protons and water are usually described by conductivity σ_m and diffusion coefficient D_l, respectively. Both σ_m and D_l are functions of membrane water content λ_w (the number of water molecules per SO_3^- group). If the membrane is thin enough and fully humidified, its resistivity is small, and the respective voltage loss can be neglected. The simplest model of a cell can thus be developed assuming ideal membrane humidification.

The main events occur in the cathode catalyst layer (CCL), where protons and oxygen meet to participate in the oxygen reduction reaction (ORR). Part of V_{oc} must be spent to activate the reaction. Furthermore, ORR requires a continuous supply of oxygen to the catalyst sites, and every transport process in a fuel cell is equivalent to a resistance which induces the respective voltage loss. Oxygen is supplied through the flow field which covers the cell active area. The geometry of the flow field may vary from a single meander channel to a complicated multi-channel fractal structure [3]. One of the tasks of theory is calculation of the expenses for oxygen transport through the flow field and GDL.

6.1.3. Cell Prototype and Quasi-2D Approach

For simplicity we will restrict our consideration to a cell equipped on both sides with a single serpentine channel. Furthermore, we will ignore the effects due to channel curvature and consider a cell with the equivalent straight channel. Clearly, the cell with meander channels can always be cut and transformed into such equivalent cell.[3] This transformation implies that we neglect oxygen "crossover" through the gas-diffusion (backing) layer between two adjacent turns of the meander. This is justified, unless a very narrow channel and/or highly permeable backing layers are used.

The prototype of the real cell suitable for analytical modelling is shown in Figure 6.3. Such a cell is essentially a two-scale system: the thickness of MEA is less than one millimeter, whereas the feed channels may be up to several metres long. Clearly, the characteristic length of variation of any parameter along the channel is much larger, than the MEA thickness. This allows us to neglect the z-component of all fluxes in MEA (Figure 6.3).

---

[3] In case of n parallel meanders this procedure gives n identical cells.

6. Analytical Models of a Polymer Electrolyte Fuel Cell

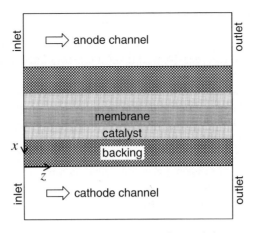

FIGURE 6.3. Sketch of the cell prototype cross section and the system of coordinates.

The transport in MEA thus occurs along the x-axis, whereas the flow in the channel transports reactants along the z-axis. The two-dimensional problem shown in Figure 6.3 can be split into two 1D problems: the channel problem along the z-axis and internal problem (in MEA) along the x-axis. The channel problem provides "boundary conditions" for the internal problem. The latter, in turn, gives local current density required to calculate the concentrations of reactants along the channel. Both problems are coupled by local current density. Following accepted terminology, this is a 1D + 1D (or quasi-2D) approach.

Physically, quasi-2D effects due to non-uniformity of oxygen concentration in the channel arise when oxygen stoichiometry λ is not large. Low λ is typical for real systems, since large stoichiometry requires a lot of energy for pumping. Many features of a cell can, however, be understood in the limit of large λ. Neglecting accumulation of water along z,[4] in the limit of large λ the cell in Figure 6.3 is a purely one-dimensional system. Solution of the respective problems give many useful insights into cell functioning and provides a basis for construction of more sophisticated quasi-2D models.

6.2. One-Dimensional Model of a PEFC

The polarization voltage of the cathode side is determined by reaction kinetics, transport of water and oxygen across the cell and by proton transport across the CCL. In this section, we will assume an ideal membrane humidification. We start with the model of the CCL.

---

[4] These effects will be considered in Section 6.4.

6.2.1. Performance of the Cathode Catalyst Layer

Many studies have been devoted to the polarization behaviour of CCL with aqueous electrolyte (see e.g. [4,5] and the literature cited therein). In these works, a general approach to the problem was formulated.

One of the first analytical models of CCL was developed by Springer and Gottesfeld [6], based on Fick's equation for oxygen transport and Tafel law for the rate of ORR. A similar approach was then used by Perry, Newman and Cairns [5] and by Eikerling and Kornyshev [7].

The aforementioned models include three governing equations: (i) mass transport equation for oxygen, (ii) proton current conservation equation with the Tafel rate of electrochemical reaction on the right side and (iii) Ohm's law, which relates proton current to the gradient of overpotential. Due to the exponential dependence of the rate of ORR on overpotential this system is strongly non-linear.

Perry, Newman and Cairns [5] obtained a numerical solution to a problem and provided asymptotic solutions for large and small proton current densities j_0. However, they did not present the expressions for the voltage current curve valid in the whole range of j_0, nor the relations for the profiles of basic parameters across the CCL. Eikerling and Kornyshev [7] used a similar approach and derived an analytical solution in the case of small overpotentials. In the general case they presented numerical results.

The CCL is a composition of carbon threads, catalyst particles, polymer electrolyte and voids for gas transport. An effective oxygen diffusion coefficient in this system is poorly known. Fortunately, modern catalyst layers are usually thin (typically less than 10 μm). This enables us to assume that the variation of oxygen concentration across the CCL is not large (the limit of fast oxygen diffusion). In this limit, Eikerling and Kornyshev obtained analytical solution in parametric form. However, their parametric solution is cumbersome and difficult to deal with. A transparent result, which reveals important features of CCL function can be obtained as follows.

6.2.1.1. Basic Equations

Consider a CCL of a thickness l_t. Let the x-axis with the origin at the membrane/catalyst layer interface be directed across the CCL (Figure 6.4). This Figure illustrates a typical profile of the overpotential $\eta(x)$ in the active layer. The proton current density entering the CCL from the membrane is j_0. The voltage loss required to convert j_0 into electron current is $\eta(0) \equiv \eta_0$ (Figure 6.4). In this section, the subscript "0" denotes the values at $\tilde{x} = 0$ (Figure 6.4). Our goal is the calculation of the polarization curve $j_0(\eta_0)$, a function which relates the boundary values of the proton current j and the overpotential η in the problem of CCL performance.

Assuming that the variation of oxygen concentration across the catalyst layer is not large, the problem is reduced to two equations:

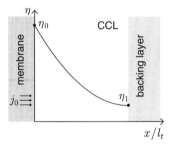

FIGURE 6.4. Sketch of the problem of CCL performance and typical distribution of overpotential across the CCL.

$$\frac{\partial j}{\partial x} = -Q(x) \quad (6.1)$$

$$j = -\sigma_t \frac{\partial \eta}{\partial x}. \quad (6.2)$$

The first equation expresses the decay of j along the x-axis at a rate $Q(x)$ due to proton consumption in electrochemical reaction. The second equation is Ohm's law, where $\eta = \varphi_m - \varphi_c$ has been substituted for φ_m, since the potential of the carbon phase φ_c is nearly constant.[5]

For the rate of ORR, we employ the Tafel law with a first-order dependence on oxygen concentration:

$$Q = i_* \left(\frac{c_t}{c_{ref}}\right) \exp\left(\frac{\eta}{b}\right) \quad (6.3)$$

Here, i_* the is exchange current density (per unit volume), c_t and c_{ref} are available in the catalyst layer and reference oxygen molar concentration, respectively, b is the Tafel slope. Introducing the dimensionless variables

$$\tilde{x} = \frac{x}{l_t}, \quad \tilde{j} = \frac{j}{j_*}, \quad \tilde{\eta} = \frac{\eta}{b}, \quad \tilde{Q} = \frac{l_t Q}{j_*} \quad (6.4)$$

where

$$j_* = \frac{2\sigma_t b}{l_t} \quad (6.5)$$

---

[5] Strictly speaking, variation of φ is negligible if $\eta \gg \varphi$. This condition is not fulfilled in the anode catalyst layer, where η can be on the order of φ. In that case η "feels" the variation of φ, which leads to two-dimensional effects [8].

is the characteristic current density (see below), Eqs. (6.1–6.3) take the form

$$\frac{\partial \tilde{j}}{\partial \tilde{x}} = -\tilde{Q} \tag{6.6}$$

$$2\tilde{j} = -\frac{\partial \tilde{\eta}}{\partial \tilde{x}} \tag{6.7}$$

$$\tilde{Q} = \tilde{k}_t \exp(\tilde{\eta}). \tag{6.8}$$

We see that behaviour of the system is controlled by a single parameter

$$\tilde{k}_t = \frac{k_t}{j_*} = \frac{i_* l_t}{j_*}\left(\frac{c_t}{c_{ref}}\right), \tag{6.9}$$

where

$$k_t = i_* l_t \left(\frac{c_t}{c_{ref}}\right). \tag{6.10}$$

Using (6.5), $\tilde{k}_t$ can be written as

$$\tilde{k}_t = \left(\frac{l_t}{l_*}\right)^2 \left(\frac{c_t}{c_{ref}}\right) \tag{6.11}$$

where

$$l_* = \sqrt{\frac{2\sigma_t b}{i_*}} \tag{6.12}$$

is the reaction penetration depth, introduced by Newman [1]. The physical meaning of l_* will be clarified below.

6.2.1.2. First integral: A Conservation Law

Multiplying together (6.6) and (6.7), and taking into account (6.8), we get

$$2\tilde{j}\frac{\partial \tilde{j}}{\partial \tilde{x}} = \tilde{k}_t \exp(\tilde{\eta})\frac{\partial \tilde{\eta}}{\partial \tilde{x}},$$

or

$$\frac{\partial \tilde{j}^2}{\partial \tilde{x}} = \frac{\partial \tilde{Q}}{\partial \tilde{x}}. \tag{6.13}$$

Integrating this equation from 0 to $\tilde{x}$ we find

$$-\tilde{Q}(\tilde{x}) + \tilde{j}^2(\tilde{x}) = -\tilde{Q}_0 + \tilde{j}_0^2, \qquad (6.14)$$

where $\tilde{Q}_0 \equiv \tilde{Q}(0)$. Eq. (6.14) is, therefore, a conservation law, which says that the sum on the left side is constant along $\tilde{x}$.

6.2.1.3. The Profile of Proton Current Across the Catalyst Layer

Taking into account (6.6), we can rewrite (6.14) as

$$\frac{\partial \tilde{j}}{\partial \tilde{x}} + \tilde{j}^2 = -\left(\tilde{Q}_0 - \tilde{j}_0^2\right) \qquad (6.15)$$

The right-hand side of (6.15) is essentially negative [9]. The solution to (6.15), subject to the boundary condition $\tilde{j}|_{\tilde{x}=1} = 0$, is

$$\tilde{j}(\tilde{x}) = \sqrt{\tilde{Q}_0 - \tilde{j}_0^2} \tan\left((1-\tilde{x})\sqrt{\tilde{Q}_0 - \tilde{j}_0^2}\right) \qquad (6.16)$$

This is the explicit form of proton current density profile across the catalyst layer. Figure 6.5 shows the profiles $\tilde{j}(\tilde{x})$ for three values of mean current density $\tilde{j}_0$. When $\tilde{j}_0$ is small, the value of $\sqrt{\tilde{Q}_0 - \tilde{j}_0^2}$ is small and the tangent function reduces to linear dependence. If $\tilde{j}_0$ is large, protons are consumed mainly near the membrane. In that case, the characteristic thickness of the current-generating domain is given by reaction penetration depth l_* (6.12), as discussed below.

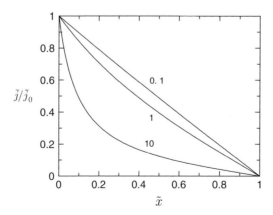

FIGURE 6.5. The profiles of proton current density across the catalyst layer for three indicated values of dimensionless current density $\tilde{j}_0$. The membrane is at $\tilde{x} = 0$.

6.2.1.4. Voltage-Current Curve

At $\tilde{x} = 0$ (6.16) gives

$$\tilde{j}_0 = \sqrt{\tilde{Q}_0 - \tilde{j}_0^2} \, \tan\left(\sqrt{\tilde{Q}_0 - \tilde{j}_0^2}\right) \qquad (6.17)$$

which together with (6.8) forms a general implicit relation for the polarization curve of the catalyst layer $\eta_0(j_0)$.

In the limits of small and large current (6.17) can be simplified. Small $\tilde{j}_0$ ($\tilde{j}_0 \ll 1$) means that the product on the right side of (6.17) is small. The function $\omega \tan \omega$ is monotonous for $\omega > 0$, thus $\tan \omega$ is also small and $\tan \omega \simeq \omega$. This gives $\tilde{j}_0 = \tilde{Q}_0 - \tilde{j}_0^2$; neglecting $\tilde{j}_0^2$ we finally obtain

$$\tilde{j}_0 = \tilde{Q}_0. \qquad (6.18)$$

With (6.8) we can rewrite (6.18) as:

$$\tilde{\eta}_0 = \ln \tilde{j}_0 - \ln \tilde{k}_t, \quad \tilde{j}_0 \ll 1. \qquad (6.19)$$

Physically, when $\tilde{j}_0 \ll 1$ the overpotential and reaction rate are almost constant across the catalyst layer (see next section), and the current density j_0 (6.18) is simply a product of reaction rate (6.3) and CCL thickness l_t.

In the opposite limit, $\tilde{j}_0 \gg 1$, the product $\omega \tan \omega$ on the right side of (6.17) is large, hence the argument of the tan-function must approach $\pi/2$. We set $\sqrt{\tilde{Q}_0 - \tilde{j}_0^2} = \pi/2$ or equivalently $\tilde{j}_0 = \sqrt{\tilde{Q}_0 - \pi^2/4}$. The latter relation shows that $\tilde{Q}_0$ is also large. We thus may neglect $\pi^2/4$ under the square root, and in the leading order we find

$$\tilde{j}_0 = \sqrt{\tilde{Q}_0}. \qquad (6.20)$$

Using (6.8) we get the voltage-current curve

$$\tilde{\eta}_0 = 2 \ln \tilde{j}_0 - \ln \tilde{k}_t, \quad \tilde{j}_0 \gg 1 \qquad (6.21)$$

Compared to (6.19), the first logarithm on the right side of (6.21) contains factor 2, which is equivalent to twice larger Tafel slope (see also [5,7]).

Relations (6.19) and (6.21) show that transition from small to large currents is accompanied by increase in apparent Tafel slope from b to $2b$. This effect is illustrated in Figure 6.6. The transition from $\tilde{j}_0 = \tilde{Q}_0$ to $\tilde{j}_0 = \sqrt{\tilde{Q}_0}$ as $\tilde{j}_0$ changes from small to large values is clearly seen in log-log coordinates. Near $\tilde{j}_0 \simeq 1$, the transition region described by exact equation (6.17) couples two limiting straight lines.[6]

---

[6] Explicit expressions $\tilde{j}_0(\tilde{\eta}_0)$, which approximate Eq. (6.17) with 2% accuracy are derived in Appendix B.

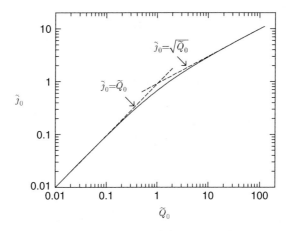

FIGURE 6.6. "Reaction rate-current" curve of the catalyst layer. Note the transition from $\tilde{j}_0 = \tilde{Q}_0$ to $\tilde{j}_0 = \sqrt{\tilde{Q}_0}$ near $\tilde{j}_0 = 1$.

It is advisable to write down polarization curves (6.19) and (6.21) in dimension form:

$$\eta_0 = b \ln\left(\frac{j_0}{l_t i_*}\right) - b \ln\left(\frac{c_t}{c_{ref}}\right), \quad j \ll j_* \tag{6.22}$$

$$\eta_0 = 2b \ln\left(\frac{j_0}{l_* i_*}\right) - b \ln\left(\frac{c_t}{c_{ref}}\right), \quad j \gg j_* \tag{6.23}$$

Equation (6.22) contains CCL thickness l_t, whereas Eq. (6.23) contains reaction penetration depth l_* (6.12).

Physically, in the low-current regime all parts of the CCL equally contribute to proton current conversion. In the large-current regime the characteristic thickness of a current-generating domain is given by internal scale of the problem l_*. In that case, proton transport gives significant contribution to voltage loss, and the reaction is concentrated close to the membrane, where protons are "cheaper". Figure 6.5 illustrates the effect.

6.2.1.5. Distribution of Overpotential and Reaction Rate Across the Catalyst Layer

All profiles across the catalyst layer depend only on one parameter (e.g., j_0). The other parameters (Q_0, η_0) may be expressed in terms of j_0 by means of voltage-current relation (6.17). Integrating (6.7) with $\tilde{j}$ (6.16), we obtain the shape of overpotential across the catalyst layer

$$\tilde{\eta}(x) = \tilde{\eta}_0 - 2\ln\left(\frac{1 + \tan(\omega)\tan(\omega\tilde{x})}{\sqrt{1+\tan^2(\omega\tilde{x})}}\right), \quad \omega \equiv \sqrt{\tilde{Q}_0 - \tilde{j}_0^2}, \quad (6.24)$$

where $\tilde{\eta}_0$ and $\tilde{Q}_0$ are related by (6.8). Equation (6.24) gives a useful relation for overpotential drop across the catalyst layer:

$$\delta\tilde{\eta} \equiv \tilde{\eta}_0 - \tilde{\eta}_1 = -2\ln\left(\cos\sqrt{\tilde{Q}_0 - \tilde{j}_0^2}\right). \quad (6.25)$$

According to (6.17), for small currents $\sqrt{\tilde{Q}_0 - \tilde{j}_0^2}$ is small, the cosine under the logarithm is close to 1 and $\delta\tilde{\eta}$ is close to zero. For large $\tilde{j}_0$, $\sqrt{\tilde{Q}_0 - \tilde{j}_0^2}$ tends to $\pi/2$, cosine tends to zero and variation of overponential tends to infinity. Figure (6.7) displays $\eta(\tilde{x})$ for small, intermediate and large $\tilde{j}_0$.

The shape of the rate of electrochemical reaction can be determined from (6.6). With $\tilde{j}$ (6.16) we get

$$\tilde{Q}(\tilde{x}) = (\tilde{Q}_0 - \tilde{j}_0^2)\left[1 + \tan^2\left((1-\tilde{x})\sqrt{\tilde{Q}_0 - \tilde{j}_0^2}\right)\right] \quad (6.26)$$

The profile $\tilde{Q}(\tilde{x})$ for $\tilde{j}_0 = 10$ is shown in Figure 6.7. The variation $\delta\tilde{Q} = \tilde{Q}_0 - \tilde{Q}_1$ of reaction rate across the catalyst layer qualitatively follows $\delta\tilde{\eta}$: it is small for small currents and tends to infinity as $\tilde{j}_0 \to \infty$.

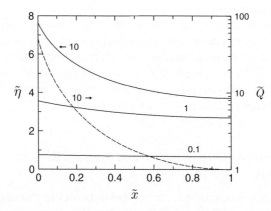

FIGURE 6.7. Profiles of dimensionless overpotential across the catalyst layer for three indicated values of dimensionless current density $\tilde{j}_0$. Dashed line - reaction rate $\tilde{Q}$ for $\tilde{j}_0 = 10$. The membrane is at $\tilde{x} = 0$.

6.2.1.6. Typical j_*

The regime of the CCL operation is determined by j_* (6.5). For typical operating conditions of a well-humidified PEFC cathode (T = 353 K, $\sigma_t \simeq 0.01 \Omega^{-1} \text{cm}^{-1}$, $l_t = 10^{-3}$ cm, $b \simeq 0.05$ V), this current is $j_* \simeq 1$ A cm^{-2}. Working current densities in a PEFC are also on the order of 1 A cm^{-2}. Thus, the CCL typically operates in the intermediate regime $\tilde{j}_0 \simeq 1$, when its voltage-current curve is given by the general relation (6.17).

6.2.2. Voltage Loss Due to Oxygen Transport

Hereafter, the subscript "0" in the symbols $\tilde{j}_0$ and $\tilde{\eta}_0$ will be omitted. The real polarization curve of the cell rapidly drops when the current reaches a certain limiting value. The limitation of current can be caused by various mechanisms. Lack of water may cause membrane drying; the relation for the respective limiting current will be derived in Section 6.4. If the cell is run at a constant inlet flow rate of oxygen, limiting current is attained when oxygen stoichiometry decreases to 1 [10,11]. Most often, however, current limitation is caused by poor oxygen transport to the catalyst sites. To calculate the respective voltage loss consider the following simple model (Figure 6.8).

Let oxygen concentration in the channel c_h be fixed. As before, we will assume that the oxygen concentration in CCL c_t is constant (Figure 6.8). Evidently, the slope of the oxygen concentration profile across the GDL depends on current density, the larger j the larger the slope. At the limiting current, c_t is zero and further growth of current is not possible.

We will describe oxygen flux through the GDL by Fick's formula with the effective diffusion coefficient D_b. Physically, in a dry GDL, D_b corresponds to a binary diffusion coefficient of oxygen corrected for porosity-tortuosity. When the GDL is partially flooded D_b should be corrected for liquid saturation. A more rigorous approach is based on Stefan-Maxwell relations, which take into account the effect of fluxes of the other mixture components on

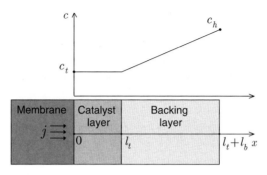

FIGURE 6.8. Sketch of the problem and the shape of oxygen concentration in the catalyst and backing layers.

oxygen transport. However, in many cases Fick's formula provides quite reasonable accuracy. Furthermore, this formula gives a transparent physical picture of the process, which is our primary goal.

Mass conservation prescribes that the diffusion flux of oxygen in the GDL be proportional to the proton current density j in the membrane (Figure 6.8). We, therefore, have

$$D_b \frac{\partial c}{\partial x} = \frac{j}{4F}. \tag{6.27}$$

Here c is the local oxygen molar concentration in the GDL. Solving this equation with the boundary condition $c|_{x=l_t+l_b} = c_h$ (Figure 6.8) and substituting $x = l_t$ into the solution, we obtain c_t:

$$c_t = c_h - \frac{l_b j}{4FD_b} = c_h \left(1 - \frac{j}{j_D}\right) \tag{6.28}$$

where

$$j_D = \frac{4FD_b c_h}{l_b} \tag{6.29}$$

is a limiting current density. Indeed, when $j = j_D$ we have $c_t = 0$.

The value of c_t appears in the expression for the rate of ORR (6.3). We, therefore, can immediately obtain the general polarization curve *of the cathode side* by simply replacing c_t in the general voltage current curve of the CCL (6.17) with (6.28). Substituting (6.28) into (6.17) and omitting the subscript "0" we get [12]

$$\tilde{j}_0 = \omega \tan \omega, \quad \omega = \sqrt{\tilde{k}_h \left(1 - \frac{\tilde{j}}{\tilde{j}_D}\right) \exp(\tilde{\eta}) - \tilde{j}^2}. \tag{6.30}$$

Here

$$\tilde{k}_h = \frac{i_* l_t}{j_*} \left(\frac{c_h}{c_{ref}}\right) \tag{6.31}$$

contains the oxygen concentration in the channel. Substituting (6.28) into (6.19) and (6.21), and omitting again the subscript "0", in the limiting cases of small and large currents, we find [12]

$$\tilde{\eta} = \ln \tilde{j} - \ln \tilde{k}_h - \ln\left(1 - \frac{\tilde{j}}{\tilde{j}_D}\right) \quad \tilde{j} \ll 1 \tag{6.32}$$

$$\tilde{\eta} = 2\ln \tilde{j} - \ln \tilde{k}_h - \ln\left(1 - \frac{\tilde{j}}{\tilde{j}_D}\right), \quad \tilde{j} \gg 1 \tag{6.33}$$

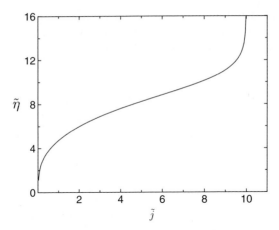

FIGURE 6.9. Polarization curve of the cathode side (6.35). Parameters: $\tilde{k}_h = 10^{-2}$, $\tilde{j}_D = 10$.

Compared to (6.19) and (6.21), Eqs. (6.32), (6.33) contain a new term $\ln(1 - \tilde{j}/\tilde{j}_D)$. Clearly, $-\ln(1 - \tilde{j}/\tilde{j}_D) \to \infty$ as $\tilde{j} \to \tilde{j}_D$, that is, at the limiting current voltage loss is infinite.

Using a simple matching function

$$\phi(t) = 1 + \frac{t}{1+t} \tag{6.34}$$

Eqs. (6.32) and (6.33) can be combined in one

$$\tilde{\eta} = \phi(\tilde{j}) \ln \tilde{j} - \ln \tilde{k}_h - \ln\left(1 - \frac{\tilde{j}}{\tilde{j}_D}\right), \tag{6.35}$$

which approximates the exact implicit relation (6.30) in the whole range of currents (Figure 6.9). The function ϕ changes from 1 to 2 as its argument varies from small to large values. If oxygen stoichiometry is large, and the cell is properly humidified, Eq. (6.35) may be used for analysis of experimental polarization curves.

Usually, however, oxygen stoichiometry λ varies between 1.5 and 3, since larger λ reduces overall efficiency of the fuel cell system. Equation (6.35) can be modified to take into account the effect of finite λ. To do this, we have to consider the second dimension of the problem: variation of oxygen concentration along the channel.

6.3. 1D + 1D Model. I. Ideally Humidified Membrane

Now we turn to the construction of quasi-2D (or 1D + 1D) models of a PEFC. The system of coordinates is shown in Figure 6.3. Following the idea discussed in Introduction, we neglect the fluxes along the z-axis in the

GDL, CCL and membrane. In the channel, we will assume plug flow conditions, i.e., well-mixed flow with constant velocity directed along the z-axis. The rationale for this assumption is as follows.

6.3.1. Flow Velocity in the Channel

Analysis of hydrodynamic equations for the flow in the fuel cell channel shows that this flow is incompressible [13]. In other words, the variation of pressure (total molar concentration) along the channel is small.[7] Consider first the case of zero water flux through the membrane. Each oxygen molecule in the cathode channel is replaced with two water molecules. Pressure is proportional to the number of molecules per unit volume. To support constant pressure, the flow velocity in the channel must increase. The growth of velocity provides expansion of elementary fluid volume; the expansion keeps pressure in this volume constant.

Crossover of water from the anode to the cathode side supplies additional water molecules to the cathode channel and induces faster growth of velocity. Let α be the overall transfer coefficient of water from the anode to the cathode side (a number of water molecules transported per each proton through the membrane, taking into account back diffusion). Calculations [13] give

$$\frac{v(z)}{v^0} = 1 + (1 + 18\alpha) \int_0^{z/h} \frac{j}{FN^0} \, d(z/h) \qquad (6.36)$$

where $j(z/h)$ the is local current density, $N^0 = \rho^0 v^0 / M_H$, v^0 and ρ^0 are inlet velocity and mass density of the flow, M_H is molecular weight of a hydrogen atom, h is the channel height.

The estimate shows that for typical conditions and $\alpha = 0$ the growth of velocity is small. For $\alpha = 0.5$ this growth is about 60%. Usually α does not exceed 0.2 [14]; thus the assumption of constant velocity is a quite reasonable approximation.

6.3.2. The Effect of Finite Oxygen Stoichiometry

Throughout this section we will assume that the cell is run in the low-current regime ($\tilde{j} \ll 1$), so that the polarization voltage of the cathode side is given by (6.32). In the limiting case of $\tilde{j} \gg 1$ and in the general case of arbitrary $\tilde{j}$ the results cannot be obtained in a closed form; the approximate relation for the general case is discussed below.

---

[7] This is correct unless a very long and/or thin channels are used.

6.3.2.1. The Shapes of Oxygen and Local Current Along the Channel

Following the general procedure, we now treat oxygen concentration in the channel c_h and current density $\tilde{j}$ in (6.32) as *local* values, which depend on $\tilde{z}$. These local values obey the following equations. c_h is governed by mass balance equation

$$v^0 \frac{\partial c_h}{\partial z} = -\frac{j}{4Fh}, \qquad (6.37)$$

which describes oxygen consumption at a rate proportional to local current density $j(z)$. The equation for $j(z)$ follows from the condition of equipotentiality of cell electrodes. In the case of ideally humidified membrane this condition reads

$$\tilde{\eta} = \tilde{E}, \qquad (6.38)$$

where $\tilde{\eta}$ is given by (6.32) and $\tilde{E} = E/b$ is constant along $\tilde{z}$.

Equation (6.38) stems from the following arguments. The cell voltage is

$$V_{cell} = V_{oc} - \eta - V_m - RJ \qquad (6.39)$$

where V_m is the voltage loss in membrane, J is the mean current density in a cell and R accumulates all contact resistances. V_{oc} and RJ in (6.39) do not depend on z. If membrane is well humidified, V_m is negligible and thus $\tilde{\eta} =$const along $\tilde{z}$.

In dimensionless variables, Eq. (6.37) is

$$\lambda \tilde{J} \frac{\partial \tilde{c}_h}{\partial \tilde{z}} = -\tilde{j}, \quad \tilde{c}_h(0) = 1 \qquad (6.40)$$

Here

$$\lambda = \frac{4F c_h^0 v^0 h}{LJ} \qquad (6.41)$$

is oxygen stoichiometry, L is the channel length,

$$\tilde{z} = \frac{z}{L}, \quad \tilde{c}_h = \frac{c_h}{c_h^0}, \quad \tilde{J} = \frac{J}{j_*} \qquad (6.42)$$

and the superscript "0" marks the values at the channel inlet (at $\tilde{z} = 0$). The solution to the system (6.40), (6.38), (6.32) is

$$\tilde{c}_h(\tilde{z}) = \left(1 - \frac{1}{\lambda}\right)^{\tilde{z}} \qquad (6.43)$$

$$\tilde{j}(\tilde{z}) = f_\lambda \tilde{J}\left(1 - \frac{1}{\lambda}\right)^{\tilde{z}}, \qquad (6.44)$$

where

$$f_\lambda = -\lambda \ln\left(1 - \frac{1}{\lambda}\right). \qquad (6.45)$$

Eqs. (6.43) and (6.44) show that $\tilde{c}_h$ and $\tilde{j}/\tilde{J}$ are universal functions of $\tilde{z}$, which are controlled by a single parameter λ. These functions are depicted in Figure 6.10. Physically, lower λ corresponds to faster oxygen consumption along $\tilde{z}$. Note, that the ratio $\tilde{j}/\tilde{c}_h = f_\lambda \tilde{J}$ does not depend on $\tilde{z}$. Furthermore, the activation and transport losses in (6.32) are constant along $\tilde{z}$. This is seen if we rewrite Eq. (6.32) as

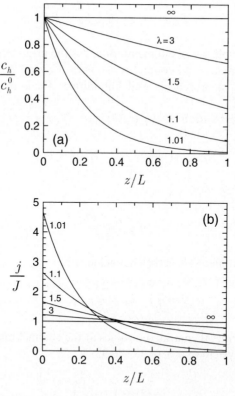

FIGURE 6.10. The profiles of (a) oxygen concentration and (b) local current density along the channel for indicated values of oxygen stoichiometry λ.

$$\tilde{\eta} = \ln\left(\frac{\tilde{j}}{\tilde{k}_h}\right) - \ln\left(1 - \frac{\tilde{j}}{\tilde{j}_D}\right). \tag{6.46}$$

Since $\tilde{k}_h \sim \tilde{c}_h$ and $\tilde{j}_D \sim \tilde{c}_h$, we see that with (6.43), (6.44) the ratios $\tilde{j}/\tilde{k}_h$ and $\tilde{j}/\tilde{j}_D$ do not depend on $\tilde{z}$, and thus both terms on the right side of (6.46) are constant along $\tilde{z}$.

Figure 6.11a shows experimental results from Kučernak's group [15,16]. The distribution of local current density along the channel is given for constant cell potential but different inlet flow rates, which correspond to different oxygen stoichiometries. Figure 6.11b shows the same results replotted in dimensionless form. The agreement with the theory is very good for the entire length of the channel. Note, that in this

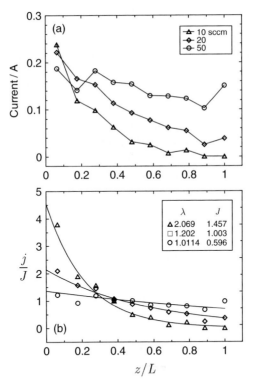

FIGURE 6.11. Experimental (points) and theoretical (curves in (b)) profiles of local current density for indicated values of oxygen stoichiometry. Experimental current distribution curves were obtained along a single flow channel (1 mm × 1 mm cross section) at a cell potential of 0.492 V and cell temperature of 30°C. For further details of experiment please see [16].

experiment special measures have been taken to keep the membrane well humidified.

6.3.2.2. Voltage Current Curve

Substitution of (6.43) and (6.44) into (6.32) yields a simple and elegant formula for polarization voltage of the cathode side [17]:

$$\tilde{\eta} = \ln(f_\lambda \tilde{J}) - \ln \tilde{k}_h^0 - \ln\left(1 - \frac{f_\lambda \tilde{J}}{\tilde{j}_D^0}\right), \quad f_\lambda \tilde{J} \ll 1 \quad (6.47)$$

where

$$\tilde{k}_h^0 = \frac{i_* l_t}{j_*}\left(\frac{c_h^0}{c_{ref}}\right) \quad (6.48)$$

$$\tilde{j}_D^0 = \frac{4FD_b c_h^0}{l_b}. \quad (6.49)$$

Comparing (6.47) and (6.32) we see that the effect of finite λ reduces to a rescaling of current by a factor f_λ. In particular, the limiting current density in Eq. (6.47) appears to be f_λ times lower than in Eq. (6.32). The function f_λ is depicted in Figure 6.12a. It tends to infinity as $\lambda \to 1$ and it tends to 1 as $\lambda \to \infty$. At large λ, $f_\lambda \simeq 1$ and Eq. (6.47) reduces to (6.32) ($\tilde{j} \equiv \tilde{J}$ in that case). However, as $\lambda \to 1$, f_λ grows and the apparent limiting current density $\tilde{j}_D^0/f_\lambda$ decreases (Figure 6.12b). The reason for this decrease is the nonuniformity of oxygen concentration along the channel (Figure 6.10a). Physically, the total limiting current density is the average of local limiting currents (6.29) over $\tilde{z}$.

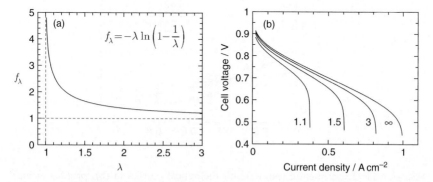

FIGURE 6.12. (a) Function f_λ, (b) the effect of indicated λ on voltage current curve. Infinite λ corresponds to uniform distribution of oxygen concentration $\tilde{c}(\tilde{z})$ along the channel. As λ decreases, the non-uniformity of $\tilde{c}(\tilde{z})$ increases (Figure 6.10), and apparent limiting current decreases by a factor f_λ.

Oxygen concentration c_h in (6.29) exponentially decreases with z, and the average value of local limiting current appears to be j_D^0/f_λ, as Eq. (6.47) shows. This can be obtained also by direct averaging of (6.29) over z with $c_h = c_h^0(1 - 1/\lambda)^{z/L}$.

In Section 6.2.2, we have shown that the general expression for local polarization curve can be approximated by Eq. (6.35). By analogy with (6.35), we may write down the following generalization of Eq. (6.47), valid for all currents in the cell:

$$\tilde{\eta} = \phi(f_\lambda \tilde{J}) \ln(f_\lambda \tilde{J}) - \ln \tilde{k}_h^0 - \ln\left(1 - \frac{f_\lambda \tilde{J}}{\tilde{j}_D^0}\right), \qquad (6.50)$$

In case of low currents ($f_\lambda \tilde{J} \ll 1$, $\phi = 1$), this equation reduces to (6.47). In case of large currents ($f_\lambda \tilde{J} \gg 1$, $\phi = 2$) and large λ, this equation reduces to (6.35), since in that case $f_\lambda \to 1$. In case of $f_\lambda \tilde{J} \simeq 1$, $\lambda \gtrsim 1$, Eq. (6.50) should be considered as a semi-phenomenological extension of (6.47).

6.3.3. Local Polarization Curves

In many situations, the PEFC is run at a constant inlet flow rate of oxygen rather than at a constant stoichiometry. In that case, it is convenient to rewrite Eq. (6.37) as

$$\beta \frac{\partial \xi_h}{\partial z} = -\frac{j}{j_D^0} \qquad (6.51)$$

Here, $\xi_h = c_h/c_{tot}$ is the oxygen molar fraction, c_{tot} is the total molar concentration of the flow and

$$\beta = \frac{l_b h v^0}{D_b L}. \qquad (6.52)$$

Equation (6.51) is convenient, since under constant v^0 parameter β is also constant. In this case, the cell performance is determined by the system of Eqs. (6.51), (6.38), (6.32), which reveals an interesting effect.

As before, solutions to this system are exponentials [16]:

$$\xi_h = \xi_h^0 \exp\left(-\frac{a_\eta \tilde{z}}{\beta}\right) \qquad (6.53)$$

$$j = j_D^0 a_\eta \exp\left(-\frac{a_\eta \tilde{z}}{\beta}\right) \qquad (6.54)$$

where a_η is a function of polarization voltage

$$a_\eta = \frac{1}{1 + \frac{1}{K}\exp(-\tilde{\eta})} \qquad (6.55)$$

and K is constant:

$$K = \frac{l_t l_b i_*}{4 F D_b c_{ref}}. \qquad (6.56)$$

Eqs. (6.54), (6.55) determine local polarization curve $\tilde{\eta}(\tilde{j})$ at $\tilde{z}$.

The set of curves for different distances along the channel is shown in Figure 6.13, together with the experimental curves (A. Kučernak) [16]. We see that this simple model qualitatively well reproduces the experimental picture. Close to the channel inlet the curves are monotonous, whereas close to the outlet they exhibit distinct maximum (Figure 6.13). These maxima result from the effect of "oxygen starvation".

Physically, operation at a constant inlet flow rate is equivalent to operation at variable $\lambda \sim 1/J$. Figure 6.10a may thus be treated as the evolution of oxygen profile under growing J (smaller λ corresponds to larger J). Consider some fixed point, e.g. $\tilde{z}_0 = 0.8$ in Figure 6.10a. As long as $\lambda \gg 1$, oxygen concentration is practically constant at this point, and the local current exponentially increases with $\tilde{\eta}$, as prescribed by Tafel law. At some η, however, λ becomes comparable to 1, and oxygen consumption *upstream from* $\tilde{z}_0$ is no longer negligible. Further growth of $\tilde{\eta}$ leads to decrease of local current at $\tilde{z}_0$ due to rapid oxygen consumption at $\tilde{z} < \tilde{z}_0$.

This S-shape effect shows up only in the case of constant inlet velocity. It is easy to show that in the case of constant λ local polarization curves are monotone.

6.3.4. *Dynamics of Cell Performance Degradation*

Fuel cells experience ageing. Proton current, flux of water and thermal effects inevitably change the structure and composition of cell components, thus causing the degradation of cell performance. The phenomenon of cell ageing is well known, though the mechanisms of ageing are not fully understood [18].

The main factors that determine cell performance are (i) kinetics of electrochemical reactions, (ii) conductivity of membrane and (iii) transport properties of the catalyst and backing layers. Ageing, therefore, may be thought of as irreversible change of one (or several) kinetic or transport parameters. In this section, we show that quasi-2D model makes it possible to predict general scenarios of the degradation process not specifying its microscopic origin.

In life test experiments and in the stacks, the cell is run under fixed total current (galvanostatic regime). In this regime, the degradation manifests itself

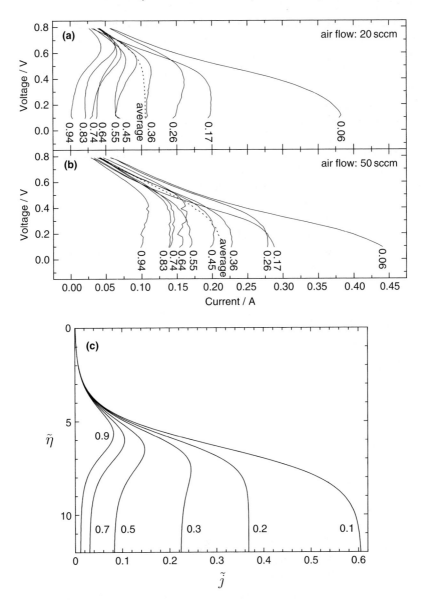

FIGURE 6.13. (a), (b) Experimental voltage current curves at different points along a single flow channel (1 mm×1 mm cross section) on the cathode side of a fuel cell, for an air flow rate of (a) 20 sccm and (b) 50 sccm. Hydrogen flow rate on anode of 25 sccm. Cell temperature of 30°C. Dimensionless distances along the channel are listed below each curve. Dotted line, average cell performance. (c) Theoretical local polarization curves for indicated dimensionless distances from the channel inlet.

as a decrease in cell voltage with time. This decrease is not linear: a gradual change is usually followed by a very rapid drop to zero (see e.g. [19]). In our experiments we observed a similar behaviour [20].

Qualitatively, we may expect that the rate of degradation is higher in the regions, where local current density j is high. We will assume that this rate is a stepwise function of j; in other words, there exists critical current density j_{crit}, at which rate of degradation jumps from zero to a certain finite value. Suppose that at time t_0 the local current density exceeds j_{crit} (Figure 6.14). The dashed domain at t_0 (Figure 6.14) is then subjected to local degradation. Let the characteristic time of local degradation be τ_d, i.e., when τ_d expires, the region where $j > j_{crit}$ does not produce current. At $t_1 = t_0 + \tau_d$, the peak of the local current density shifts to a new position (Figure 6.14), and a new domain is "exposed" to degradation. Since total current is fixed, the length of this domain increases with time (Figure 6.14). Clearly, this mechanism provides propagation of the degradation wave (DW). In this section, we will obtain an equation for the DW propagation and investigate the features of these waves [20].

6.3.4.1. Dynamics of the Degradation Wave

Let $z_w(t)$ be the instant position of the wavefront (vertical lines in Figure 6.14). Simple generalization of Eq. (6.44) for the case when current-generating domain has the length $L - z_w$ is

$$j(z) = \begin{cases} \left(\dfrac{L}{L-z_w}\right) f_\lambda J \left(1 - \dfrac{1}{\lambda}\right)^{\frac{z-z_w}{L-z_w}}, & z \geq z_w \\ 0, & z < z_w. \end{cases} \quad (6.57)$$

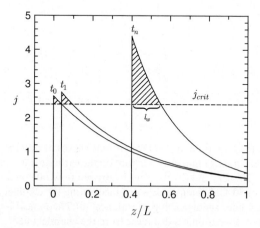

FIGURE 6.14. Sketch of the degradation wave. The domain where local current density exceeds critical value (dashed areas at $t_0, t_1, \ldots, t_n$) is subject to degradation. The degradation wave propagates toward $z/L = 1$.

Considering the practically important case of fixed total current I_{tot} and λ (galvanostatic regime under constant oxygen stoichiometry), the profile (6.57) provides invariance of I_{tot} with time. Indeed,

$$I_{tot} = d_h \int_0^L j(z)\, dz = d_h \int_{z_w}^L j(z)\, dz = d_h LJ, \qquad (6.58)$$

which does not depend on z_w. Here d_h is the in-plane width of the channel.

The length of the domain exposed to degradation l_w (Figure 6.14) is determined by condition $j(z_w + l_w) = j_{crit}$, or

$$\frac{Jf_\lambda}{1 - \tilde{z}_w}\left(1 - \frac{1}{\lambda}\right)^{\frac{\tilde{l}_w}{1 - \tilde{z}_w}} = j_{crit} \qquad (6.59)$$

where dimensionless variables are

$$\tilde{l}_w = \frac{l_w}{L}, \quad \tilde{z}_w = \frac{z_w}{L}.$$

Solving Eq. (6.59) for $\tilde{l}_w$, we get

$$\tilde{l}_w = \frac{(1 - \tilde{z}_w)\ln\left(\frac{j_{crit}(1 - \tilde{z}_w)}{f_\lambda J}\right)}{\ln\left(1 - \frac{1}{\lambda}\right)}. \qquad (6.60)$$

Since the logarithm in denominator is negative ($\lambda > 1$), the logarithm in numerator must also be negative. At $t = 0$, we have $\tilde{z}_w = 0$; the condition of DW generation is then

$$a \equiv \frac{f_\lambda J}{j_{crit}} > 1, \quad \text{or} \quad f_\lambda J > j_{crit} \qquad (6.61)$$

Putting in (6.57) $z = z_w = 0$ we get $f_\lambda J = j(0)$. Relation (6.61), therefore, is equivalent to

$$j(0)|_{t=0} > j_{crit} \qquad (6.62)$$

In other words, the wave is initiated when the maximum of local current density exceeds j_{crit}.

The velocity of the wave is $v_w = \frac{l_w}{\tau_d}$, and the wave front propagates according to the equation $\partial z_w / \partial t = v_w$ or

$$\frac{\partial \tilde{z}_w}{\partial \tilde{t}} = -(1 - \tilde{z}_w)\ln\left(\frac{1 - \tilde{z}_w}{a}\right), \quad \tilde{z}_w|_{\tilde{t}=0} = 0, \qquad (6.63)$$

where dimensionless time is

$$\tilde{t} = \frac{t}{\tau_w}, \quad \tau_w = -\tau_d \ln\left(1 - \frac{1}{\lambda}\right). \tag{6.64}$$

The solution of (6.63) is

$$\tilde{z}_w = 1 - a \exp[-\ln(a) \exp \tilde{t}]. \tag{6.65}$$

We see that in dimensionless variables, wave propagation is controlled by single parameter a.[8] The dependence $\tilde{z}_w(\tilde{t})$ is depicted in Figure 6.15. This Figure shows that when ln (a) is not small, the wave moves very fast (as a double exponent of time, Eq. (6.65)). If, however, $\ln a \to 0 (a \to 1)$ we have two distinct phases of wave propagation. Near $\tilde{z} = 0$ the wave moves very slowly; this slow phase is followed by the fast propagation phase, when velocity reaches the maximum (Figure 6.15).

The duration of the slow propagation phase can be estimated as follows. Differentiating (6.65), we get the dimensionless wave velocity $\tilde{v}_w = \partial \tilde{z}_w / \partial \tilde{t}$:

$$\tilde{v}_w = a \ln(a) \exp[\tilde{t} - \ln(a) \exp \tilde{t}] \tag{6.66}$$

At

$$\tilde{t}_{v \max} = -\ln(\ln a) \tag{6.67}$$

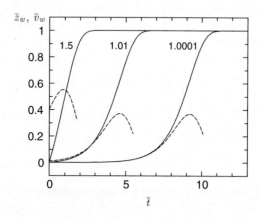

FIGURE 6.15. Dynamics of the degradation wave for indicated values of parameter a. Solid lines - position of the wavefront $\tilde{z}_w$, dashed lines – wave velocity $\tilde{v}_w$.

---

[8] In fact, wave propagation is discrete: wave front waits while the domain $j > j_{crit}$ is being degraded and then jumps to a new position $z'_w \to z_w + l_w$. Transition to continuous velocity (6.63) means that we interpolate the discrete front positions with a continuous function of time.

velocity reaches maximum

$$\tilde{v}_{max} = a\exp(-1) \simeq 0.368\,a. \tag{6.68}$$

The distance $\tilde{z}_{v\,max}$ travelled before velocity reaches the maximum is obtained by substituting $\tilde{t} = \tilde{t}_{v\,max}$ into (6.65). This gives

$$\tilde{z}_{v\,max} = 1 - \frac{a}{\exp(1)} \tag{6.69}$$

Note that $\tilde{z}_{v\,max} > 0$, if $1 < a \le \exp(1)$; however if $a > \exp(1)$, we have $\tilde{z}_{v\,max} < 0$, $\tilde{t}_{v\,max} < 0$, and the velocity has no maximum. In that case, v_w decreases with the distance, which corresponds to the onset of the fast propagation phase at $\tilde{t} = 0$.

The time (6.67) it takes for velocity to reach the maximum by the order of magnitude is the time of slow propagation. To show this we substitute $a = 1 + \varepsilon$ into (6.66) and expand the result over ε. Retaining quadratic term, we get

$$\tilde{v}_w \simeq \exp(\tilde{t})\left[1 + \left(\frac{1}{2} - \exp(\tilde{t})\right)\varepsilon\right]\varepsilon + O(\varepsilon^3). \tag{6.70}$$

Equating the expression in square brackets to zero we find

$$\tilde{t}_{slow} = \ln\left(\frac{1}{2} + \frac{1}{\varepsilon}\right) \simeq -\ln\varepsilon \tag{6.71}$$

since $\varepsilon \ll 1$. At $\tilde{t} = \tilde{t}_{slow}$, linear and quadratic in ε terms in (6.70) cancel and velocity is *very* small, $\tilde{v}_w \sim O(\varepsilon^3)$. For $\tilde{t} < \tilde{t}_{slow}$, the function $\tilde{v}_w(t)$ (6.66) monotonically increases, therefore for all $\tilde{t} < \tilde{t}_{slow}$ wave velocity is even smaller. With $a = 1 + \varepsilon$ Eq. (6.67) gives $\tilde{t}_{v\,max} = -\ln(\ln a) \simeq -\ln(\varepsilon)$, and we see that $\tilde{t}_{v\,max} \simeq \tilde{t}_{slow}$. Qualitatively, for small ε, the time of fast propagation is much smaller than the time of slow propagation, and $\tilde{t}_{v\,max}$ and $\tilde{t}_{slow}$ are close to each other.

Physically, for $\varepsilon \ll 1$ the peak $j(0)$ only slightly exceeds j_{crit} (i.e. the dashed area at t_0 in Figure 6.14 is small). In this regime, initial velocity of the wave is small, since the length of the domain exposed to degradation is small. Indeed, putting in (6.60) $a = 1 + \varepsilon$, $\tilde{z}_w = 0$ and expanding logarithm, we get $l_w = -\varepsilon/\ln(1 - \frac{1}{\lambda})$. The wave should slowly travel a certain distance until l_w becomes sufficiently large to provide further fast propagation.

6.3.4.2. Cell Potential

From (6.57), we find the instant mean current density ahead of the wave J_w:

$$J_w = \frac{1}{L-z_w}\int_{z_w}^{L} j\,dz = \frac{JL}{L-z_w} = \frac{J}{1-\tilde{z}_w} = \left(\frac{J}{a}\right)\exp\left[\ln(a)\exp\left(\frac{t}{T_w}\right)\right] \quad (6.72)$$

Clearly, J_w increases rapidly with time. When $a \to 1$, the rapid growth is preceded by the slow increase in time. To calculate the respective change in polarization voltage, we will use the large-current limit of Eq. (6.50):

$$\frac{\eta}{b} = 2\ln\left(\frac{f_\lambda J_w}{j_*}\right) - \ln\left(\frac{k_h^0}{j_*}\right) - \ln\left(1 - \frac{f_\lambda J_w}{j_D}\right). \quad (6.73)$$

Substituting J_w (6.72) into (6.73) and taking into account (6.61), we get the dependence of η on time:

$$\frac{\eta}{b} = 2\ln\left(\frac{j_{crit}}{j_*}\right) + 2\ln(a)\exp\left(\frac{t}{T_w}\right) - \ln\left(\frac{k_h^0}{j_*}\right) - \ln\left(1 - \frac{j_{crit}}{j_D^*(t)}\right) \quad (6.74)$$

where the limiting current density

$$j_D^*(t) = j_D \exp\left[-\ln(a)\exp\left(\frac{t}{T_w}\right)\right] \quad (6.75)$$

decreases with time.

The third term on the right side of (6.73) describes the effect of the limiting current density due to imperfect transport of oxygen through the backing layer. In (6.74), this term transforms to the term which limits the time of cell operation (the last term on the right side). Physically, cell potential is determined by the voltage loss due to oxygen transport in a "good" domain of a cell. This loss rapidly increases with time, since more and more oxygen must be transported through the GDL in the unspoilt domain to support total current in the load.

Equating the expression under the last logarithm on the right side of (6.74) to zero and taking into account (6.75), we get the cell life time

$$\tilde{t}_{life} = -\ln(\ln a) + \ln\left(\ln\left(\frac{j_D}{j_{crit}}\right)\right) = \tilde{t}_{slow} + \tilde{t}_D \quad (6.76)$$

where

$$\tilde{t}_D = \ln\left(\ln\left(\frac{j_D}{j_{crit}}\right)\right). \quad (6.77)$$

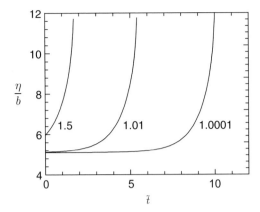

FIGURE 6.16. Cathode polarization voltage as a function of dimensionless time for indicated values of parameter a.

Physically, as $\tilde{t} \to \tilde{t}_{life}$, $\eta \to \infty$, that is, the cell potential drops to zero. The function (6.74) is shown in Figure 6.16. The catastrophe at $\tilde{t} = \tilde{t}_{life}$ is clearly seen.

Suppose that t_{slow} is positive, that is, $1 < a < \exp(1)$; we then have the two cases. If $1 < j_D/j_{crit} < \exp(1)$, the term $\tilde{t}_D$ is negative. It means that the cell potential "dies" before the end of the slow propagation phase. Physically, large transport loss in the GDL leads to a faster drop of cell voltage. If $j_D/j_{crit} > \exp(1)$ (small transport loss), $\tilde{t}_D$ in Eq. (6.76) is positive, and the time of cell operation somewhat exceeds $\tilde{t}_{slow}$.

6.3.4.3. Discussion

No assumptions were made on the nature of local degradation under the "over-current" conditions; rather, we made the phenomenological ansatz that it takes a time τ_d to "spoil" the domain in which the local current exceeds j_{crit}. The critical current density and τ_d are determined by a physical mechanism of cell degradation. If degradation is caused by thermal effects, τ_d is inversely proportional to the square of local current density. Since local current in front of the wave increases with time, τ_d would decrease with time. Qualitatively, this would lead to even faster wave propagation.

Suppose that a fresh MEA is characterized by j_{crit} associated with some physical mechanism of degradation. Depending upon λ, we may have the two regimes of cell operation:

- $\lambda - 1 \ll 1$ (λ *is close to 1*). In that case, $j(0)$ may exceed j_{crit} right upon start-up of the cell. DW then starts immediately (Figure 6.17a), and cell voltage immediately starts to decrease.

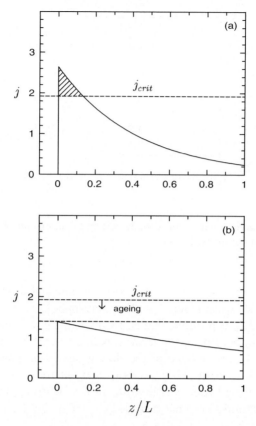

FIGURE 6.17. The two scenarios of fuel cell degradation. (a) Small $\lambda - 1$: local current density exceeds the critical level right upon start-up. The degradation wave starts immediately. (b) $\lambda - 1 \gtrsim 1$: the critical current density initially exceeds $j(0)$. Due to the slow process of ageing, j_{crit} decreases with time, and at some time moment the condition $\tilde{j}_0(0) = j_{crit}$ is fulfilled. This initiates slow and then fast propagation of the degradation wave.

- $\lambda - 1 \gtrsim 1$ (λ *is about or exceeds* 2). In that case, j_{crit} most probably exceeds $j(0)$ (Figure 6.17b). However, j_{crit} may itself decrease with time due to some slow process of ageing. At some time moment $j(0)$ would slightly exceed j_{crit} (or, equivalently, at this moment $a = 1 + \varepsilon$). This initiates slow and then fast propagation of DW, which again finally results in catastrophic drop of cell potential.

In the second scenario, we have three characteristic times: the time of ageing (slow decrease of critical current density) t_{ageing}, the time of wave slow propagation t_{slow} and the time of fast propagation t_{fast}. In view of (6.64), t_{slow} and t_{fast} are proportional to the local time of degradation τ_d.

The role of λ is twofold. For fixed mean current density, an increase in λ flattens the profile $j(z)$ and lowers $j(0)$ (Figure 6.10), thus giving more time for ageing. On the other hand, if the condition of DW generation is fulfilled, larger λ shortens the period of slow propagation (see (6.71)). At large λ, once the local current exceeds j_{crit}, the cell potential will drop very fast, on a time scale of the order of τ_d.

Non-uniformity of any parameter along the channel (e.g. partial cell flooding or membrane drying) increases non-uniformity of local current density and facilitates DW generation. If some region of the cell is flooded, local current in the non-flooded region increases and local j can exceed j_{crit}. The flooding itself may follow one of the degradation scenarios. If a local increase in current density leads to local flooding, we have the conditions for generation of the wave of flooding.

6.4. 1D + 1D Model. II. Water Management

6.4.1. Constant Oxygen Flow Rate

PEFC needs water to maintain the polymer electrolyte in a wet state. The protonic conductivity of the membrane σ_m depends linearly on the water content λ_m and at $\lambda_m = 0$, σ_m drops to zero. The proton current induces electroosmotic flux of water, which dries out the anode side of the membrane, thus increasing membrane resistance. Below, we will see that voltage loss in the membrane reduces the local polarization voltage of the cathode side and hence the local rate of oxygen consumption. Local oxygen and water fluxes across the cell are, therefore, strongly coupled. Furthermore, since oxygen is distributed non-uniformly along the MEA surface, the magnitude of these fluxes depends on position.

Analytical models of PEFCs usually ignore conjugate transport of water and oxygen across the cell. Models of water transport in the membrane [21–23] ignore transport of reactants in the catalyst and backing layers.[9] The models of the catalyst layer [5,7] disregard oxygen transport in the backing layer and water management effects. The models that take into account oxygen transport in the backing and catalyst layers [25,26] do not include water transport in membrane. Furthermore, all aforementioned models ignore along-the-channel effects.

In this section, we construct a model of a PEFC, which couples transport of water and oxygen across the cell with oxygen depletion and water accumulation along the feed channel.

---

[9] Ref. [24] contains excellent review of membrane models.

6.4.1.1. Model and Governing Equations

The main assumptions are as follows.

1. The flux of liquid water in the cathode GDL is negligible.
2. Oxygen and water vapour diffusion coefficients in the GDL coincide.
3. Diffusion coefficient of liquid water in the membrane is constant.
4. At the membrane/catalyst layer interface, the concentration of liquid water in membrane and the concentration of water vapour in the CCL are related by membrane water sorption isotherm.
5. Total flux of water in membrane is zero.

The last assumption means that local electroosmotic flux of water in membrane is exactly counterbalanced by back diffusion. Recent studies [14,27] have shown that in a wide range of operating conditions total transfer coefficient of water from the anode to the cathode does not exceed 0.2. Since electroosmotic drag coefficient in Nafion is $\simeq 1.5$ [28], we conclude that the average over the cell surface electroosmotic flux in the membrane is almost fully compensated for by back diffusion. Note that the local value of total water flux in the membrane may significantly deviate from the surface-averaged value, e.g. close to the outlet of the oxygen channel [27]. Nevertheless, assumption 5 seems to be a reasonable approximation.

Under these assumptions, the local voltage loss in the membrane is given by [29]

$$V_m = -b_m \ln\left(1 - \frac{j}{r\left(\frac{\psi_h j_D^0}{2\xi^0} + j\right)}\right), \qquad (6.78)$$

where ψ_h is local water molar fraction in the channel,

$$b_m = \frac{F D_l c_{H^+}}{\sigma_{m1} n_d} \qquad (6.79)$$

is the characteristic voltage loss, D_l the is diffusion coefficient of liquid water in membrane, σ_{m1} is the membrane conductivity at unit water content (at $\lambda_m = 1$), n_d is the electroosmotic drug coefficient and c_{H^+} is the molar concentration of protons/sulfonic groups in the membrane. The oxygen-limiting current density at the inlet j_D^0 is given by (6.49). The dimensionless parameter r in (6.78) is

$$r = \frac{D_l l_b K_\lambda c_{H^+}}{4 D_b n_d l_m c_w^{sat}}, \qquad (6.80)$$

where K_λ is the mean slope of water sorption isotherm [29], l_m is the membrane thickness and c_w^{sat} is the molar concentration of saturated water vapour.

6. Analytical Models of a Polymer Electrolyte Fuel Cell

Physically, r is proportional to the ratio of mass transfer coefficient of liquid water in membrane to mass transfer coefficient of water vapour in the backing layer. The parameter r thus describes the competition of two opposite water fluxes: back diffusion, which wets the anode side of the membrane and leakage through the backing layer to the channel, which facilitates membrane drying. Physically, r controls the local water-limiting current density (see below).

In this section, our goal is to rationalize the effects due to composition of the cathode mixture. For that reason, in all relations the dependencies on ξ_h and ψ_h will be shown explicitly. For local polarization voltage of the cathode side, we will employ low-current relation (6.32), which can be transformed to

$$\frac{\eta}{b} = \ln\left(\frac{j}{\xi_h j_T}\right) - \ln\left(1 - \frac{j}{\xi_h j_D^0}\right). \tag{6.81}$$

Here

$$j_T = l_t i_* \frac{c}{c_{ref}}.$$

The problem, hence, is governed by the following system of equations. First is the equation of oxygen mass balance in the channel (6.37). Second is the condition of equipotentiality of cell electrodes. Now, however, voltage loss in membrane is not negligible, and this condition reads

$$\eta(z) + V_m(z) = E, \tag{6.82}$$

where E does not depend on z (cf. Eq. (6.38)). Constant pressure in the channel means constant c; the sum of molar fractions is thus also invariant along z:

$$\psi_h(z) + \xi_h(z) + \zeta_h = 1, \tag{6.83}$$

where ζ_h is molar fraction of bulk gas (nitrogen). The system of Eqs. (6.37), (6.78), (6.81–6.83) define the model for cell performance.

6.4.1.2. Solution

Introducing the dimensionless variables

$$\tilde{\xi}_h = \frac{\xi_h}{\xi_h^0}, \quad \tilde{\psi}_h = \frac{\psi_h}{\xi_h^0}, \quad \tilde{V} = \frac{V}{b}, \quad \hat{j} = \frac{j}{j_D^0} \tag{6.84}$$

and substituting (6.78) and (6.81) into (6.82), Eqs. (6.37) and (6.82) become

$$\beta \frac{\partial \tilde{\xi}_h}{\partial \tilde{z}} = -\hat{j} \qquad (6.85)$$

$$\ln\left(\frac{\hat{j}}{\tilde{\xi}_h q}\right) - \ln\left(1 - \frac{\hat{j}}{\tilde{\xi}_h}\right) - p \ln\left(1 - \frac{\hat{j}}{r\left(\frac{\tilde{\psi}_h}{2} + \hat{j}\right)}\right) = \tilde{E}. \qquad (6.86)$$

As before, $\tilde{z} = z/L$, $\tilde{\eta} = \eta/b$, $\tilde{E} = E/b$. Note that in this section all current densities are normalized to j_D^0; the respective dimensionless variables are marked with the symbol "hat".

Behaviour of the system is thus governed by 4 parameters: β, q, p and r. The parameter β is given by (6.52),

$$q = \frac{l_t l_b i_*}{4FD_b c_{ref}}, \quad p = \frac{b_m}{b} \qquad (6.87)$$

and r is given by Eq. (6.80). The parameter $\tilde{E}$ determines the working point on the cell polarization curve.[10]

Equation (6.86) can be rewritten as

$$\frac{1}{\hat{j}} = \frac{1}{\tilde{\xi}_h} + \frac{1}{q\tilde{\xi}_h \exp \tilde{E}} \left[1 - \frac{\hat{j}}{r\left(\frac{\tilde{\psi}_h}{2} + \hat{j}\right)}\right]^{-p} \qquad (6.88)$$

It is seen that parameter q simply re-scales $\tilde{E}$; the shape of the solution is thus determined by 3 parameters: β, r and p. The system of Eqs. (6.85), (6.88), (6.83) determine the profiles of oxygen and local current density along the channel. In the general case, this system has to be solved numerically. However, in the case of large $\tilde{E}$ this system has explicit solutions.

6.4.1.3. Close to the Limiting Current Density ($\tilde{E} \to \infty$)

Large values of $\tilde{E}$ physically means that the cell operates close to the limiting current density. This case is of particular interest, since in this regime cell power density is close to maximal. Furthermore, the solution at the limiting current helps to understand the character of the solution in the general case.

Equation (6.86) contains two terms exhibiting limiting behaviour: the second and the third logarithms on the left side. Equating the expressions under the 2-nd and the 3-rd logarithms to zero, we find two asymptotic solutions to Eq. (6.86) at $\tilde{E} \to \infty$:

---

[10] Equation (6.85) leads to a very simple formula for oxygen utilization (see Appendix A).

$$\hat{j}^{ox}_{\lim} = \tilde{\xi}_{h\lim} \tag{6.89}$$

$$\hat{j}^{w}_{\lim} = f_r \tilde{\psi}_{h\lim}, \tag{6.90}$$

where

$$f_r = \frac{r}{2(1-r)}, \tag{6.91}$$

and the subscript "lim" denotes the values taken at $\tilde{E} \to \infty$.

The first solution corresponds to the oxygen-limiting regime, when local-limiting current is determined by oxygen transport through the backing layer. In that case, both $\hat{j}^{ox}_{\lim}$ and $\tilde{\xi}_{h\lim}$ exponentially decrease with $\tilde{z}$; this follows immediately if we substitute $\hat{j} = \tilde{\xi}_h$ into Eq. (6.85). The solution to this equation with the initial condition $\tilde{\xi}_h(0) = \tilde{\xi}_h^{in}$ is

$$\tilde{\xi}_{h\lim} = \hat{j}^{ox}_{\lim} = \tilde{\xi}_h^{in} \exp\left(-\frac{\tilde{z}}{\beta}\right) \tag{6.92}$$

Note that Eq. (6.92) can be obtained directly from Eqs. (6.53) and (6.54) in the limit of $\tilde{\eta} \to \infty (a_\eta \to 1)$. Note also that in general $\tilde{\xi}_h^{in}$ does not coincide with 1 (see below).

The second solution corresponds to water-limiting regime, when local current is limited by membrane drying. This solution arises if $f_r > 0$, or, equivalently, $r < 1$. When $r \geq 1$ this solution disappears and for all $\tilde{z}$, we have oxygen-limiting regime. Physically, if $r \geq 1$ membrane drying simply reduces cell potential without affecting the limiting current density. Substituting

$$\hat{j} = f_r\left(\frac{1}{\xi_h^0} - \tilde{\zeta}_h - \tilde{\xi}_h\right) \tag{6.93}$$

into Eq. (6.85) and solving the resulting equation with the initial condition $\tilde{\xi}_h(0) = 1$, we obtain

$$\tilde{\xi}_{h\lim}(\tilde{z}) = 1 - \tilde{\psi}_h^0\left[\exp\left(\frac{f_r \tilde{z}}{\beta}\right) - 1\right]. \tag{6.94}$$

Substituting this into Eq. (6.93), we find

$$\hat{j}_{\lim}^w(\tilde{z}) = f_r \tilde{\psi}_h^0 \exp\left(\frac{f_r \tilde{z}}{\beta}\right). \tag{6.95}$$

As discussed above, this solution arises for $r < 1$ in which case the parameter f_r is positive. We see that in the water-limiting regime local current density *increases* exponentially with $\tilde{z}$. Physically, in this case there is a plenty of oxygen everywhere, and the local-limiting current is determined by membrane drying. The growth of $\hat{j}$ with $\tilde{z}$ is then due to accumulation of water in the feed channel, which results in increasing membrane conductivity with $\tilde{z}$.

Clearly, in both oxygen- and water-limiting cases oxygen concentration in the channel monotonically decreases with $\tilde{z}$, while water concentration monotonically increases. Local current density, however, increases with $\tilde{z}$ in the water-limiting regime and decreases in the oxygen-limiting regime. We, therefore, have the three cases.

1. Oxygen-limiting regime everywhere along $\tilde{z}$ ($\hat{j}_{\lim}$ decreases with $\tilde{z}$).
2. Water-limiting regime everywhere along $\tilde{z}$ ($\hat{j}_{\lim}$ grows with $\tilde{z}$).
3. Mixed case: water-limiting close to the inlet and oxygen-limiting in the rest part of the cell (close to the inlet $\hat{j}_{\lim}$ increases with $\tilde{z}$ and then decreases).

Local current density profiles in the first and the second case are described by (6.92) with $\tilde{\xi}_h^{in} = 1$ and by (6.95), respectively. In the mixed case close to the inlet $\hat{j}_{\lim}$ is given by (6.95); along the rest part of the channel it is given by (6.92). The point $\tilde{z}_{\max}$, which in the mixed case separates water- and oxygen-limiting domains is obtained from continuity of local currents and oxygen concentrations. Equating (6.94) to (6.92) and (6.95) to (6.92), we get

$$\begin{aligned} 1 - \tilde{\psi}_h^0 \left[\exp\left(\frac{f_r \tilde{z}_{\max}}{\beta}\right) - 1\right] &= \tilde{\xi}_h^{in} \exp\left(-\frac{\tilde{z}_{\max}}{\beta}\right) \\ f_r \tilde{\psi}_h^0 \exp\left(\frac{f_r \tilde{z}_{\max}}{\beta}\right) &= \tilde{\xi}_h^{in} \exp\left(-\frac{\tilde{z}_{\max}}{\beta}\right) \end{aligned} \tag{6.96}$$

Equating the left sides of these equations we find

$$\tilde{z}_{\max} = \frac{\beta}{f_r}\left[\ln\left(1 + \frac{\xi_h^0}{\psi_h^0}\right) - \ln(1 + f_r)\right]. \tag{6.97}$$

On the right side the sign "tilde" is omitted, since $\tilde{\xi}_h/\tilde{\psi}_h \equiv \xi_h/\psi_h$.

Clearly, the mixed profile is realized if $0 < \tilde{z}_{\max} < 1$. This leads to the following condition

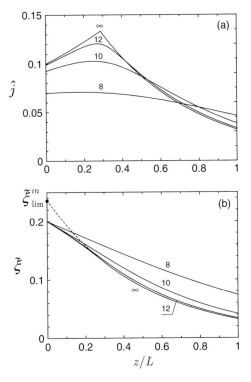

FIGURE 6.18. (a) Local current density and (b) oxygen concentration for indicated values of total voltage loss $\tilde{E}$. The other parameters are $r=0.5$, $\beta = 0.5$, $q = 10^{-3}$, $\zeta = 0.6$, $\xi_h^0 = 0.2$, $\psi_h^0 = 0.2$, $p = 1$.

$$f_r < \frac{\xi_h^0}{\psi_h^0} < (1+f_r)\exp\left(\frac{f_r}{\beta}\right) - 1 \qquad (6.98)$$

Note that the onset of the mixed regime and position of the point $\tilde{z}_{\max}$ depend on the ratio ξ_h^0/ψ_h^0, rather than on ξ_h^0 and ψ_h^0 separately. Figure 6.18 shows the profiles of $\hat{j}_{\lim}$ and $\tilde{\xi}_{h\,\lim}$ in the mixed regime (the respective curves are marked with the symbol "∞"). At $\tilde{z}_{\max}$ local current density reaches maximum. $\hat{j}_{\lim}^{\max}$ is obtained if we substitute Eq. (6.97) into Eq. (6.95). This gives

$$\hat{j}_{\lim}^{\max} = \frac{f_r}{1+f_r}\left(1 + \frac{\psi_h^0}{\xi_h^0}\right) \qquad (6.99)$$

This value does not depend on β. The increase in β simply shifts $\tilde{z}_{\max}$ towards the outlet leaving $\hat{j}_{\lim}^{\max}$ intact. The decrease in r also shifts $\tilde{z}_{\max}$ towards the outlet but $\hat{j}_{\lim}^{\max}$ then decreases. Physically, lower r means a higher rate of water removal through the backing layer to the channel which facilitates membrane drying and reduces $\hat{j}_{\lim}^{\max}$.

With Eq. (6.97) we can calculate $\tilde{\xi}_h^{in}$ from Eq. (6.96),

$$\tilde{\xi}_h^{in} = f_r \frac{\psi_h^0}{\xi_h^0} \left(\frac{1 + \frac{\xi_h^0}{\psi_h^0}}{1 + f_r} \right)^{\frac{1+f_r}{f_r}}. \tag{6.100}$$

In the pure oxygen-limiting case $\hat{j}_{\lim}^w(0) = \hat{j}_{\lim}^{ox}(0)$, thus $f_r \psi_h^0 / \xi_h^0 = 1$ and hence $\tilde{z}_{\max} = 0$, which in turn implies $1 + \xi_h^0/\psi_h^0 = 1 + f_r$. Equation (6.100) then gives $\tilde{\xi}_h^{in} = 1$. The meaning of $\tilde{\xi}_h^{in}$ is clear from Figure 6.18b.

6.4.1.4. The General Case (Finite $\tilde{E}$)

At finite $\tilde{E}$ the along-the-channel profiles of oxygen and local current density are given by the system of Eqs.(6.85), (6.86) and (6.83). Numerical solutions to this system for several values of parameter $\tilde{E}$ are depicted in Figure 6.18.[11] We see that numerical solutions qualitatively reproduce the behaviour of the

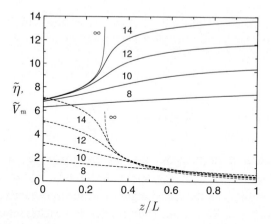

FIGURE 6.19. Dimensionless overpotential $\tilde{\eta}$ (solid curves) and voltage loss in membrane $\tilde{V}_m$ (dashed curves) for indicated values of total voltage loss $\tilde{E}$. The other parameters are the same, as in Figure 6.18.

---

[11] To simplify numerical calculations, it is useful to differentiate Eq. (6.88) with respect to $\tilde{z}$ and to transform the result into differential equation for $\partial \hat{j}/\partial \tilde{z}$ using Eq. (6.85). Initial condition $\hat{j}(0)$ is obtained from (6.86) or (6.88). Numerical solution of the system of two first-order ODEs can be obtained with any mathematical software.

limiting curve: at the inlet, current increases with $\tilde{z}$ (water-limiting domain) and then decreases (oxygen-limiting domain). As expected, with the growth of $\tilde{E}$, numerical solution tends to the limiting analytical curve.

Figure 6.18a shows that the smooth numerical plots of $\hat{j}(\tilde{z})$ at finite $\tilde{E}$ tend to a limiting curve which is non-differentiable at $\tilde{z}_{max}$. The disappearance of the derivative $\partial \hat{j}_{lim}/\partial \tilde{z}$ at $\tilde{z}_{max}$ suggests singularity at this point. The plots of $\tilde{\eta}$ and membrane voltage loss $\tilde{V}_m$ as a function of $\tilde{z}$ are shown in Figure 6.19. Although the sum $\tilde{\eta} + \tilde{V}_m$ remains constant along $\tilde{z}$, near $\tilde{z}_{max}$ both $\tilde{\eta}$ and $\tilde{V}_m$ exhibit distinct gradient, which increases with the growth of $\tilde{E}$ (Figure 6.19). Therefore, when $\tilde{E}$ is high, near $\tilde{z}_{max}$ a large $\tilde{z}$-component of proton current density arises. Clearly, operation in the mixed regime close to the limiting current requires this additional overhead.

Physically, $\tilde{z}$-gradient of $\tilde{\eta}$ is induced by the growth of membrane water content with $\tilde{z}$. This growth means the decrease in membrane voltage loss. Since $\tilde{\eta} + \tilde{V}_m$ is constant, the decrease in $\tilde{V}_m$ means the respective increase in $\tilde{\eta}$. As $\tilde{E} \to \infty$ the gradient of polarization voltage at $\tilde{z}_{max}$ becomes infinite (Figure 6.19). This means infinite z-component of proton current density in CCL and in membrane. Large $\partial \tilde{\eta}/\partial \tilde{z}$ means large oxygen consumption to support this parasitic proton current. Evidently, 1D+1D model fails to describe the details of this effect. In the vicinity of $\tilde{z}_{max}$, a true 2D distribution of local proton current arises. Accurate description of this situation requires numerical simulations.

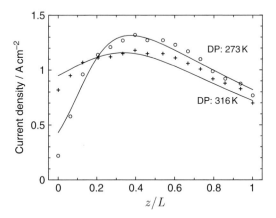

FIGURE 6.20. Along-the-channel profiles of local current density for two dew points of humidifier (two compositions of the cathode feed). Points: experimental data [27]. Circles correspond to dew point 273 K, crosses - dew point 316 K. Solid lines - theory. Parameters for the curves are $p=1$, $q = 7.6 \times 10^{-3}$, $\tilde{E} = 8.56$, $\beta = 0.880$; $r=0.935$ for d.p. 273 K and $r=0.708$ for d.p. 316 K.

6.4.1.5. Comparison with the Experiment

Recently Berg et al. [27] developed numerical 1D+1D model of PEFC and validated it against measured along-the-channel profiles of local current density. To verify the model above, we fitted the solution of Eqs. (6.85), (6.86), (6.83) to the experimental data [27]. The procedure is as follows.

Under given feed composition, the system (6.85), (6.86), (6.83) contains 6 fitting parameters: β, q, $\tilde{E}$, $\hat{j}_D^0$, p and r; for simplicity we took $p=1$. The data for dew point 273 K was fitted using the genetic algorithm [30]; the resulting parameters are $\beta = 0.880$, $\tilde{E} = 8.56$, $\hat{j}_D^0 = 9.95$ A cm^{-2} and $r=0.935$ (Figure 6.20). The curve for another feed composition (dew point 316 K) was obtained with the same β, $\tilde{E}$ and $\hat{j}_D^0$, but lower $r=0.708$ (Figure 6.20).

In both cases, local current has maximum, that is, the cell operates in a mixed regime (Figure 6.20). The growth of humidifier temperature from 0°C to 43°C means 10-fold increase in the fraction of water at the inlet. At lower water fraction, current in the water-limiting domain rises with the distance more rapidly (Figure 6.20). For further details please see [31].

6.4.1.6. Limiting Current Density, Optimal Feed Composition

Integration of $\hat{j}_{\lim}$ (6.92) and (6.95) over $\tilde{z}$ yields the respective overall limiting current density of the cell $\hat{J}_{\lim}$. For the water-limiting, oxygen-limiting and mixed cases, respectively, we write:

$$\hat{J}_{\lim}^w = \int_0^1 \hat{j}_{\lim}^w \, d\tilde{z} \tag{6.101}$$

$$\hat{J}_{\lim}^{ox} = \int_0^1 \hat{j}_{\lim}^{ox} \, d\tilde{z} \tag{6.102}$$

$$\hat{J}_{\lim}^{mix} = \int_0^{\tilde{z}_{max}} \hat{j}_{\lim}^w \, d\tilde{z} + \int_{\tilde{z}_{max}}^1 \hat{j}_{\lim}^{ox} \, d\tilde{z} \tag{6.103}$$

Polarization curves are usually measured at a constant oxygen stoichiometry ratio λ rather than at a constant β. Integrals (6.101)–(1.103) can be transformed to the case of constant λ using the identity $\beta = \hat{J}_{\lim}\lambda$, and we get [31]

$$\hat{J}_{\lim}^w = f_r \left[\lambda \ln\left(1 + \frac{s}{\lambda}\right) \right]^{-1} \tag{6.104}$$

$$\hat{J}_{\lim}^{ox} = -\left[\lambda \ln\left(1 - \frac{1}{\lambda}\right) \right]^{-1} \tag{6.105}$$

$$\hat{J}_{\lim}^{mix} = f_r \left[\lambda \left(f_r \ln\left(\frac{\lambda f_r}{s(\lambda - 1)}\right) + (1 + f_r) \ln\left(\frac{1+s}{1+f_r}\right) \right) \right]^{-1} \tag{6.106}$$

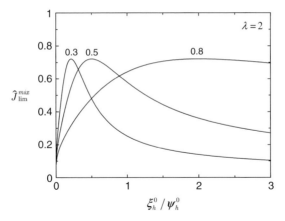

FIGURE 6.21. Limiting current density in the mixed regime as a function of the ratio of inlet oxygen to water fractions for indicated values of r. Oxygen stoichiometry $\lambda = 2$.

Here, we denote $s = \xi_h^0/\psi_h^0$. Note that Eq. (6.106) is valid only if condition of mixed regime (6.98) is fulfilled, otherwise one of the Eqs. (6.104), (6.105) should be used. In particular, one has to be careful varying parameters in Eq. (6.106), since variation of λ, f_r and s shifts position of $\tilde{z}_{\max}$.

As expected, $\hat{J}_{\lim}^{ox}$ does not depend on water content. In the water-limiting and mixed regimes, $\hat{J}_{\lim}$ depends on the ratio ξ_h^0/ψ_h^0 rather than on ξ_h^0 and ψ_h^0 separately. It follows that if we vary $\xi_h^0 \sim \psi_h^0$, the value of $\hat{J}_{\lim}$ in these regimes does not change.

In the mixed case, there exists optimal feed composition, which provides maximal $\hat{J}_{\lim}^{mix}$. This composition is a root of equation $\partial \hat{J}_{\lim}^{mix}/\partial s = 0$. Solution to this equation is $s = f_r$ or

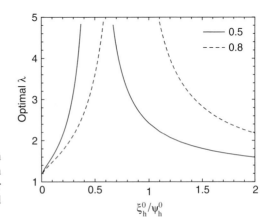

FIGURE 6.22. Optimal oxygen stoichiomtery λ_{opt} as a function of the ratio of oxygen to water inlet fractions ξ_h^0/ψ_h^0 for indicated values of f_r.

$$\left(\frac{\xi_h^0}{\psi_h^0}\right)_{opt} = \frac{r}{2(1-r)} \qquad (6.107)$$

From (6.97), it is seen that if $\xi_h^0/\psi_h^0 = f_r$, we have $\tilde{z}_{max} = 0$, or equivalently, $\hat{j}_{lim}^w(0) = \hat{j}_{lim}^{ox}(0)$. Optimal feed composition (6.107), therefore, provides equality of water- and oxygen-limiting currents at the channel inlet. This means that everywhere along $\tilde{z}$ current is not limited by drying. In other words, $\psi_h^0 = \xi_h^0/f_r$ is optimal water fraction, which provides pure oxygen- limiting regime of cell operation.

The dependence of $\hat{J}_{lim}^{mix}$ on ξ_h^0/ψ_h^0 for different r is illustrated in Figure 6.21. We see that if $r \lesssim 0.5$, the curve exhibits sharp maximum. To the left of the maximum, the cell suffers from oxygen "starvation", to the right it experiences "thirst".

If $s \neq f_r$ (i.e. the composition is non-optimal), there exists optimal oxygen stoichiometry λ_{opt}, which provides maximal $\hat{J}_{lim}^{mix}$. λ_{opt} is a root of equation $\partial \hat{J}_{lim}^{mix}/\partial \lambda = 0$. The numerical solution to this equation is shown in Figure 6.22. Note that if $s = f_r$, then λ_{opt} tends to infinity (Figure 6.22). In the case of an optimal feed composition, the curve $\hat{J}_{lim}^{mix}(\lambda)$ is monotonic, which formally is equivalent to $\lambda_{opt} \to \infty$.

Figure 6.22 shows that for $s \ll f_r$ or $s \gg f_r$, λ_{opt} is not large, typically less than 2. Physically, $\lambda = \lambda_{opt}$ provides optimal fractions of limiting current, produced in the water- and oxygen-limiting domains. When $\lambda < \lambda_{opt}$ the cell suffers from the lack of oxygen, the fraction of current produced in the oxygen-limiting domain is less than optimal. If $\lambda > \lambda_{opt}$, the rate of water removal is too high, i.e., the fraction of current produced in the water-limiting domain is less than optimal.

6.4.2. Constant Oxygen Stoichiometry

In this section, we re-formulate the model equations for the case of constant oxygen stoichiometry and obtain the criterion of ideal membrane humidification.

6.4.2.1. Modification of Model Equations

If the cell is run under constant oxygen stoichiometry λ, Eq. (6.85) should be modified. Using $\beta = \lambda \tilde{J}$ we get

$$\lambda \hat{J} \frac{\partial \tilde{\xi}_h}{\partial \tilde{z}} = -\hat{j}, \quad \tilde{\xi}_h(0) = 1 \qquad (6.108)$$

Taking into account (6.83), Eq. (6.108) yields

$$\frac{\partial \tilde{\psi}_h}{\partial \tilde{z}} = -\frac{\partial \tilde{\xi}_h}{\partial \tilde{z}} = \frac{\hat{j}}{\lambda \hat{J}} \tag{6.109}$$

For further analysis, we will convert (6.86) into differential equation for $\hat{j}$. Differentiating (6.86) with respect to $\tilde{z}$ and using (6.109) we come to

$$\frac{\partial \hat{j}}{\partial \tilde{z}} = \frac{\hat{j}^2 [-r\tilde{\psi}_h^2 + 2(1-2r)\hat{j}\tilde{\psi}_h + 2p\hat{j}\tilde{\xi}_h + 2(2-2r-p)\hat{j}^2]}{\lambda \hat{J} [r\tilde{\xi}_h \tilde{\psi}_h^2 + 2(2r-1+p)\hat{j}\tilde{\xi}_h \tilde{\psi}_h - 2p\hat{j}^2 \tilde{\psi}_h - 4(1-r)\hat{j}^2 \tilde{\xi}_h]}, \quad \hat{j}(0) = \kappa \tag{6.110}$$

An equation for local current density at the inlet κ is obtained, if we substitute inlet values $\tilde{\xi}_h = 1$, $\tilde{\psi}_h = \tilde{\psi}_h^0$ and $\hat{j} = \kappa$ into (6.86):

$$\ln\left(\frac{\kappa}{q}\right) - \ln(1-\kappa) - p \ln\left(1 - \frac{\kappa}{r\left(\frac{\tilde{\psi}_h^0}{2} + \kappa\right)}\right) = \tilde{E}, \tag{6.111}$$

where $\tilde{\psi}_h^0 = 1/\xi^0 - 1 - \tilde{\zeta}_h$.

The system (6.108), (6.110), (6.111) determines oxygen, water vapour and local current density profiles along the channel in the case of constant λ. The parameter $\tilde{E}$ is iterated so that the local current density $\hat{j}$ gives the prescribed total current density

$$\int_0^1 \tilde{J} \, d\tilde{z} = \hat{J}. \tag{6.112}$$

6.4.2.2. Condition of Ideal Membrane Humidification

When the current density is small the system (6.108), (6.110), (6.111) has an explicit solution. Expanding the right side of (6.110) with respect to $\hat{j}$ and retaining the first two non-vanishing terms we get

$$\lambda \frac{\partial \hat{j}}{\partial \tilde{z}} = -\frac{\hat{j}^2}{\hat{J}\tilde{\xi}_h}\left[1 - \frac{2p(\tilde{\xi}_h + \tilde{\psi}_h)\hat{j}}{r\tilde{\psi}_h^2}\right]. \tag{6.113}$$

The second term in the square brackets is negligible if

$$\hat{j} \ll \frac{r\tilde{\psi}_h^2}{2p(\tilde{\xi}_h + \tilde{\psi}_h)} \tag{6.114}$$

Since $\tilde{\xi}_h + \tilde{\psi}_h = \tilde{\xi}_h^0 + \tilde{\psi}_h^0 \equiv 1 + \tilde{\psi}_h^0$ and $\tilde{\psi}_h \geq \tilde{\psi}_h^0$, we can replace $\tilde{\xi}_h$ with 1 and $\tilde{\psi}_h$ with $\tilde{\psi}_h^0$ in (6.114), yielding

$$\hat{j} \ll \frac{r(\tilde{\psi}_h^0)^2}{2p(1+\tilde{\psi}_h^0)}. \tag{6.115}$$

This condition is fulfilled for sufficiently large water concentrations at the inlet $\tilde{\psi}_h^0$ and/or large parameter r. Physically, large r means a high rate of back diffusion of liquid water in membrane and/or a low rate of water vapor "leakage" to the channel. In both these cases, the upper limit of current density, which satisfies (6.115) increases.

Taking into account (6.87) and (6.49), condition (6.115) in dimension form reads

$$j \ll \frac{j_c(\psi_h^0)^2}{\psi_h^0 + \xi_h^0} \tag{6.116}$$

where

$$j_c = \frac{b\sigma_{m1} K_\lambda c_{tot}}{l_m c_w^{sat}} \tag{6.117}$$

is a characteristic current density. Note that c_w^{sat} is a function of cell temperature, whereas the inlet composition ξ_h^0 and ψ_h^0 depends on humidifier temperature.

If (6.115) is fulfilled, we can neglect the second term in the square brackets in Eq. (6.113). This equation then takes the form

$$\lambda \frac{\partial \hat{j}}{\partial \tilde{z}} = -\frac{\hat{j}^2}{\hat{J}\tilde{\xi}_h}, \quad \hat{j}(0) = \kappa. \tag{6.118}$$

The oxygen fraction $\tilde{\xi}_h$ is still governed by (6.108). The solution to (6.108) and (6.118) is

$$\tilde{\xi}_h = \exp\left(-\frac{\tilde{z}}{\mu}\right) \tag{6.119}$$

$$\hat{j} = \kappa \exp\left(-\frac{\tilde{z}}{\mu}\right), \tag{6.120}$$

where $\mu = \lambda \hat{J}/\kappa$. The local current density at the inlet κ is determined by the condition $\int_0^1 \hat{j}\, d\tilde{z} = \hat{J}$. Using (6.120) and calculating the integral we find

$$\kappa = f_\lambda \hat{J}. \tag{6.121}$$

The characteristic scale μ is then

$$\mu = -\left[\ln\left(1 - \frac{1}{\lambda}\right)\right]^{-1}. \qquad (6.122)$$

The solutions (6.119–1.122) do not contain the water management parameter r and p. Furthermore, the exponential shapes of $\tilde{\xi}_h$ and $\hat{j}/\hat{J}$ are governed by a single parameter, the oxygen stoichiometry λ. Eqs. (6.119), (6.120) with μ (6.122) and κ (6.121) coincide with the solutions (6.43), (6.44) in Section 6.3.2, where ideal membrane humidification *is assumed*. This means that the current, which obeys (6.115) does not produce any significant non-uniformity of membrane resistance along $\tilde{z}$. Equation (6.115) is thus *the condition of ideal membrane humidification*.

Physically, if the inlet water concentration is large enough and $\hat{j}$ is small, water produced by ORR has practically no effect on the profiles of oxygen concentration and local current along the channel. Membrane resistance and voltage loss in the membrane V_m are then almost constant along $\tilde{z}$. Since V_m is constant, it can be included into the "contact" resistances, and the model above reduces to that of Section 6.3.2.

This model does not take into account *transport* of liquid water in the backing layer and in the channel. If a small amount of liquid water partially fills voids of the backing layer, it simply reduces the effective diffusion coefficient of gaseous transport D_b. The level of flooding can be estimated from this model by comparing the diffusion coefficient which results from fitting of the experimental curves to the binary diffusion coefficients corrected for porosity-tortuosity.

The situation when the flux of liquid water constitutes a significant part of the total water flux in the backing layer is more difficult. This is a true two-phase regime of cell operation, which is not described by this model but is considered in Promislow's work of Chapter 7 of this book.

The model also ignores local 2D effects due to non-uniformity of oxygen and water distribution under the channel/rib [10]. This non-uniformity also reduces the effective diffusion coefficient of oxygen in 1D or 1D + 1D models. There is evidence that local 2D effects can be approximately accounted for in 1D + 1D models by simple correction of oxygen diffusion coefficient [32].

6.5. Concluding Remarks

Qualitative understanding of cell functioning is, probably, the most important outcome of the models presented above. In many situations, however, these models may offer a means for a rough characterization of cells performance. Each model considered contains several (less than 10) dimensionless parameters, which accumulate the basic transport and kinetic properties of a cell. Fitting the experimental curves to model equations provides an estimate of these properties [17, 33].

In [34], the model of Section 6.3 helps to explain the nature of high-frequency oscillations of cell voltage in the galvanostatic regime. In [33], this model was used to fit several experimental polarization curves of a cell with different oxygen concentrations at the inlet. This procedure gave quite reasonable Tafel slope, oxygen diffusion coefficient and exchange current density. The parameters resulted from fitting were then used as input data for a numerical quasi-3D simulation of a cell. Comparison of analytical, simulated and experimental curves enabled the estimation of the contribution of 3D effects to cell performance [33].

To our knowledge [33], is a first attempt to use a *hierarchy of models* rather than a single model for analysis of experimental curves. This seems to be a promising approach; the comparison of experiment with various models of differing complexity enables the evaluation of parameters which are accounted for in a high-level model and ignored in its low-level counterpart. Evidently, this idea of "models subtraction" might be useful since low-level models are fast and a large portion of comparison with the experiment can be performed with a minimal number of calculations. Analytical low-level models may thus serve as a bridge between experiment and CFD models.

Another promising application of the models above is cell optimization. Analytical solutions permit the construction of fast algorithms for optimizing the regime of cell operation. Recently the solutions of Section 6.3 were used to optimize the regime of operation of a cell with segmented electrodes [35]. Even the most sophisticated model of Section 6.4 in the general case reduces to a system of two ODEs, which can be solved very efficiently using the standard Runge-Kutta scheme. In many cases, numerical solution of two ODEs is faster than the solution of implicit algebraic equation. This model is thus also suitable for developing of optimization algorithms. Last but not least, analytical solutions presented above may serve as a basis for construction of stack models. Real PEFC stacks contain about a hundred cells, and the use of simplified models is unavoidable. Stack modelling is a large and exciting field for future analytical studies, see Chapter 9 by B. Wetton.

Acknowledgments

The author is grateful to A. Egmen, H. Scharmann, and T. Wuester for valuable experimental material. Helpful discussions with J. Divisek, H. Dohle, M. Eikerling, A. Kučernak, A. Kornyshev, J. Mergel, D. Stolten and K. Wippermann are gratefully acknowledged.

Appendix A: Formula for Oxygen Utilization

Equation (6.108) leads to a remarkably simple formula for oxygen utilization. Integrating this equation over $\tilde{z}$ we get

$$\hat{J}\lambda \int_0^1 \frac{\partial \tilde{\xi}_h}{\partial \tilde{z}} d\tilde{z} = -\int_0^1 \hat{j} d\tilde{z}$$

or

$$\hat{J}\lambda(\tilde{\xi}_h^1 - 1) = -\hat{J}, \qquad (6.123)$$

where $\tilde{\xi}_h^1$ is dimensionless oxygen molar fraction at the channel outlet. By definition, oxygen utilization $u = 1 - \tilde{\xi}_h^1$. From (6.123) we immediately obtain

$$u = \frac{1}{\lambda} \qquad (6.124)$$

Oxygen utilization is thus simply inversely proportional to stoichiometry. Note that this result follows from mass balance and does not depend on the details of ORR kinetics.

Appendix B: Accurate Explicit Approximations for General CCL Polarization Curve (6.17)

Here, we derive accurate explicit approximations for the general implicit polarization curve of the catalyst layer Eq. (6.17):

$$j = \sqrt{Q - j^2} \tan\left(\sqrt{Q - j^2}\right) \qquad (6.125)$$

Hereinafter, the sign "tilde" is omitted.

Large j

Equation (6.125) can be rewritten as

$$Q - j^2 = Q - (Q - j^2) \tan^2\left(\sqrt{Q - j^2}\right). \qquad (6.126)$$

Introducing

$$y = \sqrt{Q - j^2} \qquad (6.127)$$

Equation (6.126) becomes $y^2(1 + \tan^2 y) = Q$, or

$$y^2 = Q \cos^2 y. \qquad (6.128)$$

Taking the positive root, we get an equation suitable for iterations:

$$y_{n+1} = \arccos\left(\frac{y_n}{\sqrt{Q}}\right) \qquad (6.129)$$

We are seeking a solution to (6.125) in the limit $j \gg 1$. This means that the argument of tan-function in (6.125) must be on the order of $\pi/2$ i.e. initial guess for iterations is $y_0 = \pi/2$. Furthermore, for large j, Q must be also large, whereas $y < \pi/2$. This allows us to expand arccos in (6.129) and we get

$$y_{n+1} \simeq \frac{\pi}{2} - \frac{y_n}{\sqrt{Q}} \qquad (6.130)$$

Now we start iterations. First iteration gives

$$y_1 = \frac{\pi}{2} - \frac{\pi}{2\sqrt{Q}} = \frac{\pi}{2}\left(1 - \frac{1}{Q^{1/2}}\right)$$

Subsequent iterations give

$$y_2 = \frac{\pi}{2}\left(1 - \frac{1}{Q^{1/2}} + \frac{1}{Q}\right)$$

$$y_3 = \frac{\pi}{2}\left(1 - \frac{1}{Q^{1/2}} + \frac{1}{Q} - \frac{1}{Q^{3/2}}\right)$$

In the brackets, we get the sum $\sum_{n=0}^{\infty}(-1)^n Q^{-n/2}$, which is

$$\sum_{n=0}^{\infty}(-1)^n Q^{-n/2} = \frac{\sqrt{Q}}{1 + \sqrt{Q}}. \qquad (6.131)$$

Finally, we obtain

$$y = \frac{\pi}{2}\left(\frac{\sqrt{Q}}{1 + \sqrt{Q}}\right) \qquad (6.132)$$

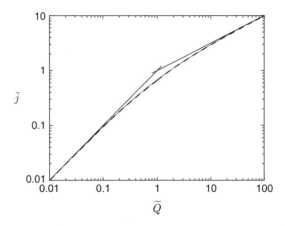

FIGURE 6.23. Thin solid line - numerical solution to Eq. (6.125), thick solid lines – leading-order asymptotic solutions $j = \sqrt{Q}$ (upper) and $j = Q$ (lower). Dashed curve - matched asymptotic solutions (6.134) ($j \leq 1$) and (6.137) ($j > 1$).

or, taking into account (6.127),

$$j = \sqrt{Q - \frac{\pi^2}{4}\left(\frac{Q}{(1+\sqrt{Q})^2}\right)} \tag{6.133}$$

A better accuracy for $j \geq 1$ gives

$$j = \sqrt{Q - \frac{\pi^2}{4}\left(\frac{AQ}{(1+\sqrt{AQ})^2}\right)}, \tag{6.134}$$

where $A = 0.82$.

Small j

In this case, we use Eq. (6.128) for iterations. Leading-order solution of Eq. (6.125) in the limit of small currents is $j = Q$. Initial guess for iterations is, therefore, $y_0 = \sqrt{Q - j^2} = \sqrt{Q(1-Q)}$. For small j, y is also small, and we can expand cosine in (6.128). This yields

$$y_{n+1} = \sqrt{Q}\left(1 - \frac{y_n^2}{2}\right) \tag{6.135}$$

Iterating, we get:

$$y_1 = \sqrt{Q}\left(1 - \frac{Q(1-Q)}{2}\right) = \sqrt{Q}\left(1 - \frac{Q}{2} + \frac{Q^2}{2}\right)$$

$$y_2 = \sqrt{Q}\left(1 - \frac{Q}{2}\left(1 - \frac{Q(1-Q)}{2}\right)^2\right) \simeq \sqrt{Q}\left(1 - \frac{Q}{2} + \frac{Q^2}{2} - \frac{5}{8}Q^3\right)$$

Solving the equation $Q - j^2 = y_2^2$ and expanding the result we find

$$j = Q - \frac{5}{8}Q^2 + \frac{87}{128}Q^3 - \frac{269}{1024}Q^4 + O(Q^5) \qquad (6.136)$$

To match solutions (6.134) and (6.136) we rewrite (6.136) as

$$j = Q - \frac{5}{8}Q^2 + aQ^3 - bQ^4 \qquad (6.137)$$

Coefficients a and b can be found from continuity of the functions (6.134) and (6.137) and their first derivatives at $Q = 1$. Simple calculation gives

$$a = 0.3800233372, \quad b = 0.08959793057 \qquad (6.138)$$

With these coefficients, accuracy of (6.134) and (6.137) is better than 2% in the whole range of current densities (Figure 6.23). Note that the second and higher derivatives $d^n j/dQ^n$ experience discontinuity at $Q = 1$.

List of Symbols

$\sim$	Marks dimensionless variables
$\hat{}$	Marks dimensionless current densities in Section 6.4
a	Dimensionless parameter (6.61)
a_η	Dimensionless parameter (6.55)
b	Tafel slope (V)
c	Oxygen molar fraction (mol cm^{-3})
c_{tot}	Total molar concentration of the mixture in the cathode channel (mol cm^{-3})
c_{ref}	Reference oxygen molar concentration (mol cm^{-3})
d_h	In-plane width of the channel
D_b	Oxygen diffusion coefficient in the backing layer (cm^{2}s^{-1})
E	Total voltage loss (V)
f_λ	Dimensionless function (6.45)
f_r	Dimensionless function (6.91)
F	Faraday constant (9.6495 × 10^4 Coulomb mol^{-1})
h	Channel height above the GDL (cm)

$\sim$	Marks dimensionless variables
i_*	Exchange current density per unit volume (A cm^{-3})
I	Total current in the external load (A)
j	Section 6.2: local proton current density in the catalyst layer; in the other sections: local current density in a cell (A cm^{-2})
j_*	Characteristic current density (A cm^{-2})
j_D	Limiting current density (A cm^{-2})
J	Mean current density in a cell (A cm^{-2})
k	Parameter (A cm^{-2})
K	Dimensionless parameter (6.56)
l	Thickness of the layer (see the subscript) in MEA (cm)
l_*	Reaction penetration depth (cm)
L	Channel length
M	Molecular weight (g mol^{-1})
N	Molar flux (mol cm^{-2}s^{-1})
Q	Rate of ORR (A cm^{-3}s^{-1})
r	Dimensionless parameter (6.80)
R	Contact resistance (Ω cm^2)
ξ_h^0/ψ_h^0	the ratio of inlet oxygen to water molar fractions.
V_m	Voltage loss in membrane (V)
v	Velocity of the flow in the channel (cm s^{-1})
v_w	Velocity of the degradation wave (cm s^{-1})
x	Coordinate across the cell (cm)
z	Coordinate along the channel (cm)
z_w	Position of the wavefront (cm)

Superscripts

0	channel inlet
1	channel outlet
mix	mixed regime
ox	oxygen-limiting regime
w	water-limiting regime

Subscripts

0	membrane/CCL interface
1	CCL/GDL interface
b	backing layer
$crit$	critical
$fast$	fast propagation of the wave
h	channel
l	liquid water
lim	limiting
max	maximal

0	membrane/CCL interface
oc	open-circuit
$slow$	slow propagation of the wave
t	catalyst layer
tot	total
v max	when wave velocity reaches maximum
w	wave

Greek Symbols

α	Total transfer coefficient of water through the membrane
β	Dimensionless parameter (6.52)
ε	Small dimensionless parameter ($\varepsilon \ll 1$)
ζ	Molar fraction of nitrogen in air
η	Polarization voltage of the cathode side (V)
λ	Stoichiometry of oxygen flow
λ_w	Membrane water content (number of water molecules per sulfonic group)
μ	Dimensionless characteristic length of oxygen consumption in the channel
ξ	Oxygen molar fraction
ρ	Mass density of the flow in the channel
σ	Electron conductivity of the carbon phase (Ω cm^{-1})
σ_m	Proton conductivity of bulk membrane (Ω cm^{-1})
σ_{m1}	Proton conductivity of bulk membrane at unit water content (Ω cm^{-1})
σ_t	Proton conductivity of a polymer electrolyte in the catalyst layer (Ω cm^{-1})
τ_d	Local time of degradation
ϕ	Potential (V)
ψ	Water molar fraction
ω	Dimensionless parameter (6.24)

References

[1] J. Newman. *Electrochemical Systems*. Prentice Hall, Inc., Englewood Cliffs, NJ 07632, 1991.

[2] A. A. Kulikovsky, J. Divisek, and A. A. Kornyshev. Modeling the cathode compartment of polymer electrolyte fuel cells: Dead and active reaction zones. *J. Electrochem. Soc.*, 146(11):3981–3991, 1999.

[3] S. M. Senn and D. Poulikakos. Laminar mixing, heat transfer and pressure drop in tree-like microchannel nets and their application for thermal management in polymer electrolyte fuel cells. *J. Power Sources*, 130:178–191, 2004.

[4] K. A. Striebel, F. R. McLarnon, and E. J. Cairns. Steady-state model for an oxygen fuel cell electrode with an aqueous carbonate electrolyte. *Ind. Eng. Chem. Res.*, 34:3632–3639, 1995.

[5] M. L. Perry, J. Newman, and E. J. Cairns. Mass transport in gas-diffusion electrodes: A diagnostic tool for fuel-cell cathodes. *J. Electrochem. Soc.*, 145(1):5–15, 1998.

[6] T. E. Springer and S. Gottesfeld. Pseudohomogeneous catalyst layer model for polymer electrolyte fuel cell. In R. E. White, M. W. Verbrugge, and J. F. Stockel, editors, *Modeling of Batteries and Fuel Cells*, volume PV 91–10 of *The Electrochemical Society Softbound Proceedings Series*, pages 197–208, 10 South Main St., Pennington, NJ 08534–2896, USA, 1991. The Electrochem. Soc., Inc.

[7] M. Eikerling and A. A. Kornyshev. Modelling the performance of the cathode catalyst layer of polymer electrolyte fuel cells. *J. Electroanal. Chem.*, 453:89–106, 1998.

[8] A. A. Kulikovsky. Quasi three-dimensional modeling of PEM fuel cell: Comparison of the catalyst layers performance. *Fuel Cells*, 1(2):162–169, 2001.

[9] A. A. Kulikovsky. Performance of catalyst layers of polymer electrolyte fuel cells: Exact solutions. *Electrochem. Comm.*, 4(4):318–323, 2002.

[10] A. A. Kornyshev and A. A. Kulikovsky. Characteristic length of fuel and oxygen consumption in feed channels of polymer electrolyte fuel cells. *Electrochimica Acta*, 46(28):4389–4395, 2001.

[11] H. Dohle, A. A. Kornyshev, A. A. Kulikovsky, J. Mergel, and D. Stolten. The current voltage plot of PEM fuel cell with long feed channels. *Electrochem. Comm.*, 3(2):73–80, 2001.

[12] A. A. Kulikovsky. The voltage current curve of a polymer electrolyte fuel cell: "Exact" and fitting equations. *Electrochem. Comm.*, 4:845–852, 2002.

[13] A. A. Kulikovsky. Gas dynamics in channels of a gas-feed direct methanol fuel cell: Exact solutions. *Electrochem. Comm.*, 3(10):572–579, 2001.

[14] G. J. M. Janssen and M. L. J. Overvelde. Water transport in the proton-exchage-membrane fuel cell: Measurements of the effective drag coefficient. *J. Power Sources*, 101:117–125, 2001.

[15] D. J. L. Brett, S. Atkins, N. P. Brandon, V. Vesovic, N. Vasileiadis, and A. R. Kućernak. Measurement of the current distribution along a single flow channel of a solid polymer fuel cell. *Electrochem. Comm.*, 3:628–632, 2001.

[16] A. A. Kulikovsky, A. Kućernak, and A. Kornyshev. Feeding PEM fuel cells. *Electrochimica Acta*, 50:1323–1333, 2005.

[17] A. A. Kulikovsky. The effect of stoichiometric ratio λ on the performance of a polymer electrolyte fuel cell. *Electrochimica Acta*, 49(4):617–625, 2004.

[18] S. Gottesfeld and T. A. Zawodzinski. Polymer electrolyte fuel cells. In R. C. Alkire, H. Gerischer, D. M. Kolb, and Ch. W. Tobias, editors, *Advances in Electrochemical Science and Engineering*, volume 5, pages 195–301. Wiley-VCH, Weinheim, 1997.

[19] S.-Y. Ahn, S.-J. Shin, H. Y. Ha, S.-A. Hong, Y.-C. Lee, T. W. Lim, and I.-H. Oh. Performance and lifetime analysis of the kW-class PEMFC stack. *J. Power Sources*, 106:295–303, 2002.

[20] A. A. Kulikovsky, H. Scharmann, and K. Wippermann. Dynamics of fuel cell performance degradation. *Electrochem. Comm.*, 6:75–82, 2004.

[21] T. Okada, G. Xie, and M. Meeg. Simulation for water management in membranes for polymer electrolyte fuel cells. *Electrochimica Acta*, 43:2141–2155, 1998.

[22] M. Eikerling, Yu. I. Kharkats, A. A. Kornyshev, and Yu. M. Volfkovich. Phenomenological theory of electro-osmotic effect and water management in

polymer electrolyte proton-conducting membranes. *J. Electrochem. Soc.*, 145:2684, 1998.

[23] D. R. Sena, E. A. Ticianelli, V. A. Paganin, and E. R. Gonzalez. Effect of water transport in a PEFC at low temperatures operating with dry hydrogen. *J. Electroanal. Chem.*, 477:164–170, 1999.

[24] A. Weber and J. Newman. Transport in polymer-electrolyte membranes I. Physical model. *J. Electrochem. Soc.*, 150:A1008-A1015, 2003.

[25] V. Gurau, F. Barbir, and H. Liu. An analytical solution of a half-cell model for PEM fuel cells. *J. Electrochem. Soc.*, 147(7):2468–2477, 2000.

[26] L. Pisani, G. Murgia, M. Valentini, and B. D'Aguanno. A new semi-empirical approach to performance curves of polymer electrolyte fuel cells. *J. Power Sources*, 108:192–203, 2002.

[27] P. Berg, K. Promislow, J. St. Pierre, J. Stumper, and B. Wetton. Water management in PEM fuel cells. *J. Electrochem. Soc.*, 151(3):A341–A353, 2004.

[28] T. F. Fuller and J. Newman. Experimental determination of the transport number of water in Nafion-117 membrane. *J. Electrochem. Soc.*, 139(5): 1332–1337, 1992.

[29] A. A. Kulikovsky. The effect of cathodic water on performance of a polymer electrolyte fuel cell. *Electrochimica Acta*, 49(28):5187–5196, 2004.

[30] T. J. VanderNoot and I. Abrahams. The use of genetic algorithms in the nonlinear regression of immittance data. *J. Electroanal. Chem.*, 448:17–23, 1998.

[31] A. A. Kulikovsky. Semi-analytical 1D + 1D model of a polymer electrolyte fuel cell. *Electrochem. Comm.*, 6:969–976, 2004.

[32] A. A. Kulikovsky. Two models of a PEFC: Semi-analytical vs numerical. *Int. J. Energy Res.* 29:1153–1165, 2005.

[33] A. A. Kulikovsky, T. Wüster, T. Egmen, and D. Stolten. Analytical and numerical analysis of PEM fuel cell performance curves. *J. Electrochem. Soc.*, 152(6):A1290–A1300, 2005.

[34] A. A. Kulikovsky, H. Scharmann, and K. Wippermann. On the origin of voltage oscillations of a polymer electrolyte fuel cell in galvanostatic regime. *Electrochem. Comm.*, 6:729–736, 2004.

[35] S. M. Senn and D. Poulikakos. Multistage polymer electrolyte fuel cells based on nonuniform cell potential distribution functions. *Electrochem. Comm.*, 7:773–780, 2005.

7
Phase Change and Hysteresis in PEMFCs

KEITH S. PROMISLOW

7.1. Introduction

A fundamental difficulty in upscaling micron-level transport parameters describing the components of polymer exchange membrane fuel cells (PEMFC) to device-level performance is the presence of fronts induced by various types of phase change. Fronts are not only a driving force behind hysteresis and slow transient behavior at the device level but also greatly complicate numerical resolution of governing models. PEMFC are typically not operated at steady-state, particularly in automotive applications, and more to the point do not necessarily perform at steady-state even when operated under steady conditions. The wide range of timescales present in PEMFC make direct simulation of comprehensive transient models impossible. Efficient transient modeling requires an identification of timescales, and particularly, a grouping of processes based upon similarity in timescale. A situation where analysis pays a handsome dividend, and which is richly exhibited in PEMFC operation, arises when time scales are widely separated, so that one group of processes may be treated at steady-state, driven adiabatically by the more slowly evolving processes [26].

We consider two examples of phase transitions in which an analytical reduction of the problem can simplify the underlying model by orders of magnitude. The first example, summarized from [28], is transient multi-phase flow in the porous electrode of the PEMFC, in which liquid water generated in the catalyst layer forms a front in the hydrophobic gas diffusion layer (GDL). Depending upon the operating conditions the front may propagate across the GDL, spilling into the channel, or stall part-way across. We derive a reduced model which captures the impact of evaporation at the front on the front motion. The second example, summarized from [3], arises in an experimental stirred-tank reactor for which, under dry inlet conditions, the membrane phase-separates into ignited and extinguished regions due to the feedback between current driven water production and membrane resistivity. Tracking the temporal evolution of the ignited and extinguished regions requires a proper accounting of the lateral flux of water within the membrane. The front motion leads to hysteresis and slow transient behavior. We finish by showing some preliminary results for partially humidified feeds in a counter-flowing PEMFC where hysteresis can arise as a transition from two-phase to single phase regimes on the cathode electrode.

7.2. Two-Phase Flow in the GDL

There has been considerable attention given to the development of 2 and 3-D numerical models to approximate the coupled two-phase mass and heat transport within unit cells, see [5, 7, 8, 15, 17, 22, 25, 29, 30, 38, 39, 37]. However, these approaches have either not fully resolved the multiphase flow, advecting the liquid as suspended droplets in the gas flow, or used isothermal models in which the vapor and liquid are treated as a combined mixture, or have experienced convergence difficulties for physically reasonable values of the parameters. The majority of the treatments are at steady-state, and none has treated the transient evolution of the free surface separating the two-phase and dry regions in the hydrophobic GDL.

The two-phase/dry interface arises for two complementary reasons. First, the liquid water transport in the hydrophobic GDL is degenerate—at low liquid volume fractions the water breaks into isolated droplets which cannot support a flux. The second is that the small pore size in porous media forces the liquid into droplets or thin rivulets with a tremendous liquid-gas surface area density. Consequently rates of phase change are extremely high, inducing boundary layers which localize the vast majority of the phase change at the two-phase/dry interface. A difficulty in resolving the evolution of two-phase fronts in the full system is the dichotomy of timescales. Even in the single phase regime, previous analysis, [27], has shown that the convective and the diffusive gas transports relax on timescales separated by 3–4 orders of magnitude. The inclusion of the slow liquid transients stretches the timescale dichotomy from 10^{-6} seconds for pressure relaxation to 10^2 seconds for motion of the free interface.

A second difficulty in numerical computations is the length scale dichotomy. The large aspect ratio of the fuel cell, roughly 1000 to 1 along-the-channel versus through-the-membrane electrode assembly (MEA), renders the resolution of yet thinner fronts within the already thin through-plane direction of the MEA a significant obstacle to numerical stability. Several authors, Berg et al. [3], Divisek et al. [9], Freunberger et al. [12], Kulikovsky [18], and Buchi et al. [6], have exploited this high aspect ratio, proposing so-called 1 + 1D models which couple 1D systems for the through-MEA transport (z direction) to 1D along-the-channel convective transport (y direction) in the flow fields; see Figure 7.2 (right). The through-MEA models provide source terms for the convective channel flow, consuming oxygen and producing water vapor as the gas flows from inlet to outlet. In turn, the gas flow provides the boundary conditions for the through-MEA codes. Such models have had success in determining the local current and voltage, and hence power generation, of the cell; however, they have not incorporated the full multiphase aspects of water management due to its significant computational complexity.

The *analytic* reduction of the two-phase front motion presented in this section reduces the complexity of computing the slow transients by several orders of magnitude, switching the timescale dichotomy from a obstacle to an asset. The inclusion of multiphase water management in models of stacks of 50–200 unit cells used for automotive applications; see [2], would seem certain to require a reduction along these lines.

The local current density, $\tilde{I}$, is the forcing term for the system, consuming oxygen, producing heat and water. However, in terms of the mass transport capacity of the gas diffusion layer, the molar and heat fluxes are small and the relaxation times are short. It is, therefore, reasonable to take these transport processes at steady-state and moreover to linearize the transport matrix about the channel values. We assume that the liquid water motion within the hydrophobic GDL is degenerate, that the water does not move unless its volume fraction reaches a percolation threshold, $\beta_* > 0$. In particular, we assume that all pores are hydrophobic. In a hydrophobic media, the capillary pressure dominates the gas pressure yielding a liquid water distribution which is a linear function of a liquid water potential.

Focusing on a 1D slice in which part of the GDL is dry and part is two-phase, as depicted in Figure 7.1, we construct a class of solutions in which the

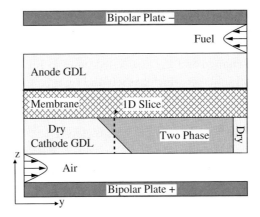

FIGURE 7.1. An along-the-channel (y-z) slice of a unit cell depicting the development of two-phase regions in the cathode GDL of a counter-flowing PEMFC unit cell. When unhumidified gases are used at the air and fuel inlets, the inlet regions of the GDL are typically dry and two-phase regions form down-stream where the gases are more humid. The goal is to predict the formation and evolution of the two-phase region within the 1D slice.

liquid water in the two-phase region is above β_*, and the gas phase is nearly saturated. The key to resolving the interface motion is to estimate the flux balance of liquid water; the liquid water generated at the catalyst layer moves into the interface, some of it evaporates, and the rest serves to move the interface. Having calculated the amount of phase change which occurs at the two-phase interface, the front motion reduces to a single differential equation, and the balance of the problem consists of inverting two, nearly singular matrices. As a by-product the analysis shows that the partitioning of product water into liquid and vapor is particularly sensitive to the thermal conductivity of the solid phase of the GDL. Increasing the thermal conductivity by factors of 2–10 has a significant impact on the water partitioning and can even *reverse* the direction of vapor phase flow, driving vapor from the channel to the cathode catalyst layer; see (68) and following discussion.

While the model presented applies to a wide range of PEM fuel cell operational regimes; there are some caveats to its application. We assume the flux of gas and heat out of the catalyst layer scale with the current density. This would not be the case, for example, with a dry anode feed and a wet cathode feed, which would generate water transport independent of the current level. We assume there is no lateral pressure gradient imposed in the x direction across the GDL such as would arise in interdigitated or serpentine flow fields: we consider straight flow fields.

7.2.1. Model and Nondimensionalization

7.2.1.1. Notation

The free boundary layer has a sub-layer structure. In the evaporative layer from (z_*^0, z_*^+), see Figure 7.2, the liquid water volume fraction is greater than the immobile volume fraction, the gas phase is under-saturated, and the amount of phase change and the associated jump in fluxes are determined. In the infinitesimal sublayer which has left and right limits z_*^- and z_*^0, respectively, the water volume fraction reaches the immobile volume fraction.

The free boundary layer is demarcated by the points z_*^-, z_*^0, z_*^+; see Figure 7.2. The evaporative boundary layer is the region (z_*^0, z_*^+), while the infinitesimal sublayer has left and right limits z_*^- and z_*^0. The jump of a quantity across the the resolved boundary layer is denoted

$$\llbracket N \rrbracket = N(z_*^+) - N(z_*^0), \tag{7.1}$$

the jump across the sublayer is denoted

$$\llbracket N \rrbracket_s = N(z_*^0) - N(z_*^-), \tag{7.2}$$

while the jump across the whole free surface is denoted

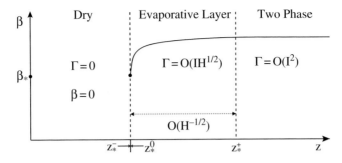

FIGURE 7.2. The variation in liquid water volume fraction β through the scaled free boundary layer. Liquid volumes below β_* are immobile. The evaporative layer has a width $O(1/\sqrt{H})$ and evaporation rate of $O(I\sqrt{H})$, where $H \gg 1$ is a rate constant for the phase change, Γ, and $I \ll 1$ is current density scaled by normalized diffusive flux. There is no phase change in the dry regime, and the phase change in the two-phase regime, outside of the evaporative layer, is $O(I^2)$. The liquid volume fraction jumps from β_* to 0 across the sub-layer.

$$[N]_f = N(z_*^+) - N(z_*^-). \tag{7.3}$$

The dimensional version of a variable has a tilde, such as the current density $\tilde{I}$, while the dimensionless version is unadorned, I. The j'th component of a vector $\vec{V}$ is denoted $\vec{V}_j$. The subscripts $\tilde{z}$ and z denote differentiation with respect to that variable. Except for the scaling parameters $t_*, I_*,$ and T_*, the sub- or superscript $*$ denotes a quantity evaluated at the two-phase boundary point z_*. The superscript t after a row-vector indicates transpose. The projections $\pi_d, \pi_w : \mathbf{R}^5 \mapsto \mathbf{R}^4$ are given by $\pi_d(x_1, \ldots x_5) = (x_1, x_2, x_3, x_4)^t$ and $\pi_w(x_1, \ldots, x_5) = (x_1, x_2, x_3, x_5)^t$. The positive z direction is from channel to membrane, positive fluxes are channel to membrane.

7.2.1.2. The GDL Model

The dimensional unknowns are the molar concentrations of oxygen, water vapor, and nitrogen, denoted by $\tilde{C}_o, \tilde{C}_v,$ and $\tilde{C}_n$, the liquid water volume fraction β, and the temperature $\tilde{T}$. The total (convective plus diffusive) flux of species α is denoted N_α. The variables represent averages over the pore-scale structure of the liquid-gas-solid domains. The gas phase occupies the fraction $1 - \beta$ of the void space, which is $\epsilon = 0.74$ of the total volume fraction. Conservation of mass and energy is expressed through the following PDEs which are coupled not only through the constitutive relations for the fluxes but also through the condensation, $\tilde{G}$, which appears as a source or sink term.

$$\epsilon \frac{\partial\left((1-\beta)\tilde{C}_o\right)}{\partial t} + (\tilde{N}_o)_{\tilde{z}} = 0, \quad (7.4)$$

$$\epsilon \frac{\partial\left((1-\beta)\tilde{C}_v\right)}{\partial t} + (\tilde{N}_v)_{\tilde{z}} = -\tilde{\Gamma}, \quad (7.5)$$

$$\epsilon \frac{\partial\left((1-\beta)\tilde{C}_n\right)}{\partial t} + (\tilde{N}_n)_{\tilde{z}} = 0, \quad (7.6)$$

$$\epsilon \frac{\partial(C_l\beta)}{\partial t} + (\tilde{N}_l)_{\tilde{z}} = \tilde{\Gamma}, \quad (7.7)$$

$$\frac{\partial(\overline{\rho c}\tilde{T})}{\partial t} + (\tilde{N}_T)_{\tilde{z}} = h_v\tilde{\Gamma}. \quad (7.8)$$

The total flux of gas species α is the sum of the convective and diffusive fluxes

$$\tilde{N}_\alpha = U_g \tilde{C}_\alpha - D\tilde{C}\left(\frac{\tilde{C}_\alpha}{\tilde{C}}\right)_{\tilde{z}}. \quad (7.9)$$

As a convenience, we have assumed a Fickian formulation for the diffusive flux, with a common diffusivity, D, for all gas species. It is straightforward to incorporate a Maxwell-Stefan formulation; see [34, 35] for the diffusive gas fluxes into the analysis presented here.

The gas pressure is given by the ideal gas law

$$\tilde{P}_g = \tilde{C}\mathcal{R}\tilde{T}, \quad (7.10)$$

where the total gas concentration $\tilde{C}$ satisfies $\tilde{C} = \tilde{C}_o + \tilde{C}_v + \tilde{C}_n$. The gas velocity is given by Darcy's law

$$U_g = -\frac{Kk_{rg}(\beta)}{\mu_g}\tilde{P}_{g,\tilde{z}}, \quad (7.11)$$

where K is the permeability, μ_g is the gas viscosity, and the gas relative permeability, k_{rg} is given by

$$k_{rg}(\beta) = (1-\beta)^3. \quad (7.12)$$

An essential feature of the multiphase model are the constitutive relations which specify the liquid relative permeability and the capillary pressure in terms of the liquid volume fractions. These relations incorporate the considerable role played by the hydrophobicity of the GDL into the model. Our approach is not new to the fuel cell literature; (see [9, 22]), and is borrowed from the hydrology literature on hydrophobic soils; (see [24] and references within). The liquid velocity is defined by Darcy's law

$$U_l = -\frac{K k_{rl}(\beta)}{\mu_l} \tilde{P}_{l,\tilde{z}}, \qquad (7.13)$$

where the liquid relative permeability k_{rl} is given by

$$k_{rl}(\beta) = (\beta - \beta_*)_+^{\frac{1}{2}}, \qquad (7.14)$$

where $(z)_+ \equiv \mathrm{Max}\{z, 0\}$. This constitutive form for the relative permeability supposes that for low liquid volume fractions, $\beta < \beta_*$, the water is contained in disconnected droplets which are immobile. For volume fractions larger than the percolation threshold β_* the droplets merge to form a locally connected domain which supports transport. A second impact of hydrophobicity is on the capillary pressure which describes the pressure jump across the pore-scale liquid-gas interfaces,

$$\tilde{P}_l = \tilde{P}_g + \tilde{P}_c, \qquad (7.15)$$

where following [19], we take the capillary pressure to have the form

$$\tilde{P}_c = S_p \gamma_s \left(\frac{\epsilon}{K}\right)^{\frac{1}{2}} \mathcal{J}(\beta). \qquad (7.16)$$

Here, γ_s is the liquid surface tension, $S_p > 0$ is a scaling constant, which includes both contact angle effects and the substantially larger scaling for capillary pressure in hydrophobic media. For the dependence of capillary pressure on liquid volume fraction, we use a simplified van Genuchten function; see [14],

$$\mathcal{J}(\beta) = \sqrt{\frac{1}{(1-\beta)^2} - 1}, \qquad (7.17)$$

which describes the increasing pressure jump required at high liquid volume fractions to force the water into increasingly smaller pores. The liquid flux, $\tilde{N}_l$, is defined in terms of the liquid velocity, the liquid volume fraction, and the constant liquid molar density, C_l, as

$$\tilde{N}_l = C_l U_l \beta. \qquad (7.18)$$

The heat flux is given by

$$\tilde{N}_T = \overline{\rho c U \tilde{T}} - \bar{\kappa} \tilde{T}_{\tilde{z}}, \qquad (7.19)$$

where c is the specific heat, ρ is the density, and κ is the thermal conductivity. The thermal conductivity is taken to be that of the solid phase, since it is dominant,

$$\overline{\kappa} = (1-\epsilon)\kappa_s, \quad (7.20)$$

while the heat capacity and convective heat flux are averaged over the phases

$$\overline{\rho c} = (1-\epsilon)\rho_s c_s + \epsilon\big((1-\beta)\rho_g c_g + \beta \rho_l c_l\big), \quad (7.21)$$

$$\overline{\rho c U} = \epsilon\big((1-\beta)\rho_g c_g U_g + \beta \rho_l c_l U_l\big). \quad (7.22)$$

The condensation rate $\tilde{\Gamma}$ is given by

$$\tilde{\Gamma} = \frac{\gamma_v h(\beta, \tilde{C}_v)}{L_p}\big(\tilde{C}_v - \tilde{C}_{sat}(\tilde{T})\big), \quad (7.23)$$

where $L_p = 1\mu m$ is the typical pore size in the GDL, and the coefficient of evaporation is approximated by Langmuir's formula; see [40],

$$\gamma_v = \sqrt{\frac{R\tilde{T}}{2\pi m_v}}, \quad (7.24)$$

where $m_v = 0.018$ kg/mole is the molecular weight of water, and the liquid-gas interface density is given by

$$h(\beta, \tilde{C}_v) = \begin{cases} \beta^{\frac{2}{3}} & \tilde{C}_v < \tilde{C}_{sat}, \\ (1-\beta)^{\frac{2}{3}} & \tilde{C}_v > \tilde{C}_{sat}. \end{cases} \quad (7.25)$$

Here the exponent 2/3 arises from the assumption of spherical water droplets and gas bubbles, with the corresponding relationship between volume and surface area. The vapor saturation concentration, $\tilde{C}_{sat}$ is given as

$$\tilde{C}_{sat}(\tilde{T}) = \frac{\tilde{P}_{sat}(\tilde{T})}{R\tilde{T}}, \quad (7.26)$$

where the saturation pressure $\tilde{P}_{sat}$ of water vapor is given from known empirical formulas, see [33].

7.2.1.3. Boundary Conditions

We prescribe fluxes at the catalyst layer-GDL boundary. At the channel-GDL boundary, we prescribe Robin-type conditions to account for boundary layers introduced in the gas channel flow by the fluxes from the GDL.

The fluxes out of the catalyst layer are described by three interrelated parameters: the current density, $\tilde{I}$, the heat flux out of the catalyst layer, $\tilde{N}_{cT}^m$, and the water (liquid plus vapor) flux out of the catalyst layer, $\tilde{N}_{cw}^m$. These last two quantities are thought of as the heat and water production of the catalyst layers, which are clearly related to the current density $\tilde{I}$, adding or subtracting heat and water flux between the membrane and catalyst layers. The heat and water fluxes between the membrane and catalyst layer depend upon the state of the membrane and the anode channel, and as such we take $\tilde{N}_{cT}^m$ and $\tilde{N}_{cw}^m$ as given quantities. In the fixed boundary layer at the catalyst layer, the phase change resolves the catalyst water flux into a vapor flux and a liquid water flux

$$\tilde{N}_{cw}^m = \tilde{N}_l^m + \tilde{N}_v^m. \tag{7.27}$$

The vapor flux will be determined by the temperature profile in the vapor-saturated two-phase region. In the event that the vapor flux $\tilde{N}_v^m$ is toward the membrane, the liquid water flux into the GDL can exceed the water flux out of the catalyst layer, $\tilde{N}_{cw}$. The phase change within the catalyst-GDL boundary layer also impacts the heat flux, $\tilde{N}_T^m$, into the GDL,

$$\tilde{N}_{cT}^m - \tilde{N}_T^m = h_v \tilde{N}_v^m. \tag{7.28}$$

We take the same mass transport parameter, $\tilde{r}$, for each of the three gas species. The heat transport is primarily through the channel landing, and we take a value of the heat transport parameter, $\tilde{r}_T$, which is consistent with a solid-solid interface; see [11]. The cathode channel total gas concentration, $\overline{C}$, satisfies

$$\overline{C} = \overline{C}_o + \overline{C}_v + \overline{C}_n = \frac{P}{\mathcal{R}\overline{T}_c} \tag{7.29}$$

and will be prescribed in place of $\overline{C}_n$. The value $\overline{C} = 102$ moles/m^3 given in Table 7.1 corresponds to a cathode pressure of 3 atmospheres at 353°K.

TABLE 7.1 Material parameters.

c_g	2.6×10^3	J/(Kg °K)	c_l	10^3	J/(Kg °K)	c_s	7.1×10^2	J/(Kg °K)
$\overline{C}$	1.02×10^2	mol/m^3	C_l	5.56×10^4	mol/m^3	D	8.0×10^{-6}	m^2/s
F	96485	C/mol	h_v	4.54×10^4	J/mole	$\tilde{I}$	1×10^4	Amp/m^2
K	1.0×10^{-12}	m^2	L	3×10^{-4}	m	L_p	1×10^{-6}	m
m_v	0.018	Kg/mole	$\mathcal{R}$	8.31	J/(mole °K)	$\tilde{r}$	8.0	m/s
$\tilde{r}_T$	1.1×10^4	J/(m^2 s °K)	S_p	100	-	$\overline{T}_c$	353	°K
β_*	0.1	-	ϵ	0.74	-	γ_s	0.0724	Kg/s^2
γ_v	161	m/s	κ_g	2.8×10^{-2}	J/(m s °K)	κ_l	0.675	J/(m s °K)
κ_s	1	J/(m s °K)	μ_g	2.24×10^{-5}	Kg/(m s)	μ_l	0.3	Kg/(m s)
ρ_g	2	Kg/m^3	ρ_l	1×10^3	Kg/m^3	ρ_s	4.9×10^2	Kg/m^3

TABLE 7.2 Boundary conditions at membrane and channel interfaces.

Region I: Membrane-GDL	Region II: Channel-GDL
$\tilde{N}_o = \frac{\tilde{I}(x)}{4F}$	$\tilde{r}(\tilde{C}_o(0) - \overline{C}_o) = -\tilde{N}_o(0)$
$\tilde{N}_v = \tilde{N}_v^m$	$\tilde{r}(\tilde{C}_v(0) - \overline{C}_v) = -\tilde{N}_v(0)$
$\tilde{N}_n = 0$	$\tilde{r}(\tilde{C}_n(0) - \overline{C}_n) = -\tilde{N}_n(0)$
$\tilde{N}_l = \tilde{N}_{cw}^m - \tilde{N}_v^m$	$\beta = 0$
$\tilde{N}_T = \tilde{N}_{cT}^m - h_v \tilde{N}_v^m$	$\tilde{r}_T(\tilde{T}(0) - \overline{T}_c) = -\tilde{N}_T(0)$

7.2.1.4. Nondimensionalization

We nondimensionalize the gas species concentrations by the channel total molar concentration. We choose the temperature scale $T_* = \frac{h_v \overline{C}}{\rho_l c_l}$ to balance the heat flux produced by evaporation against the molar fluxes of liquid and vapor water, setting the coefficient of Γ in the heat equation to one. Distance is scaled by the thickness L of the diffusion layer, and time is scaled by $\tau_* = \frac{\epsilon L^2}{D}$, the characteristic diffusive time. The dimensionless dependent and independent variables are

$$C_o = \tilde{C}_o/\overline{C} \qquad C_v = \tilde{C}_v/\overline{C} \quad C_n = \tilde{C}_n/\overline{C}, \qquad (7.30)$$
$$T = \tilde{T}/T_* \qquad \tau = t/\tau_* \qquad z = \tilde{z}/L,$$

and the dimensionless fluxes are

$$N_T = \frac{L}{h_v \overline{C} D} \tilde{N}_T, \qquad N_\alpha = \frac{L}{D\overline{C}} \tilde{N}_\alpha, \qquad N_l = \frac{L}{D\overline{C}} \tilde{N}_l,$$

where α runs over the gas species oxygen, nitrogen, and water vapor. The dimensionless current, or ion flux, is scaled by

$$I = \frac{\tilde{I}L}{4F\overline{C}D} \ll 1. \qquad (7.31)$$

A key observation is that the current, while a large charge flux, represents a weak molar flux measured against maximal GDL diffusive fluxes—the scaled value of I on the order of 10^{-2} (see Table 7.2).

Collecting the fluxes and unknowns $\vec{N} = (N_g, N_T, N_o, N_v, N_l)^t$, and $\vec{V} = (C, T, C_o, C_v, \beta)^t$, the constitutive relations for the fluxes may be written as

7. Phase Change and Hysteresis in PEMFCs   263

$$\vec{N} = \mathcal{D}(\vec{V})\vec{V}_z, \tag{7.32}$$

with the matrix $\mathcal{D}$ given by

$$\mathcal{D} = -\begin{pmatrix} R_g k_{rg} TC & R_g k_{rg} C^2 & 0 & 0 & 0 \\ \delta_l \mathcal{C}_{lg} T^2 & \delta_l \mathcal{C}_{lg} CT + R_T & 0 & 0 & R_c T\nu(\beta) \\ (R_g k_{rg} T - \frac{1}{C})C_o & R_g k_{rg} CC_o & 1 & 0 & 0 \\ (R_g k_{rg} T - \frac{1}{C})C_v & R_g k_{rg} CC_v & 0 & 1 & 0 \\ R_l \beta k_{rl} T & R_l \beta k_{rl} C & 0 & 0 & R_c \nu(\beta)/\delta_l \end{pmatrix}, \tag{7.33}$$

where $\mathcal{C}_{lg} = R_g \sigma_{lg}(1-\beta)k_{rg} + R_l \beta k_{rl}$ describes the impact of convective transport on the heat flux. The remainder of the dimensionless parameters are defined, along with typical values, in Table 7.3. The system of conservation laws takes the form

$$\mathcal{M}(\vec{V})\vec{V}_\tau + \left(\mathcal{D}(\vec{V})\vec{V}_z\right)_z = \Gamma\vec{S}, \tag{7.34}$$

where $\vec{S} = (-1, +1, 0, -1, +1)^t$ is the stoichiometeric impact of phase change on each species, the phase change term Γ takes the form

$$\Gamma(C_v) = Hh(\beta, C_v)(C_v - C_{\text{sat}}(T)), \tag{7.35}$$

where $H \gg 1$ is a key parameter whose size drives the boundary layer structure. The matrix $\mathcal{M}$ is given by

$$\mathcal{M} = \begin{pmatrix} 1-\beta & 0 & 0 & 0 & -C \\ 0 & \sigma & 0 & 0 & (1-\delta_l \sigma_{lg})T \\ 0 & 0 & 1-\beta & 0 & -C_o \\ 0 & 0 & 0 & 1-\beta & -C_v \\ 0 & 0 & 0 & 0 & \delta_l^{-1} \end{pmatrix}. \tag{7.36}$$

TABLE 7.3 Scaling constants and dimensionless parameters, at $\overline{T}_c = 353°K^\dagger$.

$\overline{C}$	$\frac{\overline{P}}{R\overline{T}_c}$	$10^{2\dagger}$ mol/m^3	I_*	$\frac{4F\overline{C}D}{L}$	1.26×10^6 C/(s m^2)	L	L	2.5×10^{-4}m
t_*	$\frac{\epsilon L^2}{D}$	5.8×10^{-3} s	T_*	$\frac{h_v \overline{C}}{\rho_l c_l}$	$4.63°K$	H	$\frac{L^2 \gamma_v}{DL_p}$	1.26×10^6
I	$\frac{L\overline{I}}{4F\overline{C}D}$	7.91×10^{-3}	r	$\frac{L\overline{i}}{D}$	2.5×10^2	r_T	$\frac{LT_* \overline{r}_T}{h_v \overline{C}D}$	0.34
R_g	$\frac{K\overline{C}RT_*}{D\mu_g}$	21.9	R_c	$\frac{\gamma_s S_p \sqrt{K\epsilon}}{D\mu_l}$	2.59	R_l	$\frac{K\overline{C}_l RT_*}{D\mu_l}$	0.891
R_T	$\frac{(1-\epsilon)\kappa_s}{D\rho_l c_l}$	3.25×10^{-2}	T_c	$\overline{T}_c/T_*$	76.2	X_o	$\overline{C}_o/\overline{C}$	[0.05-0.2]
X_v	$\overline{C}_v/\overline{C}$	0.142	X_o	$C_{\text{sat}}(0)$	0.143	X_s	$C'_{\text{sat}}(0)$	2.34×10^{-2}
δ_l	$\overline{C}/C_l$	1.83×10^{-3}	σ_{lg}	$\frac{\rho_g c_g C_l}{\rho_l c_l \overline{C}}$	2.83	σ_{ls}	$\frac{(1-\epsilon)\rho_s c_s}{\epsilon \rho_l c_l}$	0.122

7.2.2. Analysis of the Three Regimes

The nondimensionalization yields two key small parameters: the dimensionless flux density I which is the driving force for the flow and the inverse evaporation rate $1/H$ which defines the boundary layer asymptotics. Other small parameters which are tangentially involved are the liquid-gas molar density ratio, δ_l, the Lewis number, R_T, which balances thermal conductivity against molecular diffusivity, and the inverse scaled channel temperature $1/T_c$. The goal is to simplify the governing eq. (34) in each of the three regions. In the two-phase and dry regions, it will be taken at steady-state, driven adiabatically by the slowly evolving front location. The evolution of the front comes from the analysis of the full system within the boundary layer.

The analysis which follows dictates several key structures to the solution of (34); see [28] for details. The fluxes of gases, heat, and liquid are piece-wise constant to highest order in each domain. Taking the current density and the total heat and total water flux from the membrane into the GDL as given, we can compute the vapor flux in the two-phase domain from the temperature profile, and hence we can take the boundary condition at the Membrane-GDL interface as given implicity in terms of the temperature profile. If we can compute the jump in the fluxes from the two-phase to the dry region, then we know the fluxes in the dry region. From the scaling of the equations, we take a quasi-steady solution in each of the two domains, the time dependence impacts only the flux jump across the boundry layer. The solutions in the dry and two-phase regions vary linearly in space. In the dry region, the solution is given explicitly in terms of the fluxes and the channel values; see (56). The jump in the solutions across the boundary layer is $\mathcal{O}(I/\sqrt{H})$, which is lower order, while the jump in fluxes is $\mathcal{O}(I)$, see (86), which is leading order. Thus we can calculate the values of the variables on the left side of the two-phase region, and from the values of the two-phase fluxes resolve the two-phase variables; see (67). With the two-phase temperature profile, we determine the two-phase vapor flux, see (68), and so the full solution depends only upon the flux jumps across the evaporative boundary layer determined by the amount of phase change which occurs there. The analysis of the boundary layer predicts this phase change in terms of the scaled under/over saturation at the dry end of the evaporative boundary layer, again from (86). Thus we have a closed form of equation, if we know the amount of evaporation, we can compute the temperature everywhere, which in turn determines the amount of evaporation. Solving the resulting linear system of equations we obtain a closed form solution for the scaled under/over-saturation (93) which yields a simple differential equation for the front motion, see (88).

To set up the problem, we scale the fluxes at the membrane by the current density

$$\vec{N}(z=1) = \vec{F}^m I \equiv (1 + N_v^m, N_{cT}^m - N_v^m, 1, N_v^m, N_{cw}^m - N_v^m)^t I, \qquad (7.37)$$

where the scaled electrochemical heat and total water fluxes $N_{cT}^m = \frac{L\tilde{N}_{cT}^m}{h_v \overline{C}D}$ and $N_{cw}^m = \frac{L\tilde{N}_{cw}^m}{\overline{C}D}$ are prescribed while the vapor flux, $N_v^m = \frac{L\tilde{N}_v^m}{\overline{C}D}$, is to be determined. On the two-phase side of the interface we have the unknowns

$$\vec{V}(z = z_*^+) = \vec{V}_*, \tag{7.38}$$

where $\vec{V}_*$ is to be determined. On the dry side of the interface, we have the fluxes

$$\vec{N}(z = z_*^-) = \vec{F}^d I, \tag{7.39}$$

where again $\vec{F}^d$ is to be determined. Using the projection π_d defined in Section 7.2.1, we write the Robin conditions at the channel as

$$\pi_d \left(\vec{V}(z=0) - \vec{V}_c \right) = -M_r \pi_d \vec{N}(z=0), \tag{7.40}$$

and the Dirichlet condition for the liquid water as

$$\beta(0) = 0. \tag{7.41}$$

The channel values $\vec{V}_c = (1, T_c, X_o, X_v, 0)^t$ are considered known, and the 4×4 matrix M_r is diagonal

$$M_r = \text{diag}(r^{-1}, r_T^{-1}, r^{-1}, r^{-1}), \tag{7.42}$$

where $r = \frac{D}{L}\tilde{r}$ and $r_T = \frac{LT_*}{h_v \overline{C}D}\tilde{r}_T$.

7.2.2.1. Dry Regime

In the dry regime, for $z \in (0, z_*^-)$, we look for a solution as a variations from the channel values, with the variations scaling with the local current density, I. In the dry region, there is no liquid water present ($\beta = 0$), and the gas is undersaturated, so there is no phase change. The scaled unknowns are $\vec{V}^d = (C^1, T^1, C_o^1, C_v^1)^t$, where

$$\begin{aligned} C = 1 + \tfrac{C^1}{T_c} I, & \quad T = T_c + T^1 I, \\ C_o = X_o + C_o^1 I, & \quad C_v = X_v + C_v^1 I. \end{aligned} \tag{7.43}$$

The relative permeabilities simplify to $k_{rg} = 1$, and $k_{rl} = 0$, and the conservation laws (34) and boundary conditions become

$$\mathcal{M}^d \vec{V}_T^d + \left(\mathcal{D}^d(\vec{V}^d) \vec{V}_z^d \right)_z = 0, \tag{7.44}$$

$$\vec{V}^d(0) = -\mathcal{M}_r \mathcal{D}^d(\vec{V}^d)\vec{V}^d_z(0), \tag{7.45}$$

$$\mathcal{D}^d(\vec{V}^d)\vec{V}^d_z(z^-_*) = \pi_d \vec{F}^d, \tag{7.46}$$

where $\mathcal{M}^d$ and $\mathcal{D}^d$ are 4×4 matrices obtained from (36) and (33) by substituting $\beta = 0$ and eliminating the equation for β. The matrix $\mathcal{M}^d$ is independent of the channel values $\vec{V}^d$, which we take to be temporally constant, and expanding the z-derivatives we find

$$\mathcal{M}^d \vec{V}^d_\tau + \mathcal{D}^d_c \vec{V}^d_{zz} = -\Big((\mathcal{D}^d(\vec{V}^d) - \mathcal{D}^d(\vec{V}_c))\vec{V}^d_{zz} + \nabla_{\vec{V}^d}\mathcal{D}^d : \vec{V}^d_z : \vec{V}^d_z\Big)I, \tag{7.47}$$

where $\mathcal{D}^d_c = \mathcal{D}^d(\vec{V}^d_c)$. We neglect the $O(I)$ terms in (47) and find a leading order constant-coefficient parabolic system,

$$\vec{V}^d_\tau + \mathcal{K}^d \vec{V}^d_{zz} = O(I), \tag{7.48}$$

where

$$\mathcal{K}^d = \left[\mathcal{M}^d\right]^{-1}\mathcal{D}^d_c. \tag{7.49}$$

The relaxation timescale for solutions of (47) is given by the reciprocals of the eigenvalues of $\mathcal{K}^d$, which using the asymptotic relations $\delta_l, 1/T_c \ll 1$ take the form

$$\lambda_1 = \lambda_2 = -1, \tag{7.50}$$

$$\lambda_3 = -R_g T_c + \frac{R_T}{\sigma_{ls}} + O(1), \tag{7.51}$$

$$\lambda_4 = -\frac{R_g R_T T_c}{R_g T_c \sigma_{ls} + R_T} + O(1/T_c, \delta_l). \tag{7.52}$$

The fastest timescale is associated with the eigenvalue $\lambda_3 \approx -1.7 \times 10^3$ and has eigenvector $(1, 1, 0, 0)^t + O(\delta_l, 1/T_c)$, which describes the linearized dimensionless gas pressure

$$P_g = \frac{\tilde{P}_g}{CT_c\mathcal{R}} = 1 + (T^1 + C^1)I + O(I^2).$$

The smallest eigenvalue $\lambda_4 \approx -0.25$ is associated with relaxation of the temperature profile.

The two-phase interface point z_* and the dry fluxes $\vec{F}^d$ are time dependent but on a timescale several orders of magnitude longer than the relaxation times.

The corresponding solutions of (48) are therefore quasi-stationary, adiabatically driven by the slow evolution of the front and fluxes, up to the order of the timescale ratio, which is seen in (88) to be $O(\delta_l)$. In this regime, the fluxes are spatially constant to leading order, and the governing equation reduces to

$$\mathcal{D}_c^d \vec{V}_{zz}^d = O(I), \qquad \text{for } z \in \left(0, z_*^-(\tau)\right) \tag{7.53}$$

$$\mathcal{D}_c^d \vec{V}_z^d\!\left(z_*^-(\tau)\right) = \pi_d \vec{F}^d(\tau) + O(I). \tag{7.54}$$

$$\vec{V}^d(0) = -M_r \vec{F}^d(\tau) + O(I), \tag{7.55}$$

Since $\mathcal{D}_c^d$ has an $O(1)$ inverse, the system has the simple linear solution

$$\vec{V}^d(z) = \left(-M_r + z\left[\mathcal{D}_c^d\right]^{-1}\right)\pi_d \vec{F}^d(\tau) + O(I). \tag{7.56}$$

7.2.2.2. Two-Phase Regime

In the two-phase regime, the liquid water is everywhere greater than the immobile volume fraction, β_*, and the gas is saturated to leading order. As in the dry regime, the scaled variables $\vec{V}^w = (C^1, T^1, C_o^1, C_v^1, \beta^1)^t$ describe variations from the channel values, except for the scaled water vapor and water volume fractions, which describe variations from the saturation pressure and the immobile volume fraction, respectively,

$$C = 1 + C^1 I/T_c, \qquad T = T_c + T^1 I, \qquad C_o = X_o + C_o^1 I, \tag{7.57}$$
$$C_v = C_{sat}(T) + C_v^1 I/H, \qquad \beta = \beta_* + \beta^1 (I\delta_l)^{\frac{2}{3}}.$$

This change of variables renders the diffusivity matrix, $\mathcal{D}^w$, $O(1)$ with an $O(1)$ inverse. In particular, the $(I\delta_l)^{\frac{2}{3}}$ scaling for β^1 is determined from the liquid flux boundary conditions at the membrane where the effective liquid diffusivity scales as

$$\nu(\beta) = \nu^* \sqrt{\beta^1}(I\delta_l)^{\frac{1}{3}} + O(I\delta_l), \tag{7.58}$$

where $\nu^* = \nu(\beta_*) = \beta_* \mathcal{J}'(\beta_*)$. Near the membrane, neglecting the convective terms, the liquid flux has the scaling

$$N_l = -\frac{R_c}{\delta_l}\nu(\beta)\beta_z = -R_c \nu^* I(\beta^1)^{\frac{3}{2}} + O(I^{\frac{5}{3}}\delta_l^{\frac{2}{3}}), \tag{7.59}$$

which yields an $O(1)$ value for β^1 from the liquid flux boundary condition at the membrane $N_l = N_l^m I$.

In linearizing the conservation laws, it is convenient to introduce $k_{rg}^* = k_{rg}(\beta_*)$, $\sigma_* = \sigma(\beta_*)$, and $C_{lg}^* = C_{lg}(\beta_*)$. In the new variables in the two-phase region, the liquid relative permeability simplifies

$$k_{rl} = (I\delta_l)^{1/3}\sqrt{\beta^1}. \tag{7.60}$$

The dimensionless vapor saturation concentration, $C_{sat}(T) = \frac{\widetilde{C}_{sat}(\overline{T}_c + TT_*)}{\overline{C}}$ is approximated to $O(I^2)$ by linearizing about the channel temperature, T_c,

$$C_{sat}(T^1) = X_s + X_s'IT^1 + O(I^2). \tag{7.61}$$

The coefficient $X_s = C_{sat}(0)$ is the saturation mole fraction at the channel, while $X_s' = C_{sat}'(0)$ gives the linearized response of the saturation mole fraction to changes in temperature. We collect the leading order terms in the conservation laws, finding

$$\overline{\mathcal{M}}^w \vec{V}_\tau^w + (\overline{\mathcal{D}}^w(\beta^1)\vec{V}_z^w)_z = C_v^1 \vec{S} + O(I, \delta_l, 1/H), \tag{7.62}$$

where $\overline{\mathcal{D}}^w$ and $\overline{\mathcal{M}}^w$ are obtained from (36) and (33) by substituting in the values form $\vec{V}$ obtained from setting $\vec{V}^w = 0$. The matrices contain $O(H)$ terms associated with the slight variation of vapor phase from saturation; inclusion of these terms forces a boundary-layer type solution for C_v^1, which is incompatible with the boundary conditions. Neglecting these terms, the matrices $\overline{\mathcal{D}}w$ and $\overline{\mathcal{M}}^w$ are non-invertible and the leading order equation for the vapor concentration C_v^1 is degenerate; we separate this equation out,

$$(1-\beta_*)X_s'T_\tau^1 - \left((R_g k_{rg}^* - 1/T_c)X_s C_z^1 + \left(R_g k_{rg}^* X_s + X_s'\right)T_z^1\right)_z \\ = -C_v^1 + O(I, \delta_l, 1/H). \tag{7.63}$$

In the quasi-steady regime, the time derivative of T^1 is lower order, and from the analysis of the full problem we see that $C_v^1 = O(I)$, which implies that the phase change is a lower-order effect in the two-phase region, away from the boundary layer. It follows that at steady-state the vapor flux carried by the saturated gas is constant and given by the terms within the z derivative, specifically,

$$N_v^m = (R_g k_{rg}^* - 1/T_c)X_s C_z^1 + \left(R_g k_{rg}^* X_s + X_s'\right)T_z^1. \tag{7.64}$$

We drop C_v^1 from the leading order adiabatic equations in the two-phase regime. We introduce the liquid "potential" $C_\beta^1 = 2/3(\beta^1)^{3/2}$, obtaining a reduced 4×4 system for the reduced wet variables $\vec{V}^{wr} = (C^1, T^1, C_o^1, \beta^1)$,

$$\mathcal{K}^w \vec{V}_{zz}^{rw} = O(I, \delta_l, 1/H), \tag{7.65}$$

where

$$\mathcal{K}^w = [\mathcal{M}^w]^{-1} \mathcal{D}^w, \tag{7.66}$$

has a spread of eigenvalues from $O(10^3)$ for pressure relaxation to $O(10^{-2})$ for the relaxation of the liquid water profile. With a two-phase front located at $z = z_*(\tau)$, the quasi-steady solution of (65) subject to (37) and (38) is a simple linear function of position

$$\vec{V}^{wr} = \pi_w \vec{V}_* + (z - z_*)[\mathcal{K}^w]^{-1} \pi_w \vec{F}^m. \tag{7.67}$$

The analytic solution (67) determines the slope of the gas concentration and temperature in terms of the prescribed fluxes. When substituted into (64) the vapor flux carried in the two-phase region is given in terms of the other fluxes

$$N_v^m = \frac{\left(N_{cT}^m - T_c \delta_l (\frac{C_{lg}^*}{R_g k_{rg}^*} + N_{cw}^m)\right)(X_s + T_c X_s') + T_c R_T \left(1 - \frac{1}{T_c R_g k_{rg}^*}\right) X_s}{\left(1 + T_c \delta_l (\frac{C_{lg}^*}{R_g k_{rg}^*} - 1)\right)(X_s + T_c X_s') + T_c R_T \left(1 - \left(1 - \frac{1}{T_c R_g k_{rg}^*}\right) X_s\right)}. \tag{7.68}$$

The denominator is positive, while the term on the left in the numerator is negative, since the heat flux is away from the membrane $N_{cT}^m < 0$. The total water flux N_{cw}^m is dominated by the positive term $C_{lg}^*/R_g k_{rg}^*$. The term on the right in the numerator is positive, since $1/(T_c R_g k_{rg}^*) \ll 1$. Thus, we have *two competing effects driving the vapor flux*, the heat and water flux push the vapor flux negative, while the thermal conductivity, R_T, pushes it positive. The heat flux and the thermal conductivity are only partially linked; at quasi-steady state, the heat produced in the MEA must leave through either the cathode or anode side of the membrane, the total heat flux being essentially prescribed. The thermal conductivity ascribes a temperature gradient necessary to accommodate the heat flux and has only a secondary effect on the heat flux itself. For small values of R_T, as has been taken here for carbon fiber paper, the fluxes dominate and the vapor flux is negative. Indeed in the limit $R_T T_c \ll 1$, the vapor flux has the simple form

$$N_v^m = \frac{N_{cT}^m - T_c \delta_l (\frac{C_{lg}^*}{R_g k_{rg}^*} + N_{cw}^m)}{1 + T_c \delta_l (\frac{C_{lg}^*}{R_g k_{rg}^*} - 1)} < 0. \tag{7.69}$$

For a more thermally conductive GDL, intermediate between Toray carbon fiber paper and pure graphite, for example, $\kappa_s \approx 30 \, \text{J}/(\text{m s °K})$ with $R_T \approx 1$, the positive terms dominate, and the vapor flux is *toward the membrane*. In the limiting case $R_T T_c \gg 1$, the vapor flux depends only upon the channel saturation mole fraction

$$N_v^m = \frac{X_s}{1-X_s} + O(1/T_c). \tag{7.70}$$

The high thermal conductivity lowers the temperature difference between membrane and channel, forcing condensation at the GDL-catalyst boundary layer where the liquid flux

$$N_l^m = N_{cw}^m - N_v^m, \tag{7.71}$$

is more negative than the total water output of the membrane. This condensation is a sink for water vapor, and the vapor flux toward the membrane gives the oxygen a convective boost, enhancing the transport of oxygen toward the membrane. More specifically the convective boost increases the oxygen concentration at the membrane for a prescribed oxygen flux (current density), reducing oxygen reduction reaction over potential losses. The benefit comes at the cost of increased liquid water production and hence increased heat removal load for the coolant. The impact of the increased water production on flooding in the catalyst layer is an interesting modeling exercise.

7.2.2.3. The Evaporative Layer

The leading order phase change occurs in the evaporative layer, see the region (z_*^0, z_*^+) in Figure 7.2. The key to the free boundary evolution is to determine the total amount of phase change that occurs, which we will relate to Θ_*, the degree of vapor undersaturation at the two-phase point z_*^-,

$$\Theta_* = C_{sat}\left(T(z_*^-)\right) - C_v(z_*^-). \tag{7.72}$$

In the evaporative layer, we assume that the liquid water volume fraction is everywhere greater than the immobile volume fraction, β_*. We rescale the variables as in the two-phase region, except for the water vapor molar concentration whose scaling $H^{-\alpha}I$ contains a parameter α determined as part of the analysis below,

$$\begin{aligned} C &= 1 + \tfrac{C_v^1 I}{T_c}, & T &= T_c + T^1 I, & C_o &= X_o + C_o^1 I, \\ C_v &= C_{sat}(T) + \tfrac{C_v^1 I}{H^\alpha}, & \beta &= \beta_* + (\delta_l I)^{\tfrac{2}{3}} \beta^1. \end{aligned} \tag{7.73}$$

With this scaling, the local phase change takes the form

$$\Gamma = C_v^1 h_*(C_v^1) I H^{1-\alpha}, \tag{7.74}$$

where $h_*(C_v^1) = h(C_v^1, \beta_*)$.

We rescale the spatial variable within the evaporative layer, introducing $\tilde{z} = (z - z_*^0)\sqrt{H}$. The natural parabolic rescaling of time $\tau_b = H\tau$ suggests relaxation times which are yet 5–6 magnitudes of order faster than those of the two-phase regime. For this reason, we only consider the quasi-steady version of the conservation laws

$$\vec{N}_{\tilde{z}} = \frac{\Gamma}{\sqrt{H}}\vec{S} = IH^{\frac{1}{2}-\alpha}C_v^1 h_* \vec{S}. \tag{7.75}$$

The fluxes are related to the scaled variables $\vec{V}^b = (C^1, T^1, C_o^1, C_v^1, \beta^1)$ through the constitutive laws as

$$\vec{N} = I\sqrt{H}\mathcal{D}^b \vec{V}_{\tilde{z}}^b + O(I^2\sqrt{H}), \tag{7.76}$$

where $\mathcal{D}^b$ is a 5×5 matrix derived from keeping leading order terms in (33) after substituting the scalings (73).

Combining the conservation laws (75) with the constitutive relations (76), we obtain a 5×5 system

$$\mathcal{D}^b \vec{V}_{\tilde{z}\tilde{z}} = \frac{h_* C_v^1}{H^\alpha} \vec{S} + O(I). \tag{7.77}$$

A key simplification in the analysis of the evaporative layer arises from the observation that

$$\frac{1}{H^\alpha}[\mathcal{D}^b]^{-1}\vec{S} = \xi_v \vec{S}_0 + O(H^{-\alpha}), \tag{7.78}$$

where $\vec{S}_0 = (0, 0, 0, 1, 0)^t$ and

$$\xi_v = 1 - X_s + \frac{X_s + T_c X_s'}{R_T}\left(\frac{1}{T_c} + \delta_l\left(\frac{C_{lg}^*}{R_g k_{rg}^*} - 1\right)\right) + \frac{X_s}{T_c R_g k_{rg}^*}. \tag{7.79}$$

The parameter ξ_v is independent of H and strictly positive. It follows that the equation for C_v^1 uncouples at leading order,

$$(C_v^1)_{\tilde{z}\tilde{z}} = \xi_v h_* C_v^1, \tag{7.80}$$

$$\lim_{\tilde{z}\to\infty} C_v^1(\tilde{z}) = 0, \tag{7.81}$$

$$C_v^1(0) = \frac{H^\alpha \Theta_*}{I}. \tag{7.82}$$

We assume $\Theta_* > 0$, so the constant $h_* = \beta_*^{2/3}$. This equation has a classic boundary layer solution,

$$C_v^1 = C_v^1(0) e^{-\sqrt{h_* \xi_v}\, \tilde{z}}. \tag{7.83}$$

Returning to the conservation laws (77), we find that

$$\left[\!\left[\mathcal{D}^b \vec{V}_{\tilde{z}}\right]\!\right] = \int_0^\infty \mathcal{D}^b \vec{V}_{\tilde{z}\tilde{z}}\, d\tilde{z} = \int_0^\infty \frac{h_* C_v^1(\tilde{z})}{H^\alpha} d\tilde{z}$$

$$\vec{S} = \frac{h_* C_v^1(0)}{H^\alpha \sqrt{\xi_v h_*}} \vec{S} = -\frac{\Theta_* \sqrt{h_*}}{I \sqrt{\xi_v}} \vec{S}. \tag{7.84}$$

To balance the jumps in the fluxes with the magnitude of the catalyst layer fluxes suggests the scaling

$$\Theta_*^1 = \frac{\sqrt{H}}{I} \Theta_*, \tag{7.85}$$

and the associated choice of $\alpha = 1/2$ in the scaling of C_v^1. From (76) we relate the jump in the fluxes across the evaporative boundary layer to the oversaturation Θ_*^1,

$$\left[\!\left[\vec{N}\right]\!\right] = I\sqrt{H} \left[\!\left[\mathcal{D}^b \vec{V}_{\tilde{z}}\right]\!\right] = -I\Theta_*^1 \sqrt{\frac{h_*}{\xi_v}} \vec{S}. \tag{7.86}$$

7.2.2.4. The Sub-Layer and the Front Motion

The jump in the liquid volume fraction occurs at the infinitesimal sublayer delimited on the left and right by z_*^- and z_*^0. The liquid volume fraction reaches the mobile threshold, β_*, at $z = z_*^0$. Conservation of mass and energy across the interface requires that the Rankine-Hugoniot conditions be enforced. The nondegenerate variables C, T, C_o, and C_v, must be continuous across the interface, and the corresponding flux jumps must be zero. The degenerate variable, β jumps from β_* down to 0 across the inner layer, and the liquid flux is related to the interface motion and the liquid jump through a degenerate Rankine-Hugoniot relation [10, 32], which states that the liquid flux into the sub-layer must serve to move the sub-layer,

$$\frac{dz_*}{d\tau} = \frac{\delta_l [\![N_l]\!]_s}{[\![\beta]\!]_s} = \frac{\delta_l N_l(z_*^0)}{\beta_*}. \tag{7.87}$$

The jump in liquid water flux across the evaporative layer is given by (86), while the flux is constant to leading order across the two-phase region $(z_*^+, 1)$.

We express the front motion in terms of the water membrane flux and the undersaturation

$$\frac{dz_*}{d\tau} = \frac{\delta_l I\left(N_{cw}^m - N_v^m - \Theta_*^1 \sqrt{h_*/\xi_v}\right)}{\beta_*}, \tag{7.88}$$

where the vapor flux is given by (68). The front motion is on a timescale $O(10^{-5})$, which is slower than the next slowest timescale, relaxation of the β profile, by the factor $\delta_l \ll 1$, and is 8–9 orders of magnitude slower than the fastest timescale, gas pressure relaxation in the dry regime.

7.2.3. Resolution of the Front Evolution

To obtain a closed form solution for the front evolution and the slow variation of the concentrations and temperature, we glue together the dry and wet solutions developed above across the front $z = z_*(\tau)$, and impose the jump conditions

$$\left[\!\left[\pi_w \vec{V}\right]\!\right]_f = \mathcal{O}(I/\sqrt{H}), \tag{7.89}$$

$$\left[\!\left[\vec{N}\right]\!\right]_f = -I\Theta_*^1 \sqrt{\frac{h_*}{\chi_v}} \vec{S} + \frac{\beta_* }{\delta_l} \frac{dz_*}{d\tau} \vec{E}_\beta, \tag{7.90}$$

and the boundary condition on the two-phase side of the interface

$$\beta(z_*) = \beta_*. \tag{7.91}$$

Here $\vec{E}_\beta = (0, 0, 0, 0, 1)^t$. We recall that the quasi-steady solutions constructed in Sections 7.2.2.1 and 7.2.2.2 have piecewise spatially constant fluxes given by

$$\vec{N} = \begin{cases} I\vec{F}^m & \text{two} - \text{phase region}, \\ I\vec{F}^d(\tau) & \text{dry region}. \end{cases} \tag{7.92}$$

These equations lead to the following expression relating the undersaturation at the front to the scaled channel undersaturation, $\Theta_c^1 = (X_s - X_v)/I$, the vapor flux determined by (68), and the prescribed membrane fluxes

$$\Theta_*^1 = \Delta^{-1}\left(\Theta_c^1 - \frac{rX_s'(N_{cT}^m - N_v^m) - r_T N_v^m}{r_T r} - z_*\left[X_v + (1 - X_v)N_v^m\right.\right.$$

$$\left.\left. + \frac{X_v + T_c X_s'}{R_T}\left(\frac{N_{cT}^m - N_v^m}{T_c} - \delta_l \sigma_{lg}^*(1 + N_v^m)\right)\right]\right) + \mathcal{O}(1/\sqrt{H}, 1/T_c), \tag{7.93}$$

where the uniformly positive denominator Δ is given by,

$$\Delta = \sqrt{\frac{h_*}{\xi_v}\left(\frac{r_T + rX'_s}{r_T r} + z_*\left[1 + X_v\left(\frac{R_g(1 + \delta_l T_c \sigma^*_{lg}) + R_T}{T_c R_g R_T} - 1\right)\right.\right.} \\ \left.\left. + \frac{X'_s}{R_T}\left(1 + \delta_l T_c \sigma^*_{lg}\right)\right]\right). \tag{7.94}$$

This complicated but explicit formula, inserted into (88) gives the impact of each transport process on the slow front motion.

The two-phase motion problem is very stiff, with a wide separation of timescales and a transport matrix which becomes singular as the solution relaxes to its quasi-steady state. The asymptotic analysis presented eliminates the stiffness that is the bane of numerical simulations, affording computational speed-up of 3-4 *orders* of magnitude over the full system. Building this model into a unit cell simulation code promises huge reductions in computational cost and admits the possibility of performing either full stack-based calculations or doing extensive inverse calculations and parameter estimation.

In applications, the reduced system is embedded in a 1+1D computational scheme for the overall fuel cell. This includes a model of the membrane's water content and temperature, the anode GDL, and the variation of the oxygen and water vapor contents in the flow field channels in the along-the-channel direction, providing the channel conditions and fluxes which were taken as prescribed in the analysis. To present numerical results from the reduced system, we simulate this coupling by providing along-the-channel data for the oxygen and water vapor concentrations, temperature, current density, and catalyst layer production of heat and total water from a previous 1+1D computation reported in [3]. These values vary in the y direction but are constant in time and do not couple back to the reduced simulations.

The computations for two different runs are presented in Figure 7.3. On the left, the GDL thermal conductivity is taken as $\kappa_s = 1$ J/(m s °K), and the inlet and outlet ends of the channel remain dry, while the two-phase interface, plotted at 10 s intervals, grows from zero at $t=0$ to occupy the majority of the mid-channel region by $t=80$ seconds. The amount of evaporation at the two-phase boundary is an order of magnitude lower than either the liquid water or vapor fluxes from the catalyst layer. In the run on the right, with $\kappa_s = 10$ J/(m s °K), the front velocity is an order of magnitude greater, and the two-phase region occupies all but the very end section of the cathode outlet. The two-phase front is plotted at intervals of 5 seconds up to $t=15$ seconds. The dimensionless catalyst vapor flux N_v^m drops from roughly -7×10^{-3} for the low thermal conductivity case to -1.6×10^{-3}, while the liquid water flux grows by approximately an order of magnitude. For the high thermal conductivity case, the evaporation at the front is large at the cathode inlet and significantly slows the front evolution there.

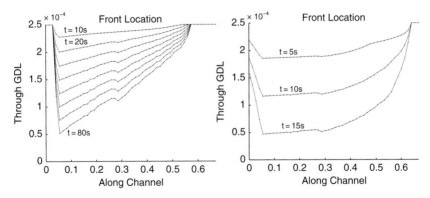

FIGURE 7.3. (Left) The front position through the GDL (not shown) are the gas flow field on the bottom, and the catalyst layer and membrane are on the top. The values of oxygen and water vapor, the current density, and catalyst layer production of heat and liquid water are prescribed in the along-the-channel direction and constant in time. Both the cathode inlet at $y=0$ and outlet at $y=0.67$ meter are in a "dry" state. The front starts at $t=0$ at the catalyst layer and is plotted every 10 seconds up to $t=80$. (Right) Front positions with the GDL thermal conductivity $\kappa_s = 10$ J/(m s °K). The front starts at the catalyst layer and is plotted every 5 seconds up to $t=15$.

7.3. Hysteresis in PEMFC

Under conditions of stress—low hydration levels, low oxygen or fuel stoichs, or high current densities—PEMFC are known to exhibit hysteresis, slow transient behavior, and even temporal oscillations. Jay Benziger studies these phenomena from an experimental perspective in Chapter 3 (the chapter authored by J. Benziger) on Reactor Dynamics, particularly his stirred tank reactor (STR) fuel cells [1, 21]. A driving force behind the hysteresis and slow transient behavior is the presence of two ignited steady-states seen in Figure 3.10 of Chapter 3, characterized as high and low water content stable states. The transition between these states was on the order of 2–4 hours, significantly longer than the relaxation timescales of the majority of processes within the cell.

A possible explanation for the observed hysteresis, summarized from [23], is a phase transition from extinguished to ignited states, which propagates though the membrane of the STR fuel cell. There is a positive feedback loop between current density, which generates water, and membrane resistance which increases with water content. The wetter parts of the membrane compete more effectively for current density, attracting yet more water. However, the membrane resistance is typically a small component of overall cell resistance. For the feedback mechanism to destabilize the cell and generate hysteresis, the membrane resistance must be particularly sensitive to water content. In PEMFC, this occurs in two situations; one, investigated in Section 7.3.1 and in Benziger's work, is when the membrane is dry, if the membrane water content falls below 3 waters/acid group then the resistance jumps dramatically, [31]; the second situation involves the transition between single-phase and two-phase operation. The Nafion absorbs significantly more water from liquid form than from vapor, even vapor approaching saturation. This is the so-called Schröder paradox, whose effect strengthens with increasing temperature. Indeed the measurements of [16] show that a liquid-equilibrated membrane's water adsorption isotherm increases with temperature, while the vapor equilibrated isotherms decrease with temperature. We present modeling data which suggest that at high current densities PEMFC can often operate in two different regimes, with different percentages of the cathode in the two-phase state, from the same operational setting.

In both these settings, we propose that the slow dynamics are accompanied by lateral diffusion of water within the membrane itself. Membranes have been made sufficiently thin to facilitate the back-diffusion of water through the membrane, which serves to keep a the membrane hydrated at the anode. However, the lateral dimensions of the membrane are much longer, and diffusion in this direction is a weak process but one which can have significant impact over long timescales. The goal is to develop models which will capture this feed-back mechanism and resolve the appreciable variation of water content in the thin through-plane direction without getting bogged down computing the fast timescale of the through-plane direction.

7.3.1. Hysteresis in STR Fuel Cells

The model STR fuel cell consists of two gas plenums which are considered well stirred, either by diffusion or by virtue of very high stoich gas feeds, see Figure 7.4. The model considers the dynamics in the anode and cathode plenums and the membrane of the fuel cell. The gas diffusion layers and catalyst layers are modeled as interfaces through effective mass transport parameters. As we are concerned primarily with relatively dry cell operation, where the water drag coefficient (waters carried per proton) for the membrane is approximately one, we assume the protons generated at the anode catalyst layer immediately form hydronium, H_3O^+ and cross the membrane vehicularly. Equating the proton and the hydronium densities within the membrane, the water drag appears as a sink term for water at the anode (water $\mapsto$ hydronium) and a modified source term at the cathode catalyst layer. This approach was applied previously, [3], yielding good agreement with experimentally measured total water cross-over.

The primary variables of the model are the water content of the membrane, M, the positive ion, i.e. the hydronium, concentration in the membrane, P, the water vapor mole fraction in the anode and cathode plenum, A_v and C_v, respectively, the oxygen concentration of the cathode plenum, C_o, and the electric potential, V. The plenums are taken to be well stirred with spatially uniform concentrations of water vapor and reactant gas. The cathode feed is

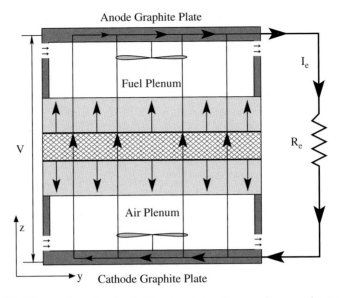

FIGURE 7.4. The anode and cathode bi-polar plates, the gas plenums, the membrane, and the gas diffusion layer. The Plenums are slowly replenished from an external feed. The black arrows indicate direction of electron flow and of proton counterflow. The external circuit between anode and cathode is completed through a load bank with variable resistance R_e. The voltage between anode and cathode plates is denoted by V and the external current by I_e.

pure O_2 and the anode feed is pure H_2. The hydrogen gas in the anode plenum is considered to be abundant, and its influence on the reaction rate is neglected. As a key reduction, the membrane water is taken at quasi-steady state in the z direction, with the z profile determined through the time-and space-dependent water contents at the anode and cathode catalyst layers, evolving in such a manner as to enforce the conservation of total water. The model is isobaric and isothermal.

The membrane water and hydronium content, M and P, are nondimensionalized by the acid molar density $a = 1200$ moles/m^3. The anode and cathode vapor molar densities, A_v and C_v, are nondimensionalized by the vapor saturation molar density, $C_{sat}(T)$; this is the principal dependence of the model upon the temperature. The cathode oxygen concentration C_o is non-dimensionalized by the inlet oxygen concentration C_o^{in}. The external voltage, V, is nondimensionalized by the open circuit voltage E_o, while the local through-plane current density $I(y)$ is nondimensionalized by $I_* = E_o \sigma_0 / H_m$, where σ_0 is a reference value for the membrane conductance and H_m is the thickness of the membrane.

A key property of an accurate set of model equations is the ability to predict the percentage of product water which remains in the membrane and the percentage which is released into the gas plenum. Based upon previous modeling efforts, [3, 13], the water flux, J, between the membrane and the gas plenum is taken to be proportional to the disequilibrium

$$J = a\gamma_m(M - M_*), \tag{7.95}$$

where the equilibrium content, M_*, of the membrane water at 80°C is given by

$$M_*(r) = 0.3 + 10.8r - 16.0r^2 + 14.1r^3, \tag{7.96}$$

with r denoting the relative humidity of the reservoir with which the membrane is in contact (see [16]) and γ_m is a mass transport coefficient which incorporates transport losses in the catalyst and gas diffusion layers. The hydronium concentration is taken to be in equilibrium with the water concentration (see [36]),

$$P(M) = -\frac{K_e M}{2} + \sqrt{\left(\frac{K_e M}{2}\right)^2 + K_e M}, \tag{7.97}$$

where $K_e = K_0 \exp\left[-\frac{H_0}{R}\left(\frac{1}{T} - \frac{1}{298}\right)\right]$. Here $K_0 = 6.2$ is a dimensionless constant and $H_0 = -52300$ J/mole is the enthalpy of hydrogen formation. The through-plane diffusion coefficients of water and hydronium in the membrane are given by

$$D_w = d_w(T)\psi_w(M), \text{ and } D_+ = d_+(T)\psi_+(M), \tag{7.98}$$

where $d_w(T) = 2.1 \times 10^{-7} \exp(-2436/T)$ m²/s, while the hydronium diffusivity is related to the membrane conductance σ_0 through $d_+(T) = \sigma_0 RT/(aF^2)$. The hydronium diffusivity is taken to vary with water concentration according to the relation

$$\psi_+(M) = \begin{cases} \frac{M}{P(M)\left(1+\beta(M-\psi_c)^{-(1+\nu)}\right)} & M > \psi_c, \\ 0 & M < \psi_c, \end{cases} \tag{7.99}$$

while the water diffusivity varies linearly with hydration level, $\psi_w(M) = M$. The values for the water self-diffusivity are taken from the literature, [4, 20], while the form of the hydronium diffusivity is taken so that the resulting membrane conductivity (see (128)) will agree with experimental values. A qualitative and rough quantitative fit to the data provided in [31] is obtained by taking $\beta = 1$ and $\nu = 0.5$.

There is little data on the diffusion of water in the in-plane direction within the membrane. It is assumed that the membrane is anisotropic with a factor α_{in} for the ratio of through-plane to in-plane diffusivities,

$$\vec{D}_+ = D_+ \text{diag}\{1, \alpha_{in}, \alpha_{in}\}, \text{ and } \vec{D}_w = D_w \text{ diag}\{1, \alpha_{in}, \alpha_{in}\}. \tag{7.100}$$

Conservation of water and hydronium lead to Nerst-Plank equations for the membrane water content, M, and the electric potential ϕ,

$$\frac{\partial P}{\partial t} - \nabla \cdot \left(a\vec{D}_+ \nabla P - \frac{aF}{\mathcal{R}T} \vec{D}_+ P \nabla \phi \right) = 0, \quad (7.101)$$

$$\frac{\partial M}{\partial t} - \nabla \cdot \left(a\vec{D}_w(M) \nabla M \right) = 0. \quad (7.102)$$

In this formulation, the absence of an explicit water drag term in (102) is compensated for by the source and sink terms which appear in eqs. (109) and (110).

The quick relaxation times in the through-plane direction as opposed to the much longer ones for the in-plane directions, motivate a model which is quasi-steady in the z direction, neglecting the influence of x and y direction fluxes on the z direction component. In particular, we assume that

$$\frac{\partial M}{\partial x}, \frac{\partial M}{\partial y} \ll \frac{\partial M}{\partial z}, \quad (7.103)$$

holds for all quantities M in the membrane. A key point is that the model remains a conservation law. We use the quasi-steady assumption to obtain a *profile* for the water concentration across the membrane, parametrized by the water concentration at the cathode and anode ends and then apply the conservation laws to this equation. We also assume, for simplicity, that the water content varies only in the y and z directions and average over the x.

7.3.1.1. Through-Plane Direction

The spatial variable y is nondimensionalized by the membrane length, L, while the through-plane variable z is scaled by the membrane thickness, H_m. Integrating the steady-state, through-plane version of eqs. (101–102) once yields

$$-\frac{aD_+}{H_m} \frac{\partial P}{\partial z} - \frac{aF}{H_m \mathcal{R}T} D_+ P \frac{\partial \phi}{\partial z} = N_+^m, \quad (7.104)$$

$$-\frac{aD_w(M)}{H_m} \frac{\partial M}{\partial z} = N_w^m, \quad (7.105)$$

where N_+^m and N_w^m are the through-plane components of the hydronium and water fluxes. The water content at the anode and cathode sides of the membrane, M_a and M_c, are determined from the in-plane membrane eqs. (109–110), which account for the influence of the fluxes. Integrating eqn. (105) yields

$$N_w^m = \frac{ad_w}{H_m}(\Psi_w(M_c) - \Psi_w(M_a)), \tag{7.106}$$

where $\Psi_w(M) = \frac{1}{2}M^2$ is the indefinite integral of ψ_w. This water profile can then be recovered in terms of the water content at the anode and cathode endpoints,

$$M(y,z,t) = \sqrt{M_c^2(y,t) - (M_c^2(y,t) - M_a^2(y,t))z}, . \tag{7.107}$$

7.3.1.2. In-Plane Direction

To develop an evolution for the membrane water content which conserves total water, we divide the membrane into an anode and a cathode side, introducing

$$\tilde{M}_c = \int_0^{\frac{1}{2}} M(y,z,t)dz \text{ and } \tilde{M}_a = \int_{\frac{1}{2}}^1 M(y,z,t)dz, \tag{7.108}$$

which represent the water content of each half of the membrane. Since M is given parametrically by M_c and M_a, imposing conservation of water in each half of the membrane yields two equations for these two unknowns,

$$\frac{\partial \tilde{M}_c}{\partial t} - \frac{\alpha_{in} d_w}{L^2} \frac{\partial^2 \tilde{\Psi}_c(M)}{\partial y^2} = -\frac{\gamma_m}{H_m}(M_c - M_*(C_v)) + \frac{3I_*I}{2FH_m a} - \frac{N_w^m}{H_m a}, \tag{7.109}$$

$$\frac{\partial \tilde{M}_a}{\partial t} - \frac{\alpha_{in} d_w}{L^2} \frac{\partial^2 \tilde{\Psi}_a(M)}{\partial y^2} = -\frac{\gamma_m}{H_m}(M_a - M_*(A_v)) - \frac{I_*I}{FH_m a} + \frac{N_w^m}{H_m a}, \tag{7.110}$$

where I is the local current production, N_w^m is the cathode to anode membrane water flux and

$$\tilde{\Psi}_c = \int_0^{\frac{1}{2}} \Psi_w(M(y,z,t))dz,$$

while $\tilde{\Psi}_a$ denotes the complementary z integral of Ψ_w. Looking at eqs. (109–110), the first term on the right-hand side represents water exchange with the gas plenums. The second term in the top equation is water gain from the combination of hydronium and electrons into water

$$O_2 + 4e^- + 4H_3O^+ \rightarrow 6H_2O.$$

The second term in the anode equation represents water loss due to conversion of water to hydronium, following the reaction

$$H_2 + 2H_2O \rightarrow 2H_3O^+ + 2e.$$

The final term represents back diffusion of water from the cathode to the anode. On the left-hand side of the equations, the $\tilde{\Psi}$ terms represent the lateral diffusive flux of water through the membrane. These conservation laws are equivalent to the full, 2D problem modulo, the assumptions of quasi-stationary in the z direction, and dominance of z flux over y flux.

The expression for $\tilde{\Psi}_c$ and $\tilde{\Psi}_a$ may be simplified using the water profile, eqn. (107),

$$\begin{pmatrix} \tilde{\Psi}_c \\ \tilde{\Psi}_a \end{pmatrix} = \mathcal{D} \begin{pmatrix} \Psi_c \\ \Psi_a \end{pmatrix}, \text{ where } \mathcal{D} = \frac{1}{8}\begin{pmatrix} 3 & 1 \\ 1 & 3 \end{pmatrix}, \qquad (7.111)$$

and $\Psi_c = \Psi_w(M_c)$, $\Psi_a = \Psi_w(M_a)$. Moreover,

$$\begin{pmatrix} \frac{\partial \tilde{M}_c}{\partial t} \\ \frac{\partial \tilde{M}_a}{\partial t} \end{pmatrix} = \mathcal{M} \begin{pmatrix} \frac{\partial M_c}{\partial t} \\ \frac{\partial M_a}{\partial t} \end{pmatrix}, \qquad (7.112)$$

where $\mathcal{M}$ is the matrix of partial derivatives of $\tilde{M}_c$ and $\tilde{M}_a$ with respect to M_c and M_a.

7.3.1.3. Gas Plenums

The slowest timescale, by several orders of magnitude, is for the diffusive motion of liquid in the lateral direction within the membrane, $\tau_{mem} = L^2/d_w = 8.3 \times 10^5$ s. On this slow timescale, we take the gas mole fractions C_v, C_o, and A_v at steady-state, driven adiabatically by the changing water flux into the plenums. The well-stirred gas plenums are assumed to have constant pressure and inlet flow rates. On the cathode side the inlet gas is pure oxygen. The total molar concentration C_T, nondimensionalized by the saturation pressure $C_{\text{sat}}(T)$, is a constant of space and time and equals the molar concentration of the inlet gases. This yields the molar balance

$$C_T = C_v + \frac{C_o^{in}}{C_{\text{sat}}} C_o, \qquad (7.113)$$

Conservation of water vapor, at the quasi-steady state, leads to

$$C_T Q_c^v + Q_c^{in} C_v^{in} = Q_c^{out} C_v, \qquad (7.114)$$

where the cathode inlet molar flux rate, Q_c^{in}, and the inlet water mole fraction, C_v^{in}, are control parameters. The vapor flux between the cathode plenum and the membrane, Q_c^v, is given by

$$Q_c^v = A\gamma_m a(\overline{M}_c - M_*^c), \tag{7.115}$$

where A is the area of the membrane, $\overline{M}_c$ is given by eqn. (118) below, and $M_*^c = M_*(C_v)$ and $M_*^a = M_*(A_v)$. The constant pressure assumption leads to the flux balance

$$Q_c^{out} = Q_c^{in} + Q_c^v - \frac{I_e}{4F}, \tag{7.116}$$

where the last term represents the consumption of oxygen. We combine eqns. (114–116) to obtain

$$\frac{AI_*}{4F}I_e C_v = A\gamma_m a(\overline{M}_c - M_*^v)(C_v - C_T) + Q_c^{in}(C_v - C_v^{in}). \tag{7.117}$$

The cathode plenum volume, V_c, equals the membrane area A, times the cathode plenum depth H_p. The total, or external, current $I_e = \int_0^1 I dy$, and the average membrane hydration, $\overline{M}_s$ for $s = a, c$ is given by

$$\overline{M}_s = \int_0^1 M_s dy. \tag{7.118}$$

On the anode side, assuming only hydrogen and vapor, we have the balance,

$$C_{\text{sat}} A_T = C_{\text{sat}} A_v + A_h^{in} A_h, \tag{7.119}$$

$$\frac{AI_*}{2F}I_e A_v = A\gamma_m a(\overline{M}_a - M_*^a)(A_v - A_T) + \frac{Q_a^{in}}{A_T}(A_v - A_v^{in}). \tag{7.120}$$

7.3.1.4. Cell Voltage and Local Current

To complete the positive feedback loop, we must include the impact of membrane hydration upon conductivity and current density. Assuming a large backing plate conductivity, the cell voltage, nondimensionalized by the open circuit voltage E_o, is a spatial constant V which must equal the voltage loss in the external resistor,

$$V = \frac{R_e I_* A}{E_o} I_e, \tag{7.121}$$

where I_* is the scaling for the external current I_e. Through each y slice of the membrane, the external voltage must balance against the open circuit voltage less the membrane losses and cathode overpotential losses,

$$E_o V = E_o - \frac{I_* I(y)}{\sigma_{\text{mea}}(M)} - \eta_* \eta, \qquad (7.122)$$

where the overpotential η is scaled by $\eta_* = \mathcal{R}T/(\alpha F)$, and the effective membrane electrode assembly (MEA) conductivity, σ_{mea}, includes the effects of the contact resistance at the membrane-catalyst layer, characterized by the area-specific conductance σ_c, and the resistance to hydronium flow in the membrane, characterized by the area-specific membrane conductance, σ_m. The overall MEA conductance follows from the series formula,

$$\sigma_{\text{mea}} = \frac{\sigma_c \sigma_m}{\sigma_c + \sigma_m}. \qquad (7.123)$$

While σ_c is taken constant, the dependence of σ_m on the membrane ignition level is obtained by integrating eqn. (104) and neglecting small terms corresponding to concentration polarizations (see [3] for details),

$$\sigma_m(M) = \left(\frac{\mathcal{R}TH_m}{\alpha F^2} \int_0^1 \frac{dz}{D_+(M)P} \right)^{-1}. \qquad (7.124)$$

Introducing the definition, eqn. (98), of the hydronium diffusivity, and using both the through-plane water profile, eqn. (107), and the equilibrium profile of eqn. (97) for the hydronium concentration, we find

$$\sigma_m(M) = \frac{\sigma_0}{H_m} \left[\Psi_c - \Psi_a \right], \qquad (7.125)$$

where

$$\sigma(M_c, M_a) = \left(\int_{M_a}^{M_c} \frac{\psi_w(s) ds}{\psi_+(s) P(s)} \right)^{-1}. \qquad (7.126)$$

Observe that in the limit $M_c \to M_a$ the conductivity reduces to the zero-water gradient formulation,

$$\sigma(M, M) = \psi_+(M) P(M). \qquad (7.127)$$

Using the formula for the proton diffusivity, the conductivity may be evaluated explicitly as

$$\sigma = \frac{M_c^2 - M_a^2}{2 \left(M_c - M_a + \frac{\beta}{\nu} ((M_a - \psi_c)^{-\nu} - (M_c - \psi_c)^{-\nu}) \right)} \qquad (7.128)$$

if both $M_c, M_a > \psi_c$, and $\sigma = 0$ otherwise.

The nondimensionalized cathode over potential, η, satisfies a Butler-Volmer relation

$$I = i_*(C_o - \delta I)\sinh(\eta), \tag{7.129}$$

where i_* represents a scaled exchange current, δ represents the mass transport limitations in the catalyst and gas diffusion layers, and Tafel slope terms in the *sinh* have been scaled into η via the nondimensionalization process. The mass transport limitation is motivated by the relation

$$C_o^{cat} = C_o - \delta I, \tag{7.130}$$

between oxygen content at the reactive catalyst cites, C_o^{cat}, and oxygen concentration in the gas plenum.

Using the definition of I_* the membrane voltage balance, eqn. (122), can be re-written as

$$\sigma(M)(1-V) = \Phi(I, M, C_o), \tag{7.131}$$

where the kinetic voltage losses Φ are given by

$$\Phi = I + \frac{\eta_*}{E_o}\sigma(M)\sinh^{-1}\left(\frac{I}{i_*(C_o - \delta I)}\right). \tag{7.132}$$

The kinetic losses are a strictly increasing function of current density $I \in [0, C_o/\delta]$, where the maximum current is dependent upon oxygen concentration and mass transport limitations. Since Φ is invertible, we formally solve the membrane voltage balance, eqn. (131), obtaining the local current $I(y)$,

$$I = \Phi^{-1}(\sigma(M)(1-V)). \tag{7.133}$$

The external voltage balance, eqn. (121), becomes a nonlinear-nonlocal equation for the cell voltage

$$V = \frac{\sigma_0 A}{\sigma_e H_m}\int_0^1 \Phi^{-1}(\sigma(M)(1-V))dy, \tag{7.134}$$

where $\sigma_e = 1/R_e$ is the conductance of the external resistor.

7.3.1.5. Full Model

Introducing the dimensionless time variable $\tau = \frac{t}{t_*}$, where $t_* = \frac{H_m}{\gamma_m}$, and taking the inlet relative humidities A_v^{in} and C_v^{in} to be zero, the mass transport equations in the along-the-membrane direction may be written as

$$\mathcal{M}\begin{pmatrix} M_c \\ M_a \end{pmatrix}_\tau - \epsilon \mathcal{D} \begin{pmatrix} \Psi_c \\ \Psi_a \end{pmatrix}_{yy}$$
$$= \begin{pmatrix} (M_c - M_*^c) - \frac{3}{2\gamma}\Phi^{-1}(\sigma(M)(1-V)) + \epsilon_\perp(\Psi_c - \Psi_a) \\ (M_a - M_*^a) + \frac{1}{\gamma}\Phi^{-1}(\sigma(M)(1-V)) - \epsilon_\perp(\Psi_c - \Psi_a), \end{pmatrix} \quad (7.135)$$

where the dimensionless parameters are given in Table 7.6. This is coupled to the quasi-steady equations for the vapor concentrations in the plenums,

$$\bar{\sigma}VC_v = \delta_c(\overline{M}_c - M_*^c)(C_v/C_T - 1) + \overline{Q}_c C_v, \quad (7.136)$$

$$2\bar{\sigma}VA_v = \delta_a(\overline{M}_a - M_*^a)(A_v/A_T - 1) + \overline{Q}_a A_v, \quad (7.137)$$

the oxygen concentration, given by $C_o = (C_T - C_v)/X_o^{in}$, and subject to the voltage balance

$$V = \bar{\rho}\int_0^1 \Phi^{-1}(\sigma(M)(1-V))dy, \quad (7.138)$$

with the current dependent voltage losses given by

$$\Phi(I) = I + \bar{\eta}\sigma(M)\sinh^{-1}\left(\frac{I}{i_*(C_o - \delta I)}\right), \quad (7.139)$$

and $I_e = V/\bar{\rho}$.

The model, as posed in eqs. (135–138), is spatially unbiased. To break symmetry, gaussian noise is added at each timestep to the discrete solution of eqn. (136). The noise is biased toward the outlet end ($y = 1$) of the cell, with the mean varying from 0.0075 at the outlet to −0.0075 at the inlet. These perturbations, on the order of 1 part in 1000, are designed to mimic the variation of humidity level within the gas plenum due to the convection, and force the ignited zone to be toward the outlet end of the cell.

7.3.1.6. Numerical Hysteresis and Transients

We compare the numerical hysteresis loop, shown in Figure 7.5, to the experimental loop in Figure 3.10 of Chapter 3. The control data for the numerical simulations are given in Table 7.4, while the material parameters and the corresponding nondimensional parameters are given in Tables 7.5 and 7.6. The numerical hysteresis loop is generated by increasing the external resistance in steps of 1.0 Ohms until the transition from uniform ignition to partial ignition is observed and then decreased in increments of 0.5 Ohm back to the minimum level of 0.5 Ohm. After each step in external load, the cell is permitted to come to

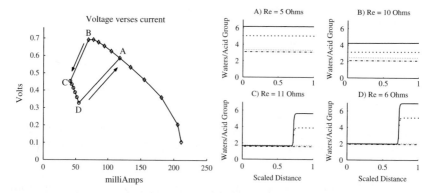

FIGURE 7.5. (left) Hysteresis in a steady-state polarization curve. The external resistance is increased from $R_e = 0.5$ Ohm to a maximum of $R_e = 11$ Ohm and back down, passing through $R_e = 5, 10, 11, 6$ Ohm at the points indicated A, B, C, and D. The diamonds correspond to a uniformly ignited membrane, while the open circles correspond to a partially ignited membrane. The corresponding data are given in Table 7.4. (right) The cathode (+) and anode (o) sides membrane water contents and equilibrium water content M_c^* (dotted line) and M_a^* (dash-dotted line) at points A-D on the polarization loop. The membrane water transitions to partial ignition at $R_e = 11$ Ohms, and the ignited area remains roughly constant as the external resistance is decreased.

steady-state over a period of 4 or more hours. The points labeled A, B, C, and D on the polarization curve, Figure 7.7 (left), indicate where the transition from uniform ignition to partial ignition occurs. The jump points from high to low polarization curves occur at similar external resistances and total currents. The lower curve on the numerical simulation has a voltage range of 0.45-0.3 volts, compared to 0.6-0.4 for the experimental curve, but both figures show a heightened slope for the lower curve, indicating a higher membrane resistance.

In Figure 7.5 (right), the water profiles of the anode and cathode sides of the membrane, and equilibrium values of membrane water content corresponding to the anode and cathode plenum relative humidities are shown for four points on the polarization curve depicted at left. At point A, $R_e = 5$ Ohms, the membrane is well ignited and the gas plenums have relatively high humidity level. At point

TABLE 7.4 Control parameters.

A_{in}	33 moles/m^3 (@1 Atm)	Anode inlet press.
C_{in}	33 moles/m^3	Cathode inlet press.
A_v^{in}	0.00	Anode inlet RH
C_v^{in}	0.00	Cathode inlet RH
Q_c	1.18 micro-moles/s = 2.4 cm^3/s	Cathode flow rate
Q_a	0.87 micro-moles/s = 1.7 cm^3/s	Anode flow rate
C_o^{in}	33 moles/m^3	Inlet O_2 molar conc.
T	368° K	Cell Temperature

TABLE 7.5 Material parameters and reference values.

a	1200	mole/m^3	membrane acid concentration
$\mathcal{A}$	10^{-4}	m^2	Membrane area
C_T	33.0	moles/m^3	inlet gas concentration
d_w	5.6×10^{-10}	m^2/s	base water self diffusivity
d_+	9.5×10^{-11}	m^2/s	base hydronium self diffusivity
E_o	1.0	Volts	open circuit voltage
F	96485	C/mol	Faraday's number
H_p	1×10^{-3}	m	Gas plenum depth
H_m	125×10^{-6}	m	Membrane thickness
H_0	-52300	J/mole	Enthalpy of hydrogen formation
K_0	6.2	-	disassociation constant for acid
L	0.01	m	Membrane length
R	8.312	J/(mole K)	Universal gas constant
R_e	1	Ω	external resistance
T	368	°K	Cell temperature
W	0.01	m	Membrane width
α_{in}	0.25	-	anisotropy of in-plane diff.
γ_m	1.1×10^{-6}	m/s	water mass transport param.
σ_0	0.35	1/(Ohm m)	Membrane conductance
ψ_c	2.8	-	conductivity threshold for membrane
I_*	$\frac{E_o \sigma_0}{H_m}$	3570 A/m^2	Current scaling
t_*	$\frac{H_m}{\gamma_m}$	118s	Time scaling
η_*	$\frac{RT}{\alpha F}$	0.063 V	Overpotential scaling

B, $R_e = 10$ Ohms, the anode side of the membrane is close to the region of reduced conductivity, the gas plenums are significantly less humid due to the decreased current production. Moving from points B to C the external load is increased to $R_e = 11$ Ohms, and the cell transitions from a uniform ignition to a partial ignition over a time period of four hours. The humidified region occupies roughly 30% of the cell area. The reduction in cell current corresponds to a reduced plenum humidity level. At point D, $R_e = 6$ Ohms, the humidified region still occupies 30% of the cell area, but the membrane hydration is higher in the

TABLE 7.6 Dimensionless parameters.

$\bar{Q}_a$	$\frac{Q_a H_m}{\gamma_m V_a \mathcal{A}^{in}}$	3.70	$\bar{Q}_c$	$\frac{Q_c^{in} H_m}{\gamma_m V_c C_{sat}}$	2.82
X_o^{in}	$\frac{C_o^{in}}{C_{sat}}$	1.00			
γ	$\frac{a \gamma_m F H_m}{E_o \sigma_0}$	0.034	δ_a	$\frac{a H_m}{C_{sat} H_p}$	0.615
δ_c	$\frac{a H_m}{C_{sat} H_p}$	0.615	ϵ	$\frac{\alpha_{in} d_w H_m}{L^2 \gamma_m}$	1.65×10^{-4}
ϵ	$\frac{d_w}{H_m \gamma_m}$	3.145	$\bar{\eta}$	$\frac{\eta_*}{E_o}$	6.34×10^{-2}
$\bar{\rho}$	$\frac{R_e \mathcal{A} \sigma_0}{H_m}$	0.178	$\bar{\sigma}$	$\frac{E_o H_m}{4 \gamma_m F V_c R_e C_{sat}}$	18.6

humidified region, and the overall gas plenum humidity levels are higher due to the higher cell current. Further decrease in external load initiates a 1 hour transition to the uniformly ignited membrane depicted in plot A.

Figure 7.6 depicts the transient response of the cell to the change in external load corresponding to the transition from point B to point C in Figure 7.5. The water profiles, Figure 7.6 (left), are snapshots of the anode and cathode membrane water content at times, $t_1 = 0.2$, $t_2 = 0.4$, $t_3 = 0.564$, and $t_4 = 4.51$ hours. Time t_1 corresponds to just before the jump in external resistance and shows the uniformly ignited membrane with its anode level above the conductivity cutoff of $\psi_c = 2.8$. Between times t_1 and t_2, the water profile remains spatially uniform but decreases in response to the lower water production. Shortly before time t_2, the water content of the left-most end of the cell passes through the minimum threshold for conductivity; the water production stops there, and a traveling wave is initiated. At time t_3, the rapid phase of the transient is over and, except for a thin boundary layer, each point y of the membrane is either extinguished or active. Over a period of 4 hours the diffusion within the membrane moves the traveling wave adiabatically through the family of quasi-steady water profiles. It is this timescale which is used to determine the anisotropy in the membrane diffusivity. Figure 7.6 (right) shows the time evolution of the current, plenum RHs, and percent of the membrane which is ignited. The relative humidity of both the cathode and anode plenums decreases by roughly 35% in response to the decreased water production, which drops from 77 milliamps to 48.5 milliamps. The percentage of the membrane area which is active decreases to 65% at time t_3 and then slowly declines to 37% at steady state.

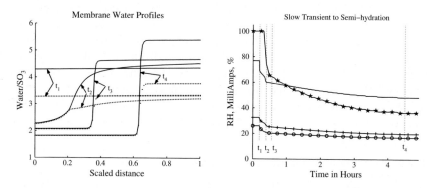

FIGURE 7.6. (left) Membrane water transients during impulsive step-up of external resistance from $R_e = 9$Ohms to $R_e = 10.5$ Ohms. The cathode (solid) and anode (dotted) side water contents of the membrane at times $t_1 = 0.200$, $t_2 = 0.400$, $t_3 = 0.564$, $t_4 = 4.51$ hours. The corresponding data is given in Table 7.3. (right) Relative humidity of the cathode (+) and anode (o) plenums, the percentage of the membrane which is ignited (star), and the total current produced (line), in milliAmps, all plotted versus time in hours, with the times $t_1, \ldots, t_4$ indicated by vertical dotted lines.

The model is quite sensitive to variations in the dimensionless parameter, γ, which governs mass transport from the membrane to the channel through eq. (95). High values of γ correspond to a catalyst layer with a large ionomer-gas pore surface area density, which facilitates transport of water into or out of the membrane, bringing it closer to its equilibrium value. Thus γ is natural bifurcation parameter, tuning the strength of the feedback mechanism between channel RH, membrane water content, and current production. This is particularly true at high temperatures, or high gas flow rates, as observed in Chapter 3, when more water vapor is required to raise the RH of the gas plenums.

Figure 7.7 shows the results of a hysteresis loop at constant external load, $R_e = 11$ Ohms. First γ is increased, in multiples of 1.03, from 0.027 up to 0.0363, and then decreased from 0.0363 down to 0.0201. In the left plot of Figure 7.7, the steady-state water profiles are shown at four values of γ on the upswing, A–D, which correspond to $\gamma = 0.027, 0.0304, 0.0342$, and 0.0363, and four on the downswing, E–H, corresponding to $\gamma = 0.0342, 0.0304, 0.027$, and 0.0247. Steady-state is achieved after 4–6 hours of clock time. In the right plot, the cell voltage, which is proportional to the total current in milliamps by a factor of 90.9, is shown versus γ for each computed value. Several features stand out; first the transition from the uniformly ignited to the partially ignited state is abrupt as γ is increased but is smooth when γ is decreased, in contrast to the variable load loop depicted in Figure 7.5, where the transition is abrupt in both directions. The abrupt transition from partially to fully ignited reflects a "run-away" ignition process, the increase in ignited cell area increases the current and the RH of the plenums, which leads to further increases in ignited area. The ignited area is quite sensitive to channel RH. Increasing the inlet RH at both anode and cathode from 0 to 5% can double the ignited area in certain regimes.

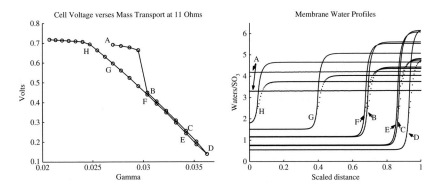

FIGURE 7.7 Sensitivity to variations in γ. At $R_e = 11$ Ohms, the dimensionless parameter γ is varied in multiples of 1.03, first increasing from 0.027 up to 0.0363 and then decreasing from 0.0363 down to 0.0201. (Left) The cell voltage as a function of γ. (Right) The water profiles at the 8 values of γ indicated in the figure at left. The profiles A and G occur at the same operating conditions.

Moreover, the closer the channel RH is to the critical value at which the reaction can spontaneously ignite, the greater the propensity for a growing ignited region to "run away". Broadly speaking, the smooth transition from partially to fully ignited operation shown in Figure 7.7 is typical of very low RH operation, while the run-away transition shown in the hysteresis loop of Figure 7.5 results from a more elevated channel RH and a channel RH which is more sensitive to membrane water production. This is reflected in the equilibrium hydration levels in Figure 7.5 (right), which are all above 2 waters/acid group, compared to those in Figure 7.7 (left), which are all below 2 waters/acid group.

7.3.2. *Hysteresis in Counter-Flowing PEMFC*

As discussed previously, the transition from a single-phase to a two-phase setting within the gas diffusion and catalyst layers can have significant impact on the membrane hydration levels and the membrane conductivity. This sensitivity of conductivity to phase change generates a positive feedback mechanism. Our numerical modeling suggests that this feedback can lead to hysteretic conditions in automotive and stationary PEM cell operation.

We consider a steady-state version of the GDL model described in Section 7.2, coupled to the membrane model and to the channel flow model developed in [3]. This a class of 1 + 1D models, which consider the lateral transport of reactants to occur only within the flow fields with the MEA supporting only through-plane transport. A 1D model for the molar fractions in the flow fields is coupled to a 1D model for through-plane transport in the MEA. The MEA model dictates the local current density and hence the consumption of reactants and production of water in liquid and vapor forms; these serve as forcing terms for the flow field equations. In turn, the flow field equations provide the boundary conditions for the MEA model, as described in Section 7.2.

In the 1 + 1D model the MEA can be in one of three distinct states, as indicated in Figure 7.1. Either the cathode GDL is entirely single phase, and the membrane water content is based upon disequilibrium from the vapor equilibrated phase, or the cathode GDL is partially or entirely two-phase, and the membrane water content is based upon disequilibrium from the liquid equilibrated phase. The switching between these states drives the hysteresis; however, lateral motion of liquid water within the membrane or within the GDL is not permitted within this model. Such motion could lead to slow transients, as found in the STR fuel cell of Section 7.3.1.

We do not present the model in detail but rather outline the main impact of the two-phase region on fuel cell performance. Considering counter-flowing unit cell operation, as outlined in Figure 7.1, we prescribe the total cell current and the coolant temperature profile and calculate the cell voltage, the local current distribution, and the distributions of reactants. In Figure 7.8, we consider a base case with inlet pressures, gas flow rates (given in terms of the Stoich), and inlet due points as listed. The cathode gas mixture is uncompressed

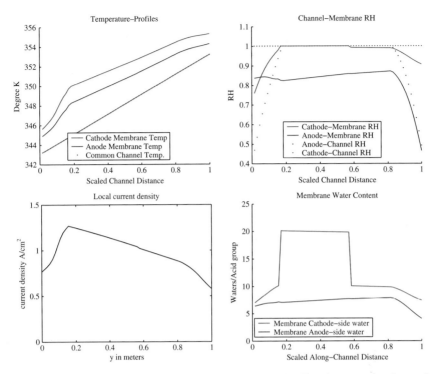

FIGURE 7.8. Temperature profile, Channel Membrane RH, Local current density, and anode-cathode membrane water contents for a counter flowing unit cell. Operating conditions are 2.2/2.0 barg An/Cat, 1.2/1.8 stoich An/Cat, 63/53C dew point An/Cat. The cell voltage is calculated at 0.6489 V at a prescribed current density of 1 Amp/cm^2.

air. The cathodic two-phase region is toward the front of the cell, indicated by the plateau for $y \in (0.2, 0.6)$ in the Membrane Water Content profile. The location and width of the two-phase region is quite sensitive to many operating conditions, such as inlet dew points, coolant temperature, thermal conductivity of the GDL, and power density. This is emphasized by the the 'Humid region' seen in the Channel-Membrane RH profile, where the RH at the cathode membrane is seen to be just slightly undersaturated in the region $y \in (.6, 0.8)$. Here the RH *falls* very slightly as one moves through the GDL from the saturated channel towards the membrane. A key factor determining whether the cathode saturates is the balance between heat flux and water vapor flux through the GDL. The local current dictates heat and vapor production; however, their removal depends upon conditions in the flow field and coolant channel. If the flow field channel is saturated than a slightly falling RH yields a dry GDL, while a slightly rising RH forces a two-phase GDL.

The temperature profile in the coolant is taken to be linear between inlet and outlet. The temperature on either side of the membrane is computed based upon the current density, the overpotential, and amount of phase change. A key feature is the inlet blip in current density, seen in the bottom left of Figure 7.8. The undersaturated air at the the cathode inlet dries the membrane, increasing resistance and decreasing current production. After the cathode flow field gas saturates, the cathode GDL goes two-phase and the membrane resistance decreases. Falling oxygen concentration leads to declining current density toward the cathode outlet. The dry gas at the anode inlet (cathode outlet) further decreases the current production at the cathode outlet. In Figure 7.9, we see that increasing the anode due point to 63C increases the width of the two-phase region, moving it yet further toward the cathode inlet. The inlet blip in current production is significantly reduced, and the cell voltage increases by about 5 millivolts. In Figure 7.10, we sweep out a polarization curve, showing that the two-phase model predicts higher cell voltages than a dry-only model, for the same operating conditions. However, the mass transport knee occurs sooner in the two-phase model. In our simulation, the inlet is coldest part of the cell, and at high current densities and larger overpotentials, the domination of heat production over vapor production moves the two-phase region toward the coldest part of the cell. This places the largest oxygen concentration and the lowest

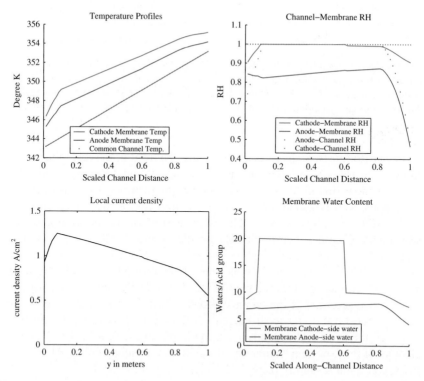

FIGURE 7.9. Same as Figure 7.8 but cathode inlet due point is 63C and cell voltage is 0.6539.

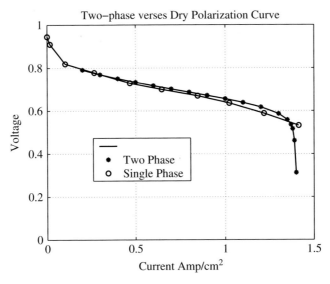

FIGURE 7.10. Polarization curve for both a single-phase and a mixed phase version of the 1 + 1D code. The wet version has higher membrane conductivity, and hence better performance at high current density, but imbalance in current distribution induced by the two-phase region leads to an earlier knee in the polarization curve.

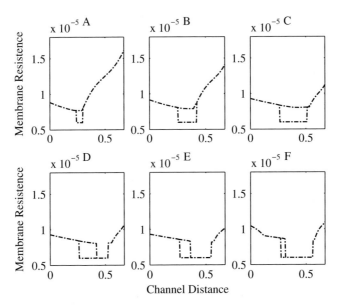

FIGURE 7.11. Membrane resistance plotted versus channel position for six different current densities, running from 1.6 A/cm^2 (A) down to 0.7 A/cm^2 (F). The dashed line and the dash-dotted line are two different steady-state configurations. The flat region with lowest membrane resistance corresponds to a liquid-equilibrated membrane.

membrane resistance at the same location, making the current distribution less level. The high current density at the cathode inlet forces the cell into the mass transport knee at lower average current density than in the flatter current profiles of the dry-cell model.

The dichotomy between dry and two-phase states within the GDL can produce hysteresis. This is depicted in Figure 7.11, where we show two possible operating states for the same external conditions. The particular operational condition obtained by the numerical model depends upon the initial guess for the iterative solver, and the solutions are continued numerically through a range of current densities. At high current density, $1.6\,\text{A}/\text{m}^2$ in subfigure (A), the cell can be either entirely dry or have a small two-phase region near the middle. As the current density is lowered, the two-phase region grows, until in subplot (D) both solutions have a two-phase region. Further lowering of the current density drives the two solutions onto each other, closing the hysteresis loop.

References

[1] Benziger, J., Chia, E., Moxley, J., & Kevrekidis, I., *Chem. Eng. Sci.*, **60** (6) 1743–1759 (2005).
[2] Berg, P., Calgar, A., Promislow, K., St.-Pierre, J., & Wetton, B. *IMA J. Applied Math.*, **71** 241–261 (2006).
[3] Berg, P., Promislow, K., St.-Pierre, J., Stumper, J., & Wetton, B. *J. Electrochem. Soc.* **151** (2004) A341–A354.
[4] Bernardi, D.M. & Verbrugge, M.W. *J. Electrochem. Soc.* **139**, 2477 (1992).
[5] Berning, T., Lu, D.M., & Djilali, N. *J. Power Sources* **106** 282–292 (2002).
[6] Bradean, R., Promislow, K., & Wetton, B. *Numerical Heat Transfer, Part A* **42** 1–18 (2002).
[7] Djilali, N. & Lu, D. *Int. J. Thermal Sci.* **41** 29–40 (2002).
[8] Divisek, J., Fuhrmann, J., Gärtner, K., & Jung, R. *J. Electrochem. Soc.* **150** A811–A825 (2003).
[9] Fife, P.C. *Dynamics of Internal Layers and Diffusive Interfaces*, SIAM, Philadelphia, PA, (1988).
[10] Fowler, A.C. *Mathematical Models in the Applied Sciences*, Cambridge Texts in Applied Mathematics, Cambridge University Press, (1997).
[11] Freunberger, S., Tsukada, A., Fafilek, G., & Buechi, F.N. Paul Scherrer Institute Scientific Report **94** 2002 ISSN 1423–7342 (2003).
[12] Freunberger, S.A., Santis, M., Schneider, I.A., Wokaun, A., & Buchi, F.N. *J. Electrochem. Soc.*, **153** (2) A396–A405 (2006).
[13] Ge, S., Li, X., Yi, B., & Hsing, I-M., *J. Electrochem. Soc.* **152** (6) A1149–A1157 (2005).
[14] van Genuchten, M. *Soil Sci. Soc. Amer. J.* **44** 892–898 (1980).
[15] He, W., Yi, J.S., & Van Nguyen, T. *AIChE J.* **46** 2053–2064 (2000).
[16] Hinatsu, J.T., Mizuhata, M., & Takenaka, H., *J. Electrochem. Soc.* **141** 1493–1498 (1994).
[17] Hsing, I.M. & Futerko, P. *Chemical Eng. Sci.* **55** 4209–4218 (2000).

[18] Kulikovsky, A. *J. Electrochem. Soc.* **150** A1432–1439 (2003).
[19] Leverett, M. *Transactions of the American Institute of Mining and Metallurgical Engineers* (*Petroleum Division*) **142** 152–169 (1941).
[20] Motupally, S., Becker, A.J., & Weidner, J.W., *J. Electrochem. Soc.*, **147** 3171 (2000).
[21] Moxley, J.F., Tulyani, S., & Benziger, J., *Chem Eng. Sci.*, **58** 4705–4708 (2003).
[22] Natarajan, N. & Van Nguyen, T., *J. Electrochem. Soc.* **148** A1324–A1335 (2001).
[23] Nazarov, I. & Promislow, K., in *Chemical Eng. Sci.* **154** B623–B630 (2007).
[24] Nieber, J., Bauters, T., Steenhuis, T., & Parlange, J.-Y. *J. Hydrology* **231–232** 295–307 (2000).
[25] Pasaogullari, U. & Wang, C.Y. *J. Electrochem. Soc.* **151** A399–A406 (2004).
[26] Promislow, K. *SIAM J. Math. Analysis* **33** (6) 1455–1482 (2002).
[27] Promislow, K. & Stockie, J. *SIAM J. Applied Math.* **62** 180–205 (2001).
[28] Promislow, K., Stockie, J., & Wetton, B., in *Proc. Roy. Soc. London: Series A* **462** 789–816 (2006).
[29] Rowe, A. & Li, X. *J. Power Sources* **102** 82–96 (2001).
[30] Siegel, N., Ellis, M., Nelson, D., & von Spakovsky, M. *J. Power Sources* **128** 173–184 (2004).
[31] Sone, Y., Ekdunge,P., & Simonsson, D., *J. Electrochem. Soc.* **143** 1254–1259 (1996).
[32] Smoller, J. *Shock waves and reaction diffusion equations*, Springer-Verlag, Fundamental Principles of Mathematical Sciences **258** (1983).
[33] Springer, T.E, Zawodzinski, T.A., & Gottesfeld, S. *J. Electrochem. Soc.* **138**, 2334 (1991).
[34] Stockie, J., Promislow, K., & Wetton, B. *Int. J. Num. Methods in Fluids* **41** 577–599 (2003).
[35] Taylor, R. & Krishna, R. *Multicomponent Mass Transfer*. Wiley Series in Chemical Engineering. John Wiley & Sons, (1993).
[36] Thampan, T., Malhotra, S., Tang, H., & Datta, R., *J. Electrochem. Soc.*, **138** 2334 (1991).
[37] Um, S. & Wang, C.Y. *J. Power Sources* **125** 40–51 (2004).
[38] Um, S., Wang, C.Y., & Chen, C.S. *J. Electrochem. Soc.* **147** 4485–4493 (2000).
[39] Wang, Z.H., Wang, C.Y., &Chen, K.S. *J. Power Sources* **94** 40–50 (2001).
[40] Zemansky, R. & Dittman M. *Heat and Thermodynamics*, McGraw Hill, (1981).

8
Modeling of Two-Phase Flow and Catalytic Reaction Kinetics for DMFCs

J. Fuhrmann and K. Gärtner

In this chapter, we discuss more deeply two aspects of a numerical model for direct methanol fuel cells published in [1]. This model describes in detail the processes in the membrane electrode assembly (MEA) of a direct methanol fuel cell (DMFC). We assume that the MEA consists of the following parts:

- an anodic porous transport layer (PTL);
- an anodic reaction zone consisting of the same material as the PTL, enhanced with catalyst and partially containing membrane material;
- a proton-conducting polymer electrolyte membrane (PEM), typically Nafion;
- a cathodic reaction zone with similar composition as the anodic reaction zone;
- a cathodic PTL with similar composition as the anodic PTL.

On the exterior, the anodic resp. cathodic PTL is equipped with electrical contacts (Anode, resp. Cathode) and interfaces to the transport channels containing a fluid – gas mixture with given phase and partial pressures.

The overall reaction taking place in the cell is the combustion of methanol

$$2CH_3OH + 3O_2 \longrightarrow 4H_2O + 2CO_2$$

which by the PEM is split into two parts

$$\begin{aligned} 2CH_3OH + 2H_2O &\longrightarrow 12H^+ + 12e^- + 2CO_2 \text{ (Anode)} \\ 12H^+ + 3O_2 + 12e^- &\longrightarrow 6H_2O \text{ (Cathode)} \end{aligned} \quad (8.1)$$

exchanging electrons through the outer circuit and protons through the PEM.

Among other processes, the model features two fundamental aspects of the processes in the MEA:

- Two-phase flow in the porous transport layers
- Catalytic reaction kinetics.

8.1. Two-Phase Flow

8.1.1. Motivation

The anodic reaction uses water and methanol to produce protons, electrons and carbon dioxide which partially dissolves in water and partially evaporates. Removing the carbon dioxide from the anodic reaction zone means that we have to support counter flow of water and gas in the porous transport layers. Methanol and water may evaporate, thus demanding to model a gas mixture.

The cathodic reaction uses electrons, protons, and oxygen and produces liquid water which at the operation temperature partially evaporates. Furthermore, liquid water may diffuse through the membrane. So, here again, we have the situation that products and educts of a reaction are available in different phases, and we have to organize a counter flow process to support the reaction.

The PTL are admixed with Teflon in order to ensure that a part of the pore space is less wettable by water, thus making it available for gas flow.

Due to the operating conditions for Nafion or other sulphonated membranes, liquid water has to be present not only in DMFC but also in H_2 PEM fuel cells; therefore, all considerations of this section apply to them as well.

When it comes to modeling these features, we can distinguish three scales:

- A microscopic scale where molecular processes take place which result in adhesion and cohesion forces.
- A mesoscopic scale corresponding to the geometry of the pore space, i.e. the "void" space not filled with substrate material. Ideally, concepts like adhesion to the pore walls, surface tension between liquid and gas phase, and contact angle are upscaled from the microscopic scale. These could be used to describe the liquid flow in the pore space, if its geometry is known in detail.
- A macroscopic scale which is based on averaging flow characteristics from the mesoscale.

The focus of this paper will be on the macroscopic scale which allows embedding of the two-phase flow model into more complex process models for the MEA.

When setting up such a model, two building blocks are available which come from different modeling fields, namely, the Stefan Maxwell equations describing the flow of a gas mixture through a porous medium and the equations of multiphase flow in a porous medium. We will shortly describe both of them. After that, we make an attempt to join both of them into a combined model. The discussion of open problems and future directions concludes the section.

8.1.2. Stefan Maxwell Model

The flow of gas mixtures in a porous medium can be modeled by the Stefan Maxwell model.

TABLE 8.1. Notations and constants for the Stefan Maxwell model as used in [1].

$\vec{N}_i^{gas}\ [mol/(m^2 s)]$	molar flux
$p_{gas_i}\ [Pa]$	partial pressure
$p_{tot} = \sum p_{gas_i}$	total gas pressure
$T\ [K]$	temperature
$y_i = p_{gas_i}/p_{tot}$	molar fractions
$B_i = D_i^K \dfrac{\omega \hat{M}_i + K_i}{1 + K_i} + \dfrac{\overline{r^2} \Psi p_{tot}}{8\eta}$	effective permeabity
$K_i = \dfrac{\lambda_i}{2\bar{r}}$	Knudsen number
$\omega = \dfrac{\pi}{4}$	slip factor
$\hat{M}_i = \sqrt{\dfrac{M_i}{\sum_i y_i M_i}}$	mean relative molecular weight
$D_i^K = \Psi \bar{r} \dfrac{2}{3} \sqrt{\dfrac{8RT}{\pi M_i}}$	Knudsen diffusion coefficient
$D_{i,j}^m = \Psi D_{i,j}^{bm}$	eff. binary molecular diffusion coefficient
$D_{i,j}^{bm} = 1.013 \cdot 10^{-2} \dfrac{T^{7/4}}{p_{tot}(\nu_i^{1/3} + \nu_j^{1/3})^2} \sqrt{\dfrac{M_i + M_j}{M_i M_j}}$	binary molecular diffusion coefficient [7]
$\lambda_i = \dfrac{T p_x}{\sqrt{2}\pi \sigma^2 p_{tot} N_A}$	mean free path
$\eta = \eta_\varsigma(T_r)\bar{f}_p^0 \dfrac{\sqrt{\bar{M}}\bar{p}_c^{2/3}}{(\bar{T}_c R)^{1/6} N_A^{1/3}}$	Viscosity
$\bar{T}_c = \sum y_i T_{ci}$	
$\bar{p}_c = R\bar{T}_c \dfrac{\sum y_i Z_{ci}}{\sum y_i \nu_i}$	
$\bar{f}_p^0 = \sum y_i f_{p,i}^0$	
$T_r = \dfrac{T}{\bar{T}_c}$	
$\sigma = 5 \cdot 10^{-10} m$	
$p_x = 8.2 \cdot 10^3\ Pa$	
$\Psi = \dfrac{\phi}{\tau}$	
$\phi\ [7]$	size of pore space
$\tau\ [7]$	tortuosity correction
$\bar{r}\ [m]$	mean pore radius
$\overline{r^2}\ [m]$	second moment of pore radius
$M_i\ [g/mol]$	molecular weight
$\nu_i\ [7]$	diffusion volume
$f_{p,i}^0\ [7]$	polarity factor
$Z_{c,i}\ [7]$	critical factor
$T_{ci}\ [K]$	critical Temperature
$R = 8.3144\ J/mol \cdot K$	gas constant
$N_A = 6.0220 \cdot 10^{23}/mol$	Avogadro number
$\eta_\varsigma(T_r) = (0.807 \cdot T_r^{0.618} - 0.357 \cdot e^{-0.449 T_r} + 0.340 \cdot e^{-4.058 T_r} + 0.018)$	

The model equation relates the gas fluxes $\vec{N}_i^{\text{gas}}$ (see Table 8.1 for notations) to the partial pressures p_{gas_i} of the components of the mixtue, respectively, their gradients:

$$RT\left(\frac{\vec{N}_i^{\text{gas}}}{D_i^K} + \sum_{j=1}^{n_{\text{gas}}} \frac{y_i \vec{N}_j^{\text{gas}} - y_j \vec{N}_i^{\text{gas}}}{D_{ij}^{bm}}\right) = \qquad (8.2)$$
$$p_{\text{tot}} \nabla y_i + y_i \left(\frac{B_i}{D_i^K} + \sum_{j=1}^{n_{\text{gas}}} \frac{B_i}{D_{ij}^m} y_j \left(1 - \frac{B_j}{B_i}\right)\right) \nabla p_{\text{tot}}.$$

The material coefficients are given by [2, 3]. The parameters of this model are derived from molecular data of the individual components and from basic parameters of the pore space ($\Psi, \bar{r}, \bar{r^2}, \tau$) (see Table 8.1).

Equation (8.2) can be solved for the molar fluxes yielding the formal expression

$$\vec{N}_i^{\text{gas}} = -\sum_{j=1}^{n_{\text{gas}}} D_{ij}^{SM}(p_1, \ldots, p_{n_{\text{gas}}}) \nabla p_{\text{gas}_j}$$

where $D_{ij}^{SM}(p_1, \ldots, p_{n_{\text{gas}}})$ is the diffusion matrix for the gas mixture, thus allowing to use the partial gas pressures p_{gas_i} as basic variables.

Hence the molar balance equations read as

$$\partial_t \left(\phi \frac{p_{\text{gas}_i}}{RT}\right) + \nabla \cdot \vec{N}_i^{\text{gas}} = R_i. \qquad (8.3)$$

Here, R_i summarizes the source/sink terms.

When used in a finite domain, this system has to be closed by boundary conditions, which in the simplest cases consist in fixing either the partial pressure p_{gas_i} or the molar flux $\vec{N}_i^{\text{gas}}$ at a given part of the boundary.

Another possible set of basic variables are the molar fractions y_i, together with the total pressure p_{tot} and the closing condition $\sum_{i=1}^{n_{\text{gas}}} y_i = 1$.

8.1.3. Extension of Darcy's Law to Two-Phase Flow

The main impact for modeling two-phase flow in porous media came from petroleum reservoir simulation [4, 5] and from hydrology [6]. Most available models are based on the concept of capillary pressure and on some upscaling of capillary information from the micro to the macro scale. We will restrict ourselves to that case.

8. Two-Phase Flow and Reaction Kinetics

In order to describe the flow of one phase, Darcy's law is at hand. It relates the rate of flow of a fluid in a porous medium to the gradient of the fluid pressure p. Formulated for molar fluxes, it reads as

$$\vec{N} = \frac{k\rho}{\mu}\nabla p.$$

In order to set up a two-phase flow model, Darcy's law for each individual phase is assumed. As the permeability for a given phase in presence of a second phase will decrease, a *relative permeability* is introduced:

$$\begin{aligned}\vec{N}_w &= -\frac{kk_{rw}\rho_w}{\mu_w}\nabla p_w \\ \vec{N}_g &= -\frac{kk_{rg}\rho_g}{\mu_g}\nabla p_g\end{aligned} \quad (8.4)$$

Here, $\cdot_w$ and $\cdot_g$ refer to the water and gas phases, respectively. Usually, a wetting and a non-wetting phase are considered, but in the context here, these denominations would be ambiguous.

The mass balances correspondingly read as

$$\begin{aligned}\partial_t(\phi s_w \rho_w) + \nabla\cdot\vec{N}_w &= R_w \\ \partial_t(\phi s_g \rho_g) + \nabla\cdot\vec{N}_g &= R_g\end{aligned} \quad (8.5)$$

Here, s_g, s_w describe the saturations of the phases, i.e. the part of pore space filled with the given phase. Assuming that the pore space is filled exclusively by the two phases leads to the condition

$$s_g + s_w = 1 \quad (8.6)$$

The saturations are expressed using the effective saturation s_e and the residual saturations s_g^{res}, s_w^{res} [6]:

$$\begin{aligned}s_w &= s_w^{res} + (1 - s_g^{res} - s_w^{res})s_e \\ s_g &= s_g^{res} + (1 - s_w^{res} - s_g^{res})(1 - s_e)\end{aligned} \quad (8.7)$$

Furthermore, the relative permeabilities are related to the effective saturations:

$$\begin{aligned}k_{rw} &= k_{rw}(s_e) \\ k_{rn} &= k_{rn}(1 - s_e)\end{aligned} \quad (8.8)$$

So far, we are left with three basic unknowns: s_e, p_w, p_g. For closing the system, we need another relation. In order to explain the idea behind, we go back to the mesoscale: Assume a thin capillary tube filled with two non-mixing fluids A and B with pressures p_A and p_B. Without surface effects, the position of the interface between them would be determined by pressure

TABLE 8.2. Two-phase flow parameters.

k	hydraulic permeability of porous medium
s_w	saturation of water
s_g	saturation of gas
p_g	pressure of gas
ρ_w	density of water
ρ_g	density of gas
μ_w	viscosity of water
μ_g	viscosity of gas
$k_{rg}(s_e)$	relative permeability for gas phase
$k_{rw}(s_e)$	relative permeability for water phase

equilibrium $p_A = p_B$. But both fluids underlay adhesion forces which attract fluid molecules to the walls and surface tension effects which stem from the disbalance of molecular cohesion forces at the fluid–fluid interface. The fluid with higher adhesion is attracted by the walls relative to that with the lower adhesion, modifying the position and the form of the interface. This process is equilibrated by surface tension which builds up a pressure drop between the liquids called capillary pressure [8]. Therefore, the amount of a given liquid in the capillary tube can be related to the capillary pressure.

A porous medium in a first approximation can be seen as a collection of capillary tubes with different and changing diameters. Using the considerations above, we therefore define the capillary pressure

$$p_c = p_g - p_w$$

as the pressure drop between gas and water phases and assume a relation

$$s_e = s_e(p_c) \tag{8.9}$$

This closes the system (8.4)–(8.9). Summarizing, to describe the two phase flow behavior of the porous medium, in addition to the porosity ϕ and the permeability k, which are characteristics of the porous medium, we introduced a capillary pressure-saturation relationship and two permeability-saturation relationships which involve the geometry of the porous medium, as well as the wetting and surface properties of the fluids. The parameters of the two-phase flow model are summarized in Table 8.2.

8.1.4. Capillary Pressure – Saturation Relationships

For general porous media, there is no practical way to obtain useful expressions for these relationships from "first principles." However there are several

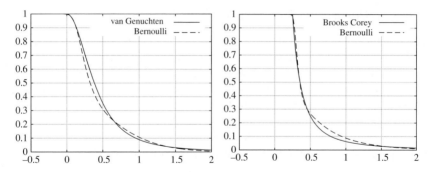

FIGURE 8.1. Pressure Saturation relationships van Genuchten: $n = 2, \alpha = 2$ [1]: $p_{\frac{1}{2}} = 0.325, \alpha = 5, \beta_n = 0.9, \beta_w = 3$ Brooks-Corey: $\lambda = 2, p_e = 0.25$ $p_{\frac{1}{2}} = 0.35, \alpha = 10, \beta_n = 0.65, \beta_w = 4$.

approaches to design expressions containing parameters, which in a coarse approximation can be related to the pore size distribution.

The most common relationships (see also Figure 8.1) are

- Brooks-Corey [9]:

$$s_e(p_c) = \begin{cases} (p_c/p_e)^\lambda, & p_c > p_e \\ 1, & p_c \leq p_e \end{cases}$$

Here, p_e is the so-called air entry pressure.
- van Genuchten [10]:

$$s_e(p_c) = \begin{cases} (1 + (\alpha p_c)^n)^{-m}, & p_c > 0 \\ 1, & p_c \leq 0 \end{cases}$$

In the vast majority of cases, $m = 1 - 1/n$

Though upscaling procedures play a large role in their derivation, and attempts to interpret the parameters in physical terms are undertaken, it turns out that the parameters of these functions have to be determined by experiments, e.g. multistep outflow, mercury porosimetry, and others, and then related to the concrete parametrization by a fitting procedure.

We note, that the Brooks-Corey function in $p_c = p_e$ and the van Genuchten function with $n < 2$ for $p_c = 0$ are not differentiable, which may cause trouble depending on the numerical realization. On the other hand, there appears to be no physical reason for having a discontinuity in the derivatives.

8.1.5. *Alternative Two Phase Flow Formulations*

The equation for the gas phase degenerates for $s_e = 1$, which is reached by both parametrizations, so well-posedness of the system cannot be guaranteed: In the case where the gas phase disappears, the system becomes indefinite. Physically, this effect describes the fact that some parts of the modeling domain are fully saturated with water. The same effect could occur – using a different parametrization – with the water phase.

There are several possibilities to circumvent this problem; in the sequel we refer shortly to two of them and then switch to a third one favorized in [1].

8.1.5.1. Phase Pressure Saturation Formulation

Introduce as new basic variables the effective saturation and the pressure of the water phase by substituting $p_g = p_w + p_c(s_e)$. The function $p_c(s_e)$ often is called the "Leverett function". The new system would read as:

$$\partial_t(\phi s_w(s_e)\rho_w) - \frac{kk_{rw}\rho_w}{\mu_w}\nabla p_w = R_w$$
$$\partial_t(\phi s_g(s_e)\rho_n) = -\frac{kk_{rg}\rho_g}{\mu_g}\nabla(p_w + p_c) = R_g$$

This modification removes the degeneration for $s_e = 1$. Furthermore, it reveals more of the mathematical nature of the problem. If we omit the capillary pressure in the second equation, which would describe the case of vanishing capillary forces, we see that the resulting equation is nonlinear hyperbolic. This observation gives rise to numerous numerical schemes using techniques from hyperbolic conservation laws.

It is straightforward that we can use as boundary conditions the phase fluxes, the water pressure, and the effective saturation.

We note that there are variants of this approach, which e.g. use the gas saturation instead of the effective saturation. Another choice would be to substitute $p_w = p_g + p_c(s_e)$ and to regard p_g, s_e as the basic variables.

8.1.5.2. Global Pressure – Saturation Formulation

A formulation removing both possible degeneracies has been introduced by [5]. This formulation has as its basic variables the wetting phase saturation and a so called *global pressure p* with values between p_g and p_w. Furthermore, it introduces a *global velocity*.

From the constitutive relationships, it is possible to derive relations between the pairs p_g, p_w, and p, s_w. For the expressions and a detailed introduction, see [11]. A drawback of this formulation is that boundary conditions now have to be given for global pressure and global velocity, which, being mathematical constructs rather than physical ones, are not directly available to physical measurements.

In the context of H_2-PEM fuel cells, this approach is taken by [12].

8.1.5.3. Pressure Pressure Formulatiom, Regularization Approach

If we can assume that always both fluids are present, or, that by regularization, we replace domains with vanishing phases by low saturation domains, we can stay with the pressure–pressure formulation which does not degenerate in this case. This approach has the advantage that boundary conditions can be set for the phase pressures and phase fluxes. Moreover, the definition of interface conditions between porous layers of different characteristics is much simpler: It is sufficient to assume continuity of phase pressures and phase fluxes, whereas in the saturation-based cases, the continuity of the capillary pressure has to be bought by a jump in the basic variable, which complicates the numerical schemes.

In [13], a mixing method is proposed which applies to porous media with mixed wettability, i.e. locally consisting of pores with different wetting characteristics. It is reasonable to assume the same for fuel cell PTLs as well. One outcome of that work is that the discontinuity in the derivative for $s \to 1$ does not occur.

In [1], the assumption of presence of both phases is made and the following parametrization is used:

$$s_e(p_c) = -B'(B(-\alpha(p_c - p_{\frac{1}{2}}))_n^\beta - B(\alpha(p_c - p_{\frac{1}{2}}))\beta_w)$$

Here, $B(x) = x/(e^x - 1)$ is the *Bernoulli function*. The parameter α controls the steepness of the curve, $p_{\frac{1}{2}}$ shifts it on the wettability scale, and β_n, β_w control the asymptotics in the saturated resp. unsaturated limit.

On one hand, it is possible to fit quite well the behavior of the Brooks-Corey and van Genuchten curves (Figure 8.1), using this relationship for a regularized pressure pressure formulation. On the other hand, fits to measurements for pore spaces of mixed wettability can be performed quite well using this ansatz [1].

8.1.6. Relative Permeability – Saturation Relationship

To the van Genuchten and Brooks-Corey parametrizations correspond relationships for the relative permeability [9, 14]. However, in many cases, a simple power law is sufficient to describe the behavior, an approach which we preferred in [1].

8.1.7. Combination of Two Phase Flow with the Stefan Maxwell Model

In order to model the two-phase transport of a liquid phase and a gas mixture, we replace the gas phase equation in the two phase flow equation by the system of Stefan Maxwell equations, modified in order to take into account the two-phase flow behavior. A first-order correction involves the factor $\Psi = \phi/\tau$ in (8.3) which is multiplied by the value of the relative permeability.

In the case of the pressure pressure formulation, this results in replacing the equation for gas flow by

$$\vec{N}_i^{gas} = -k_{rg}(s_e) \sum_{j=1}^{n_{gas}} D_{ij}^{SM}(p_1, \ldots, p_{n_{gas}}) \nabla p_{gas_j}$$

$$\partial_t \left(s_g(p_c) \right) \phi \frac{p_{gas_j}}{RT} + \nabla \cdot \vec{N}_i^{gas} = Q_i$$

$$p_c = p_{tot} - p_w.$$

A more involved procedure would allow to couple the Stefan Maxwell equations with the pressure-saturation formulation. Here, we would use the molar fractions y_i as basic variables and express $p_{tot} = p_w + p_c(s_e)$.

8.1.8. Numerical Example

We demonstrate the influence of the wettability on the fuel cell performance. We use the data from [1] and modify the parameter $p_{\frac{1}{2}}$ in the anodic and cathodic PTL, respectively.

In Figure 8.2, we see that at high metabolic rates, a less wettable anodic PTL prevents the liquid water with dissolved methanol from reaching the reaction zone in a sufficient amount. Vice versa, oxygen transport is improved with less wettability of the cathodic PTL resulting in an improved performance.

This example demonstrates that the approach taken is able to reflect well wettability effects on the fuel cell performance. Combined with the corresponding experiments to investigate the two-phase flow behavior of the PTL, it can be considered as a working, albeit pragmatic modeling approach for theses processes.

Further work should focus on improving this approach in several directions. To name only a few, one should be concerned about hysteresis in the pressure-saturation relationships, dependency on parameters influencing interface properties like temperature and methanol concentration, and extended modeling concepts like [15] which take into account the interface area between the fluid phases.

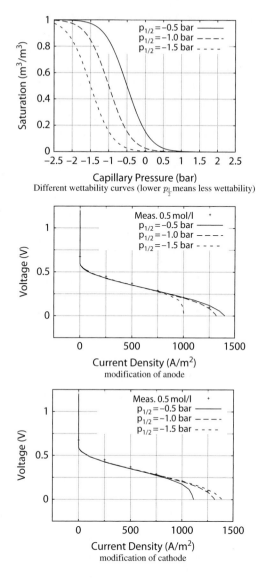

FIGURE 8.2. Influence of wettability modification on cell performance (top), (bottom).

8.2. Reaction Kinetics

8.2.1. Reaction Models

The electro-chemical reactions are assumed to take place in the fluid phase. The explicit introduction of the occupation numbers (Θ_i, $\Sigma_i \Theta_i = 1$) as variables is a natural description of the possible lack of catalytic sites and methanol cross over effects. Due to the use of expensive catalysts, electro-chemical saturation effects may be reached by fuel cells in some applications. The adsorption is described by isotherms of Temkin type and takes interaction of particles into account. The models would reduce to a Butler-Volmer-type equation, in case the occupation numbers are weakly dependent on the potential and concentrations. A natural description of the effects of methanol crossover would balance the occupation numbers for both reactions together at the cathode. In [1], the whole MEA model was under consideration; here, we would like to encourage a kinetic description of the processes and to verify different models numerically, because kinetic models are certainly within the scope of an MEA model. The results here are obtained using a specialized implementation but a much more generic one with up to twenty species seems to be an option to ease the verification of kinetic models and their use in MEA models. Specific kinetic models can be easily criticized; on the other hand, one would expect them to be the tool to work in a reasonably large range of conditions of operation and relatively close to basic concepts. The equations used follow [6, 7, 8] with respect to the indirect reduction of oxygen to water via adsorbed hydrogen peroxide and [19, 20] with respect to the anodic methanol reaction. Especially together with the constants used, one should see the results as an example and a step toward a more complete description of these processes and their related efficiency losses.

8.2.2. Oxygen Kinetics

For the oxygen kinetics, the following set of equations is used ($\phi = \phi_{e^-} - \phi_{H^+} - E_s$ empty sites):

$$O_2 + s \underset{}{\overset{k_1^\pm}{\rightleftharpoons}} O_{2,ad}$$

$$O_{2,ad} + H^+ + e^- \underset{}{\overset{k_2^\pm}{\rightleftharpoons}} HO_{2,ad}$$

$$HO_{2,ad} + H^+ + e^- \underset{}{\overset{k_3^\pm}{\rightleftharpoons}} H_2O_{2,ad}$$

$$H_2O_{2,ad} + 2H^+ + 2e^- \underset{}{\overset{k_4^\pm}{\rightleftharpoons}} 2H_2O + s$$

$$r_i = \tilde{k}_i^+ \theta_i e^{\alpha_i \gamma_i \phi} - \tilde{k}_i^- \theta_{i-1} e^{(\alpha_i - 1)\gamma_i \phi},$$

$\gamma_1 = 0, \gamma_2 = 1, \gamma_3 = 1, \gamma_4 = 2,$

$\tilde{k}_1^- = k_1^- c_{O_2}/c_{O_{2ref}}, \tilde{k}_i^- = k_i^- a_{H^+}, \; i = 2, 3, \tilde{k}_4^- = k_4^- a_{H^+}^2, \tilde{k}_i^+ = k_i^+, \; i = 1,\ldots,4$

$0 < \alpha_i < 1, (O_2)_{ad} : \theta_1, (HO_2)_{ad} : \theta_2, (H_2O_2)_{ad} : \theta_3.$

$$\theta_0 = \theta_4 = 1 - \sum_{i=1}^{3} \theta_i. \tag{8.10}$$

c_{O_2} is the concentration of oxygen dissolved in water. Instead of regarding it as a variable of the model, we assume this concentration to be in equilibrium with the local value of the partial pressure p_{O_2}. k_i^+, k_i^- denote the reaction constants for the forward and backward reactions, respectively. The reaction constants are assumed to be temperature-dependent via Arrhenius relations $k_i^\pm = k_{0,i}^\pm e^{-e_{a,i}^\pm/T}$, where $e_{a,i}^\pm = E_{a,i}^\pm/R$, $E_{a,i}^\pm$ is the activation energy. The equilibrium potential E^0 is given by the Nernst equation

$$E_0 = \frac{RT}{4F} \ln\left(\Pi_{i=1}^4 \frac{k_i^- a_{H^+}^4 c_{O_2}}{k_i^+ a_{H_2O}^2}\right),$$

which is equivalent to the requirement $\det(A_{eq}) = 0$ and a nontrivial solution to the homogeneous linear systems of equations for the occupation numbers θ_i. The reactions are assumed to be sufficiently fast (with respect to other time constants for heat and mass transfer) to describe the non-equilibrium by steady-state solutions ($S_\uparrow$: shift matrix) of

$$\frac{d\vec{\theta}}{dt} = (-A(\tilde{K})_{eq} + S_\uparrow A(\tilde{K})_{eq})\vec{\theta}$$
$$A = (-A(\tilde{K})_{eq} + S_\uparrow A(\tilde{K})_{eq}).$$

With respect to $\phi \to \pm\infty$, we find the limiting electron reaction rates:

$$r_1(\phi \to -\infty) = -k_1^- c_{O_2}/c_{O_{2ref}}$$
$$r_1(\phi \to \infty) = k_1^+.$$

The total reaction rate for electrons (in mol/l) is

$$r_{O_2} := 4r_2 A_{InnO_2},$$

where A_{InnO_2} denotes the inner surface or the number of catalytic sites per membrane area. The oxygen has to diffuse through a water film (R_1: grain thickness; $(R_1, R_2]$: film layer) with the assumption of a steady state (none of the equations has to be assumed to be in equilibrium) and mixing in the grain, the additional balance equation is:

$$c'' + \frac{n-1}{r} c' = 0$$
$$c(R_2) = c_{gas}, \quad c'(R_1) = m,$$

(r: space coordinate; m: the total reaction rate in $[0, R_1]$ divided by the grain area). With a geometric factor (f_g, $n = 1, 2, 3$ for Cartesian, cylindrical, or spherical coordinates) and r_1 as an independent variable in the first equation, the system is completed by:

TABLE 8.3. Material constants for oxygen kinetics.

A_{InnO_2}	m^2/m^3	inner surface	$9.814 \cdot 10^{+4}$
$\alpha_{O_2_2}$			0.4
$\alpha_{O_2_3}$			0.5
a_{H^+}		activity H^+	1
ϕ_{0O_2}	V	equil. potential O_2, incl. in modified $k^\pm$	1.2
D_{FilmO_2}	m^2/s	Diffusion coeff. O_2 in water film	0
r_{Film}	m	thick. water film, incl. geom. constant	0
c_{RefO_2}	mol/m^3	ref. concentration O_2	7.6

$$r_1 = k_1^+ \theta_1 - k_1^- c_{O_2}(r_1)\theta_4,$$
$$c_{O_2}(r_1) = c(R_2) + r_1 f_g.$$

Together with the explicitly used (for numerical reasons) normalization condition (8.10) these are the final equations describing the oxygen kinetics here. The notations used are found in Table 8.3 [1].

The model predicts: $O_{2,ad}$ occupies the overwhelming number of sites and the number of empty positions is small for a sufficiently large range of potentials around the equilibrium at 1.2 V. The empty sites show a reasonable dependence of the O_2 concentration. Below 0.5 V adsorption is decreasing exponentially and the electron reaction rate is saturating accordingly (see Figure 8.3).

8.2.3. Methanol Kinetics

The model equations include Pt, Ru as catalysts and balance two types of occupation numbers independently.

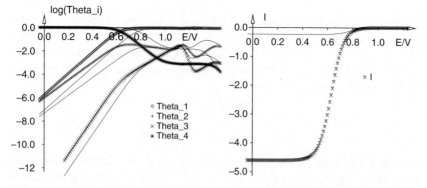

FIGURE 8.3. Occupation numbers of the oxygen kinetics (θ_4 number of empty sites, left), electron reaction rate (right, markers: 1 mol/l, lines: 0.05 mol/l).

8. Two-Phase Flow and Reaction Kinetics

$$rlCH_3OH + 2s_1 \underset{}{\overset{k_1^{\pm}}{\rightleftharpoons}} (CH_2-OH)_{ad} + H_{ad}$$

$$(CH_2-OH)_{ad} + s_1 \underset{}{\overset{k_2^{\pm}}{\rightleftharpoons}} (CH-OH)_{ad} + H_{ad}$$

$$(CH-OH)_{ad} + s_1 \underset{}{\overset{k_3^{\pm}}{\rightleftharpoons}} (C-OH)_{ad} + H_{ad}$$

$$(C-OH)_{ad} + s_1 \underset{}{\overset{k_4^{\pm}}{\rightleftharpoons}} (C-O)_{ad} + H_{ad}$$

$$(C-O)_{ad} + (OH)_{ad} \underset{}{\overset{k_5^{\pm}}{\rightleftharpoons}} (COOH)_{ad} + s_2$$

$$(COOH)_{ad} + (OH)_{ad} \underset{}{\overset{k_6^{\pm}}{\rightleftharpoons}} (CO)_2 + H_2O + s_1 + s_2$$

$$H_{ad} \underset{}{\overset{k_7^{\pm}}{\rightleftharpoons}} H^+ + s_1 + e^-$$

$$H_2O + s_2 \underset{}{\overset{k_8^{\pm}}{\rightleftharpoons}} OH_{ad} + H^+ + e^-$$

$$(OH)_{ad} \underset{}{\overset{k_9^{\pm}}{\rightleftharpoons}} O_{ad} + H^+ + e^-$$

$$(CO)_{ad} + O_{ad} \underset{}{\overset{k_{10}^{\pm}}{\rightleftharpoons}} CO_2 + s_1 + s_2$$

The first group of sites, (s_1, Θ_i), is related to the adsorption of H_{ad} and the second, (s_2, ϑ_i), are related to $(OH)_{ad}, O_{ad}$:

$$
\begin{aligned}
(CH_2-OH)_{ad} &: \Theta_1, \\
(CH-OH)_{ad} &: \Theta_2, \\
(C-OH)_{ad} &: \Theta_3, \\
(CO)_{ad} &: \Theta_4, \\
(COOH)_{ad} &: \Theta_5, \\
H_{ad} &: \Theta_6, \\
s_1 &: \Theta = \sum_{i=1}^{6} \Theta_i, \\
\bar{\Theta} &= 1 - \Theta, \\
(OH)_{ad} &: \vartheta_1, \\
O_{ad} &: \vartheta_2, \\
s_2 &: \vartheta, \vartheta = \vartheta_1 + \vartheta_2, \\
\bar{\vartheta} &= 1 - \vartheta.
\end{aligned}
$$

We obtain the rate equations:

$$r_1 = k_1^+ c_{CH_3OH} \bar{\Theta}^2 e^{\zeta_1 \Theta} - k_1^- \Theta_1 \Theta_6 e^{\xi_1 \Theta},$$

$$\zeta_1 = 2\beta g - 2g, \xi_1 = 2\beta g, i = 2,\ldots,4:$$

$$r_i = k_i^+ \Theta_{i-1} \bar{\Theta} e^{\zeta_i \Theta} - k_i^- \Theta_i \Theta_6 e^{\xi_i \Theta}, \zeta_i = 3\beta g - 2g, \xi_i = 3\beta g - g, \quad (8.10)$$

$$r_5 = k_5^+ \Theta_{i-1} \vartheta_1 e^{\beta g \vartheta} \vartheta e^{\zeta_5 \Theta} - k_5^- \Theta_i \bar{\vartheta} e^{\beta-1 g \vartheta} e^{\xi_5 \Theta},$$

$$\zeta_5 = 2\beta g - g, \xi_5 = 2\beta g - g,$$

$$r_6 = k_6^+ \Theta_{i-1} \vartheta_1 e^{\beta g \vartheta} e^{\zeta_6 \Theta} - k_6^- c X^a Y \bar{\Theta} \bar{\vartheta} e^{\beta-1 g \vartheta} e^{\xi_6 \Theta}, \zeta_6 = \beta g,$$
$$\xi_6 = \beta g - g, X = CO_2, Y = H_2O,$$
$$r_7 = k_7^+ \Theta_{i-1} e^{\zeta_7 \Theta} e^{\alpha \phi} - k_7^- \bar{\Theta} e^{\xi_7 \Theta} e^{(\alpha-1)\phi}, \zeta_7 = \beta g, \xi_7 = \beta g - g,$$
$$r_8 = k_8^+ a H_2 O \bar{\vartheta} e^{(\beta-1) g \vartheta} e^{\alpha \phi} - k_8^- \vartheta_1 e^{\beta g \vartheta} c^{(\alpha-1)\phi},$$
$$r_9 = k_9^+ \vartheta_1 e^{\alpha \phi} - k_9^- \vartheta_2 e^{(\alpha-1)\phi},$$
$$r_{10} = k_{10}^+ \Theta_4 \vartheta_2 e^{\beta g \vartheta} e \zeta_{10} \theta - k_{10}^- c C_2 \bar{\Theta} \bar{\vartheta} e^{(\beta-1)g\vartheta} e^{\xi_{10}\Theta},$$
$$\zeta_{10} = \beta g, \xi_{10} = \beta g - g.$$

$$\frac{d\Theta_i}{dt} = r_i - r_{i+1}, \quad i = 1, 2, 3, 5,$$
$$\frac{d\Theta_4}{dt} = r_4 - r_5 - r_{10},$$
$$\frac{d\Theta_6}{dt} = r_1 + r_2 + r_3 + r_4 - r_7,$$
$$\frac{d\vartheta_1}{dt} = r_8 - r_5 - r_6 - r_9,$$
$$\frac{d\vartheta_2}{dt} = r_9 - r_{10}.$$

The stationary electron reaction rate is given by

$$r_{Me} := (r_7 + r_8 + r_9) A_{InnM} = 6 r_1 A_{InnM}$$

and A_{InnM} denotes the inner surface or the number catalytic sites per membrane area at the anode. The equilibrium conditions are given by Nernst equations

$$E^0_{CH_3OH} = \frac{RT}{6F} \ln \left(\prod_{i=1}^{6} \frac{k_i^-}{k_i^+} \left(\frac{k_7^-}{k_7^+} \right)^4 \left(\frac{k_8^-}{k_8^+} \right)^2 \right)$$
$$+ \frac{RT}{6F} \ln \left(\frac{c_{CO_2}}{c_{CH_3OH} a_w} \right),$$

and

$$\bar{E}^0_{CH_3OH} = \frac{RT}{6F} \ln \left(\prod_{i=1}^{10} \frac{k_i^-}{k_i^+} \left(\frac{k_7^-}{k_7^+} \right)^3 \left(\frac{k_5^+ k_6^+}{k_5^- k_6^-} \right) \right)$$
$$+ \frac{RT}{6F} \ln \left(\frac{c_{CO_2}}{c_{CH_3OH} a_w} \right),$$

imposing the compatibility condition

$$\left(\frac{k_5^- k_6^-}{k_5^+ k_6^+} \right) \left(\frac{k_8^-}{k_8^+} \right) = \left(\frac{k_9^- k_{10}^-}{k_9^+ k_{10}^+} \right),$$

if a nontrivial equilibrium solution is required. The list of notations used is found in Table 8.4. The occupation numbers of both groups, including the number of empty sites, are shown in Figure 8.4. For small potential values CO

TABLE 8.4. Notations for methanol kinetics.

α_{Met}			0.5
β_{Met}			0.83
g		Temkin exponent	6
g_ϑ		Temkin exponent	0
A_{InnM}	m^2/m^3	inner surface	$1.186 \cdot 10^{+9}$
a_{H_2O}		activity H_2O	1
$\phi_{0\,Met}$	V	equil. pot. CH_3OH, (modified $k^{\pm}$)	0.03738
c_{RefMet}	mol/m^3	ref. concentration CH_3OH	1000
c_{RefCO_2}	mol/m^3	ref. concentration CO_2	1000

occupies the sites. The empty sites are dominant by orders of magnitude for larger potential values.

The case of the pure *Pt* catalyst can be derived by neglecting the last two reactions; this removes the second Nernst equation and the compatibility condition due to the two groups of sites. The influence of *Ru* with respect to the first group of reactions starts at 0.4 V and is shown in Figure 8.6. $(CO)_{ad}$ and $(COOH)_{ad}$ increase mainly by occupying empty sites for higher potential values, hence the electro-chemical limit of fuel cell operation, due to an electron reaction rate saturation, is shifted to higher potentials. A changed *Ru* content is modeled by a reduction of the reaction constants $k_9^+, k_9^-, k_{10}^+, k_{10}^-$ by a factor β. This yields finally a nonmonotonous reaction rate (see Figure 8.5).

In these models, all reaction rates are proportional to the number of reaction sites; hence it is natural to solve the stationary charge balance equation for the parameters applied potential difference ϕ_{appl} and the ratio of catalytic sites α.

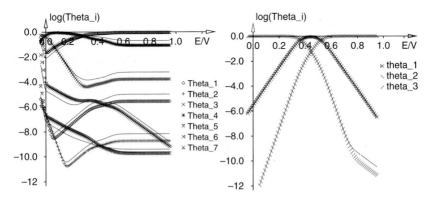

FIGURE 8.4. Occupation numbers of the methanol kinetics (θ_i left, ϑ_i right, θ_7, ϑ_3 number of empty sites, markers: 0.1 mol/l, lines: 0.8 mol/l).

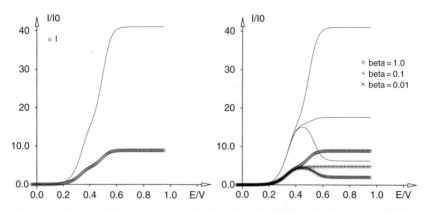

FIGURE 8.5. Reaction rates for the methanol kinetics (left) and for reduced kinetic constants $k_9^+, k_9^-, k_{10}^+, k_{10}^-$ (right, markers: 0.1 mol/l, lines: 0.8 mol/l).

$$\alpha r_{O_2}(\phi_{appl} + x) + r_{Me}(x) = 0. \qquad (8.11)$$

The only variable in equation (8.11) is x, the electro-chemical potential difference. Equation (8.11) describes an MEA with ideal transport but reaction kinetic limitations. Let $r_{O_2}(s)$ be monotone and $r_{Me}(x)$ have n_{max} local maxima, then the equation has at most $n_{max} + 1$ solution branches $r_{Me}(x)$. r_{O_2} and r_{Me} approximates for large and small ratios α an ideal ion source for the other – hence the IV–curves for the extreme values of α are close to the electron reaction rate of r_{Me} and, r_{O_2}. For intermediate values of α branches separated by cusps, complete the generic situation for one local maximum in one of the reaction rates (Figure 8.6). The study of the parameter dependence of the bifurcations is an additional source of information to verify kinetic models and related parameters.

The numerical solution of the equations describing the occupancy numbers is in general based on embedding and Newton's method. The oxygen case is just a very special reaction chain reducing to an eigenvalue problem, as long as no additional diffusion effects are taken into account. The embedding process starts from the unique equilibrium solution and the solution related to the previous boundary conditions or time step respectively. The kinetic models described here contribute not critically to the CPU time needed for the simulation of a MEA model as considered in [1].

8.3. Summary

The goal of detailed modeling of the membrane electrode assembly is to identify the essential parameters and processes to reach predictive simulation results within a parameter range of practical interest. The energetic losses due

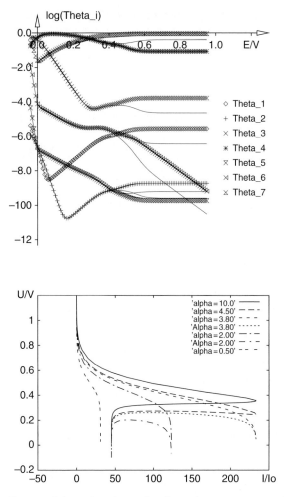

FIGURE 8.6. Influence of the reduced Ru-related reaction constants on the occupancy numbers (above, markers: $\beta = 1$, lines: $\beta = 0.01$), related IV–curves for different ratios of reaction sites α for the O_2 reaction.

to the non-ideal reaction kinetics and the different transport processes can be studied and separated. This modeling approach opens the possibility to simulate special experiments in order to obtain the necessary parameters.

Both presented parts of the DMFC model introduce details which are often neglected in computational models.

From our point of view, the limits are defined by our understanding of the processes and the availability of data. The issues like numerical approximation and stability, availability of computer resources, and programming difficulties have been at least partially solved.

Therefore, the further development of DMFC modeling depends on joint research in the fields of chemical engineering, numerical analysis, and computer science.

References

[1] J. Divisek, J. Fuhrmann, K. Gärtner, and R. Jung. *J. Electrochem. Soc.*, 150(6):A811–A825, 2003.
[2] K. Ehrhardt, K. Klusacek, and P. Schneider. *Comput. Chem. Eng.*, 12:1151–1155, 1988.
[3] D. Arnost and P. Schneider. *Chemical Eng. J.*, 57:91–99, 1995.
[4] R. Ewing. *The Mathematics of Reservoir Simulation.* SIAM, Philadelphia, 1983.
[5] G. Chavent and J. Jaffre. *Mathematical Models and Finite Elements for Reservoir Simulation.* North Holland, Amsterdam, 1978.
[6] R. Helmig *Multiphase Flow and Transport Processes in the Subsurface.* Springer, 1997.
[7] *VDI-Wärmeatlas.* Verein Deutscher Ingenieure, VDI-Gesellschaft, Düsseldorf, 1994.
[8] A. W. Adamson and A. P. Gast. *Physical Chemistry of Surfaces.* Wiley – Interscience, 1997.
[9] R. H. Brooks and A. T. Corey. In *Hydrol. Pap.*, volume 3. Colorado State Univ., Fort Collins, 1964.
[10] M. Th. van Genuchten. *Soil Sci. Soc. Amer. J.*, 44:892–898, 1980.
[11] P. Bastian. Numerical computation of multiphase flows in porous media, 1999. Habilitationsschrift, Univeristät Kiel.
[12] K. Kühn, M. Ohlberger, J. Schumacher, C. Ziegler, and R. Klöfkorn. A dynamic two-phase flow model of proton exchange membrane fuel cells. Technical Report Report 03-07, CSCAMM, 2003.
[13] P. Ustohal, F. Stauffer, and T. Dracos. *J. Contam. Hydrol.*, 33(1–2):5–38, 1998.
[14] Y. Mualem. *Water Resour. Res.*, 12(3):513–522, 1976.
[15] W. G. Gray and S. M. Hassanizadeh. *Water Resour. Res.*, 27:1855–1863, 1991.
[16] C. H. Hamann, A. Hamnett, and W. Vielstich. In *Electrochemistry*, pages 282–283. WILEY-VCH, Weinheim, 1998.
[17] M. R. Tarasevich, A. Sadkowski, and E. Yeager. In B. E. Conway, J. O'M. Bockris, E. Yeager, S. U. M. Khan, and R. E. White, editors, *Comprehensive Treatise of Electrochemistry*, pages 354–359. Plenum Press, NY and London, 1983.
[18] J. Koryta, J. Dvorak, and L. Kavan. In *Principles of Electrochemistry*, pages 358–361. John Wiley, Chichester, 1993.
[19] P. S. Kauranen, E. Skou, and J. Munk. *J. Electroanal. Chem.*, 404:1–13, 1996.
[20] A. Hamnett. *Catalysis Today*, 38:445, 1997.

9
Thermal and Electrical Coupling in Stacks

BRIAN WETTON

9.1. Overview

Many of the other chapters of this book are devoted to understanding several key aspects of PEMFCs at a fundamental but local level. These include membrane transport, the influence of catalyst layer structure on performance and the nature of two-phase flow (liquid, water, and gas) in electrodes. However, if one has accurate parametric descriptions of these phenomena, from detailed models fitted to experimental measurements, the question remains how these locally fitted models will combine with more well-understood phenomena of gas, heat, and electrical transport to determine overall *system* performance, at the unit cell or stack level. This is the question addressed in this chapter, in which a stack level computational model of a PEMFC stack is presented and discretized, and an iterative strategy is described. This computational model is capable of simulating thermal and electrical interactions of unit cells in large stacks. It is computationally efficient, requiring only a few minutes on standard, single processor machines. The reader should recognize that the model is based on very simple ideas of GDE, membrane and catalyst layer transport and electrochemical behavior in the catalyst layer. The emphasis of this chapter is placed on the structure of the robust computational framework, which can handle the addition of the more complicated models of electrodes, catalyst layers, and membranes described in other chapters. Several key simplifications are made to the model for the purposes of this exposition: operation is taken to be steady, gas channels are taken to be at constant pressure and co-flowing, the membrane resistivity is taken to be constant, and the impact of liquid water on electrode and channel transport is not taken into account. More advanced models developed in our group can handle these effects, at least individually, at the stack level.

The earliest PEMFC system models [1, 2] were for single cells at steady state, assuming isothermal and isobar conditions. Performance is averaged over the cross-channel direction, and transport in gas channels is decoupled from transport through the Membrane Electrode Assembly (MEA). The power of such system level modeling tools for scientific and commercial design was recognized. In recent years, researchers with more numerical background have returned to the development of more comprehensive reduced dimensional models [3, 4, 5]. However, the major focus for unit cell modeling has moved to 3-dimensional (3-D) models based on Computational Fluid Dynamics (CFD) codes, often commercial ones (see **e.g.** [6, 7, 8]). There are certainly valid reasons to turn to 3-D models, since there are genuine higher dimensional phenomena in fuel cells, especially gas flow effects connected with serpentine flow fields [7] and two-phase flow effects in the cross plane [8]. However, these 3-D models require significant computational resources for even unit cell studies. To study interactions in stacks that can have of the order of a hundred unit cells, reduced dimensional unit cell models must be used and their limitations taken into account when interpreting the results.

In Section 9.2 below, a summary of the nomenclature used in the chapter is given. In Section 9.3, a summary of fuel cell stack geometry, and a discussion of the dimensional reductions used in the model is given. In Section 9.4, the model of 1-D MEA transport is presented, followed by Section 9.5 on the model of channel flow for a unit cell and Section 9.6 on the electrical and thermal coupling in a stack environment. In Section 9.7, a summary of the stack model is given followed by its discretization. In Section 9.8, the iterative solution strategy for the discrete system is presented, followed by sample computational results in Section 9.9. The current state of stack modeling in this framework and future directions are summarized in the final section.

9.2. Nomenclature

9.2.1. Coordinate System

As shown in Figure 9.1
   x: down-channel coordinate.

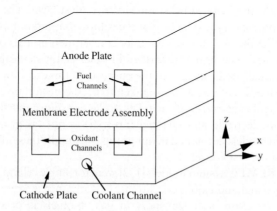

FIGURE 9.1. 3-D schematic of unit fuel cell.

y: cross-channel coordinate.
z: through-MEA coordinate.

9.2.2. Fundamental Variables

In each unit cell, the following quantities are to be determined for each x (down-channel direction):

i: local current density in A/m².

$\underline{q} = (q_o, q_c, q_n, q_h, q_a)$: Channel molar fluxes (cathode oxygen, cathode water, cathode nitrogen, anode hydrogen, and anode water) per unit orthogonal width (z) in moles/m/s.

t: average coolant channel temperature in K.

$\underline{\theta} = (\theta_l, \theta_o, \theta_c, \theta_a, \theta_h)$: temperatures in K in the plate adjacent to the coolant (l), in the cathode gas channel (o), at the cathode catalyst sites (c), anode catalyst sites (a), and anode gas channel (h).

These variables are found in each unit cell of the stack. The cell index will be denoted by a superscript j.

9.2.3. Other Quantities

References are given for fitted quantities.

a: Molar concentration of membrane acid groups, 1200 moles/m³.
A: Scaled heat transfer coefficient to coolant channel, 5600 W/m²/°K [9].
c_a: Anode channel water vapor concentration moles/m³.
c_c: Cathode channel water vapor concentration moles/m³.
c_h: Anode channel hydrogen concentration moles/m³.
c_o: Cathode channel oxygen concentration moles/m³.
c_w: Dimensionless (scaled by a) membrane water content. It varies in y.
$c_{w,a}, c_{w,c}$: Dimensionless membrane water content at anode and cathode sides of the membrane.
$c^*_{w,a}, c^*_{w,c}$: Equilibrium values of the quantities above.
C_{ref}: reference Oxygen concentration, taken to be that of pure O_2 at standard conditions (40.9 moles/m³).
d_w: Constant in the formula for D_w below, value 2.1×10^{-7} m²/s.
D_w: Diffusivity of water in membrane m²/s. An empirical function of membrane temperature T and water content c_w [10].
e: Electro-osmotic drag coefficient, taken to be 1.
E_0: Open circuit voltage 0.944 V [4].
$\mathcal{F}$: Faraday's constant 96485 C/mole.
g: Product of coolant flux (per unit orthogonal z) and the coolant heat capacity 473 W/mK.
H_{vap}: Heat of vaporization 45400 J/mole.

i_{ref}: reference current 64 A/m² [4].
i_T: Average current density for the stack A/m². A base value of 1 A/cm² is taken.
$I^{j/j+1}$: Current in the bipolar plate between cells j and j + 1, per unit orthogonal z, in A/m.
J: Membrane diffusive flux moles/s m².
L: Cell length, 1m.
L_m: Membrane thickness, 50μ.
M: Number of cells in the stack.
n: Molar ratio of oxygen to nitrogen in oxidant inlet stream.
N: Number of grid points in a discretization of the cell length.
p_a, p_c: average anode and cathode channel pressure in Pa. Base values of $p_a = 3.2 \times 10^5$ and $p_c = 3 \times 10^5$ are used.
P_{sat}: Saturation pressure in Pa as a function of temperature [11].
$\mathcal{R}$: Ideal gas constant 8.3143 J/K mole.
r_a, r_c : Anode and cathode channel relative humidity.
S_a, S_c : Anode fuel and cathode oxidant stoichiometric flow rates. Base values of $s_a = 1.2$ and $s_c = 1.8$ are taken.
T: Membrane temperature in K.
$T_{a,dew}, T_{c,dew}$: Anode and cathode dew points of inlet gases in K. Base values of 63°C are taken.
v: Cell voltage.
V_{TN}: Thermal neutral voltage of ORR to vapor 1.28 V.
w_a, w_c: Liquid water flux in anode and cathode channels per unit orthogonal width (z) in moles/m s.
α_c: Cathode transfer factor, taken to be 1 [4].
δ_o: Fitted oxygen mass transport parameter 8.993×10^{-4} moles A/m [4].
γ: Mass transfer parameter for vapor movement from catalyst sites to channel average 5.7×10^{-6} moles/s m².
γ_a, γ_c: Anode and cathode channel condensation rates, per unit orthogonal width (z) in moles/m² s.
λ: Length-specific resistivity of the bipolar plates, 4×10^{-3} Ω [13].
μ_g: Ratio of thermal conductivity to thickness of the GDE, 5×10^3 W/m² K.
μ_m: Ratio of thermal conductivity to thickness of the membrane, 1.12×10^4 W/m² K.
μ_p: Ratio of thermal conductivity to thickness of the plate, 2.3×10^3 W/m² K.
ω: membrane area-specific resistivity ($8.42 \times 10^{-6} \Omega$-m²), from the fitted model developed in [4] under saturated conditions at 348°K.

9.3. Dimensional Reduction

Unit fuel cells with straight gas channels (such as the Ballard Mk 9 design) are considered as shown in Figure 9.1 (not to scale). Oxidant gas (air), reactant gas (hydrogen), and coolant flow in straight channels in the x direction (the model assumes co-flow operation). Numerical simulation of counter-flow operation poses additional computational difficulties. Unit cells can be of the order of a meter long, but often only a few millimeters thick.

More detail of the cross-plane ($y-z$) geometry is shown in Figure 9.2 (also not to scale). The high aspect ratio of a unit cell suggests the following dimensional reduction: That the x direction transport is dominated by the gas channel and coolant flow. Furthermore, since the flow is slow enough to be laminar, these flows can be described simply with average quantities in the x direction. This leads to a "2 + 1 D" model, in which the cross-plane ($y-z$) problem can be solved for each x and connected to 1-D models for the channel flow.

Further, the cross plane ($y-z$) problem is reduced to one-dimensional averaged (over y) transport through the membrane and electrodes. A fitted averaged diffusion parameter (δ_o) is used to describe diffusive concentration differences of Oxygen from channel averages to catalyst sites. Similarly, a fitted parameter γ describes diffusive effects of water vapor from catalyst sites to channels. Temperature profiles are considered to be constant in y, and the values through the unit cell at various locations are denoted by θ, with a subscript as shown in Figure 9.2.

Unit cells (up to 100 or more) are placed in series into fuel cell stacks. The anode plate of one cell is placed next to the cathode plate of the next (so that their voltages will add in the stack). The combined plates are called *bipolar* plates. The reduced dimensional geometry of the stack model is shown in Figure 9.3.

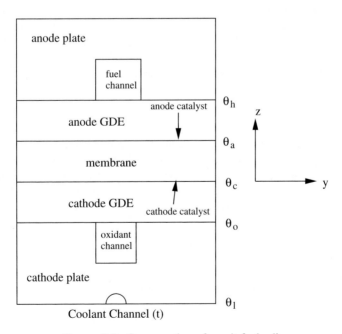

FIGURE 9.2. Cross-section of a unit fuel cell.

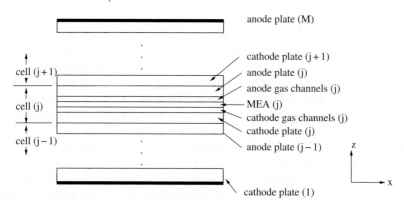

FIGURE 9.3. Cross-section of a fuel cell stack.

9.4. Simple 1-D Model of MEA Transport

At this level of the model, consider to be given the local current density i, the anode and cathode catalyst temperatures (θ_a and θ_c), the anode and cathode channel vapor concentrations (c_a and c_c), the anode channel hydrogen concentration (c_h), and the cathode channel oxygen concentration (c_o).

To be determined from the MEA model are the local cell voltage v, the diffusive water flux through the membrane J from cathode to anode, and the heat generated due to membrane resistance and cathode overpotential losses. As mentioned above, in this simplified model, the membrane resistivity is taken to be constant. Other resistances (except to in-plane currents in the bipolar plates in the stack model) are neglected.

9.4.1. Membrane Transport

The water diffusive flux through the membrane is considered following the ideas in [4]. Water content (molecule per sulphonate group) $c_w(y)$ is considered through the MEA. It is assumed to obey a simple diffusive mechanism and that the diffusive flux is not influenced by the electro-osmotic drag. The following equation is assumed:

$$-D_w \frac{dc_w}{dy} = \frac{J}{a}. \tag{9.1}$$

A fitted empirical formula is used for the diffusivity D_w [10]:

$$D_w(c_w, T) = d_w c_w \exp(-2436/T) \tag{9.2}$$

where d_w is a fitted constant and T is the membrane temperature, taken to be the average $T = (\theta_a + \theta_c)/2$. Note that the diffusivity (9.2) is linear in c_w. It is possible to solve (9.1) analytically in this case giving

$$J = ad_w \exp(-2436/T) \frac{1}{2L_m} (c_{w,c}^2 - c_{w,a}^2) \tag{9.3}$$

where $c_{w,a}$ and $c_{w,c}$ are the anode and cathode side water contents, respectively.

The anode and cathode water contents are not taken to be the equilibrium values. The equilibrium value $c_{w,c}^*$ is computed from the classic sorption curve from [12]

$$c_{w,c}^* = 0.043 + 17.81 r_c - 38.85 r_c^2 + 36 r_c^3 \tag{9.4}$$

where r_c is the cathode channel relative humidity. The same relationship is used for $c_{w,a}$ with the anode channel relative humidity r_a.

The flux of water from anode channel to catalyst sites is

$$\frac{ei}{\mathcal{F}} - J$$

and from cathode catalyst to cathode channel

$$\frac{(e+1/2)i}{\mathcal{F}} - J$$

where e is the electro-osmotic drag factor, taken to be 1. These fluxes lead to the disequilibrium in water contents using a linear relationship with fitted constant γ.

$$\frac{ei}{\mathcal{F}} - J = -\gamma \left[c_w^a - c_{w,a}^* \right], \tag{9.5}$$

$$\frac{(e+1/2)i}{\mathcal{F}} - J = \gamma \left[c_w^c - c_{w,c}^* \right], \tag{9.6}$$

For this particularly simple case, it is possible to solve analytically for the profile $c_w(y)$ and, in particular, $c_{w,a}$ and $c_{w,c}$, which determine a value of J consistent with (9.3), (9.5), and (9.6):

$$\begin{aligned} c_{w,c} &= \frac{1}{2}(\xi_+ + \xi_-) \\ c_{w,a} &= \frac{1}{2}(\xi_+ - \xi_-) \end{aligned} \tag{9.7}$$

where

$$\xi_+ = m_c^* + m_a^* + \frac{i}{2\gamma a \mathcal{F}}, \tag{9.8}$$

$$\xi_- = \frac{m_c^* - m_a^* + \frac{5i}{2\gamma a \mathcal{F}}}{1 + \frac{d_w(T)\xi_+}{L_m \gamma}}. \tag{9.9}$$

If the resulting cathode water content $c_{w,c}$ from (9.7) solution is greater than the maximum allowed at equilibrium ($c_{w,c}^*$ from (9.4) with $r_c = 1$) then an alternate solution is used. The cathode content is taken to be the maximum, and the anode content is solved to match the flux conditions there. An analytic solution to this case can also be found.

It should be noted that the analytic solutions to the membrane transport problems above and the scalar formulas for the oxygen transport and electrochemistry below are not necessary for the stack level modeling framework described in this chapter. They do allow a concise description in this expository setting and lead to fast computational methods. However, it is possible to introduce a grid in the through-MEA direction y and compute numerical approximations to more complicated models at each channel point. These more complicated relationships can be combined at the unit cell and stack levels as discussed below just as easily as the simple models presented here.

9.4.2. Oxygen Transport Losses

Assuming simple diffusive mechanisms for transport from channel average concentrations to catalyst sites, the following expression for oxygen concentration at catalyst sites can be obtained:

$$c_o - \delta_o i. \tag{9.10}$$

The oxygen concentration at catalyst sites is lower than the channel average, because the oxygen must diffuse through the GDL and catalyst layer to the active sites. The concentration reduction should be proportional to the flux, which is proportional to the local current, leading to the given form. Note that this form implies that there is a maximum local current which can be drawn, equal to c_o/δ_o.

If the diffusion occurred through a single effective media, then

$$\delta_o = \frac{L_{eff}}{4 \mathcal{F} D_{eff}}$$

where D_{eff} and L_{eff} are effective diffusivities and diffusion lengths. The minimum value of δ_o is the value from the formula above using the thickness of the

GDE and the diffusivity of oxygen in air. It was found through the fitting in [4] that a much larger value of δ_o was needed to explain experimental results. The results suggested that significant oxygen mass transport losses occur in the catalyst layer. For the purposes of this study, δ_o is taken as a fitted parameter.

9.4.3. Electrochemistry

The local cell voltage is given as

$$v = E_0 - iw - \frac{\mathcal{R}\theta_c}{4\mathcal{F}} \ln \frac{C_{ref}}{c_o - \delta_o i} - \frac{\mathcal{R}\theta_c}{\mathcal{F}\alpha_c} \ln \frac{i}{i_{ref}}, \quad (9.11)$$

where E_0 is the open circuit voltage (taken from measurements) at reference oxygen concentration C_{ref}, the second term on the right describes ohmic losses due to membrane resistivity, the third term describes Nernst losses (using the predicted catalyst oxygen concentration from (9.10)), and the final term is the overpotential that describes losses due to the irreversibility of the reaction. The parameters α_c (transfer factor) and i_{ref} (reference current) are fitted parameters.

Heat generated per unit area is $i^2 w$ from ohmic losses through the membrane and $(V_{TN} - v - iw)i$ from electrochemical losses at the cathode catalyst layer, where V_{TN} is the thermoneutral voltage of the oxygen reduction reaction with product vapor.

9.5. Channel Flow for a Unit Cell

9.5.1. Gas Concentrations in Channels

Given the local values of the channel fluxes q, the channel temperatures θ_o, θ_h, and the channel pressures p_a, p_c, it is possible to determine the channel gas concentrations c_o, c_c, c_h and c_a in moles/m³. It is assumed that the gases are ideal, obey Dalton's law, and move in the channel with a common velocity. Consider first the cathode gas channel. There are two cases to consider, depending on whether the cathode channel gases are saturated or unsaturated.

Assume first that the cathode channel is unsaturated and compute

$$c_o = \frac{p_c}{\mathcal{R}\theta_o} \frac{q_o}{q_o + q_c + q_n} \quad (9.12)$$

$$c_c = \frac{p_c}{\mathcal{R}\theta_o} \frac{q_c}{q_o + q_c + q_n} \quad (9.13)$$

where $\mathcal{R}$ is the ideal gas constant. Note that the first term on the right of both equations above represents the total molar concentration of gas at the cathode channel temperature and pressure, and the second term represents the fraction of this concentration the gas type occupies based on its fraction of the total flux. More detailed description of this flux to concentration map can be found in [4].

If the value of c_c computed above is greater than $P_{sat}(\theta_o)/(\mathcal{R}\theta_o)$, then the cathode channel gases are over-saturated. Assume that vapor will condense to prevent over-saturation and replace the concentrations c_o and c_c above with the following values:

$$c_o = \frac{p_c - P_{sat}(\theta_o)}{\mathcal{R}\theta_o} \frac{q_o}{q_o + q_n} \tag{9.14}$$

$$c_c = \frac{P_{sat}(\theta_o)}{\mathcal{R}\theta_o} \tag{9.15}$$

The cathode liquid water flux w_c can also be determined

$$w_c = q_c - \frac{c_c}{c_o} q_o \tag{9.16}$$

which would be zero in the dry case discussed in the preceding paragraph. Similarly, anode quantities c_h, c_a, and w_a can be computed.

9.5.2. Channel Flux Equations

In each cell, the following channel flux equations represent the effects of gas consumption and water production and crossover:

$$\frac{dq_o}{dx} = -\frac{i}{4\mathcal{F}} \tag{9.17}$$

$$\frac{dq_h}{dx} = -\frac{i}{2\mathcal{F}} \tag{9.18}$$

$$\frac{dq_c}{dx} = +\frac{(2e+1)i}{2\mathcal{F}} - J \tag{9.19}$$

$$\frac{dq_a}{dx} = -\frac{ei}{\mathcal{F}} + J \tag{9.20}$$

where $\mathcal{F}$ is Faraday's constant, and J is the diffusive water flux through the membrane. The cathode nitrogen flux is constant.

9.5.3. Operating Conditions

Since the unit cells of the stack are connected in series, the same total current runs through each of them. The total current is specified as an equivalent target average current density i_T. A global condition for the model is that

$$\frac{1}{L}\int_0^L i^j(x)dx = i_T \tag{9.21}$$

for each cell j.

Inlet reactant fluxes are specified as stoichiometric ratios s_a and s_c of the minimum molar flux to produce this current:

$$q_o(0) = s_c \frac{i_T}{4\mathcal{F}} L \tag{9.22}$$

$$q_h(0) = s_c \frac{i_T}{2\mathcal{F}} L. \tag{9.23}$$

The constant cathode nitrogen flux is determined by the ratio of oxygen to nitrogen in the inlet stream:

$$q_n = nq_o(0)$$

where $n \approx 0.79/0.29$ for dry air at standard conditions, which has approximately 21% oxygen and 79% nitrogen (and very small amounts of other trace elements). The channel pressures p_c and p_a must also be given.

The cathode inlet water flux is dependent on the humidification of the inlet stream, often expressed as a dew point temperature (a vapor partial pressure equivalent to saturated conditions at this temperature). If $T_{c,dew}$ is the given cathode inlet dew point, then

$$q_c(0) = q_o(0) \frac{P_{sat}(T_{c,dew})}{p_c(0) - P_{sat}(T_{c,dew})}$$

The water flux at anode inlet ($q_a(0)$) can be determined similarly from its inlet dew point $T_{a,dew}$. Note that s_a and s_c and inlet dew points can be given different values in different cells in the stack.

9.6. Stack Level Coupling

9.6.1. Thermal Coupling

Under the assumptions of dimensional reduction, the local thermal profiles through the plate and MEA are piecewise linear (quadratic through the membrane if uniform ohmic heating is assumed). The following thermal balances result:

$$\mu_p(\theta_o - \theta_l) + \mu_g(\theta_o - \theta_c) = H_{vap}\gamma_c \tag{9.24}$$

$$\mu_g(\theta_c - \theta_o) + \mu_m(\theta_c - \theta_a) = (V_{TN} - v - wi/2)i \tag{9.25}$$

$$\mu_g(\theta_a - \theta_h) + \mu_m(\theta_a - \theta_c) = \frac{1}{2}wi^2 \tag{9.26}$$

$$\mu_p(\theta_h - \theta_l) + \mu_g(\theta_h - \theta_a) = H_{vap}\gamma_a \tag{9.27}$$

$$\mu_p(\theta_o^j - \theta_l^j) + \mu_p(\theta_h^{j-1} - \theta_l^j) = A(\theta_l^j - t^j) \tag{9.28}$$

at cathode channel, cathode catalyst, anode catalyst, anode channel, and coolant channel, respectively. In the last equation, heat fluxes into the coolant come from different cells, and so the cell number superscripts are used explicitly. From Figure 9.3, it is clear that the coolant of cell j receives heat from the cathode side of cell j and the anode side of cell $j - 1$. Here, μ_p, μ_g and μ_m are the rations of thermal conductivity to thickness for the plates, electrodes, and membrane, respectively. The parameter A is a scaled Nusselt number from conjugate heat transfer theory (see [9] for details). The physical parameter H_{vap} is the heat of vaporization, and γ_a, γ_c are the condensation rates of the anode and cathode channel streams. These are given by:

$$\gamma_a = \frac{dw_a}{dx} \tag{9.29}$$

$$\gamma_c = \frac{dw_c}{dx}. \tag{9.30}$$

9.6.2. Coolant Flow

Taking into account the heat flowing into the coolant,

$$g\frac{dt^j}{dx} = \mu_p(\theta_o^j + \theta_h^{j-1} - 2\theta_l^j) \tag{9.31}$$

where g is the product of coolant heat capacity and flow rate (per unit orthogonal direction z). The scaled coolant flow rate g and its inlet temperature $t(0)$ must be given.

9.6.3. Electrical Coupling

Differences in local current densities lead to in-plane currents in the bipolar plates. The in-plane currents in the bipolar plate between cells j and $j+1$ are labeled $I^{j/j+1}$ as shown in Figure 9.4. These currents are the integrals (in z) of the x-directional current density. From current conservation, the following is obtained:

$$\frac{dI^{j/j+1}}{dx} = i^j(x) - i^{j-1}(x). \tag{9.32}$$

In-plane resistance of the bipolar plate λ (per unit length to currents per unit width) leads to changes in the cell voltage:

$$\frac{dv^j}{dx} = \lambda(-I^{j/j+1} + I^{j-1/j}). \tag{9.33}$$

Differentiating (9.33) and using (9.32), the plate currents can be eliminated, leading to the so-called *fundamental voltage* equation:

$$\frac{d^2v^j}{dx^2} - \lambda(i^{j-1} - 2i^j + i^{j+1}) = 0. \tag{9.34}$$

The boundary conditions for (9.34) are

$$\frac{dv}{dx}(0) = 0 \tag{9.35}$$

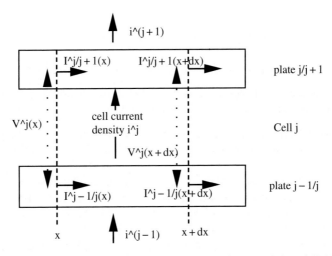

FIGURE 9.4. Diagram showing the derivation of Eqs. (9.32) and (9.33).

$$\frac{dv}{dx}(L) = 0 \tag{9.36}$$

for each cell. These represent the fact that cell ends are electrically insulated.

9.6.4. End Cell Conditions

M cells form the stack, numbered $j = 1, \ldots M$. In this paper, the end plates are taken to be infinitely electrically conductive and thermally insulated. The modifications to handle end plates of finite resistance are discussed in [13].

The electrical end plate boundary conditions can be expressed in terms of ghost values for i^0 and i^{N+1} in (9.34):

$$i^0 = i^1 \tag{9.37}$$

$$i^{M+1} = i^M. \tag{9.38}$$

Note that these conditions used in (9.34) integrated over x and using boundary conditions (9.35, 9.36) show that the total current through each cell is the same. Thus, (9.21) is a single global constraint, as expected physically.

The thermal end plate boundary conditions can be expressed in terms of ghost values for θ_h^0 and θ_o^{M+1} in (9.28) and (9.31):

$$\theta_h^0 = \theta_l^1 \tag{9.39}$$

$$\theta_l^{M+1} = \theta_h^M \tag{9.40}$$

9.7. Model Summary and Discretization

9.7.1. System Description and Important Coupling

There are two main difficulties to the fuel cell stack system presented above, besides just the complexity of detail of physical phenomena it represents:

Nonlocality: The nonlocality arises from the voltage/current coupling (9.34). Changes in conditions in one cell affect all cells at all locations through this equation.

Lack of Smoothness: The lack of smoothness in the model arises from the change between saturated and unsaturated conditions in the flux to concentration map described in Section 9.5.1. This lack of smoothness also corresponds to a change in the structure of the problem, since when the channel is saturated, derivative terms appear in (9.24, 9.27). Channel temperatures change from algebraic to differential variables.

9. Thermal and Electrical Coupling in Stacks   331

The two important couplings in the model are given below:

1. The nonlocal coupling of current density to cathode oxygen concentration through flux consumption (9.17) to concentration (9.12) or (9.14) to the concentration dependence of voltage (9.11) and its nonlocal influence back to current densities (9.34). This coupling is especially strong when the local current density is near its maximum value (i near c_o/δ in (9.11)).
2. The local coupling of channel condensation to channel temperature when the channel (anode or cathode) is saturated. Channel temperatures strongly affect channel water fluxes (9.16) through (9.15). Channel condensation is the derivative of water flux (9.30) and is a large heating source in the determination of channel temperatures (9.24).

The overall strategy taken to break the model into computationally manageable pieces is to consider the local current densities as the underlying variables. For given values of the local current densities, the channel fluxes and thermal profiles can be considered as an initial value problem. From the solution of this problem, cell voltages can be computed and residuals in (9.34) determined. In this way, the fuel cell model can be described as a nonlocal equation of nonstandard character for the current densities.

9.7.2. Discretization

The channel is discretized at $N+1$ points $k = 0, N$ with spacing $\Delta x = L/N$ corresponding to $x_k = k\Delta x$ (a regular grid). Algebraic conditions (9.10–9.16) and (9.24–9.28) apply at each grid point. First-order forward difference approximation is used for the derivatives in anode and cathode gas fluxes (9.17–9.20). First-order backward difference approximation is used for the derivatives in channel condensation (9.29, 9.30) and coolant temperature (9.31). The second order derivatives in (9.34) are approximated with centered differences, and the ghost points introduced are eliminated using the boundary conditions (9.35, 9.36). The resulting discrete approximation of the system (9.34) is consistent but rank deficient of order one. One equation of the system is replaced by a trapezoidal rule approximation to (9.21) in one cell. The expression (9.11) is regularized at currents near and past the maximum current as described in [13].

9.8. Iterative Solution Strategy

As foreshadowed in Section 9.7.1, the iterative strategy is based on an outer iteration on discrete local current densities. For given current densities, the coupled initial value problem for the channel fluxes and temperatures are solved by marching, with some terms handled implicitly. Details of the iterative strategy are given below.

9.8.1. Temperatures and Channel Fluxes

Moving from channel grid point to grid point, channel fluxes are updated with explicit approximation of (9.17-9.20) in each cell. With the channel fluxes (and the local current held fixed from the outer iteration) given at the next grid point, a nonlinear problem for θ and t in all cells remains. Given an initial guess for these quantities, and assuming for now that the channel conditions are unsaturated, (9.12), (9.13), and corresponding equations for the anode can be used to compute channel concentrations. Diffusive flux J can be determined from the algebraic relationships in Section 9.4, and the cell voltage can also be computed from (9.11) for each cell at the current channel point. In this way, residuals in (9.24–9.28) and the implicit discretization of (9.31) in every cell can be computed and θ and t for each cell adjusted to satisfy the system of equations. If at convergence the unsaturated channel assumptions are not satisfied for particular cells, the solve is conducted assuming these channels are saturated. The iteration scheme is a quasi-Newton approach, ignoring the weak dependence of cell voltage v (that enters the RHS of equation (9.25)) on the temperatures in the computation of the Jacobian. However, when channels are saturated, the dependence of water flux on channel temperature and hence its influence on condensation rate γ_a and γ_c computed by difference formulas that enter (9.24) and (9.27) must be used in the approximate Jacobian to obtain convergent iterations. At cathode inlet, a reduced system is solved, using the fixed inlet coolant temperature t.

An important aspect of the solution strategy is that channel saturation states are not changed during the quasi-Newton channel iterations. The iterations proceed to convergence, and only then are the saturation states changed if needed. In this way, only smooth problems are solved, and faster and more robust convergence using quasi-Newton methods can be obtained.

9.8.2. Current Density Update

For given discrete current densities, the problem of the temperatures and channel fluxes can be solved as discussed above. The solution of this problem provides the voltage v at all points and all the quantities in (9.11). Residuals in the discrete Eqs. (9.34) can be computed. A quasi-Newton step that preserves the stack current is used to update the discrete current densities. The approximate Jacobian used for the update uses a local approximation of voltage sensitivities to current density. Details can be found in [13]. When the local current densities are updated, the temperature profiles and channel fluxes are updated, as described in Section 8.1 above. This continues until convergence is obtained.

9.9. Computational Results

9.9.1. Unit Cell Results

A single unit cell, with thermal boundary conditions set so that it represents a cell in a uniform stack, is considered first. Base operating conditions of $s_c = 1.8$ and $i_T = 1 A/cm^2$ are taken. The inlet coolant temperature is taken to be 70° C. Other operating conditions are listed in the Nomenclature section. Representative plots for this base case are given in Figures 9.5 (local current density), 6 (coolant temperature and cathode channel temperature), and 7 (anode and cathode channel relative humidity). For comparison, a nonbase case with lower $s_c = 1.3$ is also shown. Notice that the sharp increase in cathode channel temperature in Figure 9.6 corresponds to the point at which the cathode channel saturates (see Figure 9.7), and there is additional heat generated from condensation.

Although the local current density profiles in Figure 9.5 are quite different, the cell voltage does not differ significantly: 0.688 V for the base case and 0.6602.

9.9.2. Stack Results

The stack model is now considered with $M = 13$ cells. All cells except the center cell #7 are run at base conditions. The center cell is run with the anomalous cathode stoich $s_c = 1.3$. This anomaly could arise, for example,

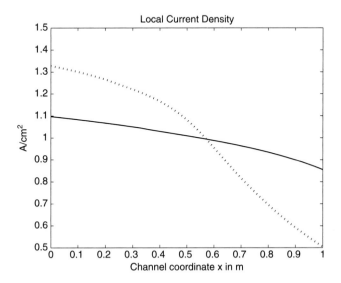

FIGURE 9.5. Local current densities for unit cell base case (solid line, cathode stoich 1.8) and anomalous case (dotted line, cathode stoich 1.3).

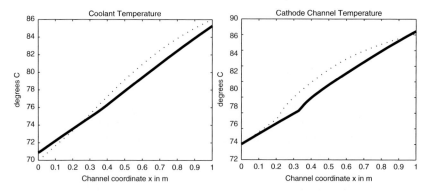

FIGURE 9.6. Coolant temperature (left) and cathode channel temperature (right) for unit cell base case (solid line, cathode stoich 1.8) and anomalous case (dotted line, cathode stoich 1.3).

from a partially blocked duct from the stack oxidant header to the inlet manifold of this cell. The unit cell results in Figure 9.5 indicate that the anomalous center cell would have quite a different local current density than its neighbors in the case of no resistivity in the bipolar plates (when there is no electrical coupling between cells in a stack). However, the bipolar resistivity is quite significant [13], and the current anomaly affects many adjacent cells as shown in Figure 9.8. The voltage is also affected as seen in Figure 9.9. Note that the anomalous cell #7 suffers a large voltage drop at outlet, but its neighbors see a slight rise. This so-called volcano effect is captured accurately by the model [14]. The cathode channel temperatures are shown in Figure 9.10. Note that the anomalous cell does run hotter than the base case, but the hottest cell is #13. This is due to end effects: cell #13 is

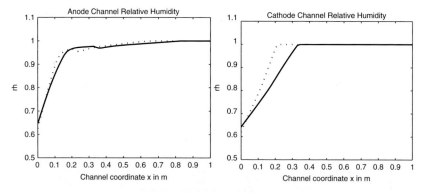

FIGURE 9.7. Anode channel (left) and cathode channel (right) relative humidities for unit cell base case (solid line, cathode stoich 1.8) and anomalous case (dotted line, cathode stoich 1.3).

9. Thermal and Electrical Coupling in Stacks   335

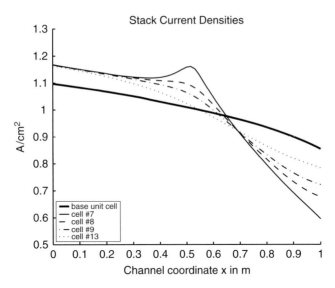

FIGURE 9.8. Local current densities for the stack computation with anomalous center cell (#7) conditions of lowered cathode stoich.

the insulated end cell with no coolant on its anode side. The end cell #1 is the coolest, since it has a coolant channel it does not have to share with a neighboring cell below.

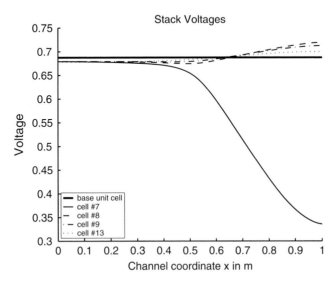

FIGURE 9.9. Local voltages for the stack computation with anomalous center cell (#7) conditions of lowered cathode stoich.

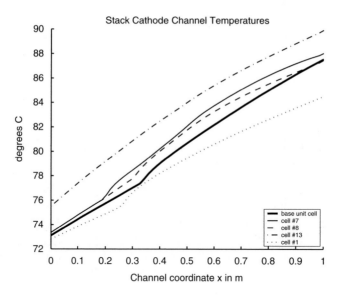

FIGURE 9.10. Cathode channel temperatures for the stack computation with anomalous centre cell (#7) conditions of lowered cathode stoich.

9.10. State of Stack Modeling

The particular model described in this chapter is not the most advanced but is representative and shows the potential as well as the limitations of the approach. Several elements of this type of stack model have been developed and validated previously in our group. A basic, reduced dimensional unit cell model was developed in [4]. Using a more detailed membrane model than that presented above, good agreement with experimental data for overall water crossover and also reduced current density near dry cathode inlets was obtained in the unit cell setting (with counter flowing gases). A simple model of stack level electrical coupling through bipolar plate resistance was given in [13]. It was later shown to accurately predict the effect of anomalous center cells and end plates compared to experimental work in [14]. Simple versions of the thermal model have been shown in [15, 9]. Comparison of thermal stack model results to experimental work is currently under way. The development of a stack model that features a more realistic description of transport in the GDE, including liquid water and its slow transients following the analytic work in [16] is currently under way.

Acknowledgments

The work described here is a summary of ideas developed by a group of academic mathematicians working with scientists from Ballard, including

Peter Berg, Radu Bradean, Atife Caglar, Paul Chang, Gwang-Soo Kim, Keith Promislow, Jean St-Pierre, John Stockie, and Juergen Stumper. The group is funded by both Ballard and the Mathematics of Information Technology and Complex Systems (MITACS) Network Centre of Excellence in Canada.

References

[1] T.F. Fuller and J. Newman, "Water and Thermal Management in Solid Polymer-electrolyte Fuel-cells," J. Electrochem. Soc **140**, 1218 (1993).
[2] T.N. Nguyen and R.E. White, J. Electrochem. Soc **140**, 2178 (1993).
[3] S. Freunberger, A. Tsukada, G. Fafilek and F.N. Buechi, "1 + 1 dimensional model of a PE fuel cell of technical size," Paul Scherrer Institut Scientific Report 2002 Volume V, article #94.
[4] P. Berg, K. Promislow, J. St-Pierre, J. Stumper, B. Wetton, "Water Management in PEM fuel cells," J. Electrochem. Soc., **115**, A341–A353 (2004).
[5] A.Z. Weber, R.M. Darling and J. Newman, "Modeling Two-Phase Behavior in PEFCs," J. Electrochem. Soc. **151**, A1715–1727 (2004).
[6] T. Berning, D. Lu and N. Djilali, "Three-Dimensional Computational Analysis of Transport Phenomena in a PEM Fuel Cell a Parametric study," J. Power Sources, **124**, 440-452 (2003).
[7] W-k. Lee, S. Shimpalee and J.W. Van Zee, "Verifying Predictions of Water and Current Distributions in a Serpentine Flow-Field PEMFC," J. Electrochem. Soc., **150**, A341–A348 (2003).
[8] D. Natarajan and T. Nguyen, "Three-Dimensional Effects of Liquid water Flooding in the Cathode of a PEM Fuel Cell," J. Power Sources, **115**, 66–80 (2003).
[9] K. Promislow and B. Wetton, "A Simple, Mathematical Model of Thermal Coupling in Fuel Cell Stacks," accepted in the J. Power Sources, **150**, 129–135. February, (2005).
[10] S. Motupally, A.J. Becker, and J.W. Weidner, "Diffusion of Water in Nafion 115 Membranes," J. Electrochem. Soc., **147**, 3171 (2000).
[11] T.E. Springer. T.A. Zawodzinski and S. Gottesfeld "Polymer electrolyte fuel cell model," *J. Electrochem. Soc.*, **138**, 2334 (1991).
[12] T. Zawodziski, C. Derouin, S. Radzinski, R. Sherman, V. Smith, T. Springer and S. Gottesfeldt, "Water-Uptake by and Transport Through Nafion 117 Membranes," J. Electrochem. Soc., **140**, 1981 (1993).
[13] P. Berg, A. Caglar, J. St-Pierre, K. Promislow and B. Wetton, "Electrical Coupling in Proton Exchange Membrane Fuel Cell Stacks: Mathematical and Computational Modelling," accepted in the IMA J. Appl. Math., March, (2005).
[14] G. S. Kim, J. St-Pierre, K. Promislow and B. Wetton, "Electrical Coupling in PEMFC Stacks", accepted in the J. Power Sources, **152**, 210–217. January, (2005).
[15] B. Wetton, K. Promislow and A. Caglar, "A Simple Thermal Model of PEM Fuel Cell Stacks," in the proceedings of the Second International Conference on Fuel Cell Science, Engineering and Technology, Rochester, June, 2004.
[16] K. Promislow, J. Stockie, B. Wetton, "A sharp interface reduction for multi-phase transport in a porous fuel cell electrode," **462**, 789–816, 2006.

Part II
Materials Modeling

Section Preface

K. D. KREUER

The fuel cell concept is actually known for more than 150 years. It was Christian Friedrich Schönbein who recognized and described the appearance of "inverse electrolysis" [1] shortly before Sir William Grove, the inventor of the platinum/zinc battery, constructed his first "gas voltaic battery" [2]. Both used platinum electrodes and dilute sulfuric acid as a proton conducting electrolyte which is chemically not very different from the materials used in modern PEM fuel cells. The most commonly used fuel cell electrolytes today are hydrated ionomers bearing sulfonic acid functional groups, and platinum or platinum alloys are still used as materials for the electrocatalysts. The following chapters focus on these two classes of materials, the main constituents of the so- called membrane electrode assembly (MEA), the heart of every PEM fuel cell. In addition, many more materials are required to construct a fuel cell, e.g. the carbon support of the electrocatalyst, the gas diffusion layer (GDL), the flow fields, current collectors, and sealings. The related material problems and recent approaches have been discussed by Steele and Heinzel [3] and descriptions of state-of-the-art materials can be found in the corresponding chapters of the *Handbook of Fuel Cell Technology Vol. 3* [4].

Electrolytes and Electrocatalysts

In contrast to aqueous sulfuric acid, where the dissociated protons and the sulfate anions (conjugated bases) are mobile, ionomers as used in modern PEM fuel cells are polymers containing covalently immobilized sulfonic acid functions. The immobilization reduces anion adsorption on the platinum cathode, which may lead to reduced exchange current densities for oxygen reduction in the case of sulfuric acid as electrolyte. Amongst the myriad of sulfonic acid-bearing polymers [5,6,7,8,9,10], the most prominent representative of this class of separators is DuPonts Nafion [11,12]. Such polymers combine, in one macromolecule, the high hydrophobicity of the backbone with the high hydrophilicity of the sulfonic acid functional group, which gives rise to a constrained hydrophobic/hydrophilic nano-separation. The sulfonic acid functional groups aggregate to form a hydrophilic domain that is hydrated upon absorption of water. It is within this continuous domain, where ionic conductivity occurs: Protons dissociate from their anionic counter ion ($-SO_3^-$) and become solvated and mobilized by the hydration water. The transport of water and the coupled transport of protonic charge carriers and water (electroosmotic drag) are also taking place within this hydrated domain (for a review on the transport in ionomers see Ref. [13]). It is interesting to note that the nature of this transport is changing from predominantly diffusional to predominantly hydrodynamic with increasing degree of hydration. The pronounced electroosmotic drag is a direct consequence of the anion immobilization and the confinement of the hydration water. In other words: transport and microstructure are closely related for this type of ionomers [14]. The limited operating temperature and the acidity of such electrolytes neccessitates the use of platinum or platinum alloys to promote the electrochemical reactions in the anode and cathode structures. But even with platinum, only rather pure hydrogen can be oxidized at sufficient rates (especially CO present in many fuels tends to poison Pt-based catalysts at low temperatures), and oxygen reduction is relatively slow under these conditions. Consequently, there have been numerous activities in the materials science community aiming at improving materials by either modifying available fuel cell materials or by trying to identify or develop alternatives.

Platinum electrocatalysts have been improved by forming alloys, e.g. with Ru to form catalysts with improved CO tolerance [15] and higher reaction rates for the direct electrochemical oxidation of methanol, or, e.g., with Co to accelerate the oxygen-reduction reaction under certain conditions [16]. While these improvements have already entered fuel cell technology, the electrolyte of almost all PEM fuel cells is still some perfluorosulfonic ionomer such as Nafion. This is surprising, considering the many attempts to develop improved proton-conducting hydrated ionomers (such as hydrocarbon membranes [5,6,7,8,9,10] and composites [17,18]) and the more radical approaches aiming at conceptually different separator materials exclusively transporting protons and being able to operate at higher temperatures in a low humidity environment [19,20,21].

The tremendous progress with respect to improving performance and long-term stability and reducing costs of different types of fuel cell, however, has essentially been achieved by sophisticated engineering efforts, making use of the few existing materials, such as Nafion and Pt-based electrocatalysts.

The Impact of Models and Simulations

Development of new and suitable materials appears to be an extremely challenging task since the applicability of the materials in a fuel cell depends on diverse properties and, with respect to many of them, established fuel cell materials already meet the given requirements. Further complications arise from the need of materials compatibilities, i.e., different fuel cell materials cannot be developed independently. As these intricacies became progessively obvious, the increasing availability of computational hardware, along with the improvement and the development of new codes, have led to a tremendeous dissemination and an increasing impact of such techniques also in the many attempts to develop improved fuel cell materials. There even had been the vision of a "virtual laboratory" substituting the "real laboratory" at some instant. The rationale behind such provocative statements is frequently the fact that, in particular *first principles* calculations are, by definition, free of assumptions and provide the full picture, in particular, details of the dynamics and chemical interactions. But there is actually nothing such as a true *ab initio* simulation. The applied techniques always rely on assumptions and approximations, which may be more or less suitable for certain types of systems (for a discussion see [13]). As a consequence, there is always a trade-off between system size (and time increment in the case of molecular dynamics (MD) simulations) and significance of the information obtained by simulations. In short: the reliability of simulation data is usually increasing with decreasing system size, while the opposite is commonly true for experiments, where sensitivity limitations frequently restricts the investigation of small systems, such as the diverse species present at the surface of an electrocatalyst under fuel cell operating conditions. In the case of more complex systems, simulations generally rely on the input of experimental data. This already makes clear that simulations may only be used in conjunction with experimental techniques, in particular, they may not be considered an independent tool in the materials development process.

What we mainly see today are attempts to reproduce experimental data by simulations of selected aspects of well-known systems. Once this has been successfully accomplished, there is a certain chance that the details produced by the simulation also have some significance. In this way, simulations may contribute to the understanding of the properties of given materials, which improves the basis for material optimizations. In this sense the application of simulation tools has already been implemented into the *"real laboratory."*

The chapters of the following section are all written in this spirit. High-level atomistic simulations of the elementary reactions taking place in the proton-conduction process and the electrochemical reduction of oxygen on a platinum surface in the presence of protons and water are presented, as well as simulations dealing with separator morphologies (the microstructure of ionomers) which strongly rely on the quality of the applied force fields (see Figure 1). Especially, the combination of simulations and the interpretation of small angle x-ray scattering (SAXS) data provide an interesting approach to mutually compensate for the restrictions of experimental and theoretical methodologies.

In this respect, simulation and modeling tools have already been established in the improvement of fuel cell materials. The complexity of the

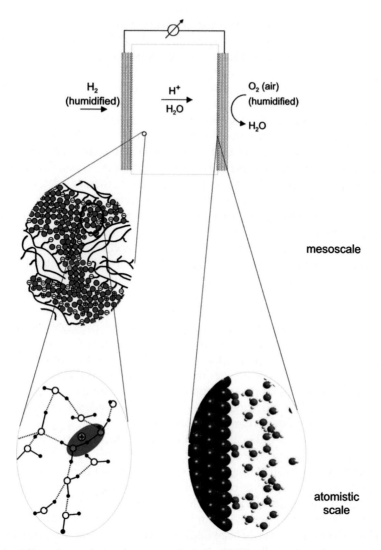

Schematic illustration of the membrane electrode assembly (MEA) of a PEM fuel cell (top) and details which have been subjected to modeling and simulation work described in the following chapters. Atomic level simulations have been performed for water and proton transport within the hydrophilic domaine of hydrated ionomers and for the electrochemical processes taking place at the electrocatalysts surfaces. The latter include the introduction of polarizable solvents and electrostatic potential variations. Mesoscale modeling is aiming at a better description and understanding of the development of ionomer microstructures.

relevant fuel cell materials such as ionomers with structural features on many length scales, and electrocatalysts with compositional and structural inhomogeneities on the nanoscale is limiting the speed of these developments at the moment, but the current progress in multi-scale modeling will surely lead to a increasing impact of modeling tools in the materials optimization process.

For more simple systems, the predictive character of ab initio and *quantum* MD simulations has already made possible the directed identification of improved proton- conducting materials. A prominent example is yttria-doped $BaZrO_3$, an oxide with the perovskite structure forming highly mobile protonic defects in the presence of water vapor [22,23]. Quantum MD simulations [24,25] have revealed details of the proton-conduction mechanism, including the critical interactions, and electronic structure calculation have helped identify the best possible dopant [26].

References

1. C.F. Schönbein, *Phil. Magazine*, **14**, 43 (1839)
2. W.R. Grove, *Phil. Magazine*, **14**, 127 (1839)
3. B.T.C. Steele, A. Heinzel, *Nature*, **414**, 345 (2001)
4. *Handbook of Fuel Cells – Fundamentals, Technology and Applications*, eds. W. Vielstich, A. Lamm, H.A. Gasteiger, John Wiley & Sons, Ltd, Chichester (2003)
5. O. Savadogo, *J. New Mat. Electrochem. Syst.*, **1**, 47 (1998),
6. Q. Li, R. He, J.O. Jensen, N.J. Bjerrum, *Chem. Mater.*, **15**, 4896 (2003),
7. M. Rikukawa, K. Sanui, *Prog. Polymer Sci.*, **25**, 1463 (2000)
8. J. Rozière, D.J. Jones, *Annu. Rev. Mater. Res.*, **33**, 503 (2003)
9. P. Jannasch, *Curr. Opin. Colloid Interface Sci.* **8**, 96 (2003)
10. K.D. Kreuer, in *Handbook of Fuel Cells – Fundamentals, Technology and Applications*, eds. W. Vielstich, A. Lamm, H. A. Gasteiger, John Wiley & Sons, Ltd, Chichester, pp. 420 (2003)
11. M. Doyle, G. Rajendran, in: *Handbook of Fuel Cells – Fundamentals, Technology and Applications*, eds. W. Vielstich, A. Lamm, H.A. Gasteiger, John Wiley & Sons, Ltd, Chichester, p.p. 351 (2003)
12. K.A. Mauritz, R. B. Moore, *Chem. Rev.*, **104**, 4535 (2004)
13. K. D. Kreuer, S. J. Paddison, E. Spohr, M. Schuster, *Chem. Rev.*, **104**, 4637 (2004)
14. K. D. Kreuer *J. Membrane Science*, **185**, 29 (2001)
15. K. Ruth, M. Vogt, R. Zuber, *Handbook of Fuel Cells – Fundamentals, Technology and Applications*, eds. W. Vielstich, A. Lamm, H.A. Gasteiger, John Wiley & Sons, Ltd, Chichester, Chap. 39, pp. 489 (2003)
16. D. Thompsett *Handbook of Fuel Cells – Fundamentals, Technology and Applications*, eds. W. Vielstich, A. Lamm, H.A. Gasteiger, John Wiley & Sons, Ltd, Chichester, Chap. 37, pp. 467 (2003)
17. G. Alberti, M. Casciola, *Annu. Rev. Mater. Res.*, **33**, 129 (2003)
18. B. Kumar, J. Fellner, P. *J. Power Sources*, **123**, 132 (2003)
19. M. F. H. Schuster, W. H. Meyer, *Annu. Rev. Mater. Res.* **33**, 233 (2003)

20 H. G. Herz, K. D. Kreuer, J. Maier, G. Scharfenberger, M. F. H. Schuster, W. H. Meyer, *Electrochim. Acta*, **48**, 2165 (2003)
21 M. Schuster, T. Rager, A. Noda, K. D. Kreuer, J. Maier, *Fuel Cells* **5**, 355 (2005)
22 K. D. Kreuer, *Solid State Ionics*, **125**, 285 (1999)
23 K. D. Kreuer, *Ann. Rev. Mater. Res.*, **33**, 333 (2003)
24 W. Münch, K. D. Kreuer, G. Seifert, J. Maier, *Solid State Ionics*, **136**, 183 (2000)
25 W. Münch, K. D. Kreuer, G. Seifert, J. Maier, *Solid State Ionics*, **125**, 39 (1999)
26 K. D. Kreuer, S. Adams, W. Münch, A. Fuchs, U. Klock, J. Maier, *Solid State Ionics*, **145**, 295 (2001)

The Membrane

10
Proton Transport in Polymer Electrolyte Membranes Using Theory and Classical Molecular Dynamics

A. A. KORNYSHEV AND E. SPOHR

10.1. The Membrane

Typical membrane materials used for polymer electrolyte fuel cells are ionomers which are composed of perfluorinated polymer backbones with side-chains containing acid groups, most commonly sulfonic acid ($-SO_3H$) end groups. In contact with water, the sulfonic acid groups dissociate and thus introduce protons as charge carriers into the membrane. At the same time, the remaining SO_3^- side-chain end groups become hydrated and form a neutralising partially ordered environment in which protons and water molecules move. Preferential interactions between water, protons and SO_3^- groups as a consequence of hydrogen bonding and ion solvation and between the hydrophobic polymer backbones lead to phase separation on the nanometer scale between polymer and aqueous phase. Protons move in the aqueous subphase, and hence the presence of water is essential for the material to become proton conducting.

The membrane in the polymer electrolyte fuel cell (PEFC) is a key component. Not by chance does the type of membrane brand the cell name – it is the most important component determining cell architecture and operation regime. Polymer electrolyte membranes are almost impermeable to gases, which is crucial for gas-feed cells, where hydrogen (or methanol) oxidation and oxygen reduction must take place at two separated electrodes. However, water can diffuse through the membrane and so can methanol. Parasitic transport of methanol (methanol crossover) severely impedes the performance of the direct methanol fuel cell (DMFC).

The reported electro-osmotic coefficient of water typically lies between 1 and 2 (but can be as high as ten at high water content, see, e.g., ref [1]), i.e., protons moving from the anode to the cathode "drag", on overage, one or two water molecules with them [2]. This leads to the obvious question how this can happen if the proton moves, as originally assumed, in a purely Grotthuss style relay fashion (in which protons "hop" from one water molecule to the next one [3, 4]) that does not require the motion of protonic aqua complexes as a whole? The answer is that in addition to the Grotthuss mechanism, there is also a "vehicle" mechanism, where the H_3O^+ or $H_5O_2^+$ ions can migrate as a whole or even with part of their solvation shell. This contributes some 20% to the overall proton mobility in water [1, 5]. In part, this water drag is compensated by back diffusion or hydraulic permeation, and for a given value of the electrical current, a gradient of water concentration will be established across the membrane [6–10]. Regions low in water content located close to the anode will not conduct protons as efficiently as water-saturated regions at the cathode side, thus creating a limiting current in the cell [11]. While the water transport problem can be solved in the PEFC by smart (but, from an engineering point of view, expensive) "water management", methanol crossover from the anode of a DMFC to the cathode is a considerable source of voltage loss [12], and such membranes are not ideal for DMFCs. This is a severe problem for further development of the DMFC, since all presently available membrane materials either suffer from the parasitic transport problem or some other problem of mechanical or chemical stability.

What, then, are the requirements for a suitable membrane? An "ideal" membrane must (i) be chemically and mechanically robust and exhibit a stable performance in the temperature range of 80–150°C with a conductivity not lower than about $0.1\ S\ cm^{-2}$ at current densities of not less than $1\ A/cm^2$; (ii) be impermeable to gases (and to methanol in the case of DMFC), and (iii) have an affordable price.

Presently the most widely used and tested polymer electrolyte membranes are various modification of Nafion® (DuPont de Nemours Co), Dow (Dow Chemicals), Flemion® (Asahi Glass) and Aciplex®, which differ in their chemical structure, their equivalent weight (EW, measured in g of dry polymer per mole of SO_3^- ion exchange functional groups) and their macroscopic thickness. Apart from the currently high price, they meet the demands of hydrogen-oxygen fuel cells. However, Nafion and Flemion (a material similar to Nafion but processed under different conditions) are not ideal for the DMFC: not only do they allow methanol permeation, they also do not sustain the high temperatures at which catalysis would be fast, and thus all the methanol supplied to the anode would have been consumed, thereby reducing the importance of the methanol permeation problem and/or reducing the amount and thus cost of precious metal catalyst. To a similar extent, this is true for the Dow and Aciplex membranes. Although the glass ("melting") transition of Nafion is close to 140°C, the standard regime for fuel cell operation with Nafion membranes lies close to 90°C, since it is difficult to keep water inside the membrane above 100°C, even under pressurisation (which is undesirable, again, from an engineering point of view). Consequently, the design of new membranes which would be able to satisfy the above-mentioned demands, particularly sustaining higher temperatures, is a subject of intensive research [13–23] and development [24–27] in many polymer laboratories world-wide. In spite of this ongoing R&D effort, the focus of activities is presently clearly on Nafion, due to its chemical stability, its availability and the wide experience in applying it in fuel cell technology. Consequently, Nafion is also studied best experimentally.

What can be expected from theory and simulation in this field? Can theory aid the design of new materials? Before this is possible, we need first of all to understand *how* protons move in the aqueous sub-phase of the membrane "interior", particularly, how does the polymer architecture (molecular structure, degree of polymerisation, arrangement of SO_3^- groups, equivalent weight, side-chain length, crosslinking etc.) as well as the dynamical properties of the polymer influence proton mobility. Here, atomistic simulations can be a valuable tool for understanding governing principles (see below). And, on the other hand, how does proton transport in the real fuel cell device act back on the molecular mechanisms of parasitic transport, microscopic polymer dynamics and macroscopic swelling of the polymer electrolyte? In this field, molecular-level simulations for open systems such as grand-canonical and Gibbs-ensemble simulations will prove very helpful, although these techniques have to date not been used in a substantial way. Only after having understood these interdependencies will theory, hopefully, be able to give some hints to polymer chemists how to change the "interior architecture" in favour of faster proton conduction and slower water/methanol diffusivity.

This is more easily said than done! First, the mobility of protons can never be higher than in pure bulk water, unless some synergetic Davydov "solitonic" mechanism [28] of collective proton transport in proton conductive "wires" exists, the subject of many speculations in "solid state protonics" [29]. It is, however, hard to envisage anything like this in a typical, disordered polymer electrolyte. Also, proton mobility is not all that needs to be maximised; rather, what needs to be maximised is the product of mobility and concentration, i.e. conductivity; thus, it is paramount to increase the concentration of proton donors as much as possible, which can be achieved by lowering equivalent weight. Hence, strategies are either to create new proton-conducting "super-ionic" materials with faster intrinsic proton mobility or to increase the concentration of protons in the material, via an understanding of how to optimise the existing materials. Particularly, Kreuer and co-workers and Paddison and co-workers investigated the prospects of the first approach experimentally, employing atomistic modelling as a tool for understanding (see their chapters in this book).

10.2. Nafion-Water-Nanostructure

Several groups, including ours, followed the latter path, studying the structural and dynamical properties of water/Nafion mixtures with atomistic simulations. Let us first briefly consider the key chemical features of the ionomer and the nature of the phase segregation. The chemical composition of Nafion is given by:

$$-(CF_2-CF_2)_{n_1}-(CF_2-CF)_{n_2}$$
$$|$$
$$O-(CF_2-CF-O)_{n_3}-(CF_2)_{n_4}-SO_3^-H^+$$
$$|$$
$$CF_3$$

($n_1 = 6-10$, $n_2 = 1$, $n_3 = 1$, $n_4 = 2$).

Apart from chemically stabilising the polymer backbone, the fluorine atoms, due to their strong electronegativity, confer on the $-SO_3H$ groups a very high acidity, similar to that of CF_3SO_3H. The dissociation products of the $-SO_3H$ groups behave very differently: SO_3^- groups stay attached to the side-chains, whereas the protons can freely move in the water sub-phase. The nanoscale segregation, due to the differences in aqueous–aqueous, polymer–polymer and aqueous–polymer interactions, into two sub-phases, can be schematically depicted as in Figure 10A.1.

The hydrophobic phase is formed by the perfluorinated polymer backbones and by the pendant side-chains except for their terminal SO_3^- groups which will seek contact with water. The hydrophilic phase is formed by water, mobile protons and SO_3^- groups; as far as this phase spans (percolates) through the sample, it is a proton conductor. At low equivalent weights, the density of mobile protons approaches the level of concentrated acids. While these basic principles are commonly accepted, there is no unified opinion on the "details" of the morphology of microphase segregation in polymer-electrolytes in spite of a large number of investigations in this area [30, 31].

There are three forms of this ionomer, sometimes labelled as shrunk (**S**), **normal** (**N**) and **expanded** (**E**) forms. The **S**-form is acquired after a treatment at high temperatures; it corresponds to an ultra-dry state, when all the residual water molecules are expelled from the polymer matrix. Since water works as a "plasticiser" for this polymer, the **S**-form is rugged and essentially

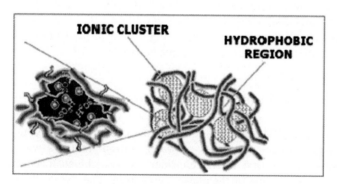

FIGURE 10.1. Cartoon of the microphase segregation in perfluorinated membranes.

ruined for membrane applications. The **E**-form is made from the polymer solution kept at high temperature and pressure. This form can have more than 50% water uptake, which is a swollen, physically crosslinked gel rather than a solid plastic material. These membranes are very permeable for water, which makes them unsuitable for fuel cell applications. The relevant **N**-form, even in its dry state (at zero extra water uptake), keeps "residual" water: between one and two molecules per SO_3^- group. The dry state may be regarded as a kind of short-range ordered crystal hydrate which is ready to absorb water; water uptake in the wet state reaches up to 30 volume percent.

Information about the membrane structure comes from various sources, including small- and wide-angle X-ray and neutron scattering, infrared and Raman spectra, time-dependent FTIR, NMR, electron microscopy, positron annihilation spectroscopy, scanning probe microscopy and scanning electrochemical microscopy (SECM) (for review of literature see Ref.[32]). The group at DuPont was first to come to a conclusion about the inverted micellar structure of the aqueous phase. The so-called Gierke model [33] considers this phase as built of approximately spherical water droplets of nanoscopic dimension, confined by anionic head groups of the side-chains. In a dry state, the diameter of these droplets is of the order of $D \approx 2$ nm, and the droplets are disconnected from each other. With water uptake the droplets grow up to $D \approx 4$ nm, and aqueous necks emerge between them at certain intermediate water content. After emergence of a critical number of necks, a continuous pathway through the water sub-phase spans through the sample, and it becomes a proton conductor.

The basic conclusions of the Gierke model were supported by the results of the DuPont [34], Kyoto [35] and Grenoble [36] groups and by many further studies (see e.g. [37]). Claims of cylindrical micelles [38] or flat lamellar structures [39–41] have also been made, but the occurrence of such structures is more typical for E-membranes. For N-membranes the concept of spherical or quasi-spherical micelles continues to be the most common conjecture about the phase segregation in the membrane, particularly in view of reports on the absence of elongated objects in the patterns of reconstruction of SAXS data [42].

However, SAXS and SANS techniques had difficulties in determining more than just the position of the Bragg or *ionomer* peak characterising the decaying periodical motif in the arrangement of the micelles. From this, we learn not about the size of the micelles but about the *distance between the centres* of the micelles. To determine the size of the micelles, one should perform the Guinier analysis, but this part of the small angle spectrum lies so close to the ionomer peak that the accuracy of such a determination is rather low. The Porod part of the spectrum which helps to evaluate the overall surface of the micelles can aid the determination of the average micelle diameter. Additional information about the size-distribution of micelles comes from capillary pressure isotherms [43] and infrared spectroscopy [44], whereas electron microscopy [45] discriminates the shape of the micelles. As to the necks, it was never clear how to detect them directly.

Beginning with a seminal paper by Eisenberg [46], a number of theoretical works focused on patterns of micro-phase segregation in ionomers (for review see [31, 47, 48]). Based on phenomenological models, this research has solidified the concepts of cluster formation and growth with water uptake. Yarusso and Cooper [49] focused on the arrangement between the polymer backbone and the aqueous aggregates, assuming that each ionic aggregate is coated with a layer of polymer backbone, and developed the so-called interparticle model that helped to rationalise SAXS data. Orfino and Holdcroft [50] utilised this model for an estimate of the length of the channels joining aqueous clusters in Nafion. However, in view of the lack of the information on the detailed structure of the hydrophobic polymer backbone, further aggravated by the frequently observed batch dependence of results, it became commonplace to consider the details of these patterns too confusing to be considered in a theoretical model.

At this point, atomistic simulations of water/Nafion mixtures can be a valuable tool to provide a molecular picture of the "environment" for proton transfer and to analyse the water dependence of this environment. Various groups have studied such systems. Vishnyakov and Neimark [51] studied alkali ion transport in aqueous and methanolic solutions, as well as in mixed solvent, in the presence of Nafion. They saw indications for the existence of a fluctuative bridging mechanism that yields a percolative, and thus proton conducting, aqueous phase "on average", in spite of the fact that at any given time not all aqueous pores are connected. Goddard and coworkers [52] investigated the influence of blocking on the nanostructure of Nafion, by comparing atomistic simulations of two different copolymerisation patterns, in order to estimate the effect of statistical vs. regular copolymerisation of the tetrafluorethylene and the sulfonated vinyl ether building blocks of Nafion. They observed more pronounced segregation in the case of the blocked polymer accompanied by dynamical changes visible in reduced water mobility for the dispersed case. The group of Khokhlov and Khalatur has also performed extensive studies of Nafion/water systems, which will be discussed in greater detail by the authors in their chapter in this book.

Recently, we studied the water dependence in mixtures of water and the protonated form of Nafion [53] using both standard force field models and an empirical valence bond model to account for the Grotthuss structural diffusion mechanism of aqueous proton transport. Results showed a transition of an irregularly shaped filamentous (cylindrical) structure in the case of low water ($\lambda = 5$, where λ is defined as the ratio of water molecules to sulfonate groups) to a structure more in accord with the above-discussed models of nano-separation, where larger clusters form which are connected by narrow bridges. A comparison of aqueous cluster sizes indicated, for a simulation time of 30 ns, no percolating clusters for $\lambda = 5$, whereas at $\lambda = 10$ most water molecules were located in a connected cluster (see [53]). Other structural

10. Proton Transport in Polymer Electrolyte Membranes 355

features such as pair correlation functions also clearly demonstrated the increased clustering of the aqueous phase for $\lambda = 10$. At the same time, proton mobility was substantially increased in the system with higher water content. Recently, Voth and coworkers [54] reached similar conclusions on the basis of a more sophisticated EVB model of a single, non-classical Grotthuss proton.

Trans-gauche motions of the Nafion side-chain were found to be extremely rare events (occurring every few nanoseconds) so that the aqueous pores can be considered quite static on the time scale of proton transfer. Nevertheless the simulations show indications of fluctuative bridging. Figure 10A.2 illustrates this by snapshots of a jelly bean representation of the aqueous (white/blue/yellow) and the polymer phases (dark red) where temporary bridges between disconnected aqueous regions (see arrow) can be clearly discerned (at times $t = 1014$ ps and $t = 1092$ ps).

Such Nafion/water simulations suffer from the fact that the time scale for polymer motions is very much longer than the typical time scales for water and proton motion. Thus, the simulation outcome is to some extent determined by the authors' choice of initial configuration and, furthermore, depends on the

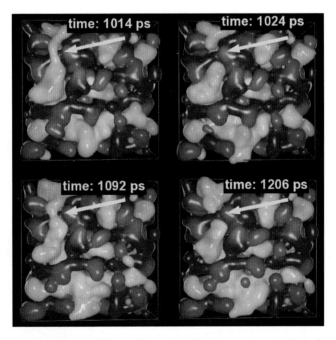

FIGURE 10.2. Snapshots at different times of a jelly-bean representation of a simulated mixture of water and Nafion in protonated form (at water content $\lambda = 5$). Dark Red regions denote polymer, grey, blue and yellow regions denote water, hydronium and sulfonate groups, respectively, in the aqueous phase. The indicated box corresponds to a length of 4.5 nm. Note that the jelly-bean surfaces hide a large number of molecules [72].

employed force field. The simulations to date thus serve more as "inspiring" guidelines than as tools capable of validating or ruling out existing models. In order to be able to study the self-organisation of aqueous pores into inverted micelles or cylinders one has to proceed to mesoscopically coarse-grained models. Recently one of us performed MD simulations using a coarse-grained model in which CF2-CF2 units, water molecules, and hydronium ions were treated as single sites [55]. With this reduction of the number of particles it becomes possible to study larger systems for a longer time. Figure 10A.3 shows some representative snapshots for three different levels of water content. Going from dry membranes ($\lambda = 3$) to intermediately humidified samples ($\lambda = 8$), the morphology changes from individual convex (spherical or cylindrical) units resembling the inverted micelles postulated by Gierke ($\lambda = 3$) to connected channel-like structures of cylindrical shape with varying diameter, which, in addition can be branched ($\lambda = 8$). At intermediate water content the convex shapes grow together and form narrow necks. The typical size of the features in Figure 10A.3 is between 2 and 4 nm, in agreement with the available experimental evidence.

FIGURE 10.3. Snapshots of simulations with a coarse-grained model at different water content l as indicated. Aqueous regions in sphere representation (blue: protons, orange: sulfonate groups), polymer in line representation. [72].

10.3. Proton Transfer

There are, however, additional sources of information associated with the membrane *performance as a proton conductor* that could help discriminating different "fine-structure" features. On a qualitative level, first steps in this direction have been made in an old conceptual review by Pourcelly and Gavach [56].

Proton conductivity has been measured as a function of water content for various membranes (for reviews see [1, 56, 57]). For Nafion the temperature dependence of conductivity has been studied, giving the free energy of activation of proton conductance. The activation energy varies dramatically between the value of 0.1 eV for a water-saturated sample and 0.35 eV for a dry sample [58, 59]. Other groups have observed less variation. The "residual" conductivity at low water content is at least three orders of magnitude smaller than the conductivity of the saturated sample. These data have been rationalised with a semi-phenomenological model of membrane swelling and effective medium percolation theory [60]. Whereas the conductivity gives information both on the global characteristics of the aqueous network inside the polymer and on the inherent proton mobility, the Arrhenius plots for different water content [59] shed light, almost exclusively, on the elementary act of proton transfer. Simultaneous analysis of both helps to narrow down the choice of the structural model of the ionomer.

In view of the time scale separation between polymer motion and elementary proton transfer steps, a theoretical analysis of proton transport best starts from the approximation that the polymer environment provides a rigid framework in which proton and water motion occurs in a static locally inhomogeneous environment. Figure 10A.4 shows the model geometries employed in our systematic studies of proton mobilities as a function of water content, temperature, sulfonate acidity and polymer equivalent weight. Proton self diffusion coefficients were calculated along the pore surface. Both Poisson-Boltzmann theory [61, 62] and molecular dynamics simulations [63, 64], including the basic modes of proton states and proton transfer in water via a simplified empirical valence bond force field [5, 65], have been applied to study proton transport in the environment of such "representative channels". The focus of these simulations was on the key parameters of the interaction of proton complexes and water with the ionised charged groups localised near the surface of a hydrophobic skeleton.

The studies revealed a number of factors determining the proton mobility, such as the width of the channel, distance between the side-chains and their flexibility. The main lessons were

- The denser the side-chains, the lower the activation energy of proton mobility.
- The narrower the channels, the lower the mobility, and, in the case of the simple Poisson-Boltzmann model, the higher the activation energy. The more realistic MD simulations showed only a minor increase in activation

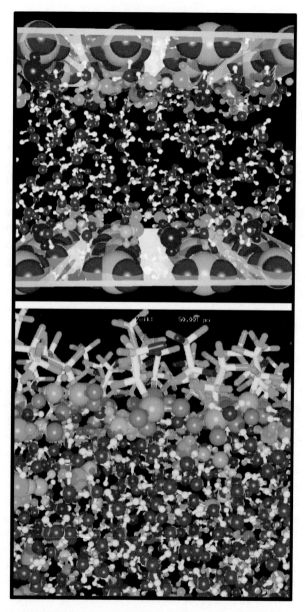

FIGURE 10.4. Model systems to study proton transport in slab pores. Top: rigid slab with embedded SO_3^- head groups. Bottom: slab boundary in a Nafion side-chain model. Proton complexes are coloured according to internal structure: blue: H_3O^+-like (Eigen complexes), green: $H_5O_2^+$-like (Zundel complexes).

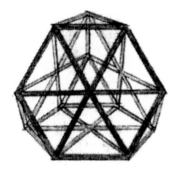 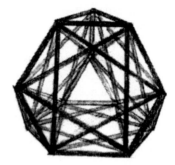

FIGURE 10.5. Two possible types of cages where the strings represent the backbones or their bundles that can provide a short-range-ordered system of fourfold, coordinated inverted micelles composed of hydrated side-chains (not shown) pointing from the strings towards the interior of the cages (taken from Ref. [32]). These micelles keep water droplets (with protons dissociated from sulfonic acid groups) encapsulated inside the cage. The channels, when they form, bridge water droplets through the windows in the cages.

energy for the single pore, supporting the notion that polymer motion may be rate determining at low water content.
- The more mobile are the SO_3^- groups about the side-chain anchoring points, the higher the mobility, but the weaker are the previous two effects.
- Longer, flexible side-chains facilitating proton conductance, impede methanol diffusion [66]. This could be a very practical lesson.

The conclusions reflect certain physics. Whereas each side-chain group donates a proton to the aqueous phase, it also contributes to the Coulomb energy landscape which protons experience due to the electrostatic potential of the array of SO_3^- groups, screened by all other protons. The farther a proton moves from the surfaces of SO_3^- groups, the stronger the screening, and the smaller the role of nonzero harmonics in a Fourier expansion of the field, which decay exponentially from the plane where the charge groups are localised. This correlation between flat density distribution and overall large proton mobility in individual pores is clearly visible in the simulation results [67, 68]. In a wide channel there are such remote regions. Attracted to the charges on the side-chains, there will be less protons there, but the mobility will be higher there. The overall channel conductance, a trade-off between mobility and concentration, will thus be higher in a wider channel. In narrow channels there will be no such remote regions.

10.4. Morphological Model of Inverted Micelles in Nafion

If the charged groups are delocalised, the Coulomb barriers will be smeared, proton mobility will be accelerated and the mentioned mechanism of proton activation energy dependence on the channel thickness will be weakened, unless these channels are much thinner than we thought. And they possibly are. Recently, a more detailed morphological model of Nafion-type ionomers was developed [32]. This model focused on the question, "how can one build an array of inverted micelles if the side-chains are attached to the backbones possibly arranged in bundles, when their persistence length is considerably larger than the size of the micelles"? A quasi-crystalline model of this arrangement has been suggested, the units cell of which looks like

This model can rationalise the observed correlation between the macroscopic swelling of the membrane and the distance between micelles [69]: it explains why in Nafion a different law is obeyed than initially expected [70]. As windows in the cages are small, the channels that can be built through these windows will also be very narrow. The whole process of membrane swelling could then be seen as follows: with initial water uptake the cages expand by strings sliding along each other. The size of the droplets grows, but each droplet is still encapsulated within each cage, disconnected from other droplets. With further water uptake, the droplet ejects water into the windows, building small cathenoids adjoining the neighbouring droplets. The analysis in [32] shows that this will take place as a first-order transition. At the transition the system may slightly shrink as some water from the droplets will be taken to build channels. With further water uptake, the system will swell again, and both the droplets and cathenoids will continuously increase in size. A theory of this phenomenon is in progress [71].

10.5. Final Remarks

How can we use our knowledge practically? This is not clear yet, but it seems that we at least understand better the nature of the proton mobility dependence on water content, temperature, equivalent weight and a few structural details of the polymer. If channels evolve initially as extremely narrow units (less than 0.5 nm radius in the narrowest part of the cathenoid) and remain narrow even in the "mature state", it is clear why the activation energy of the proton mobility, entirely controlled by the necks will depend dramatically on the water content, as is experimentally observed. The more flexible are the side-chains, the higher the proton mobility, since fluctuations of the chains will support the necks, reducing their surface tension, as well as a possible side-chain-fluctuation-promoted proton transport. Much more work still has to be done before atomistic theory and simulation on multiple time and length scales can give detailed quantitative predictions instead of qualitative guidelines for materials development of ionomer membranes suited for fuel cell applications.

References

[1] K. D. Kreuer, S. J. Paddison, E. Spohr and M. Schuster, *Chem. Rev.*, **104**, 4637 (2004).
[2] M. Ise, K.-D. Kreuer, J. Maier, *Solid State Ionics*, **125**, 213 (1999).
[3] N. Agmon, *Chem. Phys. Lett.* **244**, 256 (1995).
[4] D. Marx, M. E. Tuckerman, J. Hutter and M. Parrinello, *Nature*, **397**, 601 (1999).
[5] A. A. Kornyshev, A. M. Kuznetsov, E. Spohr and J. Ulstrup, *J. Phys, Chem. B*, **107**, 35551 (2003).
[6] T. E. Springer, T .A. Zawodsinski and S. Gottesfeld, *J. Electrochem. Soc.*, **138**, 2334 (1991).
[7] N. Nguen and R. E. White, *J. Electrochem. Soc.*, **140**, 2178 (1993).
[8] A. C. West and T. F. Fuller, *J. Appl. Electrochem.*, **26**, 557 (1996).
[9] M. Eikerling, A. A. Kornyshev and A. A. Kulikovsky, Physical modelling of polymer electrolyte fuel cell components, cells and stacks, in *Encyclopaedia of Electrochemistry*, **5** (D. Macdonald, Ed.), 2004 (in Press).
[10] R. Mosdale, G. Gebel and M. Pineri, *J. Membrane Sci.*, **118**, 269 (1996).
[11] M. Eikerling, Yu. I. Kharkats, A. A. Kornyshev and Yu. M. Volfkovich, *J. Electrochem. Soc.*, **145**, 2684 (1998).
[12] A. A. Kulikovsky, *J. Electrochem. Soc.*, **152**, A1121 (2005).
[13] K. D. Kreuer, *J. Membrane Sci.*, **185**, 29(2001).
[14] V. Arcella, A. Ghielmi and G. Tommasi, in *Advanced Membrane Technology*, **984**, New York Acad. Sci, New York, 2003, p. 226.
[15] F. Q. Liu, B. L. Yi, D. M. Xing, J. R. Yu and H. M. Zhang, *J. Membrane Sci.*, **212**, 213 (2003).
[16] M. E. Schuster, W. H. Meyer, *Ann. Rev. Mater. Res.*, **33**, 233 (2003).
[17] J. Roziere and D. J. Jones, *Ann. Rev. Mater. Res.*, **33**, 503 (2003).
[18] L. Jorissen, V. Gogel, J. Kerres and J. Garche, *J. Power Sources*, **105**, 267 (2002).
[19] G. Alberti, M. Casciola, L. Massinelli and B. Bauer, *J. Membrane Sci.*, **185**, 73 (2001).
[20] G. Inzelt, M. Pineri, J. W. Schultze and M. A. Vorotyntsev, *Electrochim. Acta.*, **45**, 2403 (2000).
[21] T. Schultz, S. Zhou and K. Sundmacher, *Chem. Eng. Technol.*, **24**, 1223 (2001).
[22] P. Staiti, A. S. Arico, V. Baglio, F. Lufrano, E. Passalacqua and V. Antonucci, *Solid State Ionics*, **145**, 101 (2001).
[23] J. S. Wainright, J. T. Wang, D. Weng, R. F. Savinell, and M. Litt, *J. Electrochem. Soc.*, **142**, L121 (1995).
[24] J. E. McGrath, M. Hickner, F. Wang, and Y.-S. Kim, in *PCT Int. Appl.* (Virginia Tech Intellectual Properties, Inc., USA), Wo. (2002). p. 46.
[25] C. Stone, A. E. Steck and R. D. Lousenberg, in *PCT Int. Appl.* (Ballard Power Systems Inc., Can.), Wo. (1996) p. 28.
[26] J. Roziere, D. Jones, L. Tchicaya-Boukary and B. Bauer, in *PCT Int. Appl.*, (Fuma-Tech G.m.b.H., Germany). Wo., 2002, p. 53.
[27] J. M. Gascoyne, G. A. Hards and T. R. Ralph, in *PCT Int. Appl.*, (Johnson Matthey Public Limited Company, UK). Wo. (2002), p. 15.
[28] A. S. Davydov, Solitons in molecular systems, 2nd ed, Kluwer, Dordrecht, London 1991.
[29] Proton transfer in hydrogen bonded systems, Ed. T. Bountis, NATO ASI Series, B: Physics 291, Plenum, New York, 1992.

[30] M. A. F. Robertson and H. L. Yeager, Structure and properties of perfluorinated ionomers, in *Ionomers. Synthesis, Structure, Properties and Applications* (M. R. Tant, K. A. Mauritz and G. L. Wilkes), Chapman and Hall, London, 1997, p. 290.
[31] K. A. Mauritz, Morphological theories, in *Ionomers. Synthesis, structure, properties and applications* (M. R. Tant, K. A. Mauritz, G. L. Wilkes), Chapman and Hall, London, 1997, p. 95.
[32] A. S. Ioselevich, A. A. Kornyshev, and J. H. G. Steinke, *J. Phys. Chem. B* **108**, 11953.
[33] T. D. Gierke and W. Y. Hsu, *ACS Symp. Ser.*, **180**, 283 (1982).
[34] T. D. Gierke, G. E. Munn and F. C. Wilson, *J. Polym. Sci. Pt. B-Polym. Phys.*, **19**, 1687 (1981).
[35] M. Fujimura, T. Hashimoto and H. Kawai, *Macromolecules*, **14**, 1309 (1981); **15**, 136–144 (1982).
[36] M. Pineri, R. Duplessix and F. Volino. *ACS Symp. Ser.*, **180**, 249 (1982).
[37] G. Xu, *Polym. J.*, **25**, 397 (1993); **26**, 840 (1994).
[38] S. F. Timashev, *Dokl. Acad. Nauk SSSR*, **283**, 930 (1985).
[39] N. A. Plate, *Brush-Like Polymers and Liquid Crystals*; Khimija: Moscow, 1980.
[40] H. W. Starkweather, Jr. *Macromolecules*, **15**, 320 (1982).
[41] M. H. Litt, *ACS Polymer Polymer Preprints*, **38**, 80 (1997).
[42] J. Halin, F. N. Buchi, O. Haas, M. Stamm, *Electrochim. Acta*, **39**, 1303 (1994); B. Dreyfus, G. Gebel, P. Aldebert, M. Pineri, M. Escoubes, M. Thomas, *J. Phys. (France)*, **51**, 1341 (1990); G. Gebel, J. Lambard, *Macromolecules*, **30**, 7914 (1997); G. Gebel, *Polymer*, **41**, 5829 (2000); J. A. Elliott, S. Hanna, A. M. S. Elliott, G. E. Cooley, *Macromolecules*, **33**, 4161 (2000); P. J. James, J. A. Elliott, T. J. McMaster, J. M. Newton, A. M. S. Elliott, S. Hanna, M. J. Miles, *J. Material. Sci.*, **35**, 5111 (2000).
[43] J. Divisek, M. Eikerling, V. Mazin, H. Schmitz, U. Stimming and Yu. I. Volfkovich, *J. Electrochem. Soc.*, **145**, 2677–2683 (1998).
[44] A. Gruger, A. Regis, T. Schmatko and Ph. Colomban, *Vib. Spectrosc.*, **26**, 215 (2001).
[45] J. A. Elliott, S. Hanna, A. M. S. Elliott and G. E. Cooley, *Macromolecules*, **33**, 4161 (2000).
[46] A. Eisenberg, *Macromolecules*, **3**, 147 (1970)
[47] A. Eisenberg and M. King, *Ion-Containing Polymers*, Academic Press, New York, 1977.
[48] K. A. Mauritz, *J. Macromol. Sci. Rev. Macromol. Chem. Phys.* **C28**, 65 (1988).
[49] D. Y. Yasusso and S. L. Cooper, *Macromolecules*, **16**, 1871 (1983).
[50] F. P. Orfino and S. Holdcroft, *J. New. Mat. Electrochem. Syst.*, **3**, 285 (2000).
[51] A. Vishnyakov and A. V. Neimark, *J. Phys. Chem. B*, **105**, 9586, (2001).
[52] S. S. Jang, Sh.-T. Lin, T. Cagin, V. Molinero and W. A. Goddard III, *J. Phys. Chem. B*, **109**, 10154 (2005).
[53] D. Seeliger, C. Hartnig and E. Spohr, *Eletrochim. Acta* **50**, 4234 (2005).
[54] M. K. Petersen, F. Wang, N. P. Blake, H. Metiu and G. A. Voth, *J. Phys. Chem. B* **109**, 3727 (2005).
[55] E. Spohr, *J. Mol. Liquids* **136**, 288 (2007).
[56] G. Pourcelly and C. Gavach, Perfluorinated membranes, in *Proton Conductors* (Ph. Colomban, Ed.), Cambridge University Press, Cambridge, 1992, pp. 294–31.

[57] S. Gottesfeld, and T. A. Zawodsinski, Polymer electrolyte fuel cells, In *Advances in Electrochemical Science an Engineering,* **5** (C. Alkire, H. Gerischer, D. M. Kolb, C. W. Tobias, Eds.), Wiley-VCH, Weinheim, Germany, 1997, p. 195–301.
[58] M. Cappadonia, J. W. Erning and U. Stimming, *J. Electroanal. Chem.,* **287**, 163 (1990).
[59] M. Cappadonia, J. W. Erning, S. M. S. Niaki and U. Stimming, *Solid State Ionics.* , **287**, 163 (1990).
[60] M. Eikerling, A. A. Kornyshev and U. Stimming, *J. Phys. Chem.,* **B 101**, 10807 (1997).
[61] M. Eikerling and A. A. Kornyshev, *J. Electroanal. Chem.,* **502**, 1-14 (2001).
[62] M. Eikerling, A. A. Kornyshev, A. M. Kuznetsov, J. Ulstrup and S. Walbran, *J. Phys. Chem,* **B105**, 3646 (2001).
[63] E. Spohr, P. Commer, and A. A. Kornyshev, *J. Phys. Chem.* **B106**, 10560 (2002).
[64] P. Commer, A. G. Cherstvy, E. Spohr and A. A. Kornyshev, *Fuel Cells*, **2**, 127 (2003).
[65] S. Walbran and A. A. Kornyshev, *J. Chem. Phys.* **114**, 1039 (2001).
[66] P. Commer, PhD thesis, Heinrich Heine Universität, Düsseldorf, 2003.
[67] P. Commer, C. Hartnig, D. Seeliger and E. Spohr, *Mol. Simulation* **30**, 755 (2004).
[68] E. Spohr, *Mol. Simulation* **30**, 107 (2004).
[69] M. Fujimura, T. Hashimoto, H. Kawai, *Macromolecules*, **14**, 1309 (1981); **15**, 136–144 (1982).
[70] C. L. Marx, D. C. Caulfield, S. L. Cooper, *Macromolecules*, **6**, 344 (1973).
[71] A. S. Ioselevich, A. A. Kornyshev, and N. Marchal, *unpublished results.*

11
Modeling the State of the Water in Polymer Electrolyte Membranes

REGINALD PAUL

11.1. Introduction

The practical details for the construction and operation of polymer electrolyte membrane (PEM) fuel cells are now reasonably familiar topics which do not require further elaboration in this chapter. It should be evident; however, that one of the features of primary importance in the operation of such a cell is the efficient conduction of the protons through the membrane.

Extensive research has gone into understanding the conduction of protons in a number of different media [1]. A typical hydrated "state-of-the-art" PEM is a system with two phases, one phase being the polymer backbone separated from a second phase composed of a random network of hydrated ion-conducting channels with strong internal electrical fields. Both, as a result of being present in a confined state and in an electrical field the water in the channels display properties that are strikingly different from those of bulk water. It is impossible for us to do justice to the abundance of literature available on the effects of confinement and electrical fields on water; however, in the following paragraphs we present a few examples that are particularly relevant to the subject of this chapter.

Gompper, Hauser and Kornyshev [2] considered water that is restricted to the space between two hydrophobic surfaces. In the vicinity of the surfaces a drop in the coordination number around each water molecules (compared to bulk water) occurs resulting in layers with low coordination numbers. When the two walls are allowed to approach each other the layers overlap filling the space with entirely low-coordinated water.

Bontha and Pintauro [3] developed a molecular level model of the partition coefficient in order to study the ability of membranes such as Nafion to absorb ionic species. Such a model requires knowledge of the difference between the electrochemical potentials of the relevant ions inside the pore and in the bulk medium, the former being strongly influenced by the electrical fields arising from the SO_3^- groups which are tethered to the pore walls. The earlier work of Guzman-Garcia et al., [4] Verbrugge and Pintauro [5] and Gur et al. [6] had shown that these fields cause the water dipole to undergo an alignment process; affecting thereby the value of the water permittivity. The authors employ the well-known Booth's equation [7] for this purpose. We will return to a more detailed consideration of Booth's work later.

Senapati and Chandra [8] used molecular dynamics to study the effects of enclosing a dipolar solvent in spherical cavities of various sizes and compared the results with the same solvent in bulk form. The molecules interacted with each other through (a) spherically symmetric 6–12 Lennard-Jones potentials and (b) long-range anisotropic potentials. In addition, the molecules were allowed to interact with the confining wall through another 3–9 Lennard-Jones potential. The density profiles were found to be highly non-uniform with pronounced orientational order close to the wall causing a lowering of the permittivity compared to bulk water (*Dielectric Saturation*).

The literature cited above is sufficient to highlight the fact that given the presence of strong electrical fields in the PEM nanopores, dielectric saturation must play a significant role in determining the properties of the contained water. The objective of the present chapter is to examine this subject in a careful manner.

11.2. Theoretical Foundations

The theoretical principles that are needed in the calculation will be presented in some detail in order to provide a coherent account of the subject [9,10,11]. For the present purpose, it is sufficient to consider a highly simplified model of a fuel cell in which the anode and the cathode are visualized as two planar and parallel surfaces with the intervening space filled by the polymeric membrane. The network of water and ion-containing domains of the membrane are visualized as a sequence of connected cylinders (nanopores) possessing arbitrary length, radius and orientation. To the walls of the cylinders are attached chains, hereafter referred to as *pendant chains or groups* protruding to a distance p into the pore interior and terminating in a $-SO_3^-$ (sulfonate) anion. In order to mathematically analyze such a complex system we consider a projection of each cylinder onto the surface normal of the electrodes and from this ensemble of projected cylinders we compute an *average* length L and radius R of an "average" cylinder (see Figure 11.1). The sulfonate ions alluded to above arise from sulfonic acid groups which are very strong acids and even a small amount of absorbed water in the membrane results in substantial ionization producing sulfonate groups which are "fixed" or tethered to the polymer backbone and the much more mobile hydrated protons (i.e. H_3O^+, $H_5O_2^+$, etc.). Owing to entropy maximization, the protons are present in a relatively more diffuse state than the $-SO_3^-$ ions. As a consequence of this spatial distribution of charges the electrical field is not homogeneous and will tend to be much stronger in the vicinity of the fixed anionic sites.

11. Modeling the State of the Water in Polymer Electrolyte Membranes

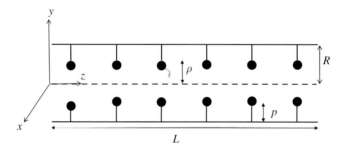

FIGURE 11.1. Cross-sectional view of the pore showing coordinate frame. ρ = distance from axis. p = protrusion length of pendant group.

The particulars of the statistical mechanical machinery are most conveniently presented in the following steps:

(1) We employ the symbols $\{x\}$ and $\{p\}$ to denote the set of coordinates and conjugate momenta for the complete specification of the phase space of all the N water molecules contained in the cylinder. In this notation, the element $x_i \in \{x\}$ refers to the variables that are required to specify the position and orientation of the ith water molecule in the pore and is thus an abbreviation of the form:

$$x_i \equiv (\mathbf{r}_i, \boldsymbol{\omega}_i).$$

Here, $\mathbf{r}_i$ is the vector position of the center of mass of the ith molecule and $\boldsymbol{\omega}_i$ is a vector, which in general, depends upon the three Euler angles required to specify the orientation of a non-linear molecule. In the present analysis, we assume that a water molecule can be treated as a "linear" molecule with a dipole moment μ and an excluded volume ϑ. Consequently only two polar angles (θ_i, γ_i) are needed in order to specify the dipole orientation and $\boldsymbol{\omega}_i$ may be expressed in terms of the unit Cartesian vectors $\mathbf{i}, \mathbf{j}$ and $\mathbf{k}$ as (see Figure 11.2):

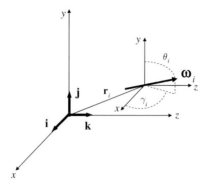

FIGURE 11.2. Diagram showing the coordinates used for the water dipole.

$$\boldsymbol{\omega}_i \equiv \mathbf{i}\sin\theta_i\cos\gamma_i + \mathbf{j}\sin\theta_i\sin\gamma_i + \mathbf{k}\cos\theta_i.$$

The elements of the set $\{p\}$ are the canonical conjugate momenta. Thus, the element $p_i \in \{\mathbf{p}_i, \mathbf{L}_i\}$ is composed of the linear momentum $\mathbf{p}_i$ of the center of mass of the ith water molecule and the angular momenta $\mathbf{L}_i$ that correspond to each angular rotation.

(2) The Hamiltonian H is now written as a sum of the energy, U_F of the internal electrical field the kinetic energy of the N water molecules, T_W and the total potential energy U_N:

$$H = U_F + T_W + U_N.$$

These three terms are further described as follows:

(a) In the present treatment we consider two electrical fields: A *weak* external field to be designated by the symbol $\mathbf{E}_e$, whose origin lies in sources that reside outside the nanopore and an *internal* field $\mathbf{E}$, which originates in the cations (H_3O^+) and anions $(-SO_3^-)$ within the nanopore, as discussed above. It is evident that if a water molecule with center of mass located at $\mathbf{r}_i$ is removed from the system then a cavity of volume ϑ will be created. The cavity will contain an amount of energy given by: $(\varepsilon_0/2)\vartheta E^2(\mathbf{r}_i)$ with ε_0 being the vacuum permittivity. This result follows from electrodynamic equations (Landau, Lifshitz and Pitaevskii [12] and Paul and Paddison [9]) according to which the energy density in vacuum is given by $(\varepsilon_0/2)E^2(\mathbf{r})$. Therefore the contribution from all the cavities $(\varepsilon_0/2)\vartheta \sum_{i=1}^{N} E^2(\mathbf{r}_i)$ must be included in the Hamiltonian. Ideally, an energy contribution from the external field should also be added; however, we assume that this is a weak field making a negligible contribution.

In order to make progress, it is necessary to adopt an explicit form for the internal field. It should be evident that with the array of charges shown in Figure 11.1 the potential energy at any point in the pore will be given by a sum of Coulombic terms. However, it is well known that such a sum converges very slowly and is fraught with computational difficulties. Recently, Grønbech-Jensen [13] et al. computed self-energies of particles in partially periodic lattices by employing Lekner summations of Coulombic potentials in three-dimensional media and were able to derive a highly convergent sum. The first term of the sum is an excellent approximation to the entire summation, and in our work we adopt this approximation as the potential due to the array of SO_3^- ions. The field from such a potential has the following explicit form:

$$\mathbf{E}(\mathbf{r}) = \frac{2ef_n n^2}{\varepsilon_0 L^2} \times \left[\mathbf{e}_\rho K_1 \left\{ \frac{2\pi n(R-p-\rho)}{L} \right\} \cos\left(\frac{2\pi nz}{L}\right) \right. \\ \left. -\mathbf{k} K_0 \left\{ \frac{2\pi n(R-p-\rho)}{L} \right\} \sin\left(\frac{2\pi nz}{L}\right) \right]. \quad (11.1)$$

11. Modeling the State of the Water in Polymer Electrolyte Membranes 369

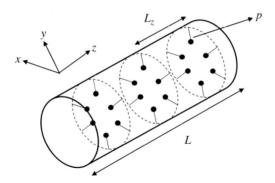

FIGURE 11.3. Axial view of the pore.

In writing this expression, we assume that the anionic groups are arranged in the form of an array of n negatively charged rings or circular arrays separated by a distance L_z and attached to the pore wall; each ring containing f_n anionic sites. The above formula has been presented in a cylindrical coordinate frame (ρ, φ, z) in which the pore axis is placed along the z axis; $\mathbf{e}_\rho$ is the unit radial vector with radial coordinate ρ. Because of the axial symmetry of the rings there is no dependence of the field on the angular variable φ (see Figure 11.3). The functions K_0 and K_1 are, respectively, the zero and first-order modified Bessel functions.

(b) Within our approximation of treating the water molecules as rigid linear molecules, the kinetic energy reduces to:

$$T_W = \sum_{i=1}^{N} \frac{p_i^2}{2m} + \sum_{i=1}^{N} \frac{1}{2I}\left(L_{\theta i}^2 + \frac{L_{\gamma i}^2}{\sin^2 \theta_i}\right). \tag{11.2}$$

Here, m is the mass of each water molecule, I is the moment of inertia, $L_{\theta i}$ and $L_{\gamma i}$ are the angular momenta of the ith molecule for rotation about the angles θ_i and γ_i, respectively.

(c) The function $U_N(\{x\}, \mathbf{E}, \mathbf{E}_e)$ that appears in H is the full N-body potential energy of the fluid within the nanopore for which an appropriate form must be chosen. We assume that, at least, three potential energy terms make significant contributions: (i) the sum of the two-body pairwise interactions between the water molecules ($U_{WW} = \sum_{i<j}^{N} \psi(x_i, x_j)$); (ii) the interaction between the internal field and all the water molecules ($U_{FW}^{(\text{int})} = -\mu \sum_{i=1}^{N} \boldsymbol{\omega}_i \cdot \mathbf{E}(\mathbf{r}_i)$); and (iii) the analogous term due to the external field ($U_{FW}^{(\text{ext})} = -\mu \sum_{i=1}^{N} \boldsymbol{\omega}_i \cdot \mathbf{E}_e(\mathbf{r}_i)$). Thus, the total potential energy will be given by:

$$U_N(\{x\}, \mathbf{E}, \mathbf{E}_e) = \sum_{i<j}^{N} \psi(x_i, x_j) - \mu \sum_{i=1}^{N} \boldsymbol{\omega}_i \cdot \mathbf{E}(\mathbf{r}_i) - \mu \sum_{i=1}^{N} \boldsymbol{\omega}_i \cdot \mathbf{E}_e(\mathbf{r}_i). \qquad (11.3)$$

The form of the function $\psi(x_i, x_j)$ will be discussed later.

(3) The total Hamiltonian may now be written as a sum of all the energy terms computed above:

$$H(\{p\}, \{x\}, \mathbf{E}, \mathbf{E}_e) = \frac{\varepsilon_0}{2} \vartheta \sum_{i=1}^{N} E^2(\mathbf{r}_i) + \sum_{i=1}^{N} \frac{p_i^2}{2m} + \sum_{i=1}^{N} \frac{1}{2I}\left(L_{\theta i}^2 + \frac{L_{\phi i}^2}{\sin^2 \theta_i}\right)$$
$$+ \sum_{i<j}^{N} \psi(x_i, x_j) - \mu \sum_{i=1}^{N} \boldsymbol{\omega}_i \cdot \mathbf{E}(\mathbf{r}_i) - \mu \sum_{i=1}^{N} \boldsymbol{\omega}_i \cdot \mathbf{E}_e(\mathbf{r}_i). \qquad (11.4)$$

(4) From the Hamiltonian the canonical partition function $Q(N, V, T, \mathbf{E}, \mathbf{E}_e)$ can be calculated by using the standard techniques of statistical mechanics:

$$Q(N, V, T, \mathbf{E}, \mathbf{E}_e) = \int \frac{d\{x\}d\{p\}}{h^{5N}} e^{-\beta H(\{p\}, \{x\}, \mathbf{E}, \mathbf{E}_e)}.$$

In writing this equation the following have been used:
(a) h is Planck's constant.
(b) $\beta = 1/kT$, where, k is the Boltzmann Constant and T is the absolute temperature.
(c) The infinitesimal elements of integration have the following interpretation:

$$d\{x\} \equiv \prod_{i=1}^{N} d\mathbf{r}_i d\boldsymbol{\omega}_i \equiv \prod_{i=1}^{N} d\mathbf{r}_i d\theta_i d\gamma_i; \quad d\{p\} \equiv \prod_{i=1}^{N} d\mathbf{p}_i d\mathbf{L}_i \equiv \prod_{i=1}^{N} d\mathbf{p}_i dL_{\theta i} dL_{\gamma i}.$$

The integrations over the center of mass variables $\mathbf{r}_i$ are limited to the volume of the nanopore; the limits of the polar angles are as usual: $0 \leq \theta_i \leq \pi$ and $0 \leq \gamma_i \leq 2\pi$ and the limits of the momentum variable components (both the linear and the angular momenta) are over all possible values, thus, for example $-\infty \leq p_{xi} \leq \infty$ and $-\infty \leq L_{\theta i} \leq \infty$. Substituting the explicit form of the Hamiltonian from Eq. (11.4), the integrations over all the momentum variables yield:

$$Q(N, V, T, \mathbf{E}, \mathbf{E}_e) = \frac{(2\pi IkT)^N (2\pi mkT)^{3N/2}}{h^{5N}} q_N(\mathbf{E}, \mathbf{E}_e), \qquad (11.5)$$

where we have:

$$q_N(\mathbf{E}, \mathbf{E}_e) \equiv \int \prod_{i=1}^{N} d\mathbf{r}_i d\omega_i \sin\theta_i e^{-\beta\left[\frac{\varepsilon_0}{2}\vartheta \sum_{i=1}^{N} E^2(\mathbf{r}_i) + U_N(\{x\}, \mathbf{E}, \mathbf{E}_e)\right]}. \quad (11.6)$$

Henceforth, the factor $\sin\theta_i$ will be incorporated within the definition of $d\omega_i$. The full N–body *spatial* distribution function: $n_N(\{x\}, \mathbf{E}, \mathbf{E}_e)$ is given by:

$$n_N(\{x\}, \mathbf{E}, \mathbf{E}_e) = \frac{1}{q_N} e^{-\beta\left[\frac{\varepsilon_0}{2}\vartheta \sum_{i=1}^{N} E^2(\mathbf{r}_i) + U_N(\{x\}, \mathbf{E}, \mathbf{E}_e)\right]}. \quad (11.7)$$

The reduced m–body generic distribution functions, with $m < N$, are obtained by integration over $N - m$ variables

$$n_m(x_1, x_2, \ldots x_m, \mathbf{E}, \mathbf{E}_e) = \frac{N!}{(N-m)!} \int dx_{m+1} dx_{m+2} \ldots dx_N n_N(\{x\}, \mathbf{E}, \mathbf{E}_e). \quad (11.8)$$

Since integration over the momentum variables has been carried out, the various functions that appear from now on will depend only on the spatial variables $x_1, x_2 \ldots$. It is expedient to introduce a further notational contraction whereby the spatial variable x_i is replaced by its subscript i.

(5) The Helmholtz free energy follows from the partition function:

$$A(\mathbf{E}, \mathbf{E}_e) = -\beta^{-1} \ln Q(N, V, T, \mathbf{E}, \mathbf{E}_e) = A^{\text{Ideal}} + A^{\text{Excess}}(\mathbf{E}, \mathbf{E}_e), \quad (11.9)$$

where:

$$A^{\text{Ideal}} \equiv -\beta^{-1} \ln\left[\frac{(2\pi IkT)^N (2\pi mkT)^{3N/2}}{h^{5N}}\right], \quad (11.10)$$

$$A^{\text{Excess}}(\mathbf{E}, \mathbf{E}_e) \equiv -\beta^{-1} \ln q_N(\mathbf{E}, \mathbf{E}_e).$$

(6) Landau, Lifshitz and Pitaevskii [14] derive from a combination of thermodynamics and the electrodynamics of continuous media an important equation:

$$\mathbf{P}(\mathbf{r}, \mathbf{E}, \mathbf{E}_e) = -\frac{\delta A(\mathbf{E}, \mathbf{E}_e)}{\delta \mathbf{E}_e}. \quad (11.11)$$

(7) Substituting Eq. (11.10) in Eq. (11.11) and using the definition of the reduced distribution function n_2 given by Eq. (11.8) we obtain for the condition $N \gg 1$:

$$\mathbf{P}(\mathbf{r}, \mathbf{E}, \mathbf{E}_e) = \frac{\mu}{N} \int d\omega_1 d\mathbf{r}_2 d\omega_2 \omega_1 n_2(\mathbf{r}, \omega_1, \mathbf{r}_2, \omega_2, \mathbf{E}, \mathbf{E}_e). \quad (11.12)$$

If the external field $\mathbf{E}_e$ is very weak then it is sufficient to consider the polarization, $\mathbf{P}(\mathbf{r}, \mathbf{E})$ after setting $\mathbf{E}_e = 0$ in Eq. (11.12)

(8) We now write $n_2(1, 2, \mathbf{E})$ as a sum of two terms [15], one of which is entirely free of the intermolecular potential $\psi(i, j)$:

$$n_2(1, 2, \mathbf{E}) \simeq n_1^{(0)}(1, \mathbf{E}) n_1^{(0)}(2, \mathbf{E})[h(1, 2, \mathbf{E}) + 1]. \quad (11.13)$$

Here:

$$n_1^{(0)}(i, \mathbf{E}) = \frac{N}{q_1} e^{-\beta \left[\vartheta \frac{\varepsilon_0}{2} E^2(\mathbf{r}_i) - \mu \boldsymbol{\omega}_i \cdot \mathbf{E}(\mathbf{r}_i)\right]}, \quad (11.14)$$

$$q_1(\mathbf{E}) = \int d\mathbf{r} d\boldsymbol{\omega} e^{-\beta \left[\vartheta \frac{\varepsilon_0}{2} E^2(\mathbf{r}) - \mu \boldsymbol{\omega} \cdot \mathbf{E}(\mathbf{r})\right]} = \int d\mathbf{r}\, e^{-\beta \vartheta \frac{\varepsilon_0}{2} E^2(\mathbf{r})} \frac{4\pi \sinh[\beta \mu E(\mathbf{r})]}{\beta \mu E(\mathbf{r})}. \quad (11.15)$$

In writing the final form for q_1, the integrations over the angular variables $\boldsymbol{\omega}$ have been carried out. The function $h(1, 2, \mathbf{E})$ is referred to as a two-body *correlation* function and carries the full effects of the two-body potential $\psi(i, j)$. The polarization is now given by the following expression:

$$\mathbf{P}(\mathbf{r}, \mathbf{E}) = \mu \int d\boldsymbol{\omega}_1 \boldsymbol{\omega}_1 n_1^{(0)}(\mathbf{r}, \boldsymbol{\omega}_1, \mathbf{E})$$
$$+ \frac{\mu}{N} \int d\boldsymbol{\omega}_1 d\mathbf{r}_1 d\boldsymbol{\omega}_2 \boldsymbol{\omega}_1 n_1^{(0)}(\mathbf{r}, \boldsymbol{\omega}_1, \mathbf{E}) n_1^{(0)}(\mathbf{r}_1, \boldsymbol{\omega}_2, \mathbf{E}) h(\mathbf{r}, \boldsymbol{\omega}_1, \mathbf{r}_1, \boldsymbol{\omega}_2, \mathbf{E}) \cdot \quad (11.16)$$
$$\equiv \mathbf{P}_F(\mathbf{r}) + \mathbf{P}_{corr}(\mathbf{r})$$

Since the angular integration in the first term of Eq. (11.16) can be carried out analytically we obtain:

$$\mathbf{P}(\mathbf{r}) = \frac{4\pi N}{\beta q_1} e^{-\beta \vartheta \frac{\varepsilon_0}{2} E^2(\mathbf{r})} \left[\frac{\cosh[\beta \mu E(\mathbf{r})]}{E^2(\mathbf{r})} - \frac{\sinh[\beta \mu E(\mathbf{r})]}{\beta \mu E^3(\mathbf{r})}\right] \mathbf{E}(\mathbf{r}) + \mathbf{P}_{corr}(\mathbf{r}). \quad (11.17)$$

The contribution of $\mathbf{P}_{corr}$ to the polarization due to the molecular correlation contains the effects of the two-body potential $\psi(i, j)$. In the appendix, we show by explicit calculations that in the PEM nanopores the internal field $\mathbf{E}$ is so strong that the interaction energy of the water dipoles with this field is greater than the intermolecular interaction energies. Under these conditions the polarization $\mathbf{P}_F$ will be the dominant term in Eq. (11.16) so that:

$$\mathbf{P}(\mathbf{r}) = \frac{4\pi N}{\beta q_1} e^{-\beta \vartheta \frac{\varepsilon_0}{2} E^2(\mathbf{r})} \left[\frac{\cosh[\beta \mu E(\mathbf{r})]}{E^2(\mathbf{r})} - \frac{\sinh[\beta \mu E(\mathbf{r})]}{\beta \mu E^3(\mathbf{r})}\right] \mathbf{E}(\mathbf{r}). \quad (11.18)$$

(9) From a practical standpoint, in the present context, the polarization is not the most convenient parameter but rather the permittivity to which it is related by simple equations derived by Landau, Lifshitz and Pitaevskii [16]. From electrodynamics these authors derive the following equation:

$$\mathbf{P}_T(\mathbf{E}) = \frac{[\varepsilon(\mathbf{E}) - 1]}{4\pi} \varepsilon_0 \mathbf{E}. \tag{11.19}$$

Here, $\mathbf{P}_T$ is the total polarization which includes both the electronic and orientational contributions. For very high frequency fields only the electrons can respond and Eq. (11.19) for the electronic contribution to the polarization may be written as:

$$\mathbf{P}_\infty = \frac{[\varepsilon_\infty - 1]}{4\pi} \varepsilon_0 \mathbf{E}.$$

$\mathbf{P}_\infty$ and ε_∞ are, respectively, the high frequency (electronic) polarization and permittivity. The latter is related to the refractive index, n by the formula $\varepsilon_\infty = n^2$. The orientational polarization is, therefore, given by the difference $\mathbf{P}_T(\mathbf{E}) - \mathbf{P}_\infty$, from which it follows that:

$$\varepsilon(\mathbf{r}) = n^2 + \frac{4\pi P(\mathbf{r})}{\varepsilon_0 E}. \tag{11.20}$$

Where E and P are the magnitudes of the vectors $\mathbf{E}$ and $\mathbf{P}$, respectively. Substituting Eq. (11.18) we obtain:

$$\varepsilon(\mathbf{r}) = n^2 + \frac{16\pi^2 N}{\beta \varepsilon_0 E^2(\mathbf{r}) q_1} e^{-\beta \vartheta \frac{\varepsilon_0}{2} E^2(\mathbf{r})} \left[\coth\{\beta \mu E(\mathbf{r})\} - \frac{1}{\beta \mu E(\mathbf{r})} \right] \sinh\{\beta \mu E(\mathbf{r})\}. \tag{11.21}$$

It is convenient to write Eq. (11.21) in a more compact form by incorporating the well-known Langevin function:

$$L(x) = \coth(x) - \frac{1}{x}.$$

Inserting this definition of L in Eq. (11.21) we obtain:

$$\varepsilon(\mathbf{r}) = n^2 + \frac{16\pi^2 N}{\beta \varepsilon_0 E^2(\mathbf{r}) q_1} e^{-\beta \vartheta \frac{\varepsilon_0}{2} E^2(\mathbf{r})} L\{\beta \mu E(\mathbf{r})\} \sinh\{\beta \mu E(\mathbf{r})\}. \tag{11.22}$$

This is the final form of the equation for the permittivity in a nanopore with an internal field and its derivation has been the primary objective of this section.

11.3. Applications

Equation (22) provides us with the means for computing the permittivity at various points **r** in the nanopore. Within a primitive model the permittivity is a measure of the ability of the medium to screen a hydrated proton located at **r** from the internal field. It follows, therefore, that regions with large values of $\varepsilon(\rho, z)$ will be the regions with the highest diffusion coefficients of these ions. Concomitantly, with the enhanced diffusion coefficient Kreuer, Paddison, Spohr and Schuster [17] point out that, compared with regions of low $\varepsilon(\rho, z)$, those with higher values display a reduction in the free energy of hydration of the protons. It follows, therefore, that in these regions there will be a higher concentration of hydrated protons. These two effects operate cooperatively to enhance the membrane conductivity.

The purpose of the present analysis is to examine the effects of the pore morphology on the permittivity which in this context is determined by the values of the six parameters L, n, N, R, p and f_n. All of these input parameters can be obtained from previous modeling studies [18, 19, 20, 21] and recent small angle x-ray scattering data; [22, 23] and are collected in the adjoining table for Nafion corresponding to three different levels of hydration expressed in terms of the parameter λ which is the number of water molecules per SO_3^- group.

The permittivity $\varepsilon(\rho, z)$ can be calculated and presented as a three-dimensional plot; however, rather than displaying such a complicated plot, it is more useful to compute an average over z and display the ρ dependence only. As discussed above the regions of high permittivity are of particular significance to the performance of the membrane and since the upper bound of the permittivity must be equal to the value of bulk water it is worth defining a volume fraction, V_F, in the following manner:

$$V_F(\lambda) = \frac{V_B}{V} \times 100. \qquad (11.23)$$

Here, V_B is the volume of water in the pore that possesses the same permittivity as bulk water, and V is the total volume of the pore.

In Figure 11.4 we show three plots of the radial variation of the permittivity and the fraction of bulk-like water corresponding to three different levels of hydration of Nafion listed in the table given above. It is evident from this diagram that the bulk-like water with relatively large permittivity is restricted to the vicinity of the pore axis and from the argument presented above the free energy of hydration of the hydrated proton is lowered in this region leading to a higher concentration of these ions.

In addition to the effects of the hydration level changes and the accompanying morphological consequences on the pore permittivity displayed in Figure 11.4 it is informative to determine, for a fixed level of hydration, the results of the surface charge density and the protrusion length of the pendant chains on the permittivity.

11. Modeling the State of the Water in Polymer Electrolyte Membranes 375

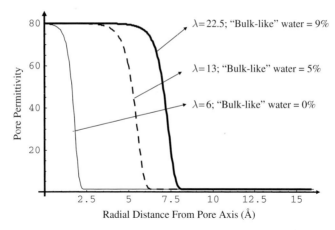

FIGURE 11.4. Plots showing the radial dependence of the permittivity in a typical nanopore of Nafion with three different levels of hydration and the volume fraction of "bulk-like water".

(1) As a means of assessing the effects of different anionic charge densities and distributions on the permittivity of the water, the number of radially symmetric arrays (i.e. rings) of $-SO_3^-$ groups can be varied while keeping the total amount of charge fixed for a hypothetical membrane pore with given water content. We select, for example, a pore with dimensions similar to Nafion with $\lambda = 6$ (see Table). If the number of rings are varied from n = 4, corresponding to a very inhomogeneous distribution of sulfonate groups along the length of the pore to n = 9, where the charge is uniformly (i.e. smoothly) distributed then Figure 11.5 is obtained. These plots clearly demonstrate that by only increasing the uniformity in the sulfonate distribution, the ordering and dielectric saturation of the water can be substantially decreased from pores showing no "free" or "bulk-like water" (i.e. for n = 4–6) to pores with as much as 4% (see Eq. (11.23)).

	L, Å	n	N	R, Å	p, Å	f_n
$\lambda = 6$	30	6	216	8	4	6
$\lambda = 13$	56	8	1001	14	4	9.5
$\lambda = 22.5$	64	8	1800	16	4	10

(2) The effects of changing the length of intrusion of the fixed sites within the membrane pore on the dielectric properties of the water is also a matter of interest and is motivated by the observed differences in the transport properties of PEMs with distinct side chains (i.e. length and chemistry).

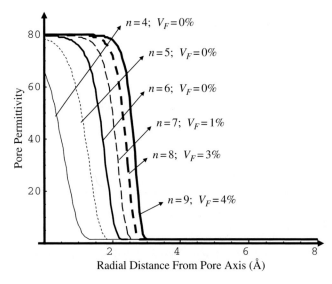

FIGURE 11.5. The effects of increasing the "smoothness: of anionic charge distribution in Nafion with $\lambda = 6$.

The calculations can be conducted in a pore with fixed size, volume, and total number of sulfonate groups found, for example, in Nafion with $\lambda = 6$. All the pore parameters except the intrusion length p are held fixed. The permittivity of the water is computed as a function of the radial distance for two values of the intrusion lengths: p = 1.5A and p = 3.0A. The relevant plots are shown in Figure 11.6. These results show that as the anionic sites penetrate further into the pore interior they are brought into closer proximity to the region where bulk-like water persists leading to a more dramatic display of dielectric saturation. It is also interesting to witness the effects of what might be perceived as only a slight or insignificant change in side chain length (i.e. an intrusion increase of only 1.5A) to the nature of the water.

11.4. Connection with Older Results

Equation (22) is a result that has been specifically derived for PEM nanopores in which the internal field $\mathbf{E}(\mathbf{r})$ is the dominant force acting within the volume. In most of the traditional theories, it is the interaction between the water molecules that constitutes the leading influence, which would imply that in Eq. (11.17) $\mathbf{P}_{corr}$ plays the dominant role. However, the calculation of $\mathbf{P}_{corr}$ requires the full apparatus of many-body theory to be employed which, unfortunately, has not yet been successfully developed, and the authors of the conventional theories resort to some form of a mean field theory. Within a mean field approach Eq. (11.18) once again becomes the central object but it

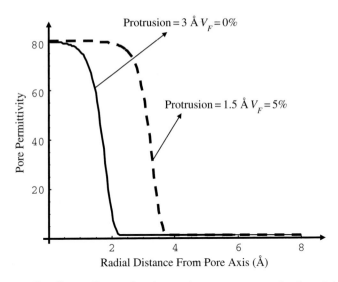

FIGURE 11.6. The effects of increasing the pendant group protrusion length in Nafion with $\lambda = 6$.

appears with new multiplicative constants. The values of these constants depend upon the precise manner in which the mean field calculation is implemented and the many-body effects approximated.

We begin by making contact with the work of Booth [7] to which reference has already been made. Booth makes use of earlier models derived independently by Kirkwood, Onsager and Fröhlich [24,25,26] and the most convenient place to introduce the work is by starting from Eq. (11.18) and proceeding as follows:

(1) Since no internal field is present it follows that in the Hamiltonian $U_F = 0$ and the resulting effect can be realized in Eq. (11.18) by setting $\vartheta = 0$. Furthermore, in employing Eq. (11.18) we are effectively dealing with a single dipole (hereafter referred to as the "dipole of interest") in vacuum which can be duly recognized by replacing μ by μ_v with the subscript v referring to vacuum.

(2) The field $\mathbf{E}(\mathbf{r})$ is replaced by the constant external field $\mathbf{E}_e$ which implies that the integration in Eq. (11.15) can be carried analytically to give:

$$q_1 = V \frac{4\pi \sinh(\beta\mu_v E_e)}{(\beta\mu_v E_e)}. \tag{11.24}$$

(3) Substitution in Eq. (11.18) results in a very simple form for the polarization:

$$\mathbf{P} = \frac{N\mu_v}{V} \mathbf{e} L(\beta\mu_v E_e). \tag{11.25}$$

Here, $\mathbf{e}$ is a unit vector parallel to the field $\mathbf{E}_e$.

(4) From Eq. (11.20) the permittivity is given by:

$$\varepsilon = n^2 + \frac{4\pi N \mu_v}{\varepsilon_0 V E_e} L(\beta \mu_v E_e). \tag{11.26}$$

This is one of earliest equations of permittivity and was derived by Debye [27], for a very special case in which only a single dipole is considered in the field $\mathbf{E}_e$ and the other $N-1$ molecules are ignored.

(5) Onsager [25] started from Eq. (11.25) and introduced two constants A and B to take into consideration the effects of the $N-1$ molecules that were ignored in the Debye model:

$$\mathbf{P} = \frac{N(A\mu_v)}{V} \mathbf{e} L[\beta(A\mu_v)(BE_e)].$$

The constant A is introduced to correct for the fact that the dipole is not in a vacuum but in a medium while the constant B is a correction factor that recognizes the screening effects of the medium on the field $\mathbf{E}_e$. In evaluating these constants, the $N-1$ molecules are considered to be in the form of a continuum that surrounds a cavity in which the dipole of interest is placed. The details of the calculations of these constants are well known in literature, and a very readable account can be found in Scaife [28]. The final forms are as follows:

$$A = \frac{n^2 + 2}{3}; \; B = \frac{3}{2}.$$

If the polarization vector $\mathbf{P}$ is first calculated with these constants and the result substituted in Eq. (11.20) the well-known Onsager's equation for the permittivity is obtained:

$$\varepsilon = n^2 + \frac{4\pi N (n^2 + 2) \mu_v}{3\varepsilon_0 V E_e} L\left[\frac{\beta(n^2 + 2)\mu_v E_e}{2}\right].$$

(6) Kirkwood extended the work of Onsager by recognizing the local structure of water around the dipole of interest. It was assumed that this dipole is surrounded by other water molecules in a tetrahedral arrangement and that outside this immediate neighborhood a continuum can be assumed. For such a local arrangement around the dipole of interest Kirkwood obtains the following modification of Eq. (11.26) which is adopted by Booth in his work on dielectric saturation:

$$\varepsilon = n^2 + 28\left(\frac{N}{V}\right)\frac{\pi\mu}{\sqrt{73}E_e}L\left[\frac{\sqrt{73}}{3}\mu\beta E_e\right]. \qquad (11.27)$$

It is important to realize that Eq. (11.27), which is popularly referred to as the Booth equation, is strictly only valid when the electrical field causing the polarization is a constant because it is only then that Eq. (11.24) is valid. In order to extend the range of applicability of Eq. (11.27) to PEM membranes, Bontha and Pintauro [3] replaced the constant field magnitude, E_e by the spatially dependent internal field magnitude $E(\mathbf{r})$. Such a-posteriori introduction of a functional dependence is in contradiction to the principles embodied in Eq. (11.15) where the spatially dependent field is present within an integrand. If the field is generated by an external system of macroscopic electrodes, then it is possible to envisage a situation where the field varies slowly over space and the Bontha-Pintauro approximation justified. In the present case, however, the field has a local molecular origin, arising as a consequence of the ionization of sulfonic acid groups and it seems very unlikely that slowly varying fields would arise.

If the field $\mathbf{E}$ is allowed to be spatially dependent and Eq. (11.15) used for the evaluation of the permittivity then in the absence of the internal field energy $U_F = 0$ pathologically divergent results are obtained.

In our earlier work, [29,30] we attempted to go beyond the mean field theory of Onsager. Because of the inherent assumption of a constant field it is clear that all the molecules in the tetrahedral unit must experience the same electrical field; which would not be true if the spatial dependence of the field is recognized. We employed a calculation where it was assumed that if the dipole of interest experiences a field $\mathbf{E}(\mathbf{r})$ given by Eq. (11.1) then those in the first coordination layer must experience the fields $\mathbf{E}(\mathbf{r} + \mathbf{t}_i)$ with $i = 1, 2, 3$ and 4. Here, $\mathbf{t}_i$ is the vector from the dipole of interest to each of the tetrahedral apices. Unfortunately we had not yet recognized the significance of the internal field energy in the Hamiltonian, and this lead to divergences in the terms that correct the mean field result.

11.5. Closing Remarks

In this chapter, we have presented a molecular, statistical mechanical model for the computation of the spatially dependent water permittivity in the pores of hydrated polymer electrolyte membranes. We summarize as follows some of the fundamental features of the calculation:

(1) The method relies very heavily on the molecular details of the pore morphology and utilizes the tools of statistical mechanics with an overall emphasis on compact and analytical mathematical forms for the results. The latter emphasis provides a practitioner in the field with a means for rapid

computation of the numerical value of the permittivity without resorting to expensive computer techniques.

(2) As has been pointed out in Section 11.4 of this chapter, many of the traditional methods show unphysical divergences when inhomogeneous electrical fields are present. In contrast, the method presented in this chapter overcomes this problem by correctly including the field energy in the Hamiltonian.

(3) Given the emphasis on closed analytical results it becomes necessary to categorize, at an early stage, the relative importance of various factors and include those that are deemed more important. In a competing calculation using extensive numerical methods more factors may be included; however, the results become very specific to the pore under investigation and difficult to apply to new situations.

(4) The results presented can be easily applied to obtain the effects of the field due to the charged anionic sites on the transport coefficients of the hydrated protons and also provide valuable insight into those domains in the pore where these hydrated species are most likely to occur. In determining the effects of the anionic sites the formula (22) very easily reveals the influence of such morphological features as the pendant chain length and the surface charge density.

(5) The final Eq. (11.22) is, strictly speaking, applicable to those materials in which the hydrated ion containing nanopores are characterized by internal fields whose interaction with the water molecules is stronger than the intermolecular interactions between the water molecules.

Appendix

We assume the interaction energies $\psi(i,j)$ to be composed of:

(A) A hard-core interaction with an exclusion radius σ.
(B) A dipolar interaction v_{dd}.

Two cases are now considered:

(1) Shown in Figure 11.7 in which the anionic sites in the pore are replaced by a "*fixed*" dipole, attached to the pore wall at $z = 0$ with its axis parallel to the pore axis. A "test" dipole is introduced with its axis parallel to the pore axis in order to maximize the interaction energy. For such a configuration the mathematical form of the dipole–dipole potential assumes a relatively simple form:

$$v_{dd}(\rho, z) = \frac{\mu^2\left[(R-\rho)^2 - 2z^2\right]}{4\pi\varepsilon_0\left[(R-\rho)^2 + z^2\right]^{5/2}}.$$

11. Modeling the State of the Water in Polymer Electrolyte Membranes   381

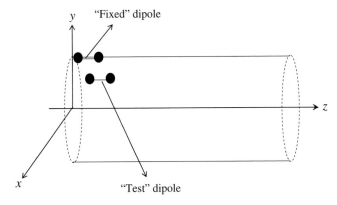

FIGURE 11.7. Anionic sites replaced by a "Fixed" dipole and interacting with a "Test" dipole.

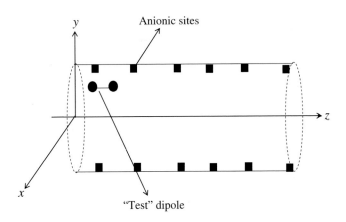

FIGURE 11.8. Anionic sites interacting with a "Test" dipole.

(2) Shown in Figure 11.8 the "test" dipole is introduced with its axis parallel to the pore axis in a pore with the anionic sites intact but the fixed dipole removed. In this case the interaction potential is:

$$\Psi_{pd} = -\mu \cdot \mathbf{E}.$$

Here, **E** is given by Eq. (11.1).

In Figure 11.9 four plots comparing the otentials experienced by the "test" dipole in a Nafion membrane with $\lambda = 6$ are shown. In (a) case (1) is considered with the "test" dipole placed at a distance of 2.6 Å from the pore

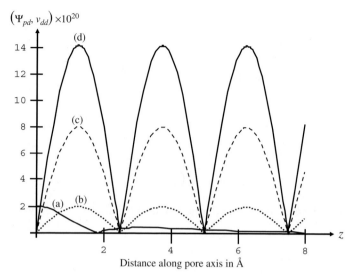

FIGURE 11.9. Diagram comparing the interaction potential between the internal field and a dipole with the potential between two dipoles. See text for details.

wall while (b), (c) and (d) case (2) is considered with the "test" dipole placed at the distances of 4 Å, 3 Å and 2.6 Å respectively from the pore wall. These plots show that at the pore axis the dipole-dipole interactions dominate resulting in bulk-like water; however the assumption $\mathbf{P}_F > \mathbf{P}_{corr}$ is reasonable at points removed from the center.

References

[1] K. D. Kreuer, *Chem. Mater.*, **8**, 610 (1996).
[2] G. Gompper, M. Hauser and A. A. Kornyshev, *J. Chem. Phys.*, **101** (4), 3378 (1994).
[3] J. R. Bontha and P. N. Pintauro, *Chem. Eng. Sci.*, **49** (No. 23), 3835 (1994).
[4] A. Guzman-Garcia, P. N. Pintauro, and M. W. Verbrugge, *AIChE J*, **36**, 1061 (1990).
[5] P. N. Pintauro and M. W. Verbrugge, *J. Memb. Sci.*, **44**, 197 (1989).
[6] Y. Gur, I. Ravina, and A. Babchin, *J. Colloid Interface Sci.*, **64**, 333 (1978).
[7] F. Booth, *J. Chem. Phys.*, **19**, 391 (1950).
[8] S. Senapati and A. Chandra, *J. Chem. Phys.*, **111**, 1223 (1999).
[9] R. Paul and S. J. Paddison, *J. Phys. Chem. B*, **108**, 13231 (2004).
[10] R. Paul and S. J. Paddison, *Solid State Ionics*, **168**, 245 (2004).
[11] R. Paul and S. J. Paddison, "*Development of Thermodynamic Methods for the Study of Nanopores*", in Nanotech 2002: 2nd International Conference on Computational Nanoscience and Nanotechnology, Puerto Rico April 2002.

[12] L. D. Landau, E. M. Lifshitz and L. P. Pitaevskii, *Electrodynamics of Continuous Media*, (Pergamon Press, New York, 1984), Chapter: 2 p. 44.
[13] N. Grønbech-Jensen, G. Hummer, and K. M. Beardmore, *Mol. Phys.*, **92**, 941 (1997).
[14] L. D. Landau, E. M. Lifshitz and L. P. Pitaevskii, *Electrodynamics of Continuous Media*, (Pergamon Press, New York, 1984), Chapter: 2 p. 49.
[15] D. A. McQuarrie: *Statistical Mechanics* (Harper Collins Publishers, New York, 1973) Chapter 13.
[16] L. D. Landau, E. M. Lifshitz and L. P. Pitaevskii, *Electrodynamics of Continuous Media*, (Pergamon Press, New York, 1984), Chapter: 2 p. 36.
[17] K. D. Kreuer, S. J. Paddison, E. Spohr, and M. Schuster, *Chem. Rev.*, **104**, 4637 (2004).
[18] S. J. Paddison, L. R. Pratt, T. A. Zawodzinski, and D. W. Reagor, *Fluid Phase Equilibria*, **150**, 235 (1998).
[19] S. J. Paddison and T. A. Zawodzinski Jr., *Solid State Ionics*, **115**, 333 (1998).
[20] S. J. Paddison, L. R. Pratt, and T. A. Zawodzinski Jr., *J. New Mater. Electrochem. Syst.*, **2**, 183 (1999).
[21] S. J. Paddison, L. R. Pratt, and T. A. Zawodzinski Jr., in *Proton Conducting Membrane Fuel Cells II*, S. Gottesfeld and T. F. Fuller, Editors, PV 98-27, p. 99, The Electrochemical Society Proceedings Series, Pennington, NJ (1999).
[22] S. J. Paddison, R. Paul, and B. S. Pivovar, in *Direct Methanol Fuel Cells*, S. Narayanan, S. Gottesfeld and T. A. Zawodzinski, Editors, PV 01-04, The Electrochemical Society Proceedings Series, Pennington, NJ (2001).
[23] M. Ise, Ph.D.-Thesis, University Stuttgart, 2000.
[24] J. Kirkwood, *J. Chem. Phys.*, **7**, 911 (1939).
[25] L. Onsager, *J. Am. Chem. Soc.*, **58**, 1486 (1936).
[26] H. Fröhlich: *Theory of Dielectrics* (Oxford University Press, New York, 1949) Chapter 11.
[27] P. Debye, *Physikalische Zeitschrift*, **13**, 97 (1912).
[28] B. K. P. Scaife: *Principles of Dielectrics* (Clarendon Press, Oxford, 1989) Chapter 2.
[29] R. Paul and S. J. Paddison, *J. Chem. Phys.*, **115**, 7762 (2001).
[30] R. Paul and S. J. Paddison, in *Advances in Materials Theory and Modeling – Bridging Over Multiple-Length and Time Scales*, edited by V. Bulatov, L. Colombo, F. Cleri, L. J. Lewis, and N. Mousseau (Materials Research Society, Warrendale, Pennsylvania, 2001), pp. A7.16.1–7.16.7.

12
Proton Conduction in PEMs: Complexity, Cooperativity and Connectivity

S. J. PADDISON

12.1. Introduction

The pursuit to develop improved high performance polymer electrolyte membranes (PEMs) for fuel cells operating at temperatures above 100 °C has motivated extensive study of existing materials such as the archetypical perfluorosulfonic acid (PFSA) membrane Nafion® [1]. Central to both experimental and theoretical investigations of these ionomers have been efforts aimed at elucidating how protons move in these materials [2]. Although it is widely recognized that the proton conductivity of PFSA membranes is highly dependent on the water content, exactly how polymer chemistry and morphology determine proton conductivity is not understood.

As these polymers are composed of both hydrophilic (i.e. the sulfonic acid groups) and hydrophobic (typically the backbone) components, with the absorption of water these materials phase separate. The water and ions (both the hydrated protons and conjugate tethered sulfonate groups) coexist in regions of only a few nanometers in dimension and are distributed through inhomogeneous regions some of which are crystalline consisting of the polytetrafluoroethylene (PTFE) backbone and portions of the pendant perfluorinated side chains (see the chapters in this volume for further description concerning modeling of membrane morphology). Hence, the solvated protons are confined to very small and/or narrow regimes that are of non-uniform and complex shape [3]. This "exotic" environment makes understanding the underlying physical and chemical processes which determine the proton conductivity difficult. In recent years, multi-scale modeling has been undertaken to understand proton acidity, hydration, diffusion, and confinement in PEMs [4]. This chapter presents a summary of the insight into understanding proton conduction within a framework consisting of: *complexity*, *connectivity*, and *cooperativity*.

12.1.1. Types of PEMs

DuPont's Nafion® is still considered the state-of-the-art membrane material for PEM fuel cells despite exhibiting sufficient proton conductivity only at high degrees of hydration [5]. This hydration requirement results in a problematic operating temperature limited to the boiling point of water (i.e. $T \leq 100°C$ at 1 atm) and significant water and/or methanol (if used as the fuel at the anode) "cross-over" due to permeation and electro-osmotic drag: the transport of water coupled to the protonic current. The constraint of low operating temperature necessitates the use of expensive platinum or platinum alloy catalysts to promote electrochemical oxidation and reduction at the anode and cathode, respectively [6]. Notwithstanding the use of platinum, the hydrogen must be of very high purity to be converted at sufficient rates and, with direct methanol fuel cells, sufficient oxidation is impossible at typical operating temperatures of 80°C. If a hydrogen-rich reformate (e.g. that produced by steam reforming of methanol or methane) is used as the fuel, then the presence of even trace amounts of CO will poison platinum based catalysts through adsorption and consequent blocking of reaction sites. Hence, the humidification requirements coupled with the high electro-osmotic drag in presently available membrane materials severely complicates both the water and heat management of the fuel cell, leading to detrimental chemical short-circuiting through parasitic chemical oxidation of methanol at the cathode.

Several different approaches have been used in attempts to improve proton conduction in PEMs [7]. These include the use of alternative fluids (e.g. phosphoric acid and polybenzimidazole [8], phosphonic acid [9], and imidazole [10]) to replace the function of water in the membrane, the addition of inorganic particles [11] (e.g. silica, heteropolyacids) into polymeric conductors that purportedly allow proton conduction along the inorganic surface or maintain the water content of the membrane by adding an additional hydrophilic component and, as alluded to above, the preparation of various alternative proton-conducting polymers, such as bis(sulfonylimide) analogues of Nafion®, sulfonated poly (arylphosphazenes), sulfonated poly (aryl ketones) and poly (aryl sulfones), sulfonated poly (phenylene oxide) and sulfonated polystyrenes. Along with the synthesis and testing of entirely novel PEMs has come modification of specific molecular features of presently available PFSAs. These include the use of alternate protogenic groups (e.g. phosphonic acid), and distinct backbone and/or side chain chemistry. With respect to the latter, Figure 12.1 highlights the potentially substantial differences that both the choice of the polymeric backbone and the side chain length make on the proton diffusion or conductivity [12]. Proton and water diffusion coefficients, as determined from NMR measurements of Nafion and sulfonated polyether ether ketone (i.e. PEEK) membranes at 300 K are plotted as a function of the water volume fraction in Figure 12.1(a) and dramatically show that the diffusion rates are much greater at low to intermediate water contents in the membrane with the PTFE backbone when compared to the aromatic

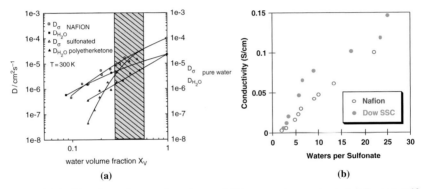

FIGURE 12.1. The dependence of proton mobility on water content. (a) Proton self-diffusion coefficients (Dσ) of Nafion and sulfonated polyetherketone membranes at 300 K plotted as a function of the water volume fraction showing the substantially greater proton mobility in the PFSA membrane as low to intermediate water contents. Taken from Ref. 13. (b) Proton conductivity of Nafion and low EW ($\approx$800) Dow SSC membranes plotted as a function of the water content expressed as the number of water molecules per sulfonic acid group showing substantially higher conductivity at intermediate hydration levels. Taken from Ref. [12].

material [13]. However, when the two membranes are fully hydrated the proton self-diffusion coefficients are very similar to one another. A further underscoring of the differences that polymer chemistry may make on proton transport is seen in Figure 12.1(b) where the proton conductivity is plotted for Nafion (again) and the short side chain (i.e. SSC) PFSA ionomer (originally synthesized by the Dow Chemical Company [14]) and clearly shows that the membrane with the shorter side chain (i.e. $-OCF_2CF_2SO_3H$) has much higher conductivity at intermediate water content (i.e. from 4–18 H_2O/SO_3H). The reasons for these substantial differences in proton mobilities undoubtedly have to do with the density of the hydrated protons but is not fully understood and therefore impetus for molecular-level structure/function modeling.

12.1.2. Motivation for Modeling

It is widely appreciated that during operation the PEM in a fuel cell is not uniformly hydrated, with the membrane only partially hydrated on the anode side and typically fully hydrated (and often 'flooded') on the cathode side. This is despite the humidification of the H_2 gas stream and removal of water at the site of reduction. This necessitates that under operation the flux of protons occurs across a gradient in the concentration of the water which impacts the rate of transport (e.g. see Figure 12.1) and consequently the mechanism whereby the proton traverses the electrolyte. Experimental studies also indicate that even relatively subtle changes in either the backbone or side chain chemistry may substantially affect the proton mobility (to be discussed below). Hence, the goal of theoretical and computational studies must be to illuminate mechanistic differences in the transport of protons in PEMs as a function of both polymer chemistry and water content [15].

12.1.3. Goals and Challenges for Modeling

The general goals, therefore, in modeling proton movement in PEMs are (1) to obtain a *molecular-level* understanding of transport mechanisms in hydrated PEMs; (2) to *connect* and *correlate* this structural information with the function and properties of the materials; and (3) to design in collaboration with synthetic and experimental characterization efforts *improved materials*. As has been pointed out in recent reviews [4,15], early modeling investigations of fuel cell materials were not molecular and therefore could not really connect polymer chemistry with measurable properties including conductivity. However, more recently a wide range of methods that treat the polymer and or water atomistically including classical molecular dynamics (MD) simulations [16,17,18,19,20], ab initio MD simulations [21], molecular orbital calculations [22,23,24], statistical mechanical models [25,26,27,28,29], and empirical valence bond (i.e. EVB) simulations [30,31]. It is through these types of methods that theoretical work will impact the understanding and design of novel high performance fuel cell materials.

12.2. Proton Conduction

As introduced earlier, examination of experimental and theoretical studies has suggested a useful framework of guiding principles for understanding proton conductivity in hydrated ionomers consists of three aspects or ingredients: *complexity*, *connectivity*, and *cooperativity*. The present understanding derived from both experimental and theoretical studies of the mechanisms of proton transport in hydrated PEMs was recently reviewed by several authors [4,15]. Although a substantial amount of effort has been undertaken over the past two decades, much remains to be understood in terms of how structural features of the hydrated membrane morphology determine key molecular processes, such as the transport of protons, water, and other species present in PFSA and other fuel cell electrolyte membranes. The mobility of protons in both the solid state [32,33] (*e.g.* cubic perovskite-type oxides) and in bulk water [34,35,36,37] provide a framework for understanding proton transport, as to some extent these systems may be viewed as limiting or extreme cases in the range of hydration of the electrolyte (*i.e.* from dry to fully hydrated). Important common features include the observation that the excess protonic charge follows the symmetry of the dynamic and fluctuating hydrogen bond coordination pattern [38]. With solids, this is typically within direct proximity of the mobile proton, while in water the proton-hopping is coupled to the dynamics of the hydrogen bonds in at least the second hydration shell [35,36,39,40]. The transport of protons in a hydrated ionomer is, however, considerably more complex due to the heterogeneous nature of the polymer (*i.e.* co-organized crystalline and ionic domains [41,42,43]) and the nano-confinement of the water around the tethered anionic groups (*e.g.* $-SO_3^-$) and hydrated protons [44,45]. As indicated earlier, the hydrated morphology of PFSAs has been difficult to elucidate experimentally, with the result that much of the early modeling of proton transport in these materials was either empirical or largely phenomenological in nature. Since these studies, along with others, were reviewed elsewhere, we limit the focus of the following discussion to molecular level modeling of transport, and principally that which avoids much phenomenology.

12.2.1. Complexity

Complexity in the conduction of protons encompasses: (1) dissociation of the proton from the acidic site; (2) subsequent transfer of the proton to the first hydration shell water molecules; (3) separation of the hydrated proton from the conjugate base (*e.g.* the sulfonate anion); and finally (4) diffusion of the protons in the media consisting of confined water and tethered sulfonates within the polymeric matrix. Hence, we will endeavor to discuss the insight from theoretical modeling into these four aspects.

12.2.1.1. Dissociation

In an early effort to understand proton dissociation and the stabilization of the dissociated proton in the first hydration shell and subsequent separation of the hydrated proton, Paddison et al. [22,23] conducted a series of explicit water electronic structure calculations with both trifluoromethane and *para*-toluene sulfonic acids, the former to serve as a benchmark for ionization of PFSA ionomers and the latter for PEEK membranes. Starting from the fully optimized geometry of a single isolated acid molecule, water molecules were systematically added to B3LYP/6-31 G** minimum energy conformations to obtain successively larger water clusters of each acid (*i.e.* $-SO_3H + xH_2O$, where $1 \leq x \leq 6$). The resulting equilibrium structures for trifluoromethane-sulfonic acid with from 1 to 5 water molecules are depicted in Figure 12.2 and reveal that as the number of water molecules in the cluster around the acid is increased, proton dissociates occurs only after 3 water molecules are added (Figure 12.2(c)), forming a hydronium ion-triflate anion pair. With the addition of the fourth and fifth water molecules the $-SO_2O^- \bullet\bullet\bullet OH_3^+$ distance increases slightly: from 2.56 to 2.69 Å, but the H_3O^+ remains hydrogen bonded to the $-SO_3^-$. This contact ion pair disappears when the sixth water molecule is added, Figure 12.3(a), whereupon the hydrated proton separates from the anion forming a solvent-separated pair as an Eigen (i.e. $H_9O_4^+$) cation and sulfonate anion. A qualitatively similar result was observed for the

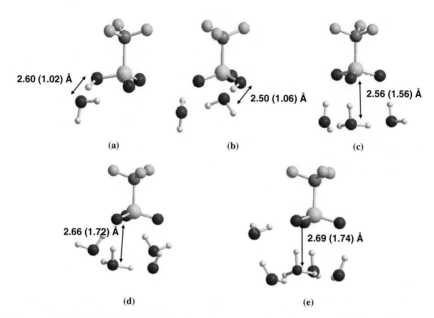

FIGURE 12.2. Global minimum energy structures determined by full optimization at the B3LYP/6-31G** level of a trifluoromethanesulfonic acid molecule with different numbers of water molecules, showing the O–O and O–H (in brackets) distances, and revealing that with: (a) 1 and (b) 2 water molecules, the proton does not dissociate; with (c) 3, (d) 4, and (e) 5 water molecules, the proton is dissociated forming a contact hydronium ion-sulfonate pair. Redrawn from results first presented in Ref. [23].

aromatic sulfonic acid (see Figure 12.3(b)) with the important distinction that the Eigen cation resides 0.3 Å closer (as measured by the $-SO_2O^- \bullet\bullet\bullet OH_3^+$ distance) to the sulfonate anion. The authors concluded that the self-dissociation of the proton in sulfonic acids is a function of the strength of the Lowry-Brønsted acid, which is equivalent to the stability of the conjugate base (*i.e.* the more stable its conjugate base, the stronger the acid) [46]. These electronic structure calculations of water clusters of CF_3SO_3H and $CH_3C_6H_5SO_3H$ acids indicate that the former is the stronger acid. Calculation of the total atomic charge residing on the sulfonate oxygens with the CHelpG scheme confirm that there is consistently more negative charge residing on the oxygen atoms on the *p*-toluene sulfonate suggesting that the aromatic ring is not as good an electron withdrawing group as the trifluoromethyl group. Examination of Figure 12.4 shows that for both acids the electron density residing on the sulfonate oxygen atoms initially increases as the first hydration shell is formed, and then decreases once the hydrated proton separates from the conjugate base (*i.e.* following the demise of the contact ion pair). Furthermore, these molecular orbital calculations indicated that Zundel-like cations, Figures 12.2 and 12.3, initially form when only a few water molecules are associated with the acids but,

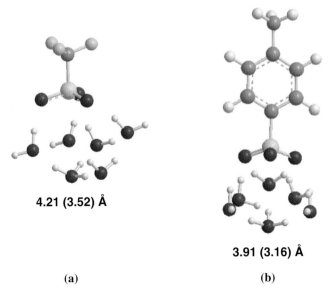

FIGURE 12.3. Global minimum energy structures determined by full optimization at the B3LYP/6-31G** level of a trifluoromethanesulfonic acid molecule (a) and a *para*-toluene sulfonic acid molecule (b) each with 6 water molecules. Both acids show the dissociated proton exists as a solvent separated pair consisting of what resembles an Eigen cation and sulfonate ion. The O–O and O–H (in brackets) distances for each acid reveal that aromatic anion is the stronger base. First presented in Ref. [23].

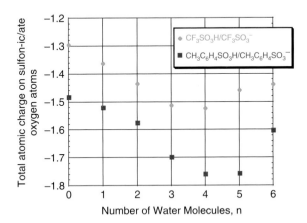

FIGURE 12.4. Total atomic charge as computed with the CHelpG routine of G03 residing on the oxygen atoms of trifluoromethanesulfonic acid and *para*-toluene sulfonic acid as a function of the number of hydrating water molecules. The greater negative charge residing on the sulfonic/sulfonate oxygen atoms of the aromatic molecule indicates that more electron density has been stabilized on the fluorine atoms than on the π-ring.

after separation, the proton is stabilized as an Eigen-like cation. It is important, however, to recognize that these results are only of isolated single acid molecules and hence their applicability to the conditions in condensed phase PEMs is to be used with caution.

These studies were extended recently by Paddison and Elliott in electronic structure calculations of fragments of the SSC ionomer consisting of two pendant side chains possessing different three different separations of the side chains: $CF_3CF(-O(CF_2)_2SO_3H)-(CF_2)_n-CF(-O(CF_2)_2SO_3H)CF_3$, where $n = 5, 7$ and 9 with the addition of from 1 to 7 water molecules [47]. Fully optimized conformations of the fragments with the explicit water molecules at the B3LYP/6-311 G** level revealed that the separation of the side chains affects the minimum amount of water necessary to observe the dissociation of proton(s) to the first hydration shell. Dissociation of both protons occurred in the SSC fragment with five difluoromethylene groups separating the pendant side chains after only 5 water molecules were added and the global minimum energy structure is displayed in Figure 12.5(a). The larger

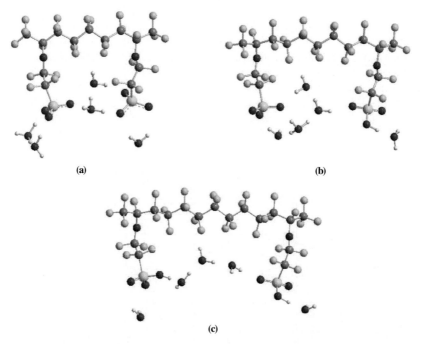

FIGURE 12.5. Global minimum energy structures determined by full optimization at the B3LYP/6-31G** level of oligomeric fragments with 5 water molecules of the SSC PFSA membrane consisting of two side chains, but with differing number (5, 7, and 9) of difluoromethylene groups in the backbone separating the side chains. The fragment in (a) shows both protons are dissociated; in the larger fragment in (b) only one proton is dissociated; and in (c) neither proton dissociates in the fragment with the largest separation of the sulfonic acid groups. Taken from Ref. [47].

fragments with $n = 7$ and 9 CF_2 groups shown in Figures 12.5(b) and (c), however, showed dissociation of only one proton and neither protons, respectively, with a similar number of water molecules hydrating the oligomeric fragment. The addition of seven water molecules to each of the fragments resulted in increased separation of the dissociated protons from their conjugate bases in the two smaller fragments (Figures 12.6(a) and (b)) but in the fragment with 7 CF_2 groups no dissociation of the second proton was observed (Figure 12.6(c)). All the SSC fragments with differing separation of the side chains confirm the importance and seeming preponderance of the Zundel-like cation ($H_5O_2^+$) in the first hydration shell under the minimally hydrated (i.e. $\leq 3\,H_2O/\text{–}SO_3H$) conditions examined. These calculations suggest that despite the hydrogen bonding of the water molecules with the conjugate base (i.e. $-SO_3^-$), transfer of protons may be partially facilitated

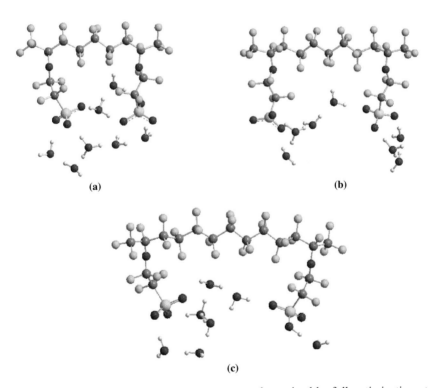

FIGURE 12.6. Global minimum energy structures determined by full optimization at the B3LYP/6-31G** level of oligomeric fragments with 7 water molecules of the SSC PFSA membrane consisting of two side chains, but with differing number (5, 7, and 9) of difluoromethylene groups in the backbone separating the side chains. The fragment in (a) suggests the dissociated protons exist in Zundel-like contact cations; in the larger fragment in (b) both proton are (now) dissociated; and in (c) only one of the protons dissociates in the largest fragment. Taken from Ref. [47].

through some kind of hopping mechanism, though clearly it is different from that observed in bulk water or in fully hydrated PEMs. Finally, these calculations have demonstrated that side chain separation affects the establishment of a hydrogen bond network of the sulfonic acid groups in the SSC PEM under minimal hydration conditions.

12.2.1.2. Transfer and Separation

The effects of both the neighboring chemistry to the acid (i.e. α-group to the $-SO_3H$) and the number of hydrating water molecules on the separation of the proton from the conjugate base have been described in the previous section but only as they pertain to a single isolated acid molecule. Clearly the close proximity of additional acid groups as is observed in PFSA membranes (dry or at least partially hydrated) will impact proton dissociation. Paddison and Elliott, in a very recent subsequent study [48], examined the effects of conformational changes of the PTFE backbone in the $CF_3CF(-O(CF_2)_2SO_3H)-(CF_2)_7-CF(-O(CF_2)_2SO_3H)CF_3$ fragment of the SSC PFSA membrane on proton transfer and separation. Extensive searches for minimum energy structures (at the B3LYP/6-311 G(d,p) level) of this oligomeric fragment with from 4 to 7 explicit water molecules revealed that the perfluorocarbon backbone may adopt either an elongated geometry (as shown in Figures 12.5(b) and 6(b), with all carbons in a *trans* configuration, or a folded conformation where at least two of the carbons exhibit a *gauche* configuration. This seemingly slight 'kink' to the backbone results in more than just a reduction in the separation of the sulfonic acid groups. Fully optimized structures of the oligomeric fragment displaying the latter backbone geometry with 5 and 7 water molecules are displayed in Figures 12.7(a) and (b), respectively. At both of these hydration levels the conformational changes in the PTFE backbone bring the sulfonate groups more than 3 Å closer to one another, all water molecules form a single hydration cluster, and increased separation of the acidic protons is observed. Of additional significance is the observation that the structures in Figures 12.7(a) and (b) are 9.7 and 12.7 kcal/mol lower in energy (B3LYP/6-311 G**) than their counterparts in Figures 12.5(b) and 12.6(b) suggesting that aggregation of the $-SO_3H$ groups is strongly preferred even under low levels of hydration. Furthermore, these electronic structure calculations show that the fragments displaying the 'kinked' backbone possess stronger binding of the water to the dissociated sulfonic acid groups and this is clearly related to the enhanced transfer and separation of the proton with fewer hydrating water molecules. Specifically, when the counterpoise method of Boys and Bernardi was implemented to correct for basis set superposition error in these calculations the binding of the water to the fragment was determined to be 14.8 and 14.5 kcal/mol per H_2O molecule for conformations depicted in Figures 12.7(a) and (b), respectively, whereas the

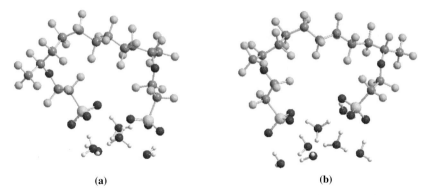

FIGURE 12.7. Global minimum energy structures determined by full optimization at the B3LYP/6-31G** level of the $CF_3CF(-O(CF_2)_2SO_3H)-(CF_2)_7-CF(-O(CF_2)_2SO_3H)CF_3$ oligomeric fragment of the SSC PFSA membrane displaying "kinked" backbones. The fragment in (a) shows that closer proximity of the sulfonic acids achieved by the alternation of the conformation of the backbone results in dissociation of both protons with only 5 water molecules. All 7 water molecules in (b) form a single cluster and both protons are dissociated as Zundel-like cations bridging both sulfonate groups. Taken from Ref. [59].

binding energies of their counterparts with the extended backbone (i.e. Figures 12.5(b) and 6(b)) were only 13.1 and 12.6 kcal/mol. Hence, despite the enhancement in the dissociation of the protons with closer proximity of the sulfonic acid groups, this increased binding of the molecules to the sulfonate groups may result in trapping or pinning of the hydrated protons, thereby impeding the mobility of the protons

12.2.1.3. Diffusion

Various physical and theoretical models, largely phenomenological in nature, have been developed in an effort to describe and understand the transport of various species in ionomer membranes. Typically these models involve a substantial degree of 'coarse-graining' in the form of phenomenological parameters and macroscopic descriptions of the underlying processes, and hence the molecular details have been to a large extent "averaged out" or ignored. An earlier chapter in this book reviews much of this work, and the reader is referred to this chapter and other articles in the peer-reviewed literature [4]. In this section, we focus on exclusively on discussion of a molecular-level model of proton transport applicable to hydrated ionomers based on a nonequilibrium statistical mechanical framework [25] which affords the calculation of proton self-diffusion coefficients in single water and ion containing domains (i.e. pores or channels) of the membrane.

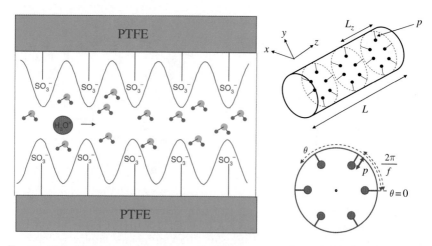

FIGURE 12.8. The assumed structure of a PEM pore or channel showing the request nomenclature. The pore has a length L, Cartesian frame showing the pore orientation with the $-SO_2^-$ groups located at the termini of pendant groups of length p which are axially arranged in n segments separated by a distance L_z. The cross sectional view of the pore shows the f pendant groups arranged in a circular array. Taken from Refs. [25] and [29].

This model does possess the clearly phenomenological assumptions that the water and ion domains are regular cylindrical channels of constant diameter with uniformly distributed sulfonic acid groups, and the features of such an 'idealized' channel along with request nomenclature are presented in Figure 12.8. These are clearly gross approximations as PFSAs are typically random copolymers where the shape of either the PTFE portion of the polymer or the water and ion domains are anything but regular [42]. The model also assumes that the diffusion of a proton is to a large extent coupled to the diffusion of a water molecule; another simplification that for membranes at high degrees of hydration is not tenable due to the observed increase in the mobility of protons over that measured for water molecules [4,13]. The observation that the self-diffusion coefficient of the hydrated protons increases more rapidly than the diffusivity of the water molecules with increasing water content has been attributed to an increase in the contribution of a Grotthuss-type hopping or shuttling mechanism in the mobility of the hydrated protons. This structural diffusion of the proton is taken into account by the model but not in a manner that allows for an identification of the relative contribution of hopping to the net mobility of the hydrated protons.

The model computes an effective 'friction coefficient' of a hydrated proton (assumed to be a structureless classical hydronium ion and labeled α) in a single pore of length L and cross-sectional radius R, filled with N water molecules each possessing a dipole moment μ. All sulfonic acid groups are assumed to be dissociated and periodically distributed along the length of the pore, each with a charge of $-e$ protruding a distance p from the wall of the pore. The system energy is taken to consist of classical two-body potentials describing: (1) interaction of the hydrated proton with all the water molecules and sulfonate groups; (2) interaction of all water molecules with the sulfonate groups; and (3) water-water interactions. The average force on the hydrated proton, $\langle \mathbf{F}(\mathbf{r}_\alpha, \mathbf{v}_\alpha) \rangle$, is calculated by performing both time and phase space integrations of the mean force with a time-dependent nonequilibrium distribution function, $f(\mathbf{r}_\alpha, \mathbf{v}_\alpha, \{\mathbf{r}\}, \{\mathbf{v}\}, t)$ according to:

$$\langle \mathbf{F}(\mathbf{r}_\alpha, \mathbf{v}_\alpha) \rangle = - \int_0^\infty dt \int_\Omega d\{\mathbf{r}\}d\{\mathbf{v}\} f(\mathbf{r}_\alpha, \mathbf{v}_\alpha \{\mathbf{r}\}, \{\mathbf{v}\}, t) \nabla_\alpha V(\mathbf{r}_\alpha, \{\mathbf{r}\})$$

where $\mathbf{r}_\alpha$ and $\mathbf{v}_\alpha$ denote the position and velocity of the hydrated proton; $\{\mathbf{v}\}$ and $\{\mathbf{r}\}$ denote the velocity and position of all other species; $V(\mathbf{r}_\alpha, \{\mathbf{r}\})$ is the potential energy of the hydrated proton within the channel; and Ω is the pore volume. Intrinsic to this equation is the assumption that a non-equilibrium stationary state moving with the proton contributes significantly to this average force and is obtained in the limit of $t \to \infty$. In general, the time evolution of this distribution function is governed by the continuity (i.e. Liouville) equation:

$$i \frac{\partial f(\mathbf{r}_\alpha, \{\mathbf{r}\}, \{\mathbf{v}\}, t)}{\partial t} = L_0 f(\mathbf{r}_\alpha, \{\mathbf{r}\}, \{\mathbf{v}\}, t)$$

where L_0 is the Liouville operator for the system with a coordinate reference system moving with an assumed constant velocity, $\mathbf{v}_\alpha$, of the hydrated proton. A formal solution of this equation is utilized according to:

$$f(\mathbf{r}_\alpha, \{\mathbf{r}\}, \{\mathbf{v}\}, t) = e^{-iL_0 t} f(\mathbf{r}_\alpha, \{\mathbf{r}\}, \{\mathbf{v}\}, 0) = e^{-iL_0 t} f_{eq}(\mathbf{r}_\alpha, \{\mathbf{r}\}, \{\mathbf{v}\})$$

where $f_{eq}(\mathbf{r}_\alpha, \{\mathbf{r}\}, \{\mathbf{v}\})$ is the distribution function for the system at equilibrium ($t = 0$) and the time displacement operator, $e^{-iL_0 t}$, constructed from the Poisson bracket:

$$L_0 = i\{H_0(\mathbf{r}_\alpha, \{\mathbf{r}\}, \{\mathbf{v}\}), \}$$

where $H_0(\mathbf{r}_\alpha, \{\mathbf{r}\}, \{\mathbf{v}\})$ is the Hamiltonian for the pore. As indicated earlier, the total energy of the pore consists of the kinetic energy of all the water molecules and the net potential energy, $V(\mathbf{r}_\alpha, \{\mathbf{r}\})$, due to two-body interactions described above:

$$H_0(\mathbf{r}_\alpha, \{\mathbf{r}\}, \{\mathbf{v}\}) = \sum_{i=1}^N \frac{m(\mathbf{v}_i + \mathbf{v}_\alpha)^2}{2} + V(\mathbf{r}_\alpha, \{\mathbf{r}\})$$

where m is the mass and $\mathbf{v}_i$ the velocity of the ith water molecule. The potential for this system is assumed to consist of the following four terms:

$$V(\mathbf{r}_\alpha, \mathbf{r}) = -\sum_{i=1}^{N} \frac{\mu^2 e^2}{48\pi^2 \varepsilon^2 kT} \frac{1}{|\mathbf{r}_\alpha - \mathbf{r}_i|^4}$$
$$+ \Psi_0 \cos\left(\frac{2\pi n z_\alpha}{L}\right) + \sum_{i<j}^{N} \frac{2\mu^4}{3(4\pi\varepsilon)^2 kT} \frac{1}{|\mathbf{r}_i - \mathbf{r}_j|^6}$$
$$- \sum_{i=1}^{N} \frac{2\pi \mu \Psi_0 n}{eL} \sin\left(\frac{2\pi n z_i}{L}\right)$$

where ε is the permittivity of the water in the pore, k the Boltzmann constant, T the temperature, and Ψ_0 the 'amplitude of the potential energy' due to interaction of the hydronium ion with the $-SO_3^-$ groups. The form of this latter function (i.e. Ψ_0) received careful consideration by the authors and two different forms were eventually derived, and further details are provided later. A power series expansion of the mean force experienced by the hydrated proton where only the linear term is retained allows evaluation of the friction coefficient, ζ, as a sum of four terms:

$$\zeta_\alpha = \frac{\beta}{3} \int_0^\infty dt \left(\langle \mathbf{F}_{\alpha s} e^{-iL_0 t} \mathbf{F}_{\alpha s} \rangle_0 + \langle \mathbf{F}_{\alpha s} e^{-iL_0 t} \mathbf{F}_{ps} \rangle_0 + \langle \mathbf{F}_{\alpha p} e^{-iL_0 t} \mathbf{F}_{ps} \rangle_0 + \langle \mathbf{F}_{\alpha p} e^{-iL_0 t} \mathbf{F}_{\alpha s} \rangle_0 \right)$$
$$= \zeta_1 + \zeta_2 + \zeta_3 + \zeta_4$$

where $\beta = 1/kT$, and $\mathbf{F}_{\alpha s}, \mathbf{F}_{ps}$, and $\mathbf{F}_{\alpha p}$ are the forces between the hydronium ion and the water molecules, the fixed sites and the water molecules, and the hydronium ion and the fixed sites, respectively. This result is deemed to be quite important as it partitions the friction experienced by the hydrated proton into four distinct contributions ($\zeta_1, \zeta_2, \zeta_3,$ and ζ_4); thereby allowing not only a comparison of the magnitude of the various terms but identification of the dominant contribution for a particular pore. A pictorial representation of these four force-force correlation functions is given in Figure 12.9 and illustrates the 'source of the friction' in each term. The first term, ζ_1, is the friction coefficient experienced by the hydrated proton in the absence of the SO_3^- pendant groups and was not explicitly evaluated by the authors but was rather assumed to be the friction coefficient of either an hydronium ion in bulk water (computed via the Stokes relation: $\zeta_1 = 6\pi\eta a$; where a is the radius and η the viscosity), or that of a proton in bulk water (derived from the experimental self-diffusion coefficient, $D_{exp} = 9.311 \times 10^{-5} cm^2 s^{-1}$, from $\zeta_1 = \frac{kT}{D_{exp}}$). The choice of which of the two values to implement was selected based on the nature of the water in the pore: at low hydration levels where the water molecules are quite constrained the value corresponding to the friction

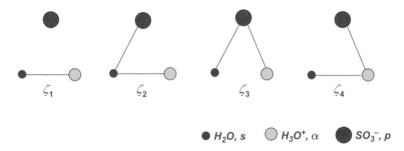

FIGURE 12.9. A pictorial representation of the four contributions (ζ_n) to the friction coefficient of the hydrated proton showing the two body interactions involving the selected proton, α, the water molecules, s, or the sulfonate groups, p, in each force–force correlation function. First shown in Ref. [25].

coefficient of a hydronium ion was used; but at high hydration levels the value inclusive of proton hopping was used. The latter three terms were explicitly evaluated and describe the contribution to the 'hydrodynamic friction' of the solvated proton due to interaction with the water and sulfonate groups either through direct electrostatic interaction with the sulfonate groups (i.e. ζ_3 and ζ_4) or more synergistically through the constrained water molecules (as in ζ_2).

Prior to discussing the results for the diffusion coefficients computed with this model we return to the potential energy implemented in the model, specifically to the derivation of the term describing the collective effective of all of the SO_3^- groups in the pore, Ψ_0. As indicated earlier, the authors used two different forms for this potential; the first, which was strictly speaking only applicable for periodic systems of infinite length, had the form [25,26]:

$$\Psi_0(r_{xy}, z) = \frac{q_1 q_2}{\pi \varepsilon_0 \varepsilon_r L_z} K_0(2\pi \frac{r_{xy}}{L_z}) \cos(2\pi \frac{z}{L_z})$$

where r_{xy} is the vertical component of the separation distance of the SO_3^- groups with net charge q_1 from a charge q_2 (e.g. if hydrated proton then $q_2 = +e$). It should be pointed out that this potential is a truncation of a potential derived by Grønbech-Jensen et al using Lekner summations of Coulombic interactions in infinite 3-dimensional systems having periodicity in one or two dimensions [49]. With this potential and experimentally-derived parameters for the dimensions of the pore,[13] along with the density and distribution of water molecules and sulfonate groups, the authors computed diffusion coefficients over a wide range of water contents (i.e. from $\lambda = 6$–30) in both Nafion [25,27] and S-PEEK[28] (sulfonated poly(arylene ether ether ketone)) membranes, the results which are compared to experimentally measured values from pulsed-field gradient NMR measurements in a single plot in Figure 12.10. The agreement to the experimental results is remarkable with all 6 computed values uniformly lower than their experimental counterpart by

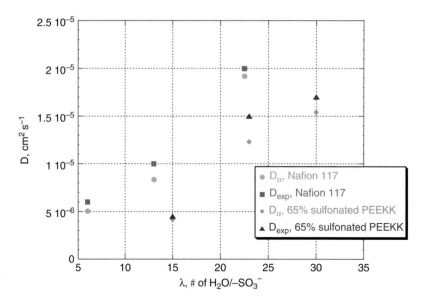

FIGURE 12.10. Computed and experimentally determined proton self-diffusion coefficients in Nafion 117 and 65% sulfonated PEEKK membranes plotted as a function of water content expressed as the number of water molecules per sulfonic acid group. The comparison to the experimental measurements indicates the generally remarkable agreement (within 15%) range of membrane hydration. Taken from Ref. [23].

no more than 15%. The authors reasoned that as the diffusivities vary over nearly an order of magnitude this level of agreement indicates substantial predictive capability of the model. The parametric sensitivity of the model was subsequently investigated [26] and in a particularly insightful study they calculated diffusion coefficients for a PEM pore with fixed dimensions ($R = 8$ and $L = 30$), constant hydration level (6 H_2O/SO_3^-), and fixed number of sulfonate groups, varying both the intrusion and the 'uniformity' of the axial distribution of the fixed sites independently. The computed diffusion coefficients are plotted as a function of the length of the side chain (i.e. intrusion as measured from the pore wall) for various distributions of the fixed sites in Figure 12.11. This plot clearly shows the substantial sensitivity of the diffusion coefficient to the intrusion of the sulfonate groups; where for a distribution of the sulfonate groups on only 4 axial arrays (the most heterogeneous distribution examined), the diffusion coefficient varies over more than three orders of magnitude. However, when the sulfonate groups are distributed uniformly across the length of the pore (i.e. for n = 9), the diffusion of the proton is only affected when the sulfonate groups are almost in the center of the pore. Their study suggests that both the intrusion of the side chains and uniformity of the distribution of the anionic groups have a substantial impact on proton diffusion.

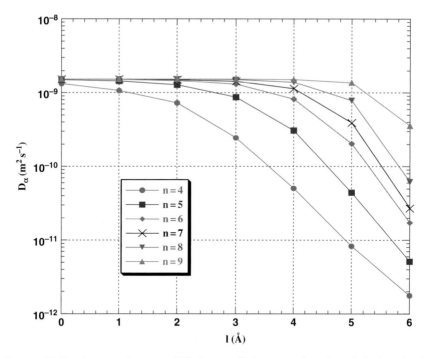

FIGURE 12.11. Computed proton diffusion coefficients as a function of the length of intrusion of the side chain (l) and as a function of the number (n) of axially positioned arrays of fixed sites for an arbitrary membrane pore with fixed: length, diameter, total number of anionic groups, and water content. The plots show the substantial sensitivity (i.e. varying over more than 3 orders of magnitude) of the computed diffusion coefficient to these parameters and suggest that the most uniform distribution of anionic groups (i.e. where $n = 9$) results in the highest proton diffusion notwithstanding the protrusion of the anionic groups. Taken from Ref. [26].

Despite the seeming success of their model in matching experimental diffusion coefficients the authors acknowledged two important features that had been neglected: (1) many-body effects due to the presence of the complementary set of protons; and (2) the effects of the internal field due to the anions on the permittivity of the water within the pore [50,51]. Both of these features were dealt with by first deriving a potential that corrected accounted for only a finite number of periodic distributed sulfonate anions [29]:

$$\Psi(\rho, z) = \frac{q_1 q_2 n f}{8\pi^2 \varepsilon_0 \varepsilon_r L_z} c_2 - \frac{q_1 q_2 n f}{2\pi \varepsilon_0 \varepsilon_r L_z} K_0 \left[\frac{2\pi}{L_z} r_{xy} \right] \cos\left(\frac{2\pi}{L_z} z\right)$$

which has an additional contribution from a nonperiodic component (the term containing c_2), determined from the boundary conditions. Implementing a recent work of Kjellander and Mitchell [52], who used the Ornstein-Zernicke equation to derive an integrodifferential for a screened potential in terms of quasiparticles or "dressed" ions, they obtained a potential that included the screening due to the presence of the hydrated protons of the form:

$$\Psi_{sc}(\rho, z) = \frac{q_1 q_2 nf}{4\pi\varepsilon_0\varepsilon_r L_z}\lambda_1 - \frac{q_1 q_2 nf}{2\pi\varepsilon_0\varepsilon_r L_z} K_0\left[\left(\frac{4\pi^2}{L_z^2} + \kappa^2\right)^{1/122} r_{xy}\right]\cos\left(\frac{2\pi}{L_z}z\right)$$

where κ is a screening parameter. Two different forms were examined for this parameter; the first, the commonly employed Debye-Hückel parameter, κ_D, applicable to dilute solutions:

$$\kappa_D^2 = \frac{f^2 e^2 n_{H_3O^+}}{4\pi\varepsilon_0\varepsilon_r kT}$$

and a screening constant derived by Attard [53,54] that excludes a hard-core diameter d for the ions which in terms of κ_D has the form:

$$\kappa^2(1 + \kappa d) = \kappa_D^2\left[1 + \kappa d + (\kappa d)^2/2 + (\kappa d)^3/6\right]$$

Using this extended model they computed the radial dependence of the proton diffusion coefficient for a Nafion membrane pore at a hydration level of 6 H_2O/SO_3^- with inclusion of the two different formulations for the shielding. The results are shown in Figure 12.12 and show that the Attard formulation for the screening parameter results in higher proton diffusion particularly in the center of the pore (i.e. more efficient screening of the sulfonate anions). By comparing it to the experimental diffusion coefficient (6.0×10^{-10} m^2s^{-1}) the authors estimated that the hydrated protons reside (on average) about 3.2 Å (with the Attard screening parameter) from the sulfonate groups which appears to be consistent (albeit slightly shorter) with results from electronic structure calculations [23] (discussed earlier) that indicate the formation of solvent-separated sulfonate-hydrated proton pairs at this hydration level. This finding of the propensity of the hydrated protons to separate from their conjugate bases was also observed in recent MD simulations performed on Nafion "pockets" using a multi-state empirical valence bond (MS-EVB) methodology [31]. These dynamical simulations revealed a sulfonate oxygen–hydronium oxygen radial distribution function with two peaks corresponding to a contact ion pair at 2.6 Å and a solvent (i.e. water) separated ion pair at approximately 4.4 Å. Separation of the hydrated proton from the sulfonate groups, therefore, appears to be an important feature in proton diffusion in these membranes particularly under minimally hydrated conditions.

12.2.2. Connectivity

Connectivity in the conduction of protons encompasses hydrogen bonding of the of the protogenic groups to one another and/or water molecules. On

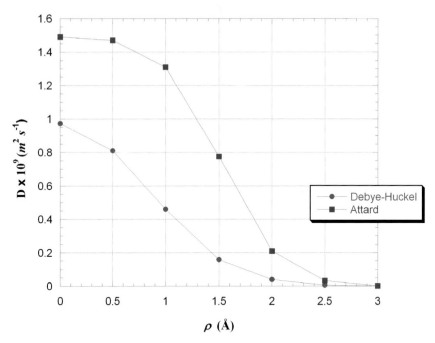

FIGURE 12.12. Variation of the diffusion coefficient as a function of the distance (ρ) from the center of the pore (i.e. $\rho = 0$ Å) to within 1 Å of the $-SO_3^-$ groups. The calculation is for Nafion with 6 $H_2O/-SO_3^-$ with inclusion of either Debye–Hückel or Attard models for the shielding. At the center of the pore the Attard formulation results in the most significant screening of the sulfonate groups and hence the higher proton self-diffusion coefficient.

greater length scales it also includes the connection of the water and ion domains to one another within the polymeric matrix.

12.2.2.1. Dry Connectivity

In an effort to determine the hydrogen bonding among neighboring (i.e. next to each other on the same macromolecule) sulfonic acid groups in PFSA membranes Paddison and Elliott undertook extensive electronic structure calculations of oligomeric fragments of the SSC ionomer consisting of two pendant side chains of different equivalent weight: $CF_3CF(-O(CF_2)_2SO_3H)-(CF_2)_n-C\ F(-O(CF_2)_2SO_3H)CF_3$, where $n = 5$, 7 and 9.[47] Full optimizations of each of these fragments beginning with the side chains and backbone in various configurations (including those with the sulfonic acid groups in close proximity to one another) revealed that in all cases the preferred energetic state was with the side chains well separated from one another and the difluoromethylene groups staggered along the backbone giving a typical helical pitch as observed in PTFE. Figure 12.13 displays the

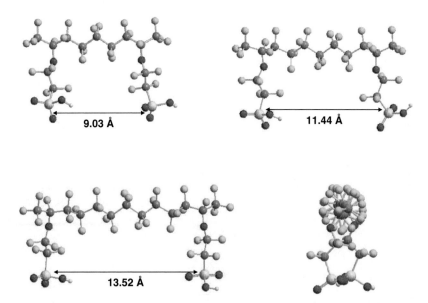

FIGURE 12.13. Fully optimized (B3LYP/6-311G**) global minimum energy structures of isolated two side chain fragments of the SSC PFSA membrane with differing number (5, 7, and 9) of difluoromethylene groups showing the side chains on the same "side" of the backbone and indicating that the separation of the sulfonic acid groups scale with the number of CF_2 groups in the backbone.

global minimum energy conformations of each of the SSC ionomer fragments along with a typical view down the backbone and indicate that the separation of the sulfonic acid groups scales with the number of CF_2 groups separating the pendant side chains: ≈ 9.0 Å with 5 CF_2 groups; nearly 11.5 Å with 7 CF_2 groups; and ≈ 13.5 Å with 9 CF_2 groups. In contrast to these results of well separated sulfonic acid groups, earlier reported electronic structure calculations at the same level of theory (i.e. B3LYP/6-31 G**) for a smaller fragment of the SSC ionomer, $CF_3CF(-O(CF_2)_2SO_3H)-(CF_2)_4-CF(-O(CF_2)_2SO_3H)CF_3$ did determine that a conformation where the sulfonic acid groups are doubly hydrogen bonded to one another is as much as 12.2 kcal/mol lower in energy than those with separated sulfonic acid groups [15]. It should be pointed out, however, that this fragment with an even (i.e. 4 CF_2) number difluoromethylene groups between the side chains is not realizable with either of the synthetic routes developed thus far for this ionomer.

12.2.2.2. Connectivity Through Hydration

To assess the impact of water molecules to the hydrogen bonding of neighboring sulfonic acid groups in PFSA membranes, Paddison and Elliott [47] continued their study with the two side chain fragments of the

SSC ionomer with different equivalent weights by adding water molecules to the 'dry' fragments (described in Section 12.2.2.1) to determine the minimum number needed to effect the connectivity of the groups. Full optimizations where undertaken on each of the $CF_3CF(-O(CF_2)_2SO_3H)-(CF_2)_n-CF(-O(CF_2)_2SO_3H)CF_3$ (n = 5, 7, 9) fragments after adding water molecules from different locations around the fragment, and it was found that the number of H_2Os necessary to connect or bridge the SO_3H groups was proportional to the separation of the side chains. The resulting B3LYP/6-311 G** global minimum energy structures are shown in Figure 12.14 and when compared to their counterparts in Figure 12.13 reveal that while the distance between the tertiary carbon atoms (i.e. backbone carbon where side chain is attached) in all three fragments has essentially remained unchanged (indicating a very similar conformation of the backbone), the sulfonic acid groups are closer to one another with the –S ••• S– distance approximately 2 Å shorter in each fragment. Specifically it was observed in these minimally 'connected' fragments that the distances between the sulfonic acid groups were approximately 7.2, 9.4, and 11.5 Å for n = 5, 7, and 9, respectively.

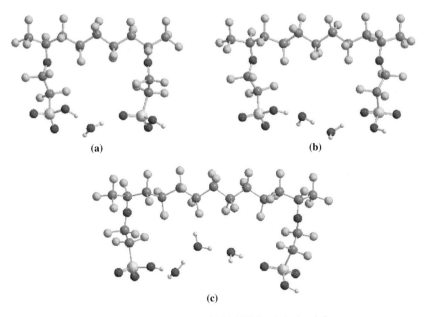

FIGURE 12.14. Fully optimized (B3LYP/6-311G**) global minimum energy structures of two side chain fragments of the SSC PFSA membrane showing the connectivity of the sulfonic acid groups with water molecules: (a) 1 H_2O connects the $-SO_3H$ groups in the fragment with five $-CF_2-$ backbone units; (b) 2 H_2Os connect the $-SO_3H$ groups in the fragment with seven $-CF_2-$ backbone units; (c) 3 H_2Os connect the $-SO_3H$ groups in the fragment with nine $-CF_2-$ backbone units.

It should be realized that closer proximity of the sulfonic acid groups may be achieved due to conformational changes in the backbone (e.g. the structures in Figure 12.7(a) and (b) where the –S ••• S– distances are 5.3 and 6.8 Å, respectively) but was only observed with the addition of a greater number of water molecules.

The effects of a high density and thereby close proximity of perfluorosulfonic acid groups to one another was investigated in a very extensive and to some extent ground-breaking investigation of trifluoromethanesulfonic acid monohydrate through ab initio molecular dynamics (AIMD) simulations [21]. Although this system is actually a solid at room temperature (i.e. m.p. = 309 K) it is nevertheless an important model system for proton transfer in PFSAs applicable to a minimal water content regime and high ion exchange capacity. The crystal structure provides an appropriate basis for controlled simulations as it possesses four formal $CF_3SO_3^-H_3O^+$ moieties in the unit cell and hence allows for studies of the extended system through periodic boundary conditions in all dimensions. Hence, this investigation has the distinct advantage over the previously described electronic structure calculations in that entropic effects at finite temperatures are explored through the MD simulations. The VASP Density Functional Theory based total-energy code [55,56,57] was used under the PW91 Generalized Gradient Approximation [56] with ultrasoft pseudopotentials [58] for all atoms and a plane wave energy cutoff was 40 Rydberg and a Γ-point sampling of the Brillioun. Beginning from the experimental structure at 298 K optimizations followed by several decades of both constant temperature and constant energy MD runs were performed with a step size of 0.2 fs. This provided for a verification of the chosen methodology and structural data for the native structure a snap shot which is displayed in Figure 12.15(a). This structure shows all protons dissociated forming hydronium ions that are hydrogen bonded to three different sulfonate ions with SO_2O ••• H–O distance varying from 2.5 to 2.75 Å. Thus, the sulfonate groups are very close to one another with sulfur–sulfur distances between 4.6 and 5.0 Å. This system therefore demonstrates that high density of perfluorosulfonic acid groups accompanied with only minimal (i.e. $\lambda = 1$) hydration demonstrates that close connectivity may bring about complete dissociation. Perturbation (removal and subsequent return of a single proton in the unit cell separated by extensive MD runs) brought about the transition to a stable defect state in the solid shown in Figure 12.15(b). One of these protons gives rise to a Zundel ion ($H_5O_2^+$), whereas the other one is accommodated between two of the oxygen atoms of a pair of triflate anions. The formation of the latter sulfonate O•••H•••O complex required a considerable rearrangement of the crystal structure so that the oxygens can approach each other at the hydrogen bond distance. The formation of the Zundel cation is critical in this defect state and the free energy of formation was calculated to be 0.2 eV from a quasi-harmonic model based upon statistical determination of normal modes. This energy difference corresponds well with experimental measurements of the activation energy

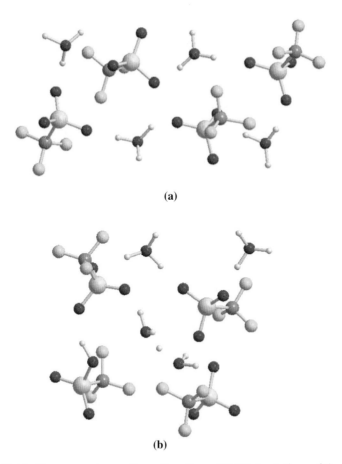

FIGURE 12.15. Representative configurations from AIMD simulations of the trifluoromethanesulfonicacid monohydrate solid showing: (a) the native solid with each hydronium ion hydrogen bonded to three different $CF_3SO_3^-$ anions; and (b) a defect state of the solid with two of the four protons existing as Zundel-like cations: one with two water molecules and the other with two $CF_3SO_3^-$ anions.

for proton transport in minimally hydrated Nafion. The plot of the spectral density in the defect state showed the expected increase in soft modes associated with the presence of the delocalized protons. An observed 'drift' of the Zundel ion suggests that it may function as a mediator of proton transfer, working as a relay group between hydronium ions or sulfonate anions. The observations of the concerted dynamics in the solid indicate that an appropriate flexibility of anionic side chains is an important ingredient of proton transport in PFSAs under conditions of minimal water content and high anion concentrations. This latter result is consistent with the electronic structure results discussed earlier [23].

12.2.3. Cooperativity

Cooperativity in the conduction of protons in hydrated PEMs encompasses effects including the mobility of protons via a flux of water molecules, the amphotericity (i.e. the ability to act as both a Lowry-Brønsted proton donor and acceptor) of the protogenic groups, and the motion of either the protogenic group or side chain that facilitates the 'hand off' or net transport of a proton. In this section we exclude our discussion to only the latter flexibility of the side chains of PFSA membranes.

Paddison and Elliott, in parallel to their studies of hydration and proton dissociation in two side chain fragments of the SSC PFSA membrane, also performed extensive electronic structure calculations on a (dry) fragment to determine rotational potential energy profiles of a CF_2-CF_2 bond along the backbone and each of the bonds along the length of one of the side chains [59] The bonds in the $CF_3CF(-O(CF_2)_2SO_3H)-(CF_2)_7-CF(-O(CF_2)_2SO_3H)CF_3$ fragment for which rotational potential energy surfaces were computed at the B3LYP/6-31 G(d,p) level of theory are highlighted in Figure 12.16. The potential energy profile for the CF_2-CF_2 bond in the center of the backbone is displayed in Figure 12.17(a) and indicates that although the barrier for complete rotation is a substantial 6.9 kcal/mol there is quite a bit of flexibility with smaller barriers of only about 3.7 kcal/mol between the *trans* and *gauche* states. Although it was widely accepted that the ether linkage(s) in PFSA membranes provide enhanced rotational degrees of freedom to the side chains and hence should be incorporated into novel ionomers, the electronic structure calculations indicated that the stiffest portion of the side chain is at its attachment to the backbone. The torsional profiles for the FC–O and O–CF_2 bonds are plotted in Figure 12.17(b) and (c) and these reveal barriers of 9.1

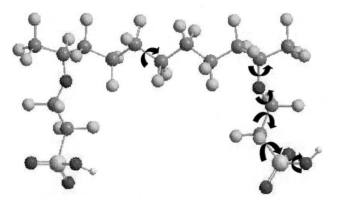

FIGURE 12.16. Fully optimized (B3LYP/6-311G**) oligomeric fragment of the SSC PFSA polymer with seven –CF_2– units in the backbone separating the side chains. Arrows indicate the six different bonds for which torsional energy profiles were calculated (see Figs. 17 and 18).

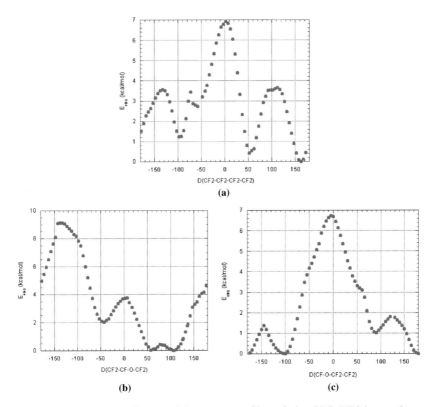

FIGURE 12.17. Computed potential energy profiles of the SSC PFSA membrane fragment in Fig. 16 for rotation about the: (a) F_2C_2–CF_2 bond in the center of the backbone; (b) FC–O and (c) O–CF_2 bonds located at the attachment of the right side chain to the backbone in the oligomeric fragment.

and 6.7 kcal/mol. respectively. The barriers are due to the eclipsing of the O–C bond with the C–F bond and the C–O bond with the C–C, respectively; but despite the significant barrier to full rotation, there is a substantial amount of the surface below 4 kcal/mol and the electronic structure calculation of the oligomeric fragment with from 5 to 7 water molecules revealed conformations of the side chains with dihedral angles within these higher energy regions (i.e. distinct from the global minima). The most flexible portion of the side chain is the terminal section consisting of the carbon-carbon and carbon-sulfur bonds and the corresponding potential energy profiles for rotation about both of these bonds along with the sulfur-oxygen (proton bearing oxygen) bond are displayed in Figure 12.18. Comparison of the torsional profile for the C–C on the side chain (Figure 12.18a) with its counterpart on the backbone indicated that although the three peaks are present in both plots there is nevertheless substantial differences in the energy to rotate about both bonds. The former exhibits essentially three-fold degeneracy with a maximum barrier of slightly

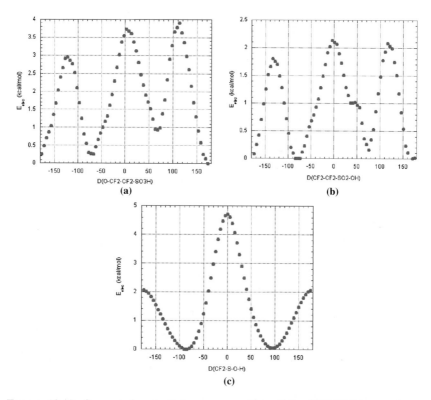

FIGURE 12.18. Computed potential energy profiles of the SSC PFSA membrane fragment in Fig. 16 for rotation about the: (a) F_2C–CF_2 and (b) F_2C–S bonds (b) FC–O of the right side chain; and (c) he S–OH bond at the termini of the side chain.

less than 4 kcal/mol. The F_2C–SO_3H bond is even more labile and similar shaped-profile but characterized by substantially lower barriers of only about 2 kcal/mol. The torsional profile shown in Figure 12.18(c) for the O_2SC–OH bond exhibits two physically equivalent minima and two distinctly different maxima. The highest barrier is nearly 4.8 kcal/mol and is due to rotation that eclipses the O–H bond with the S–C bond. The lower barrier of about 2 kcal/mol is due to rotation that places the acidic proton away from the side chain.

12.3. Conclusions

Although a considerable effort has been undertaken to understand proton conduction in PEMs, much remains to be understood in terms of how molecular chemistry and hydrated morphology dictate fuel cell performance. Molecular modeling of acidic functional groups, polymeric fragments, proton diffusion, and dielectric properties of the confined water in several different PEMs has suggested that the critical ingredients of proton conduction include: *complexity*;

connectivity, and *cooperativity*; and that furthermore, the underlying chemical and physical processes need to be examined across diverse length and time scales. This chapter has sought to review what is presently understood in terms of these three aspects. The *complexity* of proton conduction encompasses dissociation of the proton from the acidic site, subsequent transfer of the proton to the aqueous medium, separation of the hydrated proton from the conjugate base (*e.g.* the sulfonate anion), and finally diffusion of the proton in the confined water within the polymeric matrix. The *connectivity* involves not only hydrogen bonding of the water to the protogenic groups but also greater length scales including connection of the water domains within the polymeric matrix. *Cooperative* effects include the amphotericity of the protogenic groups and also the flexibility of the side chains and/or backbone. These concepts should prove helpful to those performing both computational and experimental studies in the development of high performance electrolytes for fuel cells.

References

[1] Mauritz, K. A.; Moore, R. B. *Chemical Reviews* **2004**, *104*, 4535.
[2] Kreuer, K. D. *Solid State Ionics* **2000**, *136*, 149.
[3] Gebel, G. *Polymer* **2000**, *41*, 5829.
[4] Kreuer, K. D.; Paddison, S. J.; Spohr, E.; Schuster, M. *Chemical Reviews* **2004**, *104*, 4637.
[5] Kreuer, K. D. *Chemphyschem* **2002**, *3*, 771.
[6] Steele, B. C. H.; Heinzel, A. *Nature* **2001**, *414*, 345.
[7] Kerres, J. A. *Journal of Membrane Science* **2001**, *185*, 3.
[8] Wainright, J. S.; Wang, J. T.; Weng, D.; Savinell, R. F.; Litt, M. *Journal of the Electrochemical Society* **1995**, *142*, L121.
[9] Allcock, H. R.; Hofmann, M. A.; Ambler, C. M.; Lvov, S. N.; Zhou, X. Y. Y.; Chalkova, E.; Weston, J. *Journal of Membrane Science* **2002**, *201*, 47.
[10] Kreuer, K. D.; Fuchs, A.; Ise, M.; Spaeth, M.; Maier, J. *Electrochimica Acta* **1998**, *43*, 1281.
[11] Alberti, G.; Casciola, M. *Annual Review of Materials Research* **2003**, *33*, 129.
[12] Zawodzinski, T. A.; Derouin, C.; Radzinski, S.; Sherman, R. J.; Smith, V. T.; Springer, T. E.; Gottesfeld, S. *J. Electrochem. Soc.* **1993**, *140*, 1041.
[13] Kreuer, K. D. *Journal of Membrane Science* **2001**, *185*, 29.
[14] Ezzell, B. R.; Carl, W. P.; Mod, W. A.; The Dow Chemical Company: U.S., **1982**.
[15] Paddison, S. J. *Annu. Rev. Mater. Res.* **2003**, *33*, 289.
[16] Vishnyakov, A.; Neimark, A. V. *Journal of Physical Chemistry B* **2000**, *104*, 4471.
[17] Vishnyakov, A.; Neimark, A. V. *Journal of Physical Chemistry B* **2001**, *105*, 9586.
[18] Urata, S.; Irisawa, J.; Takada, A.; Shinoda, W.; Tsuzuki, S.; Mikami, M. *Journal of Physical Chemistry B* **2005**, *109*, 4269.
[19] Urata, S.; Irisawa, J.; Takada, A.; Shinoda, W.; Tsuzuki, S.; Mikami, M. *Journal of Physical Chemistry B* **2005**, *109*, 17274.
[20] Spohr, E.; Commer, P.; Kornyshev, A. A. *Journal of Physical Chemistry B* **2002**, *106*, 10560.
[21] Eikerling, M.; Paddison, S. J.; Pratt, L. R.; Zawodzinski, T. A. *Chemical Physics Letters* **2003**, *368*, 108.

[22] Paddison, S. J.; Pratt, L. R.; Zawodzinski, T.; Reagor, D. W. *Fluid Phase Equilibria* **1998**, *151*, 235.
[23] Paddison, S. J. *Journal of New Materials for Electrochemical Systems* **2001**, *4*, 197.
[24] Eikerling, M.; Paddison, S. J.; Zawodzinski, T. A. *Journal of New Materials for Electrochemical Systems* **2002**, *5*, 15.
[25] Paddison, S. J.; Paul, R.; Zawodzinski, T. A. *Journal of the Electrochemical Society* **2000**, *147*, 617.
[26] Paddison, S. J.; Paul, R.; Zawodzinski, T. A. *Journal of Chemical Physics* **2001**, *115*, 7753.
[27] Paddison, S. J.; Paul, R. *Physical Chemistry Chemical Physics* **2002**, *4*, 1158.
[28] Paddison, S. J.; Paul, R.; Kreuer, K. D. *Physical Chemistry Chemical Physics* **2002**, *4*, 1151.
[29] Paul, R.; Paddison, S. J. *Journal of Chemical Physics* **2005**, *123*, 224704.
[30] Spohr, E. *Molecular Simulation* **2004**, *30*, 107.
[31] Petersen, M. K.; Wang, F.; Blake, N. P.; Metiu, H.; Voth, G. A. *Journal of Physical Chemistry B* **2005**, *109*, 3727.
[32] Slade, R. C. T.; Omana, M. J. *Solid State Ionics* **1992**, *58*, 195.
[33] Rosso, L.; Tuckerman, M. E. *Solid State Ionics* **2003**, *161*, 219.
[34] Tuckerman, M.; Laasonen, K.; Sprik, M.; Parrinello, M. *Journal of Chemical Physics* **1995**, *103*, 150.
[35] Marx, D.; Tuckerman, M. E.; Hutter, J.; Parrinello, M. *Nature* **1999**, *397*, 601.
[36] Lapid, H.; Agmon, N.; Petersen, M. K.; Voth, G. A. *Journal of Chemical Physics* **2005**, *122*, 014506.
[37] Woutersen, S.; Bakker, H. J. *Physical Review Letters* **2005**, *96*, 138305.
[38] Kreuer, K. D. *Solid State Ionics* **1997**, *94*, 55.
[39] Agmon, N. *Chemical Physics Letters* **1995**, *244*, 456.
[40] Day, T. J. F.; Schmitt, U. W.; Voth, G. A. *Journal of the American Chemical Society* **2000**, *122*, 12027.
[41] Rubatat, L.; Rollet, A. L.; Gebel, G.; Diat, O. *Macromolecules* **2002**, *35*, 4050.
[42] Rubatat, L.; Rollet, A. L.; Diat, O.; Gebel, G. *Journal De Physique Iv* **2002**, *12*, 197.
[43] Rubatat, L.; Gebel, G.; Diat, O. *Macromolecules* **2004**, *37*, 7772.
[44] Paddison, S. J.; Reagor, D. W.; Zawodzinski, T. A. *Journal of Electroanalytical Chemistry* **1998**, *459*, 91.
[45] Paddison, S. J.; Bender, G.; Kreuer, K. D.; Nicoloso, N.; Zawodzinski, T. A. *Journal of New Materials for Electrochemical Systems* **2000**, *3*, 291.
[46] Paddison, S. J.; Pratt, L. R.; Zawodzinski, T. A. *Journal of Physical Chemistry A* **2001**, *105*, 6266.
[47] Paddison, S. J.; Elliott, J. A. *Journal of Physical Chemistry A* **2005**, *109*, 7583.
[48] Paddison, S. J.; Elliott, J. A. *Solid State Ionics* **2005**, in press.
[49] Gronbech-Jensen, N.; Hummer, G.; Beardmore, K. M. *Molecular Physics* **1997**, *92*, 941.
[50] Paul, R.; Paddison, S. J. *Journal of Chemical Physics* **2001**, *115*, 7762.
[51] Paul, R.; Paddison, S. J. *Journal of Physical Chemistry B* **2004**, *108*, 13231.
[52] Kjellander, R.; Mitchell, D. J. *Journal of Chemical Physics* **1994**, *101*, 603.
[53] Attard, P. *Physical Review E* **1993**, *48*, 3604.
[54] Attard, P. *Advances in Chemical Physics* **1996**, *92*, 1.
[55] Kresse, G.; Hafner, J. *Physical Review B* **1994**, *49*, 14251.
[56] Kresse, G.; Hafner, J. *Journal of Physics-Condensed Matter* **1994**, *6*, 8245.
[57] Kresse, G.; Furthmuller, J. *Physical Review B* **1996**, *54*, 11169.
[58] Vanderbilt, D. *Physical Review B* **1990**, *41*, 7892.
[59] Paddison, S. J.; Elliott, J. A. *Physical Chemistry Chemical Physics* **2006**, *8*, 2193.

13
Atomistic Structural Modelling of Ionomer Membrane Morphology

J. A. ELLIOTT

13.1. Introduction

The aim of this chapter is to provide a concise review and synthesis, intended mainly for the non-specialist, of some of the more recent (post-1997) applications of *atomistic* computer simulation techniques, chiefly classical molecular dynamics (MD) and Monte Carlo (MC) methods, with some discussion of quantum chemical methods, including semi-empirical (SE) and ab initio molecular orbital (MO) and density functional (DFT) approaches, to the study of perfluorosulphonate ionomer (PFSI) membrane morphology at the level of the fluorocarbon matrix and ionic 'clusters', and the influence that this has on ion transport. The focus will be mainly on perfluorosulphonate systems, since these are of most widespread industrial interest, but some of the results are illustrated by comparison with those from membrane systems involving hydrocarbon-based ionomers. Since they are covered elsewhere in this volume, we specifically exclude from consideration in this chapter those atomistic modelling results focusing principally on the nature of water and proton transport (see chapters 10, 11), and ab initio MD (see chapter 14). There are clearly strong links to microscopic and mesoscale modelling of morphology (chapter 15) and aspects of proton conduction relating to membrane morphology (chapter 12), but it is hoped that any overlap will be beneficial in aiding the understanding of non-specialists by the consideration of these studies from an alternative point of view.

13.1.1. Structure and Morphology of PFSI Membranes

For the purposes of delineating the scope of the membrane systems considered in this chapter, the generic PFSI chemical structure is defined in Figure 13.1, following the notation used by Tant et al. [1]. All the structures have a common feature –they consist of a (presumed) random copolymer of tetrafluoroethylene and perfluorovinyl ether monomer units – but differ in terms of the length and distribution of the ionic side groups along the main

$$-(CF_2CF_2)_x\text{-}(CF_2CF)_y\text{-}$$
$$|$$
$$(OCF_2CF)_m\text{-}O\text{-}(CF_2)_n\text{-}SO_3^-$$
$$|$$
$$CF_3$$

FIGURE 13.1 Generic chemical formula of perfluorosulphonate ionomer (PFSI), with variable indices m and n. For Flemion: $m = 0–1$, $n = 1–5$; Nafion: $m = 1$, $n = 2$; Dow: $m = 1$, $n = 1$; Aciplex: $m = 0–2$, $n = 1–4$. Typical molecular weights are estimated to be in the range 10^5–10^6 g mol^{-1} (Mauritz and Moore [2]), which would correspond a random copolymer with $x = 595$–5945, $y = 91$–909 for 1100 EW molecules of Nafion.

fluorocarbon chain (the "backbone"). It should be remembered that the quoted values of equivalent weight, (EW), i.e. the mass of dry polymer per mole of sulphonate groups, for PFSIs are always statistical averages and that not much is truly known about the distribution of comonomers along the chain, although it is often presumed that this distribution is uniformly random [2]. However, in section 13.2.1.4, the effects of changing monomeric sequence are discussed. It is immediately clear by inspection from the chemical structures of the PFSIs that there will be a large separation of length scales between the chain length (as measured along the chain) and radius of gyration of the fluorocarbon backbone and the spatial extent of the ionic side group, and thus also the time scale over which these entities can relax to an equilibrium state. This is an important consideration when undertaking atomistic modelling, as we shall discuss later in section 13.2.1.3.

A very large amount of experimental work has already been carried out over the past two decades in an attempt to characterise the morphology of PFSIs at the atomistic scale, using small-angle X-ray and neutron scattering, microscopy (TEM, AFM) and various spectroscopic techniques (IR, near-IR and solid state NMR). We will not attempt to review this work here, but we refer interested readers to an early work by Eisenberg and Yeager [3], and more recent reviews by Heitner-Wirguin [4] and by Mauritz and Moore [2]. Despite an overall consensus that the morphology of hydrated PFSIs consists of nanophase-separated ionic domains embedded in a largely amorphous fluorocarbon matrix, there remain many unanswered questions that continue to cause controversy in the literature. Perhaps foremost among these is the question of the precise nature of ionic 'clusters' in the PFSIs that can clearly be observed with microscopic techniques [5, 6, 7, 8, 9] and give rise to X-ray and neutron scattering [10, 11, 12, 13, 14, 15, 16, 17, 18, 19] by virtue of their contrast against the fluorocarbon matrix. Some authors regard these entities as hard, spherical objects dispersed in a paracrystalline fashion [20], giving rise to scattering by interference between particle centres [21]. Others maintain that the interface between the ionic and fluorocarbon material is much more diffuse [22], with scattering taking place from contrast within individual clusters, and that considerable rearrangement of side groups occurs during swelling and dehydration [23]. Even the overall shape of the ionic domains is disputed; in addition to spherical clusters [23], lamellar [24] rod-like [25, 26] and ribbon-like [27] structures have also been proposed. Much of the remaining confusion in the literature derives from the fact that morphological information from scattering studies is of limited predictive capability due to the difficulty in interpretation of diffuse data and requirement for use of a priori structural models. With atomistic modelling techniques, it is possible to build structures using the bare minimum of physical assumptions, bringing some of the machinery of statistical mechanics and thermodynamics to bear on whatever morphological prejudices one might have to begin with, and to interrogate the resulting models in a level of detail that would not be possible from an experimental study alone. Atomistic modelling, therefore, has an important role to play both in its own right and also as a tool for improving understanding and interpretation of experimental data on membrane morphology.

In addition to the nature of the ionic aggregates, another important unanswered question is the state of aggregation of the fluorocarbon backbone and how this relates to the behaviour of the ionic aggregates on swelling and dehydration, if at all. Most morphological models for ion clustering simply ignore the presence of small amounts of fluorocarbon crystallites or treat them as a small perturbation to the swelling process. Since the fraction of crystallites is rather low, typically 8–12 wt% for Nafion [21], it is hard to obtain detailed information from either microscopic or scattering studies, although it is known that the crystallites are around 4 nm in size both parallel and perpendicular to the chain axis [28] (cf. average distance between side groups along the backbone of around 1.4 nm in 1100 EW Nafion). There is also strong evidence from small-angle X-ray scattering of long-range correlations (of order 10 nm [21]) between the crystallites, which display similar changes on swelling to those of ionomeric domains. A definitive answer as to whether the crystalline domains were intimately associated with the ionic clusters would be extremely useful in helping to resolve some of the controversies concerning membrane morphology currently still discussed in the literature. Again, atomistic modelling techniques can be applied to address the state of aggregation of fluorocarbon material and how this affects, and is affected by, changes in the ionic phase induced by swelling by solvents or changes to the molecular architecture of the polymer.

Lastly, although this chapter focuses on modelling of membrane morphology, the issue of the extent to which the side groups in PFSIs interact and move within the bulk ionic phases is nevertheless of critical importance to understand proton transport and conduction within the membranes. It is now well accepted that there are at least two different environments for water in the membranes [29]: that which is strongly associated with cations or complexes "bound" to the sulphonate groups on the side group and that which behaves essentially as "free water" in the bulk. Clearly, the nature of membrane morphology will have a strong influence on the structure of the water channels, and hence on the proton dissociation and diffusion via both structural ("Grotthuss" [30]) and vehicular (i.e. bulk diffusion) mechanisms. The problem here lies with the very great separation in time and length scales of the molecular relaxations of the polymer backbone, ionic side groups and water/cation complexes. In order to circumvent this difficulty, many existing analytical models for PFSIs, which are usually designed to predict macroscopic transport properties (see e.g. studies by Paddison et al. [31], Paul and Paddison [32], and Bontha and Pintauro [33]), use rather idealised geometries, such as slabs, cylinders or spherical pores, to study the motion of ions. Although such models are capable of yielding the correct qualitative trends, linking fully atomistic models of membrane morphology to predictive models for ion transport would be a very major step forward in realising a multi-scale hierarchy to link chemical and molecular structure to functional properties at the macroscale.

13.1.2. Application of Atomistic Modelling to PFSI Membrane Morphology

While it might be admitted, as Mauritz and Moore [2] claim, that "... there have been no fundamental principles-based ... model for Nafion that has predicted significantly new phenomena or caused property improvements", it is nevertheless the case that computer simulation has an important role to play in establishing an unambiguous connection between molecular-scale phenomena in PFSI sand macroscopic properties relevant to engineering applications. At their best, molecular simulations can offer novel and unexpected predictions concerning membrane and device properties in a fraction of the time and cost of trial-and-error experimentation. However, there is always a lingering possibility for abuse if they are simply used to provide post hoc justifications for already well-known results. This can normally be avoided by validating models against experimental data, abstracting from the model more general hypotheses with which to make new predictions, and finally testing these predictions against new experimental data or analytical theory. For the reasons outlined in section 13.1.1, there are many areas in the area of PFSI morphology where these methods have already been usefully applied, and where future work is urgently required.

In the remainder of this chapter, we will attempt to summarise the main results published so far from atomistic approaches to computational modelling of PFSI membranes. Crudely speaking, fully atomistic simulations can be classified by whether they take account of electronic degrees of freedom (i.e. solve the Schrödinger equation subject to some set of approximations) or just evolve the nuclear co-ordinates according to some classical semi-empirical potential field (which is known generally as the molecular mechanics [34] approach). There are deficiencies with both methods as applied to simulation of ionomer morphology. As highlighted by Mauritz and Moore [2], these include a lack of chemical specificity, inability to simulate both short-range and large scale structures in same model and therefore difficulty with coupling to experimental probes of structure such as SAXS or TEM. Broadly speaking, quantum mechanical methods can take account only of local structures at the level of several ionic side groups, but are capable of good treatment of phenomena such as charge transfer and changes in chemical bonding, and have very few free parameters (and therefore a large predictive capability). On the other hand, molecular mechanics approaches are now currently able to simulate something more resembling a realistic membrane morphology, at least at the level of clusters of ionic side groups and counter ions, but are unable to capture more detailed chemical effects and require careful parameterisation to yield accurate and meaningful results.

The approximate regimes of utility of various computational methods are summarised in Figure 13.2 [35], in particular their links to processing modelling techniques such as finite element analysis via mesoscale modelling techniques (see chapter 15). More recently, there have been attempts to bridge the gaps in the capability at the molecular scale by use of empirical valence bond (EVB) force fields that allow for changes in chemical bonding (see e.g. work by Voth and co-workers [36] and Kornyshev and co-workers [37] on the solvation of an excess proton in water) and hybrid QM-MM techniques that embed quantum clusters (described, say, using molecular orbital theory) within molecular mechanics simulations of macromolecules or periodic solids. One such QM-MM method, known as ONIOM [38] and

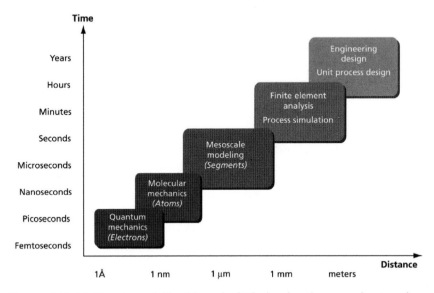

FIGURE 13.2 Multi-scale modelling hierarchy [35], showing the approximate regimes of time and length scales over which atomistic modelling techniques (quantum and molecular mechanics) can usefully be applied and how these link with process methods via mesoscale modelling.

implemented in the GAUSSIAN code, is based on the definition of interconnected "layers" within a simulation that can be treated with different levels of molecular orbital theory and/or molecular mechanics and has been very successful in applications such as enzyme-facilitated catalysis where the regions of quantum mechanical activity within a system can clearly be spatially delineated. However, its application to systems where such separations are less obvious is still problematic. Although we will not discuss QM-MM methods explicitly in this chapter, it is becoming increasingly clear that their application to PFSI materials is now both technically feasible and potentially useful.

13.2. A Review of Recent Atomistic Computational Studies of PFSIs

We will now proceed to discuss some of the existing atomistic studies of PFSI morphology in greater detail, starting with and focusing mainly on results obtained with purely classical molecular mechanics and molecular dynamics methods. There will be a brief discussion of EVB methods for treating excess protons in section 13.2.2 and finally as consideration of molecular orbital and DFT simulations on atomistic fragments of PFSIs.

13.2.1. Classical Molecular Mechanics Models

13.2.1.1. Polyelectrolyte Models

The earliest molecular dynamics studies (MD) of PFSIs were based on only partially atomistic models, usually consisting of atomistic side groups confined in some specified geometry, e.g. Dyakov and Tovbin [39] (slit-shaped pores) and Din and Michaelides [40] (cylindrical pores with negatively charged walls). Although interesting in the context of studying ion transport, by definition they cannot provide much information on membrane morphology. The first truly all-atom MD simulations of ionomers were actually based on polyelectrolyte analogues: sulphonic acid anions similar to Nafion side groups by Elliott and Hanna [41], and polyethylene oxide sulphonic acid anions by Ennari et al. [42]. The main reason for this simplification was that the computer technology of the time was incapable of simulating a realistic polymer morphology on a reasonable time scale, at least using brute force methods alone (i.e. canonical MD with standard valence potentials). Nevertheless, despite the absence of a polymeric backbone, some interesting phenomena of relevance to real PFSI systems were revealed. Using an ensemble of 26 sulphonate anions (modelling using augmented DREIDING [43] force field, with AM1 partial charges), 74 TIP3P [44] water molecules and 26 hydronium ions (H_3O^+, made by adding proton to TIP3P and distributing excess positive charge over hydrogens), Elliott and

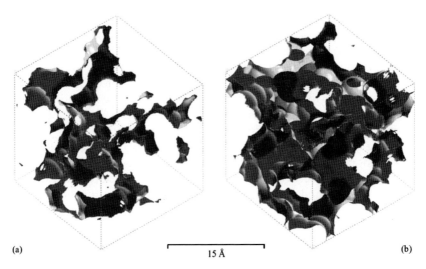

FIGURE 13.3 Visualisation of the water-filled channels in polyelectrolyte model for Nafion, as simulated by Elliott and Hanna [41]. The atomic fragments have been deleted for clarity, replaced by three-dimensional Connolly surfaces produced from models containing: (a) 6.3% and (b) 15.9% water by mass. In the original paper, the surfaces were coloured by proximity to polar and non-polar regions.

Hanna found that the system spontaneously assembled from a random initial state into a stable phase-separated morphology with water-filled channels that appeared to swell on addition of further water molecules. A visualisation of the morphologies for the ambient (6.3 wt% water) and swollen (15.9 wt% water) states is shown in Figure 13.3. These structures are highly reminiscent of the cluster-network model of Hsu and Gierke [23], although the length scales of the channels are rather smaller, probably due to the absence of the fluorocarbon backbone. Of particular interest, however, was the fact that the channels were shown to be selectively conductive to cations when hydroxyl ions (together with a commensurate number of neutralising hydronium ions) were added to the simulation. It was shown that the mechanism of selective conductivity was jump diffusion of H_3O^+ between sulphonate groups, with OH^- confined to the bulk water phase where it remained relatively immobile. Of course, this mechanism is naïve in the sense that it ignores completely the proton transfer between water molecules and assumes 100% ionisation of the sulphonic acid groups when it is known that this is also a function of membrane water content. Nevertheless, it demonstrates that even such a simple model system can reproduce correctly many of the qualitative features of morphology and associated ion transport in PFSIs.

The polyelectrolyte system modelled by Ennari et al. consisted of a single PEO sulphonate dianion (modelled by augmented PCFF forcefield [45, 46]), a lone proton, a single hydronium ion and 220 water molecules. The proton was created with non-bonding terms so that it was free to jump between water molecules during the simulation, driven by electrostatic interactions. It was found that the hydronium ion was strongly co-ordinated to one or two of the sulphonate groups on the dianion, whereas the proton was co-ordinated to the only one end of the anion or moved freely. The diffusion coefficients of the proton, hydronium ion and water were calculated, and the conductivity of the system was found to be of the correct order of magnitude when compared to experimental values for H^+ in water at infinite dilution. Since just two sulphonate groups were present in their model, only limited information about membrane morphology can be inferred, but in general the pair distribution functions between the water oxygens and the sulphonate dianion show the usual bound and free water peaks, as also found by Elliott and Hanna [41].

13.2.1.2. Oligomeric Models

More recently, Vishnyakov and Neimark described the use of force field models capable of describing oligomeric PFSI systems, some of which could reasonably be termed as "polymeric". In the first of their studies [47], they simulated the molecular dynamics of perfluorosulphonate oligomers consisting of 10 monomer units solvated in water and methanol, which revealed a noticeable difference between the geometries of the fluorocarbon backbone in different solvents. In particular, the fluorocarbon chain in water was substantially more folded than in methanol. The ionic side group of the Nafion oligomers was found to be rather stiff, and only a few conformational transitions were detected. A similar simulation methodology was used recently by Rivin et al. [48] to investigate the interactions of Nafion and dimethylmethylphosphonate (DMMP), revealing that DMMP segregates to the hydrophilic interface in solvent mixtures with water, although no further new insights into membrane morphology were discussed.

In a later study [49], Vishnyakov and Neimark extended their force field model, which was based on DREIDING [43] augmented by bond distances and angles from experimental and ab initio data and backbone dihedral angles calculated using DFT, to study Nafion oligomers with 4 ionic side groups and 1164 EW (corresponding to $x=7$, $y=4$ in Figure 13.1, but with an alternating comonomer sequence) swollen in equimolar mixtures of methanol and water using Na^+ as the counterion. By way of comparison, force fields for unsubstituted fluoroalkanes such as PTFE have been available for some time (see e.g. work of Holt et al. [50]), which can accurately reproduce the rotational chain conformations in the crystalline state. Recently, Jang et al. [51] undertook a more detailed ab initio study of perfluoroalkanes, combined with input from X-ray crystallographic data and experimental compressibility curves, and concluded that electrostatic forces play an important role in their helical chain conformations at low temperatures.

The initial condition for the MD simulations of Vishnyakov and Neimark was one four-unit oligomer surrounded by 1024 solvent molecules (water being represented by the SPC/E model of Berendsen et al. [52], and methanol by the three-centre potential of van Leeuwen and Smit [53]) on a lattice at low density (0.2 g cm^{-3}), which was then thermalised and pressurised to standard conditions ($T = 298$ K, $p = 1$ atm) by an alternating series of isobaric-isothermal (NpT) and canonical (NVT) MD simulations until an equilibrium density of 0.946 g cm^{-3} was reached. The reasons for this time-consuming and careful equilibration procedure were to ensure a relaxed chain configuration and low residual stresses in the equilibrated structures. As later studies would show, the problem of achieving equilibrated macromolecular structures is exacerbated as the degree of polymerisation of the PFSI chains increases. From their simulations of oligomeric fragments in equimolar methanol/water mixtures, Vishnyakov and Neimark found a surprisingly high mobility for the fluorocarbon backbone: the number of conformational transitions was three times as large as for the simulations in pure water. This has interesting ramifications for the rationalisation of exaggerated PFSI swelling behaviour in solvent mixtures [54], where it has been hypothesised that the addition of less polar solvents to water can allow greater swelling due to plasticization of the fluorocarbon matrix [55]. As in their previous study of smaller oligomeric PFSIs [47], Vishnyakov and Neimark found that the ionic side groups on the oligomer were extremely stiff, and no appreciable clustering was observed, probably due to the very small number of sulphonate groups present in their simulations.

13.2.1.3. Polymeric Models

The first study to describe atomistic simulations of what could realistically be called "polymeric" PFSIs was published by Vishnyakov and Neimark [56] as a development from their work on oligomeric systems [47, 49], although the molecular weights used (around 1.2×10^4 g mol^{-1}) were still at least in an order of magnitude smaller than those found in real PFSI membranes. Nevertheless, over the time scales capable of being simulated using classical MD, such entities can be considered to be macromolecular to a fairly good approximation. The Nafion "polymer" was represented by 15 polymer chains each containing 10 side groups (corresponding to a EW 1164 membrane) at three different water contents: 5.0 wt% (ambient humidification), 12.5 wt% (saturation) and 17.0 wt% (an unphysically high water content) using K$^+$ as the counterion. In order to reduce the computational workload, the force field developed for the oligomeric fragments described in previous studies [47, 49] was coarse-grained by using a united atom (UA) representation for the CF$_2$ and CF$_3$ groups. This required re-parameterisation of the torsional potentials for the ionic side group, for which no previously published data were available, although UA models for perfluoroalkanes are readily available in the literature [57]. As in the previous study using longer oligomers [49], the equilibration procedure was not straightforward, and consisted of building an amorphous structure from the components at low density (0.2 g cm^{-3}), compressing via NpT simulations at elevated pressures (10 MPa) to intermediate densities (1.5 g cm^{-3}), before reaching the final density via an extended NpT simulation at 1 atm. It was claimed that a fully relaxed state had been reached by monitoring the profiles of equilibrium properties of the system, such as density and contributions to the potential energy. However, the authors acknowledged that this might not be sufficient to establish that equilibrium has been achieved, and in fact other more stringent tests that are well known in the context of simulating glassy polymers include monitoring of the equilibrium torsional distributions and chain end-to-end distances. In particular, methods for preparation of relaxed polymer melts for use in computer simulation studies are discussed by Brown et al. [58] and more recently Auhl et al. [59], which aim to produce dense ensembles of polymer chains with low residual stresses and correct conformation statistics. To date, these methods have not been applied directly to PFSIs, but work is ongoing in this area.

Nevertheless, Vishnyakov and Neimark [56] undertook a detailed structural analysis of their "polymeric" models, including a discussion of the morphology of microsegregation between a hydrophilic (aqueous) phase formed by water and counter ions, and a hydrophobic (organic) phase formed by the fluorocarbon backbone. The evidence from RDFs suggested a segregation that became more pronounced with increasing water content. Although no direct visualisations were published, the structure qualitatively resembled that displayed in previous studies using polyelectrolyte or oligomeric models. The authors found that the hydrophilic phase consisted of disconnected clusters, containing up to 100 water molecules. There was no evidence of a percolation transition from a collection of isolated clusters to a fully connected network, as envisaged in the cluster-network model of Hsu and Gierke [23]. Instead, the authors proposed that water and counter ion transport occurred by the formation of short-lived channels between clusters, leading to a redistribution of water molecules between clusters over the course of a 1 ns MD simulation. The time scale for the formation and break-up of these connective channels was found to be of the order of 100 ps, giving an effective mobility of water molecules in the channels approximately the same as in bulk water. However, it was clear from measurements of the mean-squared displacements of the water molecules and counter ions that the phase segregated structure was still evolving significantly over the time scale of the simulation. This made it impossible to distinguish between motions of water molecules within a cluster, and between different clusters. Also, the size of the simulation cell was not sufficient to give a clear picture of transport processes in real Nafion membranes. Furthermore, the force field used does not allow for a detailed study of the structure of the fluorocarbon matrix, in particular whether crystallinity is present or not. Nevertheless, it is clear from molecular simulations that the phase segregated morphology of PFSIs is highly labile at the atomic scale and that considerable structural rearrangements of the hydrophilic phase and associated ionic side groups can take place even with the stereochemical constraints imposed by the polymeric backbone.

In a related but more recent study, Urata et al. [60] performed classical molecular dynamics on model PFSI polymers with a very similar chemistry to those considered by Vishnyakov and Neimark [56]. However, whereas the latter approximated the entire polymer using united atoms, Urata et al. [60] used an explicit all-atom model for the ionic side groups, with torsional potentials and partial atomic charges calculated both using hybrid density functional theory (DFT) at the B3LYP/6-31G* level, and molecular orbital (MO) theory at the MP2/6-31G* level. UA sites on the backbone were set to be charge neutral, which was considered to be acceptable since such a model can reproduce the vapour–liquid equilibria for fluorocarbon molecules. NpT dynamics simulations were run with a Nosé-Hoover chain thermostat [61] and an Andersen barostat [62] on systems at 358.15 K and 0.1 MPa with four different water contents: 5, 10, 20 and 40 wt%, with hydronium as the counterion. The results were qualitatively similar to those described in previous MD studies, in that a nanophase-separated structure was observed, comprising clusters of sulphonate groups with spacings between 4.6–7.7 Å, together with bound water strongly associated with each sulphonate (and not around the ether oxygen of the side group, due to a reduction of the proton affinity on fluorination [63]). However, perhaps with the benefit of their all-atom representation of the ionic side group, the authors were able to observe that the side groups tended to align perpendicular to the water-ionomer interface, with long-range correlations in the orientation emerging as the water content increased. They also noted that the water dynamics was highly restricted at lower water contents due to the strong interactions with the sulphonate groups. By contrast, at higher water contents, even water near the sulphonate groups was relatively mobile (although some of their water contents were unphysically high). Finally, Urata et al. [60] calculated simulated structure factors $S(q)$ from their RDFs to compare with experimental observations from scattering studies. These showed a smaller ionic cluster size in the simulations, which was attributed to the shorter polymer chain (i.e. lower molecular weight) used. However, it is also likely to be a model-size effect and illustrates the limitations of MD for looking at large-scale structures in PFSIs. It is clear that even more coarse-grained methods than classical MD may be of benefit in simulating larger scale polymeric structures.

Between the appearance of the MD studies by Vishnyakov and Neimark [56], and that by Urata et al. [60], Khalatur, Khokhlov and co-workers published two coarse-grained molecular modelling studies of Nafion [64, 65] that utilised rather different approaches to studying the morphology of PFSIs. The first of these [64] was based on a hybrid Monte Carlo/reference interaction site model (MC/RISM). The principle behind this method was to use MC simulations, based on the rotational isomeric state (RIS) approximation [66] with short-range interactions obtained from semi-empirical quantum mechanics (AM1), to generate conformations of a single polymer chain that were used to calculate intramolecular distribution functions. These distribution functions were used to find the partial intermolecular distributions by solving a series of coupled integral equations, which in turn define an effective potential in which the original single chain conformations could evolve. The process was then iterated until self-consistency between the intra- and intermolecular distributions was achieved. Rather than using an all-atom model, a representation of the Nafion chain based on a united atom (UA) approximation using hard core repulsive particles with attractive tails (specifically a Yukawa-type potential parameterised from AM1 results and excluded volume arguments) was used. There were 9 UA units (corresponding to CF_2 or CF_3 moieties) between each ionic side group (defined by 7 UA units), giving an EW of approximately 900. An attempt was made to simulate the differing spatial orientations of ionic side groups along the backbone by pre-averaging the structure factors from RIS conformational statistics before MC/RISM calculation, but there was no account taken of the statistical distribution of the groups along the chain. It should also be noted that the densities of the models studied (ca. $1.5\,\mathrm{g\,cm^{-3}}$) were 10–15% lower than for a typical hydrated membrane, for technical reasons.

The main conclusion of this work was that the water and polar sulphonic acid groups (there was no attempt to simulate dissociation of the acid) segregate into a three-layer structure with a central water-rich region, an outer layer of side groups strongly associated with water molecules and a "corona" of the less polar regions of the side groups which are immersed in the fluorocarbon matrix. These results were very similar to those obtained from earlier MC studies of other ionomeric systems in the presence of water obtained by Khalatur et al. [67], indicating that such morphologies are rather a generic feature of polymer systems containing both repulsive and attractive interactions. Nevertheless, in their study of Nafion, Khalatur, Khokhlov and co-workers found that the characteristic length scale of the structural segregation, as measured from partial structure factors for sulphur units and water molecules, increased perfectly linearly with water content (as opposed to with the one-third power, as might be expected for spherical aggregates). They rationalised this observation by invoking the irregular nature of the interface between the polar and non-polar regions in the phase-segregated morphology, and the presence of water adsorbed in the fluorocarbon matrix. However, such a linear dependence of microscopic swelling on water content has also been observed experimentally [14, 21, 68] and attributed variously to the expansion of 2-dimensional (lamellar) aggregates [24] or structural reorganisation during growth of 3-dimensional (spherical) clusters [14, 68]. In their simulations, Khalatur, Khokhlov and co-workers observed that incorporation of water molecules inside the aggregates resulted in both an increase in their stability and also in the number of associating groups in a stable aggregate. This would seem to be strongly supportive of view that X-ray scattering is produced by interference between spherical ionic clusters that reorganisation on swelling with water, and it is significant that such a result can emerge from a fully polymeric, albeit coarse-grained, molecular model containing no a priori assumptions about the membrane morphology. However, it should be noted that the model perpetuates the idea of long continuous channels required for ionic conductivity in Nafion, at even very low water contents, despite Vishnyakov and Neimark's findings [56] that water and counter ion transport can occur by the formation of short-lived channels between clusters. Therefore, especially in view of the very simplistic treatment of the hydrated sulphonic acid groups, and the high degree of structural coarse-graining of the side groups, the implications of the morphological findings for ion transport should be interpreted with caution.

In the second study using coarse-grained methods [65], Mologin et al. used a lattice molecular dynamics (LMD) approach (which they referred to as a cellular automaton-based simulation) that can give convergence to limiting distribution functions nearly an order of magnitude faster than typical MC lattice-based simulations of polymers, such as the bond fluctuation model (BFM) [69]. The LMD algorithm is based on a coarse-grained model for polymer chains with variable bond lengths between connected nodes, similar to the BFM, but with deterministic rules for propagation of polymer segments on the lattice as opposed to stochastic evolution in the standard MC approach. In this sense, the simulations are analogous to lattice gas automata models used for studying fluid flow (for a comprehensive description of such techniques, see e.g. Rothman and Zaleski [70]). A coarse-grained representation of Nafion as a comb-like copolymer was used, with sulphonic acid side groups spaced at regular intervals down the backbone. There were approximately 9 lattice units (corresponding to CF_2 or CF_3 moieties) between each ionic side group (defined by 7 lattice units), giving an EW of approximately 900. Systems of 30 or 240 macromolecules were run at differing water contents, using realistic densities for hydrated membranes of ca. 2.0 g cm^{-3} (cf. previous MC/RISM study of Khalatur et al. [67], which used densities 25% lower). Semi-empirical quantum mechanical calculations (AM1) were used to set pairwise nearest neighbour interaction terms for the occupied lattice sites, such that there were attractive interactions between sulphonic acid groups, water, and between water and sulphonic acid groups. All other interaction terms were set to zero.

Using their LMD model, Mologin et al. [65] found that there was a decrease in the radius of gyration of the fluorocarbon backbone together with a corresponding stretching of the side groups on changing from a completely athermal system (i.e. with all interaction terms set to zero) to a system with pairwise attractive terms between sulphonic acid groups and water. Both of these were monotonically increasing functions of water content. The attractive interactions drove an aggregation of the water and side groups to produce a familiar structure with spherical agglomerates of polar material embedded in a fluorocarbon matrix. From the dependence of the radius of gyration and moments of inertia of the aggregates on water content, it was concluded that there are large changes in the internal structure and shape as the number of water molecules increases. The static structure factors computed from all particles in the system showed the presence of a peak attributed to the average interdomain distance (as opposed to cluster size), which shifted to lower angles and became broader with increasing water content in a similar fashion to the "ionomer" peak observed in experimental SAXS studies. The network of water-filled channels was visualised using three-dimensional Connolly surfaces, and a selection of these are shown in Figure 13.4 using a probe with radius of twice the lattice spacing. Figure 13.4(a) shows that sulphonic acid groups are grouped into spherical aggregates that are highly inhomogeneously distributed throughout the model, and Figure 13.4(b) shows that the water network consists also of discrete cluster linked by narrow connective channels. Mologin et al. rationalised the structure at this point in terms of a disordered bicontinuous network where the polar component, Figure 13.4(b), is close to the percolation threshold. They hypothesised that the rapid onset of ionic conductivity in membranes at water contents of 3–4 H_2O molecules per sulphonic acid group is due the formation a fully connected bicontinuous structure, which was consistent with preceding MC/RISM calculations. However, since it is well known that dissociation of the sulphonic acid group (which itself is a function of water content) is required for proton conductivity in the acid form of the membrane, and other studies have shown that a statically continuous network is not required for ion conduction, it is probable that the direct link with onset of percolation is just coincidence.

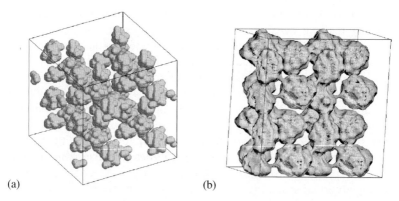

FIGURE 13.4 Connolly surfaces, taken from LMD simulations by Mologin et al. [65], showing (a) surface of associating sulphonic acid groups at a water content of 4 H_2O molecules per sulphonic acid group and (b) surface of solvent system from a different configuration of a simulation with the same water content. The probe radius was twice the lattice spacing.

Mologin et al. [65] also made some direct comparisons between the structure of the polar aggregates in their simulations and some phenomenological models used historically to analyse SAXS data, such as the core-shell [11, 21, 71] (CS) and depleted-zone core-shell [11, 72] (DZCS) models, concluding that neither model accurately represents the internal structure of the agglomerates. Furthermore, the results shed a disfavourable light on lamellar models for ionic aggregates, such as that proposed by Litt [24], since such high aspect ratio domains were not observed during the simulations.

13.2.1.4. Effect of Monomeric Sequence in Polymeric Models

So far, none of the atomistic models for polymeric PFSIs discussed have made any explicit mention of the effect of monomeric sequence on the morphological properties. As mentioned in the introduction, it is often presumed that the distribution of ionic side groups along the chain is random, however this has not been definitively established by experiment and it is clearly likely that there will be statistical variations which could possibly be exploited to improve membrane properties. To this end, Jang et al. [73, 74] recently published two related studies using molecular dynamics to investigate the effect of monomeric sequence in Nafion-type PFSIs on nanostructure and water transport. They investigated two extreme cases of ionic side group distribution on a backbone with EW 1150 and molecular weight 1.15×10^4 g mol^{-1} (i.e. almost equivalent to that used by Vishnyakov and Neimark, but still an order of magnitude smaller than those found in real PFSI membranes). In one case, the side groups were distributed uniformly randomly along the chain (i.e. a random copolymer) and, in the other, they were grouped into two distinct blocks (i.e. a diblock copolymer). The authors stressed that neither case was representative of a true Nafion chain, rather that their aim was to demonstrate an effect of monomeric sequence on properties using two extreme examples.

The systems consisted of four Nafion chains with 560 F3C [75] water molecules and 40 hydronium ions, corresponding to 15 water molecules per sulphonate group (around 20 wt% water). Two independent samples with different initial configurations were studied so that the sensitivity to statistical variations could be assessed. The initial amorphous polymer configurations were generated using a Monte Carlo growth technique into a periodic cell at high density (1.8 g cm^{-3}) before insertion of water and hydronium fragments into the larger voids. The intra- and inter-molecular interaction potentials were described with DREIDING [43], augmented with a more detailed force field [51] for the fluorocarbon portions of the chain. A rather involved equilibration procedure, consisting of a series of expansions and compressions using NpT molecular dynamics (Nosé-Hoover method [76, 77, 78]) at elevated temperatures (600 K) followed by a final relaxation at 300 K, was used to produce periodic structures with final densities of 1.67 g cm^{-3} for the diblock copolymer and 1.60 g cm^{-3} for the random copolymer. The authors claimed that the explanation for the calculated densities being around 5–10% lower than experimental membranes with equivalent water content was that their model systems were purely amorphous. However, since the crystallinity of real membranes is typically 8–12 wt% for Nafion [21] and the density of crystalline PTFE is only about 6% higher than "dry" Nafion, this explanation cannot account for more than a 1% decrease in the model density compared to experiment. It is most likely that the remaining discrepancy is due to either incomplete relaxation of the amorphous polymer or details of the force field parameterisation, but little data on the backbone chain relaxation were presented (although, later in the paper, it is stated that simulated structure factors from the models are unchanged after the equilibration period). It is possible, therefore, that at least some of the difference between the densities of the diblock and random copolymers could be due to kinetic effects, although it would seem reasonable that the diblock copolymer density is higher since it is well known that random copolymers are typically amorphous.

In addition to the variation in density between the two models with different monomer sequence distributions, Jang et al. [73] observed a difference between the nanophase-separated morphologies formed in each case. In the diblock copolymer, the sulphonate groups were more clearly aggregated into clusters, whereas in the random copolymer they were more uniformly distributed throughout the system. However, the sulphur-sulphur distances in each case were almost independent of monomeric sequence. The authors calculated simulated structure factors, $S(q)$, from their models to enable a comparison with experimental SAXS and SANS studies. These confirmed that the diblock copolymer had a more strongly segregated morphology than the random copolymer. They obtained $S(q)$ profiles with an "ionomer" peak at low q which, for the largest systems studied (comprising 8 independent configurations of the smaller systems combined into a single cell) corresponded to a characteristic spacing of around 50 Å for the diblock copolymer, which is comparable to values obtained from experiment for membranes with similar water contents. In contrast, the equivalent structure formed by the random copolymer had a smaller characteristic spacing of around 30 Å. Jang et al. [73] concluded from this that the degree of "blockiness" in real Nafion membranes is intermediate between the two extreme cases they studied but closer to the diblock copolymer. Of course, as in the case of work by Urara et al. [60], it is dangerous to extrapolate from periodic structures with cell sizes of the same order as the characteristic spacing of interest, but the conclusion that increased blockiness of monomeric sequences leads to larger characteristic spacings for the "ionomer" peak seems reasonable given the stereochemical constraints imposed on the sulphonate groups in the random copolymer case. Nevertheless, it is worth noting that even the largest systems considered contain no more than two or three "clusters", so these can hardly be considered representative.

Jang et al. [73] analysed their model structures further in terms of the heterogeneity of the water-ionomer interface by partitioning the Connolly surface of the water phase into hydrophilic patches interspersed on a hydrophobic background. A hydrophilic patch was defined by the locus of points on the Connolly surface lying within the centre of a sphere centered on each sulphur atom. Figure 13.5 shows the decomposition of the interface into hydrophilic and hydrophobic domains, and it should be noticed that the diblock copolymer, Figure 13.5(b), has a larger proportion of bigger hydrophilic patches than the random copolymer, Figure 13.5(a). The authors took this to be further evidence that the interface between the water and ionomer phases is more segregated as monomeric sequence becomes blockier. Finally, Jang et al. [73] analysed the transport properties of water and hydronium ions in their structure. Both diffusion coefficients and activation energies for calculated for the water molecules were broadly consistent with experimental measurements, although the protonic diffusion is slower due to neglect of the structural diffusion mechanism. Interestingly, they showed that while water transport is inhibited in random copolymer relative to the diblock case, the hydronium diffusion coefficients are largely unaffected by monomeric sequence within statistical errors. The authors therefore concluded that experimentally observed differences in proton diffusion must be accounted for by "proton hopping". It is not clear whether they were referring to protons hopping between water molecules from sulphonate groups to water molecules, or from sulphonate to sulphonate. However, it is obvious that in order to explain important details of the relationship between morphology and ion

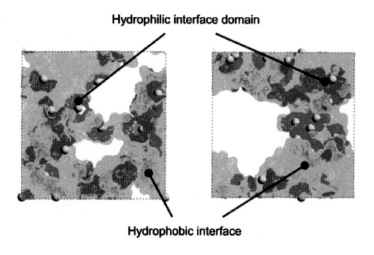

FIGURE 13.5 Connolly surfaces, taken from MD simulations by Jang et al. [73], showing the regions of hydrophilic and hydrophobic interface between the water and ionomer phases in polymeric models for Nafion with different monomeric sequences: (a) random copolymer and (b) diblock copolymer.

13.2.2. Empirical Valence Bond Models

Clearly, a full-scale ab initio MD simulation of proton transport in an atomistic model for PFSIs is out of the question at the time writing. However, one method for taking into account the proton mobility in a more realistic way in a classical MD framework is to use an empirical valence bond (EVB) approach, in which co-ordinated clusters of water with can exchange an excess proton in quasi-quantum mechanical fashion by approximating the motion of protons as classical on the time-dependent ground-state potential energy surface. Despite the approximations involved and technical difficulties in implementation, such EVB approaches have enjoyed reasonable success in the area of simulating excess proton transport in water (see, e.g. work of Schmitt, Voth, et al. [79, 80, 81]) with a very large saving in computational effort compared to full ab initio MD simulations. In the following discussion, we will necessarily restrict our attention to studies that use EVB simulations in conjunction with an atomistic model for PFSI morphology and refer the reader to chapter 10 for fuller discussion of modelling proton transport in general.

Spohr et al. [37] investigated proton transport in concentrated aqueous proton solutions in slab pores as model systems for Nafion-like PFSIs. They used the single-state EVB model of Walbran and Kornyshev [82] to incorporate a Grothuss-type proton transfer mechanism into their classical MD simulations, in which the proton delocalises over just two water molecules. The fluorocarbon portion of the Nafion was replaced by a wall potential defining a two-dimensional slab, to which a variety of different chemical representations of the ionic side groups were attached: (i) homogeneous negative charge, (ii) static point charges, (iii) static SO_3^- groups, (iv) harmonically tethered SO_3^- groups and (v) fully atomistic Nafion side group (based on that used by Vishnyakov and Neimark in their studies of oligomeric Nafion [49]). The authors found that proton mobility generally increased with increasing delocalisation of the negative charge on the "side groups", which were modelled in the various ways described above. It was found that changing the side group model from a simple point charge to a sulphonate group produced the most drastic reduction. On the other hand, side group conformational motion (as allowed in the fully atomistic case) led to an increased mobility, which demonstrates the importance of atomistic treatment of the side groups. In general, any motion of the sulphonate groups led to increased mobility of the protons, which was interpreted as a lowering of the Coulombic barrier between adjacent bound proton-water clusters. The general conclusions of the work are that the precise arrangement and atomistic structure of the ionic side groups in PFSIs can have a profound effect on the proton mobility when the structural diffusion mechanism is allowed.

In a more recent study, Petersen et al. [83] used a second-generation multistate EVB (MS-EVB2) model produced by Day et al. [36] to study the solvation of a proton in a hydrophilic "pocket" of Nafion 117. In contrast to Spohr et al. [37], only one delocalised proton was considered, however the authors argued that the MS-EVB2 model is superior to the EVB model used previously since it can consider states in which the proton is delocalised over many different water molecules. Nevertheless, they acknowledged that further work is required to enable all excess protons to be delocalised, especially at low levels of hydration. The simulations were based on an all-atom model for 1100 EW Nafion containing 4 chains each with 10 equally spaced ionic side groups. The polymeric portion of the chain was described by the force field parameters of Jang et al. [73], whereas the water and excess proton was described by the MS-EVB2 model [36]. Two systems with differing degrees of hydration were considered: one with 7 water molecules per sulphonate group ($\sim$10 wt% water) and the other with 15 ($\sim$20 wt% water). Having prepared amorphous structures by compressing periodic simulation cells from reduced densities (50% of experimental density) at elevated temperature, Petersen et al. [83] performed constant temperature Nosé-Hoover MD at 300 K on 10 distinct configurations in which a classical hydronium ion was replaced at random by a hydronium treated by MS-EVB2. From these trajectories, a total of 2 ns of simulation data at each hydration level were collected for further analysis.

From their simulations, Petersen et al. [83] found a very similar nanophase-separated morphology for both of their hydrated models to that observed in previous MD studies [41, 56, 73]. However, the solvation structure around the sulphonate anions showed that the proportion of protons lying close to the sulphonates is reduced by a factor of roughly one half compared to that found using purely classical hydronium ions (i.e. with structural diffusion mechanism switched off) for the 10 wt% hydration model. This ratio approached 1:1 as the hydration level was increased to 20 wt%, which the authors attributed to the greater affinity for water of the MS-EVB2 hydronium in the first solvation shell. Furthermore, they found a reduced diffusion coefficient for the proton at the lower hydration level, which was attributed to trapping of the MS-EVB2 proton by "cages" of classical hydronium.

In conclusion, it would appear that a more realistic treatment of proton transport than simple bulk diffusion of H_3O^+ ions is required to connect bulk PFSI model morphologies to their effects on macroscopic properties. In particular, the static distribution of protons around sulphonate groups using EVB approaches is quite different from that predicted by classical MD, especially in the regime of minimal hydration. Since it is precisely this regime that is most interest for fuel cell applications, we will conclude our review of atomistic simulation methods by considering how far explicit quantum mechanical modelling using molecular orbital and density functional theories can contribute to the understanding of membrane morphology emerging from classical and semi-classical simulation work.

13.2.3. Molecular Orbital and Density Functional Models

Due to the extreme computational demands of full ab initio calculations, some of the earliest work in this area focused exclusively on modelling of side group conformation and interactions with water in PFSIs. Paddison and Zawodzinski [84] used a combination of Hartree-Fock (HF), Møller-Plesset perturbation theory (MP2) and hybrid DFT (B3LYP) with 6–31 G** split valence basis sets to examine the energy minimised conformations of triflic acid (CF_3SO_3H), the di-trifluoromethane ether linkage (CF_3OCF_3) and the entire Nafion side group ($CF_3OCF_2CF(CF_3)OCF_2CF_2SO_3H$). Since this study, and much of the subsequent work by Paddison and co-workers [85, 86], is discussed more fully in chapter 12, we will mention only very briefly here the results obtained on the flexibility and hydrophobicity of the side group. Paddison and Zawodzinski [84] found that the ether portion of the side group is hydrophobic and stiff, but that the sulphonate groups is very hydrophilic and relatively flexible. This conclusion is also supported by more recent work by Urata et al. [63] who modelled ionic side group interactions with water using MO and DFT. With only very small amounts of water, Paddison and Zawodzinski [84] found that the side group conformation was tightly coiled, with an energy barrier to unfolding of between 3 and 4 kcal mol^{-1}. More recent work by Paddison and Elliott [87], who calculated minimum energy structures of the shorter Dow side group (see Figure 13.1, $n = 1$) attached to a fluorocarbon backbone using hybrid DFT at the B3LYP/6-31 G** level, revealed that the conformation of the side group and, in particular, the connectivity of the sulphonic acid groups is highly influenced by the water content of the membrane. They also demonstrated that connectivity and amount of water required to produce dissociation of the acidic proton is governed by the spacings between side groups along the chain. Of course, the global conformations of the side group are intimately related to the backbone conformation [88], but to date there have been no published studies that combine both large-scale classical models of fluorocarbon matrix (such as those of Vishnyakov and Neimark [56], Jang et al. [73] and Petersen et al. [83]) with detailed quantum mechanical calculations of side group conformation and water interactions (rather than resorting to empirical methods). Such work is currently ongoing, and will hopefully lead to further improvement in our understanding of the connection between morphology and ion transport in PFSI systems.

13.3. Conclusions

The theme of this chapter has been to demonstrate how state-of-the-art molecular simulation methodologies can be applied to tackle the simulation of PFSI systems in order to link chemical structure to macroscopic ion transport properties. It is clear that over the last six years, the scale of systems that can be studied in atomistic detail has increased from simple polyelectrolyte solutions and single ionic side groups, through oligomeric species, up to macromolecules more closely approximating real polymers. It is now possible to explore the effect of structural changes in the PFSI polymers, such as monomeric sequence, side group length and flexibility, and acidity of the proton. The resulting challenges for atomistic modelling approaches are now to tackle the separation in length and time scales between ionic motion and local rearrangement of the charged groups on the polymer and the more global relaxation times of the fluorocarbon backbone. While empirical valence bond techniques give some hope of being able to incorporate quantum mechanical features of proton transport into a classical model framework, their accuracy still leaves something to be desired and one of the key challenges in the future will be the integration of molecular orbital and hybrid density functional calculations of local ion transport models with realistic calculations of bulk membrane morphology, perhaps using QM-MM techniques. Also, the use of mesoscale modelling to accelerate the generation of relaxed conformational states of the polymeric backbone will help to complete the multi-scale modelling chain, illustrated in Figure 13.2, between the level of atomistic structure and engineering properties, allowing fundamental advances in PFSI membrane technology.

Acknowledgments

JAE would like to acknowledge helpful discussions with S. J. Paddison and K.-D. Kreuer during the writing of this chapter.

References

[1] M. R. Tant, K. A. Mauritz and G. L. Wilkes (eds.) *Ionomers : Synthesis, Structure, Properties and Applications*, (Blackie Academic & Professional, Glasgow, 1997).
[2] K. A. Mauritz and R. B. Moore, Chemical Reviews **104**, 4535 (2004).
[3] A. Eisenberg and H. L. Yeager, *Perfluorinated ionomer membranes* (American Chemical Society, Washington, DC, 1982).
[4] C. Heitner-Wirguin, Journal of Membrane Science **120**, 1 (1996).
[5] S. Rieberer and K. H. Norian, Ultramicroscopy **41**, 225 (1992).
[6] M. Chomakova-Haefke, R. Nyffenegger and E. Schmidt, Applied Physics A **59**, 151 (1994).
[7] Z. Porat, J. R. Fryer, M. Huxham, et al., Journal of Physical Chemistry **99**, 4667 (1995).
[8] A. Lehmani, S. Durand-Vidal and P. Turq, Journal of Applied Polymer Science **68**, 503 (1998).
[9] P. J. James, J. A. Elliott, T. J. McMaster, et al., Journal of Materials Science **35**, 5111 (2000).

[10] E. J. Roche, M. Pineri, R. Duplessix, et al., Journal of Polymer Science **19**, 1 (1981).
[11] M. Fujimura, T. Hashimoto and H. Kawai, Macromolecules **15**, 136 (1982).
[12] G. Gebel and J. Lambard, Macromolecules **30**, 7914 (1997).
[13] B. Loppinet, G. Gebel and C. E. Williams, Journal of Physical Chemistry B **101**, 1884 (1997).
[14] J. A. Elliott, S. Hanna, A. M. S. Elliott, et al., Macromolecules **33**, 4161 (2000).
[15] H. G. Haubold, T. Vad, H. Jungbluth, et al., Electrochimica Acta **46**, 1559 (2001).
[16] L. Rubatat, A. L. Rollet, O. Diat, et al., Journal De Physique Iv **12**, 197 (2002).
[17] S. K. Young, S. F. Trevino and N. C. B. Tan, Journal of Polymer Science Part B-Polymer Physics **40**, 387 (2002).
[18] V. Barbi, S. S. Funari, R. Gehrke, et al., Polymer **44**, 4853 (2003).
[19] G. Gebel, O. Diat and C. Stone, Journal of New Materials for Electrochemical Systems **6**, 17 (2003).
[20] C. L. Marx, D. F. Caulfield and S. L. Cooper, Macromolecules **6**, 344 (1973).
[21] M. Fujimura, T. Hashimoto and H. Kawai, Macromolecules **14**, 1309 (1981).
[22] M. Falk, Canadian Journal of Chemistry **58**, 1495 (1980).
[23] W. Y. Hsu and T. D. Gierke, Journal of Membrane Science **13**, 307 (1983).
[24] M. Litt, Abstracts of Papers of the American Chemical Society **213**, 33 (1997).
[25] E. M. Lee, R. K. Thomas, A. N. Burgess, et al., Macromolecules **25**, 3106 (1992).
[26] B. Loppinet and G. Gebel, Langmuir **14**, 1977 (1998).
[27] L. Rubatat, A. L. Rollet, G. Gebel, et al., Macromolecules **35**, 4050 (2002).
[28] H. W. Starkweather, Macromolecules **15**, 320 (1982).
[29] R. Duplessix, M. Escoubes, B. Rodmacq, et al., in *Water in Polymers*, edited by S. P. Rowland (American Chemical Society, Washington, D.C., 1980), p. 487.
[30] D. Marx, M. E. Tuckerman, J. Hutter, et al., Nature **397**, 601 (1999).
[31] S. J. Paddison, R. Paul and T. A. Zawodzinski, Journal of Chemical Physics **115**, 7753 (2001).
[32] R. Paul and S. J. Paddison, Solid State Ionics **168**, 245 (2004).
[33] J. R. Bontha and P. N. Pintauro, Chemical and Engineering Science **49**, 3835 (1994).
[34] A. R. Leach, *Molecular Modelling: Principles And Applications* (Prentice-Hall, Englewood NJ, 2001).
[35] J. A. Elliott, in *Introduction to Materials Modelling*, edited by Z. H. Barber (Maney Publishing, London, UK, 2005), p. 120.
[36] T. J. F. Day, A. V. Soudackov, M. Cuma, et al., Journal of Chemical Physics **117**, 5839 (2002).
[37] E. Spohr, P. Commer and A. A. Kornyshev, Journal of Physical Chemistry B **106**, 10560 (2002).
[38] S. Dapprich, I. Komaromi, K. S. Byun, et al., Journal of Molecular Structure-Theochem **462**, 1 (1999).
[39] Y. A. Dyakov and Y. K. Tovbin, Russian Chemical Bulletin **44**, 1186 (1995).
[40] X. D. Din and E. E. Michaelides, Aiche Journal **44**, 35 (1998).
[41] J. A. Elliott, S. Hanna, A. M. S. Elliott, et al., Physical Chemistry Chemical Physics **1**, 4855 (1999).
[42] J. Ennari, M. Elomaa and F. Sundholm, Polymer **40**, 5035 (1999).
[43] S. L. Mayo, B. D. Olafson and W. A. Goddard, Journal of Physical Chemistry **94**, 8897 (1990).

[44] W. L. Jorgensen, J. Chandrasekhar, J. D. Madura, et al., Journal of Chemical Physics. **79**, 926 (1983).
[45] H. Sun, S. J. Mumby, J. R. Maple, et al., Journal of the American Chemical Society **116**, 2978 (1994).
[46] H. Sun, S. J. Mumby, J. R. Maple, et al., Journal of Physical Chemistry **99**, 5873 (1995).
[47] A. Vishnyakov and A. V. Neimark, Journal of Physical Chemistry B **104**, 4471 (2000).
[48] D. Rivin, G. Meermeier, N. S. Schneider, et al., Journal of Physical Chemistry B **108**, 8900 (2004).
[49] A. Vishnyakov and A. V. Neimark, Journal of Physical Chemistry B **105**, 7830 (2001).
[50] D. B. Holt, B. L. Farmer, K. S. Macturk, et al., Polymer **37**, 1847 (1996).
[51] S. S. Jang, M. Blanco, W. A. Goddard, et al., Macromolecules **36**, 5331 (2003).
[52] H. J. C. Berendsen, J. R. Grigera and T. P. Straatsma, Journal of Physical Chemistry **91**, 6269 (1987).
[53] M. E. Van Leeuwen and B. Smit, Journal of Physical Chemistry **99**, 1831 (1995).
[54] J. A. Elliott, S. Hanna, A. M. S. Elliott, et al., Polymer **42**, 2251 (2001).
[55] X. Gong, A. Bandis, A. Tao, et al., Polymer **42**, 6485 (2001).
[56] A. Vishnyakov and A. V. Neimark, Journal of Physical Chemistry B **105**, 9586 (2001).
[57] S. T. Cui, J. I. Siepmann, H. D. Cochran, et al., Fluid Phase Equilibria **146**, 51 (1998).
[58] D. Brown, J. H. R. Clarke, M. Okuda, et al., Journal of Chemical Physics **100**, 6011 (1994).
[59] R. Auhl, R. Everaers, G. S. Grest, et al., Journal of Chemical Physics **119**, 12718 (2003).
[60] S. Urata, J. Irisawa, A. Takada, et al., Journal of Physical Chemistry B **109**, 4269 (2005).
[61] G. J. Martyna, M. E. Tuckerman, D. J. Tobias, et al., Molecular Physics **87**, 1117 (1996).
[62] H. C. Andersen, Journal of Chemical Physics **72**, 2384 (1980).
[63] S. Urata, J. Irisawa, A. Takada, et al., Physical Chemistry Chemical Physics **6**, 3325 (2004).
[64] P. G. Khalatur, S. K. Talitskikh and A. R. Khokhlov, Macromolecular Theory and Simulations **11**, 566 (2002).
[65] D. A. Mologin, P. G. Khalatur and A. R. Kholhlov, Macromolecular Theory and Simulations **11**, 587 (2002).
[66] P. J. Flory, *Statistical Mechanics of Chain Molecules* (Oxford University Press, Oxford, 1989).
[67] P. G. Khalatur, A. R. Khokhlov, D. A. Mologin, et al., Macromolecular Theory and Simulations **7**, 299 (1998).
[68] T. D. Gierke, G. E. Munn and F. C. Wilson, Journal of Polymer Science. **19**, 1687 (1981).
[69] I. Carmesin and K. Kremer, Macromolecules **21**, 2819 (1988).
[70] D. H. Rothman and S. Zaleski, Reviews of Modern Physics 66, 1417 (1994).
[71] W. J. MacKnight, W. P. Taggart and R. S. Stein, *Journal of Polymer Science.: Polymer Symposia*, **45** 113, (1974).
[72] D. J. Yarusso and S. L. Cooper, Macromolecules **16**, 1871 (1983).

[73] S. S. Jang, V. Molinero, T. Cagin, et al., Journal of Physical Chemistry B **108**, 3149 (2004).
[74] S. S. Jang, V. Molinero, T. Cagin, et al., Solid State Ionics **175**, 805 (2004).
[75] M. Levitt, M. Hirshberg, R. Sharon, et al., Journal of Physical Chemistry B **101**, 5051 (1997).
[76] S. Nose and M. L. Klein, Molecular Physics **50**, 1055 (1983).
[77] S. Nose, Molecular Physics **52**, 255 (1984).
[78] S. Nose, Molecular Physics **57**, 187 (1986).
[79] U. W. Schmitt and G. A. Voth, Journal of Physical Chemistry B **102**, 5547 (1998).
[80] U. W. Schmitt and G. A. Voth, Journal of Chemical Physics **111**, 9361 (1999).
[81] T. J. F. Day, U. W. Schmitt and G. A. Voth, Journal of the American Chemical Society **122**, 12027 (2000).
[82] S. Walbran and A. A. Kornyshev, Journal of Chemical Physics **114**, 10039 (2001).
[83] M. K. Petersen, F. Wang, N. P. Blake, et al., Journal of Physical Chemistry B **109**, 3727 (2005).
[84] S. J. Paddison and T. A. Zawodzinski, Solid State Ionics **115**, 333 (1998).
[85] S. J. Paddison, L. R. Pratt and T. A. Zawodzinski, Journal of New Materials for Electrochemical Systems **2**, 183 (1999).
[86] S. J. Paddison, Journal of New Materials for Electrochemical Systems **4**, 197 (2001).
[87] S. J. Paddison and J. A. Elliott, Journal of Physical Chemistry A **109**, 7583 (2005).
[88] S. J. Paddison and J. A. Elliott, Solid State Ionics, **177**, 2385 (2007).

14
Quantum Molecular Dynamic Simulation of Proton Conducting Materials

G. Seifert, S. Hazebroucq and W. Münch

14.1. Introduction

Computer simulations are powerful for the understanding of properties, reactions, and processes. Due to the improvement of both computer capabilities and new methods or algorithms, the investigated systems become more and more complex. Nowadays, these tools can be applied to any domain of chemistry to bring fundamental information concerning structures, reactivity, and properties on the components of the system of interest. The fuel-cell applications bring newer challenges, due to the complexity of model systems. The study of proton transport through polymeric membranes is not a routine task, due to the large timescale and the variety of possible mechanisms. The choice of theoretical chemistry allows to avoid any need for assumptions concerning the system.

The dynamical processes can be investigated in two different ways: the adiabatic processes, where the system remains in the electronic ground state and the nonadiabatic processes where electronic excitation, ionization or charge transfer occur. Only the ways to study adiabatic phenoma will be described here.

In order to explore the proton conductivity, quantum molecular dynamics methods are extensively used. There are two different types of ab initio molecular dynamics based on two theories: Born-Oppenheimer and Car-Parrinello. Both of them use an explicit treatment of the electron interactions. The bonding state of the system can change along the simulation. Therefore, these methods are of great interest to obtain information concerning the transfer of protons. They play a central role in the description of dynamic phenomena, using ab initio or empirical level of theory.

14.2. Basic Theory

The basic equation of quantum mechanics is the Schrödinger equation

$$\hat{H}\Psi = E\Psi \qquad (14.1)$$

where $\hat{H}$ is the hamiltonian operator, E the energy eigenvalue, and Ψ the wavefunction of the system. The hamiltonian operator is written as

$$\hat{H} = \hat{T} + V \qquad (14.2)$$

The solutions are stationary solutions of the system. To this non-relativistic approximation, Born and Oppenheimer added in 1927 the decoupling of the movements between nuclei and electrons [1]. Therefore, the nuclei are considered fixed for the computations, and the electronic hamiltonian operator is written as (in atomic units)

$$\hat{H}_{el} = -\frac{1}{2}\sum_{i=1}^{n}\nabla_i^2 - \sum_{i=1}^{n}\sum_{K=1}^{N}\frac{Z_K}{r_{iK}} + \sum_{i=1}^{n}\sum_{j>i}^{n}\frac{1}{r_{ij}} \qquad (14.3)$$

The terms are, respectively, the kinetic energy of the electrons $\hat{T}_e$, the electron-nuclei attraction $\hat{V}_{eN}$, and the electron–electron repulsion $\hat{V}_{ee}$, with i running over the n electrons and K over the N nuclei.

The Schrödinger equation is not exactly soluble for polyelectronic systems due to the electron–electron interaction. In order to solve approximately the Schrödinger equation, different methods are available. The most traditional way is based on the Hartree-Fock (HF) method. An alternative treatment is based on the Density-Functional Theory (DFT).

14.2.1. HF and Post-HF Methods

In these methods, each electron is moving in a mean-field potential due to the other electrons. The total wavefunction Ψ is written as a determinant, the Slater determinant, of spin-orbitals.

$$\Psi(1\cdots n) = \frac{1}{\sqrt{n!}}\begin{vmatrix} \phi_1(1) & \phi_2(1) & \cdots & \phi_{n-1}(1) & \phi_n(1) \\ \cdots & \cdots & \ddots & \cdots & \cdots \\ \phi_1(n) & \phi_2(n) & \cdots & \phi_{n-1}(n) & \phi_n(n) \end{vmatrix} = |\phi_1\cdots\phi_n| \qquad (14.4)$$

The spin-orbital ϕ is a product of an orbital, the space function and α or β, the spin function. The variational principle allows to determine the "optimal" spin-orbitals through a self-consistent field (SCF) procedure minimizing the HF energy.

In the HF method, the electron correlation is missing. For the post-HF methods, the correlation may be considered in different ways. In the configuration interaction (CI) treatment, excited state determinants are constructed from the SCF solution, and the total wave function is written as a superposition of these determinants. The coefficients of the determinants in these expansions are optimized via a variational principle. Instead of the CI expansion also many body perturbation theories can be used to consider the electron correlation. The Møller Plesset method [2] in the second order of the perturbation expansion (MP2) is the most frequently used perturbative method.

14.2.2. Density Functional Theory

The electronic density $\rho(r)$ of a system is defined as

$$\rho(\vec{r}) = \int \cdots \int |\Psi(\vec{r_1} \cdots \vec{r_n})|^2 ds_1 d\vec{r_2} \cdots d\vec{r_n} \qquad (14.5)$$

It represents the probability of finding one of the n electrons of the system in the volume element $d\vec{r_1}$. Systems are described through their density function $\rho(r)$ and not anymore through the wave function. Then, for an n electron system, instead of a function of $3n$ space variables, the density is function of only 3 space variables. Hohenberg and Kohn give the two fundamental theorems of DFT [3]. The first theorem states that all ground state properties – including the corresponding potential – are determined by the ground state electron density ρ alone in a unique way. The second theorem provides the variational principle. If $\rho(r)$ is the exact density, then $E[\rho(r)]$ is minimal and exact.

$$E[\rho] = F[\rho] + \int \rho(r) v_{ext}(r) dr \qquad (14.6)$$

The functional $F[\rho]$ is universal because it is not dependent on the external potential. The universal functional remains unknown, and in order to find an approximate expression, it has been divided into the same components as the hamiltonan operator

$$E_{el} = T_e[\rho] + V_{ee}[\rho] + V_{ne}[\rho] \qquad (14.7)$$

Kohn and Sham [4] proposed an ansatz for the density $\rho(r)$

$$\rho(r) = \sum_i^N |\phi_i(r)|^2 \qquad (14.8)$$

A variation of the energy with respect to the Kohn-Sham orbitals $\phi_i(r)$ leads to effective single particle equations, the Kohn-Sham equations

$$h^{KS} \phi_i(\vec{r}) = \epsilon_i \phi_i(\vec{r}) \qquad (14.9)$$

where

$$h^{KS} = -\frac{1}{2}\nabla^2 + v^{KS}(\vec{r}) \tag{14.10}$$

From these equations, the total energy is

$$E = T^{KS}[\rho] + J[\rho] + V_{Ne}[\rho] + E_{xc}[\rho] \tag{14.11}$$

$$E = T^{KS}[\rho] + \frac{1}{2}\int \frac{\rho(\vec{r_1})\rho(\vec{r_2})}{|\vec{r_1} - \vec{r_2}|}d\vec{r_1}d\vec{r_2} + \int \rho(\vec{r_1})v(\vec{r_1})d\vec{r_1} + E_{xc}[\rho] \tag{14.12}$$

The exchange-correlation term E_{xc} is defined as the exact kinetic energy $T[\rho]$ minus the kinetic energy in the KS representation $T^{KS}[\rho]$, and the exact exchange-correlation energy $V_{ee}[\rho]$ minus the "mean field" Coulombic interaction energy $J[\rho]$

$$E_{xc} = (T[\rho] - T^{KS}[\rho]) + (V_{ee}[\rho] - J[\rho]) \tag{14.13}$$

The potential v^{KS} is then defined by

$$v^{KS} = v_{eN} + \frac{\partial J[\rho]}{\partial \rho} + \frac{\partial E_{xc}[\rho]}{\partial \rho} \tag{14.14}$$

The theory is precisely defined, but an expression for the exchange-correlation energy remains unknown. Generally, it is written as

$$E_{xc} = \int \rho(\vec{r})\epsilon_{xc}(\rho)d\vec{r} \tag{14.15}$$

where ϵ_{xc} is the exchange-correlation energy per particle.

14.2.2.1. Functionals in DFT

In the most simple approximation for the exchange correlation energy, the Local Density Approximation (LDA), ϵ_{xc} is taken from the value of a uniform electron gas of a density ρ. The exchange part is equal to [5]

$$\epsilon_x = -C_x \rho^{\frac{1}{3}} \tag{14.16}$$

The correlation energy for a uniform electron gas can be obtained from quantum Monte Carlo calculations. A widely used functional, derived from such calculations, is that one from Vosko, Wilk, and Nusair [6]. Combined with the exchange funtional of Slater, it is the so-called LDA SVWN functional.

The LDA can be modified by consideration of the gradient of the density

$$\epsilon_{xc} = F[\nabla \rho, \rho] \tag{14.17}$$

Based on the so-called generalized gradient approximation (GGA) [7] numerous functionals are used today, for example LYP [8], B88 [9], or PW91 [10].

Because the GGA methods are often not sufficiently accurate, several corrections have been proposed. Becke introduced a part of exact exchange [11] from Hartree-Fock exchange into the exchange-correlation energy. These hybrid functionals may partially be justified by the so-called adiabatic connection principle [12, 13, 14, 15]. The so-called meta-GGA methods [16] are other possibilities to correct the inaccuracy of the LDA or GGA aproximations, due to the locality of the E_{xc} term, adding a dependence on the laplacian of the density: $\nabla^2 \rho$ or on the density of the kinetic energy of the occupied orbitals.

14.2.2.2. Approximate Treatment

Although the computers are increasingly powerful, the ab initio methods cannot be performed on large systems. It is even more limiting for dynamic computations where long time simulations are needed. A compromise between computational effort and accuracy can be reached within an approximate KS-DFT scheme with an LCAO representation of the KS orbitals – a density-functional based "tight-binding" (DFTB) method. Formally, it can be derived within a variational treatment of an approximate KS energy functional given by second-order perturbation with respect to charge density fluctuations around a properly chosen reference density [17]. But it may also be related to cellular Wigner-Seitz methods [18] and to the Harris functional [19]. For an overview and detailed discussion of the method, see e.g. [20]. Based on earlier work [18], the DFTB method [21, 22] was developed as a method, which avoids any empirical parametrization by calculating the Hamiltonian and overlap matrices out of a DFT-LDA-derived local orbital (atomic orbitals – AOs) and a restriction to only two-center integrals. Therefore, the method includes ab initio concepts in relating the Kohn-Sham orbitals of the atomic configuration to a minimal basis of the localized atomic valence orbitals of the atoms, which are determined self-consistently within the local-density approximation (LDA), together with the corresponding atomic KS potentials. For the many-atom configuration the effective one-electron potential in the Kohn-Sham Hamiltonian is approximated as a sum of the atomic potentials. Consistent with this approximation, the Hamiltonian matrix elements can strictly be restricted to a two-center representation. This leaves to solve a general eigenvalue problem for determining the KS-like eigenvalues of the molecule, cluster or solid. Finally, taking advantage of the compensation of the so-called "double counting terms" and the nuclear repulsion energy in the DFT total energy expression, the energy may be approximated as a sum of the occupied KS single-particle energies and a repulsive energy E_{rep}, which can be obtained from DFT calculations in properly chosen reference systems [21, 22]. This relates the method to common standard "tight-binding -TB" schemes, which are well known in solid state physics. This approach defines the density-functional tight-binding (DFTB) method in its original (non–self-consistent) version. The DFTB method was further developed, including self-consistency, and in this way also a more general derivation within the DFT was given [17].

In the following, this DFT foundation of the method is outlined briefly.

Within density functional theory (DFT), the total energy of a system can be expressed as a functional of a charge density ρ - see above.

Decomposing the electron density into a sum of a reference density and a density fluctuation, $\rho = \rho_0 + \delta\rho$ the energy at the reference density $\rho = \rho_0$ can be expanded up to the second order in the density fluctuations $\delta\rho$. For the DFTB approximation, the Hamiltonian matrix elements $\langle \psi_i | \hat{H}^0 | \psi_i \rangle$ are represented in a minimal basis of optimized pseudo-atomic orbitals φ_μ.

$$\psi_i = \sum_\mu c_{\mu i} \varphi_\mu (\vec{r} - \vec{R}_\alpha). \tag{14.18}$$

As Hamiltonian matrix elements $H^0_{\mu\nu}$ in this basis are calculated within a two-center approximation [21]

$$H^0_{\mu\nu} = \langle \varphi_\mu | \hat{T} + V_{eff}[\rho^0_\alpha + \rho^0_\beta] | \varphi_\nu \rangle \quad \mu \in \alpha, \nu \in \beta. \tag{14.19}$$

The charge density fluctuations are approximated by monopolar charge fluctuations at the atom α, $\Delta q_\alpha = q_\alpha - q^0_\alpha$. The second order term in $\delta\rho$ in the energy expression is then approximated as

$$\frac{1}{2} \int \int \left(\frac{1}{|\vec{r} - \vec{r}'|} + \frac{\delta^2 E_{xc}}{\delta\rho\delta\rho'}|_{\rho_0,0} \right) \delta\rho \, \delta\rho' d^3r d^3r'$$
$$\approx \sum_{\alpha\beta} \gamma_{\alpha\beta}(|\vec{R}_\alpha - \vec{R}_\beta|) \Delta q_\alpha \Delta q_\beta.$$

The remaining "double counting terms" in the energy expression and the nuclear repulsion energy ($E_{\alpha\beta}$) are summarized into a short-range repulsive energy $E_{rep} = \sum_{\alpha \neq \beta} U[R_{\alpha\beta}]$, consisting of atom-type specific pair potentials $U[R_{\alpha\beta}]$. These are constructed as the difference between the total energy versus distance calculated in DFT and the corresponding electronic energy derived within the DFTB approach for properly chosen reference systems – see [21].

Variation of this approximate Kohn-Sham energy expression with respect to the minimal basis yields single-particle "Kohn-Sham-like" equations

$$\sum_\nu c_{\nu i} (H_{\mu\nu} - \varepsilon_i S_{\mu\nu}) = 0 \quad \forall \mu, i, \tag{14.21}$$

where the Hamiltonian matrix elements are given by

$$H_{\mu\nu} = H^0_{\mu\nu} + \underbrace{\frac{1}{2} S_{\mu\nu} \sum_{\zeta}^{N} (\gamma_{\alpha\zeta} + \gamma_{\beta\zeta}) \Delta q_{\zeta}}_{H^1_{\mu\nu}} \qquad (14.22)$$

Analytical interatomic forces can easily be calculated by differentiating the total energy with respect to the nuclear coordinates,

$$\vec{F}_\alpha = -\sum_i n_i \sum_\mu \sum_\nu c^*_{\mu i} c_{\nu i} \left(\frac{\partial H^0_{\mu\nu}}{\partial \vec{R}_\alpha} - \left(\varepsilon_i - \frac{H^1_{\mu\nu}}{S_{\mu\nu}} \right) \frac{\partial S_{\mu\nu}}{\partial \vec{R}_\alpha} \right)$$
$$- \Delta q_\alpha \sum_\zeta \frac{\partial \gamma_{\alpha\zeta}}{\partial \vec{R}_\alpha} \Delta q_\zeta - \frac{\partial E_{rep}}{\partial \vec{R}_\alpha}. \qquad (14.23)$$

The pseudo-atomic basis functions $|\varphi_\mu\rangle$ are obtained by solving the Kohn-Sham equation for a spherical symmetric spin-unpolarized neutral atom selfconsistently. From this procedure, we obtain for each atom type optimized atomic basis sets $\{\varphi_\mu\}_\mu$ and atomic densities ρ^0_α, which are used to calculate the matrix elements of the zeroth order Hamiltonian $H^0_{\mu\nu}$ in a two-center approximation. The integrals $\gamma_{\alpha\beta}, \alpha \neq \beta$ are calculated analytically from the Coulomb interaction of two atom-centered spherical charge distributions [17]. The method has been extended also to the consideration of spin-polarized systems; see [20].

The DFTB method has been applied for a large variety of molecules, clusters and condensed systems, ranging from medium size organic molecules over biomolecules, fullerenes, nanotubes to metallic liquid alloys – for an overview see [20].

14.3. Molecular Dynamics

The time evolution of an isolated system has to obey three conditions:

- energy conservation
- momentum conservation
- time reversibility

The particles trajectory of a system is governed by the second law of Newton or *fundamental law of dynamics*. This law describes the motion of the particles as a function of time: For a body of constant mass m, the undergone acceleration is proportional to the sum of the forces and inversely proportional to its mass m

$$\sum_i \vec{F}_i = m\vec{a} = -\frac{d\vec{p}}{dt} \qquad (14.24)$$

Because the atoms exert forces on each others, the degrees of freedom of the system are all coupled, and an analytical solution is impossible. The integration of these equations is done by dividing the trajectory in a series of discrete states separated by short time intervals. Concerning the ab initio molecular dynamics, two methods will be explained here: the Born-Oppenheimer and the Car Parrinello techniques.

14.3.1. Born-Oppenheimer Molecular Dynamics

In this technique, at each step the electronic structure is computed according to the static coordinates of the nuclei, in other words, solving for each new positions the time-independent Schrödinger equation. Between two steps, the nuclei are moved following the rules of classical mechanics. The BO dynamics can be realized with any ab initio method: Hartree-Fock, post-HF, DFT, or DFTB. Each step brings the new positions of the nuclei; the minimum energy is computed following the choosen method and then the forces which are used in the determination of the displacements of the atoms. In this way, the system is kept on the BO hyper-surface all along the simulation.

14.3.2. Car-Parrinello Molecular Dynamics

The CPMD [23] approach exploits in another way the separation of fast electronic and slow nuclear motions: A fictitious wave function dynamics is used together with the classical dynamics in a single Lagrangian

$$\begin{aligned}L_{CP} = &\sum_j \frac{1}{2} M_j \dot{\vec{R}}_j^2 - E\left[\{\psi_i(\vec{r})\}, \{\vec{R}_i\}\right] \\ &+ \sum_i \frac{\mu}{2} \int d^3r \, |\dot{\psi}_i(\vec{r})|^2 \\ &+ \sum_{i,j} \Lambda_{ij} \left(\int d^3r \, \psi_j^*(\vec{r})\psi_i(\vec{r}) - \delta_{ij}\right).\end{aligned} \qquad (14.25)$$

The terms in the first line describe the kinetic energy of the atomic motion and the DFT energy functional. The second line stands for the kinetic energy of the orbitals with a fictitious mass μ. The sum in the third line guarantees the orthonormality of the $\{\psi_i\}$ as a constraint for the Lagrangian multipliers Λ_{ij}. The variation of the action corresponding to (14.25) yields the classical equations of motion for the nuclei (14.24) and, in addition, equations of motion for the electronic orbitals

$$\mu\ddot{\psi}_i(\vec{r},t) = -\frac{\delta E}{\delta \psi_i^*(\vec{r},t)} + \sum_j \Lambda_{ij}\psi_j(\vec{r},t). \qquad (14.26)$$

These equations describe a fictitious dynamics of the electronic system, which should not to be mixed up with the real electronic dynamics, rather it considers the time evolution of the KS wave functions connected with the atomic motion. Concerning a detailed discussion of this method and its applications, see e.g. [24, 25, 26].

14.3.3. Data analysis

Simulations using BOMD or CPMD give as result a set of snapshots of the system, as coordinates, velocities, and forces. Exploitation of this information allows to know statistical quantities as well as dynamic quantities. As an example, the radial distribution function gives the probability to find a pair of atoms a distance r apart, relative to the probability for a random distribution at the same density [27]

$$g(r) = \frac{V}{N^2} < \sum_i \sum_{j \neq i} \delta(r - r_{ij}) > \qquad (14.27)$$

The angular distribution can also be computed following the same rules.

Diffusion coefficients D can be obtained after Einstein's relation on the mean square displacement

$$2Dt = \frac{1}{3} < |r_i(t) - r_i(0)|^2 > \qquad (14.28)$$

Using the least square method, we can compute the mean slope of the square displacements all along the simulation [27].

14.4. Applications

In this section, we discuss the results of some MD studies of proton transport systems, relevant for fuel cell applications. The first two examples are water and imidazole. In the condensed phase, these systems are characterized by intermolecular hydrogen bonds, and the dynamics of the hydrogen bond network is significant for the proton transport. They are homogeneous media, and the protogenic group is also the protonic charge carrier, because the molecules undergo a self-dissociation. For these two examples, the proton transport can result from two mechanisms: the structure diffusion and the mass diffusion. In the last example, perovskites, the protons are not constituents of the crystal but the results of absorption of water from the gas phase onto defects in the solid.

Other systems investigated in the literature with MD simulations concerning the proton conductivity are e.g. methanol [28], ammonia [29], or water [30, 31, 32] in particular cases: along linear chains of water, and in the presence of an electric field, or considering the triplet state of the ion complexes. These studies, although not directly related to fuel cells, provide interesting information concerning proton transfer and play a significant role in the understanding of the proton transport mechanism.

Ab initio molecular dynamics simulations allow to identify the limiting step of the proton transport, between the transfer and the reorganization. An understanding of the transport mechanisms helps design new materials, whose structure is based on well-known materials, but whose proton conductivity may be improved considerably.

14.4.1. Water

Aqueous proton transport is a fundamental process in chemistry. The very high proton mobility cannot be explained by the protonic defects, the H_3O^+ and OH^- species, and the hydrodynamics diffusion. It has been demonstrated that the structure diffusion explains the anomalous mobility of protons [33, 34] in water. Two structural models were proposed: In the Eigen complex $H_9O_4^+$ [35, 36], the proton carrier H_3O^+ is stabilized by three surrounding water molecules, in the Zundel complex [37, 38] the proton is delocalized over two water molecules to form the complex $H_5O_2^+$ (see Figure 14.1). A series of CPMD computations [39, 40] and NMR investigations [41] showed that both complexes are involved in the proton transport and the protonic charge follows the existing hydrogen bond network. All water molecules are acceptors and donors of protons. The breaking and forming of bonds happen in the outer region of the complexes and in a concerted way, allowing the excess charge to displace along the hydrogen bond structure, forming alternatively the one or the other complex. These two complexes represent only some extreme configurations, and many other intermediates are involved in the transport [42]. The mechanism associated with the structure diffusion, almost barrier free, is the most favorable one and is now considered as the appropriate model for the proton conductivity in water and may serve also as a good reference for other materials.

14.4.2. Imidazole

Imidazole is a heterocycle, exhibiting a high proton mobility, similar to water. Because of its properties as an amphoteric molecule, the low hydrogen

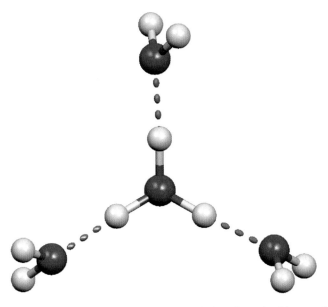

FIGURE 14.1 Illustration of the Zundel (*left*) and the Eigen (*right*) complexes. The excess proton is, respectively, delocalized on two or four water molecules.

bonding barrier between the highly polarizable nitrogen atoms, imidazole was investigated as a proton solvent in separator materials for fuel cells [43]. An important result of these investigations was the high difference between the proton mobility and the molecular diffusion coefficient. This is an indication for a structure diffusion, which was demonstrated by CPMD studies, for imidazole by itself [44], as well as for imidazole-based compounds [45].

In order to observe a proton transfer between imidazole molecules within the simulation time, an excess proton was added to a molecule, acting therefore as a proton donor for its two nearest neighbors (see Figure 14.2). The system can be represented as Im – Im H^+ – Im with hydrogen bonds slightly

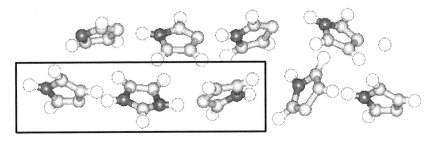

FIGURE 14.2 Illustration of the imidazole chain. The protonated molecule is in the center of the frame, surrounded by its two nearest neighbors.

shortened compared to the average hydrogen bond length of the system. The imidazoles linked in this way represent a chain-like configuration, where the polarization of the hydrogen bonds has two possible orientations within segments, which are separated by imidazoles bent out of the chain. Their proton is then directed out of the chain and could also link to a non-protonated nitrogen from another strand of imidazole. The MD simulation showed that some imidazole molecules close to the protonic defect in hydrogen bond patterns, rapidly change by hydrogen bond breaking and forming processes. As for the case of water, a complete proton transfer can occur because the excess proton is shifted within this region, and the turned imidazole is propagating in the same time as the proton.

The studied system is somewhat artificial because there is no counter charge compensating the excess proton. In such a model system (8 molecules in a box), it is unlikely that self-dissociation occurs within a simulation time of 10 ps. The absence of counter charge could explain the difference in the results obtained experimentally by ^{1}H PFG NMR [46]. In the liquid, some molecules undergo the self-dissociation, and therefore some regions contain an excess proton and some are proton deficient. These regions attract each other depending on their distance and the dielectric constant of the liquid. Defects are created by transfers of a proton from one molecule to another. Logically, the donor molecule should retrieve the proton, but after a reorientation, the acceptor imidazole has two equivalent protons giving rise to nonequivalent transfer. If the same proton is transferred back, there is no contribution to conductivity and diffusion. If the second proton is transferred, there is a contribution to the diffusion and not to the conductivity. Such a mechanism has not been proved to happen in liquid imidazole, but this cyclic intermolecular proton transfer is known to occur in certain organic pyrazole-containing complexes [47] and proton diffusion in hydroxides [48, 49].

14.4.3. Perovskites

Some simple cubic perovskites have been found to have a high thermodynamic stability and a high proton conductivity [50, 51, 52]. This conductivity is the result of protonic defects in the structure, more precisely a hydroxide ion residing on an oxide ion site. Therefore, these materials have been tested for proton transport in fuel cell applications. The protonic defects are the results of absorption of water molecules, which dissociate into a hydroxide ion and a proton, the hydroxide ion filling an oxide ion vacancy and the proton bonding to an oxide ion to form another hydroxide. This reaction creates two protonic defects. The hydroxide anion is stabilized by one of the eight possible orientations of the hydrogen bond with one of the eight nearest oxygen neighbors. As shown by DFTB and CPMD simulations, the proton transport is based on two steps: a rotational diffusion of the protonic defect and transfer of the proton onto a neighbor oxygen [53, 54, 55, 56, 57] (see

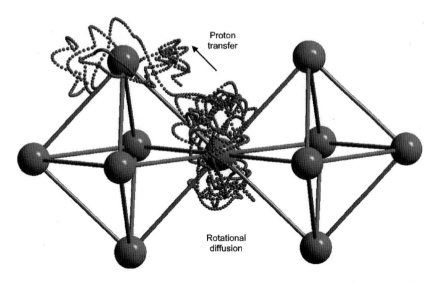

FIGURE 14.3 Illustration of the trajectory of a proton transfering inside the perovskite lattice.

Figure 14.3). The rotational diffusion is fast with a low activation barrier, as demonstrated experimentally [58, 59, 60] and theoretically [53, 54, 55, 56]. Spectroscopic data (IR spectra [61] and neutron diffraction experiments [62]) indicate strong hydrogen bonds, giving favor to a fast proton transfer rather than a fast reorientation step, which needs breaking of the hydrogen bond.

Due to the distance between two oxide ions (larger than 290 pm), there is a competition between the energy gained by the formation of a hydrogen bond and the energy necessary to the distortion of the lattice because strong hydrogen bonds appear at shorter distances. The corresponding free energies are of the same order of magnitude for distances of 250–300 pm [46]. Thus, short oxygen distances favoring proton transfer, and large oxygen distances allowing rapid bond breaking, result in similar free energies of the entire system. Both situations can then be found with the same probability. Indeed, the protonic defect is found to form rapidly, but hydrogen bonds are formed temporarily with all eight nearest oxygen neighbors. Finally, the hydrogen bond is soft and the protonic defect acts almost like a free OH group with small stretching amplitudes, compared to the extended stretching vibrations in the hydrogen bond state [63]. Following the thermodynamics of hydrogen bonds, the activation enthalpy should be around 0.15 eV and not 0.4–0.6 eV [61] which is found for perovskites. This suggests that the activation is controlled by another phenomenon, which could be the repulsion between the proton and the B-site cation in the center of the lattice. In many configurations produced by the simulation, the proton is not found between the two oxygens on the edge of the octahedron but outside the BO_6 octahedron as part of a strongly bent hydrogen bond [56] that still possesses some barrier for proton transfer. The H/B repulsion is reduced in the transition states by the elongation of B–O bonds, and the proton being transferred is on the edge of the distorted octahedron [63]. This B–O elongation gives the main contribution to the reduction of the proton transfer barrier. The H/B repulsion may be used as an estimate of the upper limit for these contributions to the activation enthalpy. This is demonstrated by the activation barrier of cubic perovskites with pentavalent B-site cations (I-V perovskites), which is much higher than for the perovskites with tetravalent B-site cations (II-IV perovskites) [51, 63].

14.5. Conclusions

In general, molecular dynamics simulations, in the framework of the Born-Oppenheimer or Car-Parrinello approximation, are of great importance for the understanding of materials dedicated to proton transport. Especially for materials, where interactions are dominated by covalent or hydrogen bonds, ab initio molecular dynamics provide a proper description. The results obtained by such methods give details at the atomic level, which are not accessible by experimental investigations. Nevertheless, the choice of the model system has to be done in a very careful way in order to consider the manifold possibilities of structures and mechanisms.

The coupling of ab initio and experimental techniques provides a powerful tool for the understanding of proton conductivity and the development of new materials for fuel cells.

References

[1] M. Born, J. R. Oppenheimer, Ann. Phys. 84, 457 (1927)
[2] C. Møller, M. S. Plesset, Rev. Phys. 46, 618 (1934)
[3] P. Hohenberg, W. Kohn, Phys. Rev. 136, B 864 (1964)
[4] W. Kohn, L. J. Sham, Phys. Rev. 140, A1133 (1965)
[5] J. C. Slater, Phys. Rev. 81, 385 (1951)
[6] S. J. Vosko, L . Wilk, M. Nusair, Can. J. Phys. 58, 1200 (1980)
[7] J. P. Perdew, Phys. Rev. Lett. 55, 1665 (1985)
[8] C. Lee, W. Yang, R. G. Parr, Phys. Rev. B 37, 785 (1988)
[9] A. D. Becke, Phys. Rev. A 38, 3098 (1989)
[10] J. P. Perdew, Y. Wang, Phys. Rev. B 45, 13244 (1992)
[11] A. D. Becke, J. Chem. Phys. 98, 548 (1993)
[12] J. Harris, R. O. Jones, J. Phys. F 4, 1170 (1984)
[13] O. Gunnarson, B. T. Lundquist, Phys. Rev. B 13, 1274 (1976)
[14] J. Harris, Phys. Rev. B 29, 1648 (1984)
[15] A. D. Becke, J. Chem. Phys. 88, 1053 (1988)
[16] J. P. Perdew, S. Kurth, A. Zupan, P. Blaha, Phys. Rev. Lett. 82, 2544 (1999); 82, 5179(E) (1999)

[17] M. Elstner, D. Porezag, G. Jungnickel, J. Elsner, M. Haugk, T. Frauenheim, S. Suhai, G. Seifert, Phys. Rev. B 58, 7260 (1998)
[18] G. Seifert, H. Eschrig, W. Bieger, Z. Phys. Chem. (Leipzig) 267, 529 (1986)
[19] J. Harris, Phys. Rev. B 31, 1770 (1985)
[20] Th. Frauenheim, G. Seifert, M. Elstner, T.A. Niehaus, C. Köhler, M. Amkreutz, M. Sternberg, Z. Hajnal, A. di Carlo, S. Suhai, J. Phys. C 14, 3015 (2002)
[21] D. Porezag, T. Frauenheim, T. Köhler, G. Seifert, R. Kaschner, Phys. Rev. B 51, 12947 (1995)
[22] G. Seifert, D. Porezag, T. Frauenheim, Int. J. Quant. Chem. 58, 185 (1996)
[23] R. Car, M. Parrinello, Phys. Rev. Lett 55, 2471 (1985)
[24] R. Car, M. Parrinello, Simple Molecular Systems at Very High Density, NATO Advanced Study Institute, Series Plenum Press, 1988, p. 455
[25] G. Pastore, E. Smargiassi, F. Buda, Phys. Rev. A 44, 6334 (1991)
[26] M. E. Tuckerman, J. Phys. Condens. Matter 14, 1297 (2002)
[27] M. P. Allen, D. J. Tildesley, Computer Simulations of Liquids, (Clarendon Press, Oxford, 1989)
[28] J. A. Morrone, M. E. Tuckerman, J. Chem. Phys. 117, 4403 (2002)
[29] V. Zoete, M. Meuwly, J. Chem. Phys. 120, 7085 (2004)
[30] S. Scheiner, J. Am. Chem. Soc. 103, 315 (1981)
[31] A. A. Tulub, J. Chem. Phys. 120, 1217 (2004)
[32] K. Drukker, S. W. De Leeuw, S. Hammes-Schiffer, J. Chem. Phys. 108, 6799 (1998)
[33] M. Eigen, L. De Mayer, Proc. R. Soc. (London) Ser. A 247, 505 (1958)
[34] M. Eigen, Angew. Chem. 75, 489 (1963)
[35] E. Wicke, M. Eigen, Th. Ackermann, Z. Phys. Chem. 1, 340 (1954)
[36] M. Eigen, Angew. Chem. Int. Edn Engl. 3, 1 (1964)
[37] G. Zundel, H. Metzger, Z. Physik. Chem. 58, 225 (1968)
[38] G. Zundel, The Hydrogen Bond-Recent Developments in Theory and Experiments. II. Structure and Spectroscopy (eds P. Schuster, G. Zundel, C. Sandorfy) 683–766 (North-Holland, Amsterdam, 1976).
[39] M. Tuckerman, K. Laasonen, M. Sprik, M. Parrinello, J. Chem. Phys. 103, 150 (1995)
[40] M. E. Tuckerman, D. Marx, M. L. Klein, M. Parrinello, Science 275, 817 (1995)
[41] N. Agmon, Chem. Phys. Lett. 244, 456 (1995)
[42] D. Marx, M. E. Tuckerman, J. Hütter, M. Parrinello, Nature 397, 601 (1999)
[43] K.-D. Kreuer, A. Fuchs, M. Ise, M. Spaeth, J. Maier, Electrochim. Acta 43, 1281 (1998)
[44] W. Münch, K.-D. Kreuer, W. Silvestri, J. Maier, G. Seifert, Solid States Ionics 145, 437 (2001)
[45] M. Iannuzzi, M. Parrinello, Phys. Rev. Lett. 93, 25901 (2004)
[46] K.-D. Kreuer, S. J. Paddison, E. Spohr, M. Schuster, Chem. Rev. 104, 4637 (2004)
[47] F. Toda, K. Tanaka, C. Foces-Foces, A. L. Llamas-Saiz, H. H. Limbach, F. Aguilar-Parrilla, R. M. Claramunt, C. Lòopez, J. Elguero, J. Chem. Soc. Chem. Commun. 1139 (1993)
[48] M. Spaeth, K.-D. Kreuer, T. Dippel, J. Maier, Solid State Ionics 97, 291 (1997)
[49] M. Spaeth, K.-D. Kreuer, J. Maier, J. Solid State Chem. 148, 169 (1999)
[50] K.-D. Kreuer, Annu. Rev. Mater. Res. 33, 333 (2003)
[51] T. Norby, Y. Larring, Curr. Opin. Solid State Mater. Sci. 2, 593 (1997)

[52] K.-D. Kreuer, Solid State Ionics 125, 285 (1999)
[53] W. Münch, G. Seifert, K.-D. Kreuer, J. Maier, Solid State Ionics 86–88, 647 (1996)
[54] W. Münch, G. Seifert, K.-D. Kreuer, J. Maier, Solid State Ionics 97, 3399 (1997)
[55] W. Münch, K.-D. Kreuer, St. Adams, G. Seifert, J. Maier, Phase Trans. 68, 567–586 (1999)
[56] F. Shimojo, K. Hoshino, H. Okazaki, J. Phys. Soc. Jap. 66, 8 (1997)
[57] F. Shimojo, K. Hoshino, Solid State Ionics 145, 421 (2001)
[58] M. Pionke, T. Mono, W. Schweika, T. Springer, T. Schober, Solid State Ionics 97, 497 (1997)
[59] T. Matzke, U. Stimming, C. Karmonik, M. Soetramo, R. Hempelmann, F. Güthoff, Solid State Ionics 86–88, 621 (1996)
[60] R. Hempelmann, M. Soetramo, O. Hartmann, R. Wäppling, Solid State Ionics 107, 269 (1998)
[61] K.-D. Kreuer, Solid State Ionics 97, 1 (1997)
[62] N. Sata, K. Hiramoto, M. Ishigame, S. Hosoya, N. Niimura, S. Shin, Phys. Rev. B 54, 15795 (1996)
[63] K.-D. Kreuer, Solid State Ionics 136, 149 (2000).

15
Morphology of Nafion Membranes: Microscopic and Mesoscopic Modeling

DMITRY GALPERIN, PAVEL G. KHALATUR AND ALEXEI R. KHOKHLOV

15.1. Introduction

Polymer electrolyte fuel cells (PEFCs) have attracted much interest as one of the most promising nonpolluting power sources capable of producing electrical energy with high thermodynamic efficiencies. The key element of PEFCs is a polymer electrolyte membrane (PEM) that serves as proton conductor and gas separator [1, 2, 3, 4]. The membrane commonly employed in most PEFC developments is based on Nafion, which represents a family of comb-shaped ionomers with a perfluorinated polymeric backbone and short pendant (side) chains having sulfonic acid end groups

$$-(CF_2-CF_2)_n-CF_2-CF\diagdown_{O-CF_2-CF(CF_3)-)_m-O-CF_2-CF_2-SO_3H}$$

where n and m are small numbers (usually, n is approximately five-six times the value of m and $m = 1-2$). The Nafion-based PEFCs offer several advantages over other fuel cell membrane types (e.g., phosphoric acid, alkaline, and molten carbonate).

In the wet state, the SO_3H groups of Nafion dissociate and deliver protons as charge carriers. These groups and water molecules are known to organize into hydrophilic nanodomains in the chemical stable hydrophobic fluorocarbon matrix [5, 6, 7, 8, 9]. As a result, there are two distinct regions in the system, hydrophilic and hydrophobic. This leads to unique permeability characteristics of the Nafion membranes: good proton conductivity, good barrier properties for hydrogen and oxygen, optimum dependence of the proton conductivity on water content in the membrane, etc. Also, it is known that conductivity starts to increase considerably already at very small amounts of water uptake [1, 2, 3, 4]. The latter fact suggests that the ionic channels for conductivity exist already in the structure of "dry" Nafion membranes. This is in spite of the relatively low contents of immobile ionic groups when one should rather expect isolated ionic aggregates, not channels. The structural reason for the existence of ionic channels at low concentrations of water and charged groups in ionomers is unknown and a molecular level understanding of such features in Nafion is not available yet.

Despite a long history of research and a number of successful industrial applications of Nafion membranes, their morphology remains the focus of debate in the current literature. The specific organization of the microphase-separated morphology of Nafion is basically corroborated by the existence of a small-angle scattering (SAXS and SANS) maximum, called "ionomer" peak [5, 6, 7, 8, 9], and by TEM data. Unfortunately, scattering techniques provide only indirect information on the 3D organization of the material. Furthermore, a hysteresis of properties of the same sample with hydration/dehydration cycles is typically observed. This indicates that the membrane structure is usually not an equilibrium structure. One may regard the ionomer membrane rather as a soft "glassy" nonequilibrium composite with water as one of its components. Thus, by fitting to the SAXS/SANS results, a variety of models appeared which relatively accurate describe the experimental data but conflict with each other. There are a number of recent reviews that cover this subject (see, e.g., Refs. [3, 10, 11, 12, 13, 14]).

In these circumstances, when the microphase-separated morphology is clearly evidenced by numerous experiments, which can be however ambiguously interpreted, modeling and simulation attract much attention as a means to clarify and extend the concept of Nafion morphology and ionic conductivity. Such studies provide a valuable tool for a deeper understanding of the complex properties of water-containing Nafion systems at a fundamental level.

The conventional theoretical approach to determine the three-dimensional organization of (co)polymers at equilibrium is based on minimizing a suitable Ginzburg-Landau free energy functional with respect to the corresponding lattice parameters for some a priori presupposed structure of ideal symmetry (e.g., body centered cubical, hexagonal, lamellar, etc.) [15, 16]. Because of its inherent assumed symmetries, this approach fails to predict the irregular and "defect" microstructures. Also, the computations of the free-energy coefficients within the Leibler-like or random phase approximation (RPA) treatment [16] do not take into account directly the interaction potentials of polymer segments and the system-specific features of polymer architecture on short length scales.

Although various analytical approaches were advanced considerably and for simple models of complex polymer systems accurate results can be obtained, in general, the problems related to the prediction of the system morphology are still analytically (exactly) intractable. As always, in these circumstances, an alternative to analytical theories is numerical methods, which are designed to obtain a numerical answer without knowledge of an analytical solution.

In principle, Nafion morphologies that develop as a result of molecular properties and processing conditions can be modeled on three different levels. On one end of the scale are the microscopic models, which are based on detailed molecular descriptions. Atomistic simulation of a molecular system involves the generation of a statistically representative set of samples of the system from which the properties of interest can be estimated. There are two principal simulation methods – molecular dynamics (MD) and Monte Carlo (MC) – which correspond to two different approaches in statistical mechanics [17]. The main weakness of MD and MC is an inability to access long time and length scales. Such limitations become particularly significant in the context of modeling inhomogeneous self-assembled phases, wherein the length scale of self-assembly morphology corresponds to many atomic lengths. On the other end of the scale are the macroscopic approaches, which are often based on equations of state or constitutive equations, the latter being continuum partial differential equations solved by conventional finite difference or finite element methods. This is the domain of the chemical engineering. In between, in both length and time scales are the mesoscopic approaches. They include coarse-grained (particle-based) models and simulation techniques and statistical continuum methods, with the possibility of the latter using local concentration fields as collective variables to describe self-organizing structures. Typical length scales are $10-10^3$ nm and typical time scales are up to milliseconds or even seconds. In this article, we will mainly focus on the mesoscopic approaches.

Various simulation techniques, as applied to PEFCs, differing in the degree of coarseness and the basic equations solved have been reviewed [10–14]. In contrast to these reviews, in which the transport phenomena in ionomer membranes are discussed in detail, we will consider the problems related to the modeling of Nafion morphology, with the emphasis on the role of water content.

15.2. Coarse-Grained Particle-Based Simulations

For computational simplicity, the chemical groups of a complex polymer can be represented as single interaction sites with suitable nonbonded interaction parameters. In other words, one can coarse-grain the structure of a polymer molecule. Coarse-grained models simplify the problem by combining atoms into effective united atoms (UAs). In this way, only significant microscopic information – "essential features" of the real system – is retained. The unit can represent a chemical group of a few atoms, a monomeric unit in a polymer, groups of monomeric units, or chain segments of various lengths. Certainly, whether a coarse-grained description is adequate for understanding a particular polymer system depends very much on the problem being studied. The challenge lies in selecting just the right amount of atomistic detail to build coarse-grained models with maximum generality. In some sense, of course, that is precisely what physics is about. Therefore, the question is how can we construct copolymer models that are sufficiently realistic to capture the essential features of real macromolecules yet simple enough to allow large-scale computations of polymer conformation and dynamics?

15.2.1. Dissipative Particle Dynamics

Dissipative particle dynamics (DPD) is a simulation technique initially developed for the simulation of complex fluids [18] and later extended for polymers. The DPD model consists of pointlike particles interacting with each other through a set of prescribed forces [19]. From a physical point of view, each dissipative particle is regarded not as a single atom or molecule but rather as a collection of atomic groups (molecules) that move in a coherent fashion.

Actually this method can be considered a mesoscopic realization of the conventional MD since the time evolution of the system is obtained by solving Newton's equation of motion $d\mathbf{p}_i(t)/dt = \mathbf{f}_i(t)$, where $\mathbf{p}_i$ and $\mathbf{f}_i$ are the impulse of the particle (UA) and the force applied to it. In contrast to atomistic MD, the DPD particles are soft beads connected by springs. The attention is focused now on the center-of-mass motion of these coarse-grained particles while their internal degrees of freedom are completely ignored. Instead of interatomic forces calculated in conventional MD, the forces $\mathbf{f}_i$ applied to the DPD particle i are of three types: a conservative force, a dissipative force that tries to reduce velocity differences between the particles, and a stochastic force directed along the line joining the center of the particles. The amplitude of these forces is dictated by a fluctuation-dissipation theorem [19]. The forces are modulated with a weight function (usually a Mexican hat function) that specifies the range of interaction between dissipative particles and renders this interaction local. Just by changing the conservative interactions between the particles, one can easily construct polymers, colloids, amphiphiles, and mixtures. In particular, by joining consecutively sets of particles with springs, which are usually harmonic, one can model polymers. Also, the conservative force is responsible for a soft pairwise repulsion/attraction and is determined for polymer models by Flory-Huggins interaction parameters, χ.

Recently, Yamamoto and Hyodo have employed the DPD method for studying Nafion membranes [20]. The systems considered in this study were built using two distinct molecular species, denoted "comb-shaped polymer" (p) and "water" (w). The polymer was presented as a branched sequence of beads. It consisted of a main chain (backbone) of $N_b = 20$ effective monomer units ($-CF_2CF_2CF_2CF_2-$) linked with $n_s = 5$ short side chains of $n = 2$ units [$-OCF_2C(CF_3)FO-$ and $-CF_2CF_2SO_3H$]; the total number of interaction sites in the macromolecule was $N_p = N_b + n \times n_s = 30$. A water-like particle was modeled as the same size as the units of the Nafion fragment ($\sigma = 6.1$ Å) and represented four water molecules. The χ parameters were found using an atomistic calculation. The DPD simulation was performed for water volume fractions of $\varphi_w = 0.1, 0.2$, and 0.3 that corresponds to the ratio of water (n_w) to sulfonate groups (n_s) $h = n_w/n_s \approx 3.3, 6.6$, and 10, respectively.

It was shown that water-like particles and hydrophilic segments of Nafion side chains form separated hydrophilic clusters which are embedded in the hydrophobic phase of the Nafion backbone. The radial distribution function of water particles, $g_w(r)$, calculated at $\varphi_w = 0.2$ for the structure at equilibrium had sharp peaks at $r = 5.2$ Å and $r = 10.8$ Å. These peaks originate from the first and second nearest water particles in a water cluster. A second peak was observed at $r = 41.5$ Å, which corresponds to the first nearest water cluster. As the hydration level increased from $h = 3.3$ to $h = 10$, the average cluster size changed from 32 to 52 Å, and the cluster center-to-center spacing changed from 36 to 60 Å, in good agreement with existing experimental data [6]. The study revealed that a sponge-like (or a bicontinuous microphase) structure of hydrated Nafion might exist, even at very low hydration levels, $h < 4$. Although this morphology was essentially identical to that predicted by the cluster-network model in experimental studies [6], the shape of water clusters was not spherical but irregular, and the water regions were indistinguishable structures of water clusters and their connected channels.

In principle, using the DPD results [20], it was possible to mimic SAXS and SANS experiments to discriminate the structural models used for interpretation of the experimental data [20]. Also, Hyodo [21] proposed a hierarchical procedure for calculating the electronic states of a hydronium ion in a hydrated Nafion membrane via the mesoscopic structure predicted by DPD [20]. A mixed basis function method was introduced for electronic state calculations in inhomogeneous fields as a combination of Gaussian basis functions and shape functions of the finite element method for expressing electronic wave functions.

15.2.2. Cellular-Automaton-Based Simulation

For studying dense polymeric systems, one of the best coarse-grained MC models is the bond-fluctuation model (BFM) [22]. The model combines many of the advantages of lattice methods with those of the off-lattice (continuous) models. Generally, however, the continuum MD simulations are more flexible compared to MC. Also, they can predict transport properties. Therefore, it is instructive to combine the best features of lattice BFM and dynamic simulations in one algorithm. In Refs. [23,24,25], the corresponding approach – a cellular-automaton-(CA) based "lattice molecular dynamics" (LMD) – was proposed. It turned out that in some cases, this coarse-grained model requires about one order of magnitude less computer time than conventional MC/BFM simulations. A CA-based algorithm for simulating two-dimensional polymers was also presented by Koelman [26].

The BFM is a lattice model with a variable bond length b [22]. In three dimensions, each chain segment occupies eight neighboring lattice sites of a simple cubic lattice, and each lattice site can only be part of one segment.

In the CA-based polymer model [23, 24], it is assumed that chain segments move on a cubic lattice with unit velocity **v** between a given lattice cell occupied by segment i and one of its six nearest neighbors. The changes in velocities are caused by "elastic collisions" between pairs of segments i and j. The collision laws of the discrete coarse-grained model give the transition rules for the internal states of each cell occupied by segments. The choice of the collision laws is somewhat arbitrary. In Ref. [24], the following highly simplified collision law $\mathbf{v}_\alpha \leftrightarrow \mathbf{v}_\beta$ was used for any two colliding segments i and j. In other words, the motion of the chain segments is such that if their relative positions do not satisfy the excluded-volume condition and the bond-length restrictions the segments move freely; in the opposite case, elastic collision takes place between the corresponding pair of segments and results in interchanging of their velocities. Therefore, linear momentum and kinetic energy are conserved in all collisions. At each time step, the segments attempt to move in the corresponding directions, whereas the trajectories between collisions are straight lines. When the collision occurs, the velocities of the two colliding segments are interchanged, but these segments are not moved until the next time step. One can also introduce a variant of the CA-based model in which the coupling to a heat bath is simulated [24]. In a straightforward application of the Metropolis algorithm [17] to the lattice system with attracting particles, a trial configuration is accepted with the transition probability defined by the energy of this configuration.

Using the CA-based algorithm, Mologin et al. [27] simulated the structure of hydrated Nafion membranes. Each polymer molecule consisted only of two types of monomer units: hydrophilic sulfonate groups located at the end of side chains and remaining hydrophobic units. The system was modeled as an ensemble of n_p polymers (up to 240 macromolecules) and n_w water-like one-site particles, at different water contents. Nafion is characterized by its equivalent weight, EW, which is defined as the mass of "dry" (water-free) polymer containing one mole of terminal sulfonate groups (e.g., for the Nafion 117 membrane, EW = 1100 g/equiv, corresponding to 13 CF_2 groups on average between two successive lateral chains of length n). Therefore, for the Nafion oligomer (with N_b = 68, n_s = 8, and n = 7) studied in [27], the value of EW can be estimated as EW $\approx$ 700 g/equiv.

At low water contents, the simulation [27] predicted the formation of specific microphase-separated morphologies in which sulfonate groups formed a distinct population of large but submacroscopic aggregates ("multiplets") that were wrapped by hydrophobic polymer sections connecting these aggregates and playing the role of peculiar bridges. It was found that as the level of polymer hydratation increases the aggregates grow in size, coalesce, and can form a percolating network-like structure which runs through the whole computational box. The associations between the sulfonate groups and the water-like particles resulted in a swelling of the main and side chains. This behavior is typical for copolymer systems in the region of microphase separation transition [28, 29]. However, the equilibrium conformational structure predicted for the side chains was distinctly different from the fully stretched conformation suggested in the so-called lamellar model of Litt [30].

The simulation predicted that the mean aggregate sizes are nearly linear functions of the hydratation level when the number of solvent molecules per sulfonate group $h \lesssim 4$ [27]. This indicates that practically all the solvent particles added are included in the mixed, solvent-containing aggregates as previously shown by the SANS contrast variation method [31]. These aggregates with significant amount of incorporated small solvent molecules had a clathrate-like structure consisting of an inner region, where the solvent molecules dominate, the first outer layer of the sulfonate groups mixed with the solvent, and an outer shell formed by hydrophobic polymer sections, which are immersed into external polymer matrix. There is a diffuse interface between the inner and first outer regions inside the aggregate core, as well as a quite sharp interface between the hydrophilic mixed core and the hydrophobic environment. In the case of sufficiently large excess of solvent molecules over sulfonate groups, it was found that the solvent particles began to "stick" to the surface of the mixed aggregates causing some erosion of the intramicellar interface. Thus, no homogeneous structure of the mixed aggregates with SO_3H groups homogeneously distributed inside the aggregate core was observed. One may say that a microphase separation of absorbed water and ionic groups took place inside a core although both core-forming components were completely compatible. There are experimental data supporting the simulation results [32, 33, 34, 35, 36]. Also, such organization has been already observed in micellar systems [37, 38].

The cluster configurations were actually quite rough and irregular. The cluster surface area to volume ratio, which is directly affected by the degree of surface irregularities, was unusually high. Such a geometry is favorable to the formation of long channels of connected water-containing aggregates. The same conclusion has been drawn from the N.M.R. data obtained for the Nafion/water systems. [39]

At sufficiently high hydration levels ($h \gtrsim 3$–4), the aggregation numbers began to grow very rapidly, implying the coalescence of the aggregates [27]. The existing structural models of hydrated Nafion (in particular, the so-called local order model [40]) do not reproduce such satisfactory behavior. The contradiction is mainly due to the fact that in the local order model [40] the effective cluster radius is supposed to be fixed while the simulation predicts that the swollen clusters can coalesce as the hydration level increases. The transition itself can be relatively sharp. Indications from the results presented in Ref. [27] are that the intramolecular chemical constraints, i.e., the connectivity of the side chains having terminal cluster-forming sulfonate groups and complex molecular architecture of comb-shaped copolymers, play an important role, promoting the coalescence. Due to the attachment of ionic groups to the polymer chains, the existence of an ionic cluster increases the probability of finding other clusters in its close vicinity and thus promotes the formation of channels constructed from these neighboring clusters. This supports the notion of extended, ion-conductive structures or aggregated clusters.

The coalescence of hydrophilic clusters can lead to the formation of a bicontinuous network-like morphology, in which both the polar (water-containing) component and the apolar polymer sections are found on the continuous (effectively infinite) domains. To visualize the three-dimensional structures of these microphases more clearly, the density of hydrophilic particles (water and sulfonate groups) was measured on a three-dimensional grid. In the hydrophilic domains, the density of polar components is high, and in the hydrophobic domains this density is low (~ 0). The dividing surface between the hydrophilic and hydrophobic domains can be represented by the isosurface shown in Figure 15.1 for $h = 4$. For this relatively high solvent

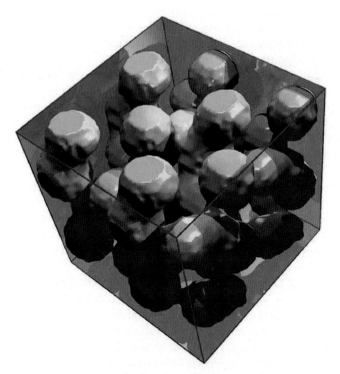

FIGURE 15.1. Isosurface between the hydrophilic and hydrophobic domains showing a microphase-separated water-containing Nafion system at hydration level corresponding to $h = 4 \, H_2O$ molecules per SO_3H group. There are continuous passes in the water-rich regions from any side of the system to any other its side. Adapted from Ref. [27].

concentration, essentially not isolated, more or less large hydrophilic clusters are present in the system. Of course, the hydrophobic component is also found in a continuous region that serves as a background for the polar species. As a result, it is observed rather distinctly a bicontinuous structure that can be pictured as an intertwisted sponge-like network where two effectively infinite domains are present simultaneously. The visual analysis of the simulated structures shows that there are continuous passes in the water-rich regions from any side of the system to any other its side (Figure 15.1). The reader may compare these results with those of Yamamoto and Hyodo [20]. Despite the differences in the CA-based model [27] and the DPD model [20], the essential results are practically the same. In the vicinity of the percolation threshold ($h \gtrsim 3$), when the episodical formation of the mixed aggregates running through the whole computational box took place, the maximum of topological complexity of cluster morphologies and the maximum of their diversity were observed [27].

15.3. Mesoscopic Continuum Modeling

Mesoscopic statistical-mechanical polymer theories include the polymer RISM (PRISM) theory [41, 42], which is an extension to long-chain polymers of the integral equation reference interaction site model (RISM) theory of Chandler and Andersen [43], density functional (DF) theory [44], dynamic mean-field density functional approach [45], and self-consistent mean-field (SCMF) methods [46, 47]. The main goal of these approaches is to calculate the density correlation functions and to determine the relation between the polymer architecture and the intermolecular correlations. All these methods coarse-grain the familiar atomistic representation of the (macro)molecule to gain orders of magnitude in both length and time scale. The chemistry of the system is captured through effective potentials and intramolecular correlations. In this section, we will only address PRISM and SCMF methods.

15.3.1. Integral Equation Theory

The polymer RISM theory relates the set of total correlation functions $\{h(r)\} = \{g(r)–1\}$ to the set of site-site direct correlation functions, $\{c(r)\}$, and the set of intramolecular distribution functions, $\{w(r)\}$, via the nonlinear site-site Ornstein-Zernike-like (SSOZ) integral matrix equation [41]

$$\mathrm{H}(r) = \int_{(\mathbf{r}')} \int_{(\mathbf{r}'')} \mathrm{W}(|\mathbf{r}-\mathbf{r}'|) C(|\mathbf{r}'-\mathbf{r}''|)[\mathrm{W}(\mathbf{r}'') + \rho \mathrm{H}(\mathbf{r}'')] d\mathbf{r}' d\mathbf{r}'' \quad (15.1)$$

where H, C, and W are the square matrices constructed from the functions $h(r)$, $c(r)$, and $w(r)$, respectively, and ρ is the system density.

For a copolymer system, the aim is to find the partial total, $h_{\alpha\beta}(r)$, and direct, $c_{\alpha\beta}(r)$, correlation functions, where the subscripts α and β indicate the type of the monomer species. From the complete set of these correlation functions, one can combine the square matrices $\mathbf{H}_p(r) = [\rho_\alpha \rho_\beta h_{\alpha\beta}(r)]$ and $\mathbf{C}_p(r) = [c_{\alpha\beta}(r)]$, where $\rho_\alpha = f_\alpha \rho_p$ is the site density of species α, ρ_p is the total monomer number density, and f_α denotes the fraction of the corresponding units in a macromolecule composed of N_p units.

Chemical structure is taken into account via the structural matrix, $\mathbf{W}_p(r)$, composed of the partial intrapolymer site-site distribution functions, $w_{\alpha\beta}(r)$. The elements of $\mathbf{W}_p(r)$ are given by $W_{\alpha\alpha}(r) = \rho_\alpha w_{\alpha\alpha}(r)$ and $W_{\alpha\beta}(r) = \rho_p w_{\alpha\beta}(r)$, where

$$w_{\alpha\alpha}(r) = \frac{1}{f_\alpha N_p} \sum_{i \in \alpha} \sum_{j \in \alpha} w_{ij}(r), \quad w_{\alpha\beta}(r) = \frac{1}{N_p} \sum_{i \in \alpha} \sum_{j \in \beta} w_{ij}(r) \qquad (15.2)$$

The correlation function $w_{ij}(r)$ characterizes the distribution of chain segments i and j belonging to the corresponding species inside a copolymer.

For the bicomponent system "copolymer plus N_w-atomic solvent" (e.g., water) at a given solvent number density ρ_w, we can write the analogous matrices belonging to the solvent subsystem; below they are denoted as $\mathbf{H}_w(r)$, $\mathbf{C}_w(r)$, and $\mathbf{W}_w(r)$. Using these matrices, the following square supermatrices are constructed

$$\mathbf{H}(r) = \begin{bmatrix} \mathbf{H}_p(r) & 0 \\ 0 & \mathbf{H}_w(r) \end{bmatrix}, \mathbf{C}(r) = \begin{bmatrix} \mathbf{C}_p(r) & 0 \\ 0 & \mathbf{C}_w(r) \end{bmatrix}, \mathbf{W}(r) = \begin{bmatrix} \mathbf{W}_p(r) & 0 \\ 0 & \mathbf{W}_w(r) \end{bmatrix} \qquad (15.3)$$

where 0 is the matrix with zero elements. Then the RISM Eq. (15.1) is written in Fourier transform space as

$$\mathbf{H}(q) = \mathbf{W}(q)\mathbf{C}(q)[\mathbf{W}(q) + \mathbf{H}(q)] \qquad (15.4)$$

Here, the matrices $\mathbf{H}(q)$, $\mathbf{C}(q)$, and $\mathbf{W}(q)$ contain the functions $h(q)$, $c(q)$, and $w(q)$ which are the Fourier transformations of the corresponding correlation functions with wave vector $\mathbf{q}$. Having these functions, one can find the partial static structure factors, $S_{\alpha\beta}(q)$, which are the Fourier transformed density-density fluctuation correlation functions and are proportional to the scattering intensities observable in experiments. They are defined as

$$S_{\alpha\beta}(q) = [\mathbf{E} - \mathbf{W}(q)\mathbf{C}(q)]^{-1} W_{\alpha\beta}(q) \qquad (15.5)$$

where $[\bullet]^{-1}$ denotes the matrix inverse, and $\mathbf{E}$ is the diagonal unity matrix.

The actual solution of Eqs. (15.1) or (15.4) can be obtained if the relation between direct correlation functions, total correlation functions, and site-site potentials, $\{u(r)\}$, is known. Several such closure relations are known, e.g., the Percus-Yevick (PY) closure or the hypernetted chain (HNC) approximation [48]. Substitution of the Fourier transform $c(r) \xrightarrow{F} c(q)$ of suchf relation into Eq. (15.4) gives a closed system of nonlinear integral equations with respect to the functions of interest $\{h(r)\}$. In general, this system can be solved only numerically. The PY and HNC equations are typical examples of "atomic" closure relations, which are widely used in statistical physics of simple monoatomic and low-molecular-weight liquids. However, as was found [49], these equations do not correctly describe the behavior of polymer systems in the presence of attractive interactions between polymer segments. The reason is that the relation between intrapolymer and interpolymer correlations is not taken into account. To eliminate this inconsistency, the so-called molecular closures were proposed [49], which explicitly include the intramolecular correlations. If the site-site potentials $u_{\alpha\beta}(r)$ are composed of an attractive tail interaction and a short-range repulsive hard core [i.e., $u_{\alpha\beta}(r) < 0$ at $r > \sigma_{\alpha\beta}$ and $u_{\alpha\beta}(r) = \infty$ at $r < \sigma_{\alpha\beta}$, where $\sigma_{\alpha\beta} = (\sigma_\alpha + \sigma_\beta)/2$, σ_α and σ_β are the effective hard core diameters of the monomer species of the type α and β], the molecular closures can be written in the following general form [49]

$$[\mathbf{W} * \mathbf{C} * \mathbf{W}(r)]_{\alpha\beta} = [\mathbf{W} * (\mathbf{C}_0 + \Delta\mathbf{C}) * \mathbf{W}(r)]_{\alpha\beta}, \quad r > \sigma_{\alpha\beta} \quad (15.6)$$

where the asterisks "*" denote convolution integrals, the subscript "0" refers to direct correlation functions for a hard core reference system (without attractive interactions), and the matrix $\Delta\mathbf{C}$ characterizes attractive interactions. Usually, the exact core condition $h_{\alpha\beta}(r) = -1$ for $r < \sigma_{\alpha\beta}$ is used. This corresponds to the usual PY approximation for the reference direct correlation function $c_{(0)\alpha\beta}(r)$, that is, $c_{(0)\alpha\beta}(r) = 0$ at $r > \sigma_{\alpha\beta}$. The attraction between interaction sites can be introduced, e.g., through the reference molecular mean-spherical approximation (RMMSA) [49]

$$\Delta c_{\alpha\beta}(r) = -u_{\alpha\beta}(r)/k_B T, \quad r > \sigma_{\alpha\beta} \quad (15.7)$$

where $u_{\alpha\beta}(r)$ is the model site-site potential between species α and β.

In the RISM theory of small rigid molecules [43], the intramolecular correlation functions $w_{ij}(q)$ are well defined and serve to specify the positions of the atoms on the molecule: $w_{ij}(q) = \sin(qr_{ij})/qr_{ij}$, where $r_{ij} = |\mathbf{r}_i - \mathbf{r}_j|$ is the fixed distance between pairs of atoms i and j. In the case of flexible polymers having a large number of internal degrees of freedom, rigorous definition of $w_{ij}(q)$ is a complicated theoretical problem. Moreover, in dense medium the equilibrium *intra*molecular and *inter*molecular structures should be formed self-consistently. The polymer RISM theory can be implemented in a self-consistent manner when the intermolecular correlation functions are used to find a medium-induced intramolecular interaction, $\Psi(r)$, between any two polymer segments and, in its turn, the intramolecular pair correlation function is determined by $\Psi(r)$. Briefly, some MC simulation method is applied to generate the configurations of a single polymer. Using the coordinates of chain segments, the averaged intrapolymer correlation functions are obtained. Then, solving the coupled polymer RISM equations, one can find the partial intermolecular correlation functions for a given polymer density. It yields the medium-induced intrapolymer potential and the corresponding effective intramolecular energies, which are used in the standard Metropolis MC procedure. The structural properties of the system are computed by averaging over the statistically representative set of configurations. As a result of many such iterations, the intramolecular and intermolecular structure is determined self-consistently.

For a single-site solute in a solvent, the potential of mean force (PMF) between two interaction sites i and j of type α and β is given by $\Psi_{\alpha\beta}(r_{ij}) = -k_B T \ln[h_{\alpha\beta}(r_{ij})+1]$ or $\Psi_{\alpha\beta}(r_{ij}) = u_{\alpha\beta}(r_{ij}) + \Delta\Psi_{\alpha\beta}(r_{ij})$, where the function $\Delta\Psi_{\alpha\beta}(r_{ij})$ denotes the correction to the bare site-site pair potential $u_{\alpha\beta}(r_{ij})$ due to the effect of medium. This medium-induced contribution is the excess chemical potential for the two particles (at a given distance r from each other) due to their interactions with other surrounding particles. Once the direct correlation function is known, the medium-induced potential can be obtained from [50]

$$\Delta\Psi_{\alpha\beta}(r) = -k_B T \rho_\alpha c_{\alpha\beta}(r)^* S_{\beta\alpha}(r)^* c_{\alpha\beta}(r) \tag{15.8}$$

The numerical self-consistent (SC) MC/RISM procedure [51, 52] employed to solve the matrix polymer RISM equation (4) with the RMMSA closure relation (6)–(7) was used in Ref. [53] to study water-containing Nafion systems. The single-chain MC simulation was based on the realistic rotational-isomeric-state (RIS) model [54], in which the short-range intramolecular interactions depending on the details of chemical structure were taken into account via appropriate matrices of statistical weights [54].

To give a visual impression of the simulated system, Figure 15.2 presents a typical snapshot of single macromolecule generated in the curse of the SC-MC/RISM calculations [53]. The macromolecule consists of a fluorocarbon backbone of $N_b = 98$ effective UA monomer units CF_2 (or CF_3) linked to $n_s = 10$ side chains $OCF_2CF(CF_3)OCF_2CF_2SO_3H$ consisting of $n = 7$ UA units in the main chain. The stereochemistry of the comb-like polymer was explicitly included in the UA site representation. It was taken into account in the single-chain RIS conformational statistics used as input to the SC-MC/RISM calculations in that macromolecules of differing tacticity had different statistical weights. Because of statistical distribution of pendant chains in Nafion, the intramolecular structure factors of combs with differing microstructures (with randomly generated space orientations of the pendant chains) were averaged together before performing the SC-MC/RISM computations [53]. Such "preaveraging" implies that the single-polymer structure factors are averaged together to form the intramolecular structure factor for

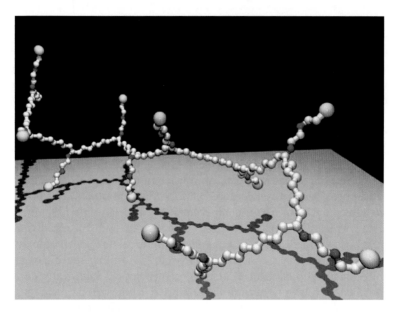

FIGURE 15.2. Typical snapshot of single Nafion macromolecule generated in the curse of the SC-MC/RISM calculations [53]. The macromolecule consists of a fluorocarbon backbone of 98 effective units (CF_2 or CF_3) linked to 10 side chains $OCF_2CF(CF_3)OCF_2CF_2SO_3H$ consisting of 7 units in the main chain.

an effective "atactic" comb-shaped copolymer with pendant chains randomly oriented in space. The interaction between the effective UA units was described by Yukawa-type site-site potential. Semiempirical molecular orbital calculations were performed in order to determine site-site interaction parameters, as well as the equilibrium, energy-minimized conformations of molecular fragments.

The results [53] establish molecular scale information necessary to understanding the equilibrium structure and thermodynamics of water-containing Nafion, as well as water distribution in hydrated Nafion membranes. In particular, it was concluded that water molecules absorbed by the membrane are preferentially distributed near the terminal parts of the side chains. The geometry of comb-shaped Nafion macromolecules suggests that the outer SO_3H groups of side chains are more accessible to each other than the inner polymer groups, since the latter are shielded from other macromolecules by the former. Another trend was that the hydrophobic polymer units are weakly hydrated, i.e., the solvent access to the side chain and perfluorinated backbone is hindered; such a behavior followed from the inspection of the polymer-water correlation function $g_{pw}(r)$, as well as from the quantum mechanical calculation [53]. Therefore, the ether groups of the side chains and the perfluorinated backbone should be considered hydrophobic.

The partial static structure factors calculated for sulfonate groups and water molecules, $S_s(q)$ and $S_w(q)$, had sharp small-angle ("ionomer") peaks at some wave number q^* which corresponds to a Bragg spacing of r^* ($=2\pi/q^*$) ≈ 30–50 Å, characteristic of a system containing hydrophilic domains (clusters) within a hydrophobic matrix. The observed spatial distribution of the hydrophilic regions is not, by any means, crystalline-like or uniform. The addition of even relatively small amount of water (≈6 wt-%, $h ≈ 4$) led to the strong intensification of aggregative processes observed for polar sulfonic acid groups. All these results are in accordance with most experimental studies performed on bulk or solvent-swollen ionomer membranes [5,6,7,8,9,10,11,12,13,14].

Water absorbed in Nafion was found to be in the well-ordered state. The theoretical prediction of well-ordered state for incorporated water is in qualitative agreement with existing experimental (N.M.R.) data [39]. Furthermore, the water absorbed in Nafion demonstrated the peculiar ordered structure both on small (local) scales and on intermediate (mesoscopic) scales. From these results it was concluded that water is forming specific cage-like structures (similar to clathrates) of many molecules. At high water content, the so-called free water (noticeably separated from sulfonic acid groups) exists in water clusters. Note that compound water within hydrated Nafion has been observed experimentally using IR methods [55] (see also the review of Paddison [11]), as well as in the MD simulations of Elliott et al. [56] and Urata et al. [57].

It is interesting to monitor the influence of the level of polymer hydration on the structural reorganizations of the system. Some typical data for the partial structure factors $S_s(q)$ and $S_w(q)$, as well as for the SO_3H-SO_3H and SO_3H-water pair correlation functions, $g_{ss}(r)$ and $g_{sw}(r)$, are plotted in Figure 15.3 at different densities of water, ρ_w, and $T = 300$ K. It is seen that the main peak of $S_s(q)$ shifts toward small q with increasing hydration level. From these data, one can conclude that the characteristic distance, r^*, is an increasing function of the moisture content. The functions $g_{ss}(r)$ and $g_{sw}(r)$ also show visible composition dependence. A general tendency is that the first and second peaks of $g_{ss}(r)$ and $g_{sw}(r)$ become more intensive when water content increases, indicating that an increase in the content of physisorbed water promotes the aggregation of SO_3H groups and leads to the formation of more perfect local structure in hydrophilic regions of the model ionomer membrane. On the other hand, it was observed that there was a moderate decrease in the local ordering of sulfonate groups and their hydration with increasing temperature. It is interesting to note that the structural reorganizations observed for the water and SO_3H subsystems go on synchronously as hydration level is increased (see Figure 15.3). This implies that both the subsystems are bonded intimately to each other.

In considering the ionomer peak observed for hydrated Nafion to be an intercluster origin, the cluster center-to-center spacing r^* was found to be a linear function of hydration level [53]. Generally, CA-based simulation [27] predicts the same behavior. An interesting result was that the value of r_s* (for

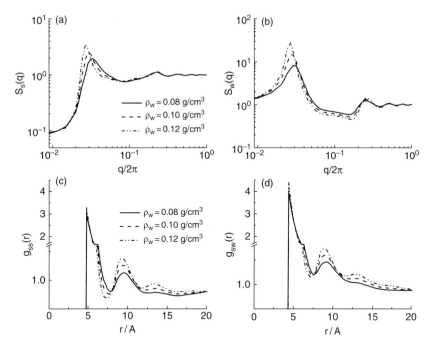

FIGURE 15.3. (a) The static structure factors for sulfonate groups and water molecules, $S_s(q)$ and $S_w(q)$, and (b) the SO$_3$H-SO$_3$H and SO$_3$H-water pair correlation functions, $g_{ss}(r)$ and $g_{sw}(r)$, for different partial densities of water, ρ_w, at $T = 300$ K. Adapted from Ref. [53].

SO$_3$H groups) was smaller than r_w^* (for water) for the wide range of hydration levels ($\lesssim 7.5$ wt% H$_2$O). From simple geometrical arguments, it might be expected that for a system containing nearly spherical aggregates the value of r^* should be proportional to $n_w^{1/3}$, where n_w is the total number of water molecules in the system. This is in contrast with the linear dependence of r^* on the hydration level predicted in Refs. [27, 53]. This fact may be explained by the irregular nature of the surface of the water-containing clusters. As previously hypothesized [39], the surface area to volume ratio of the hydrophilic aggregates within Nafion is unusually high; from the N.M.R. data, the cluster surface area was found to be proportional to $R^{2.5}$, where R is the effective cluster size. This supports the view of the irregularly shaped cluster surfaces – the "fuzzy" spheres, which have a ratio of surface area to volume considerably higher than for a perfect solid sphere. The corresponding tendency should dominate as water content in the system is increased.

The most important achievement of the work [53] was the implementation of SC-MC/RISM technique to probe the continuum percolation in the water subphase of Nafion. The solution of such a problem is of interest for molecular-level understanding of the mechanism of ionic conductivity. To this end, the formalism of so-called pair-connectedness correlation functions was employed that enabled to study the effect of interaction between the molecular species on association equilibrium in the system and, in particular, to determine the mean aggregation number of water clusters, m_w, and their mean size. The details of the underlying theory are beyond the scope of this review and so will be omitted here.

Some of the results for absorbed water are shown in Figure 15.4 as a function of temperature. As seen, for a given hydration level, the average number of molecules in a water cluster increases when temperature decreases. This is quite an expected behavior: as temperature is reduced, the association equilibrium in the physisorbed water phase shifts toward the formation of larger water-rich domains. In the case when $\rho_w \lesssim 0.04\,\text{g/cm}^3$, the size of the clusters is small at any reasonable temperature. In the region of room temperatures, the clustering is sharply dependent on the level of hydration, showing the divergence of m_w that corresponds to the percolation transition for $\rho_w \gtrsim 0.06$–$0.07\,\text{g/cm}^3$, i.e., for a surprisingly low hydration level of $h \approx 3$.

This estimate of the percolation threshold is consistent with existing theoretical predictions known for simple percolating (lattice and off-lattice) systems [58–60] and with DPD and CA-based simulations [20, 27]. There are several experimental measurements that certainly corroborate the findings discussed above. Zawodzinski et al. [61] showed that reasonable conductivities are possible in Nafion membranes with very little water present (as low as 1 H_2O per SO_3H). From the recent conductivity data of Weber and Newman [62], it follows that the percolation threshold practically occurs around $h = 2$.

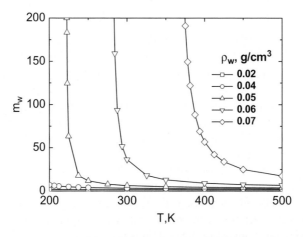

FIGURE 15.4. Mean aggregation number of hydrophilic (water) clusters, m_w, as a function of temperature, at different hydration levels. The clustering is sharply dependent on hydration, showing the divergence of m_w that corresponds to the percolation transition. Adapted from Ref. [53].

Edmondson and Fontanella studied proton conductivity and water diffusion in diverse membrane materials [63]. They established the existence of a universal percolation threshold at a water volume fraction of ~5 % ($h \approx 2$). The results of the MD simulations of Urata et al. [57] and Jang et al. [64] suggest the formation of a stable water percolated structure in hydrated Nafion at significantly higher water content (above ≈ 20 wt-%, i.e., at $h \approx 15$) and dynamic formation and destruction of continuous water clusters at lower hydration. However, these numerical results must be regarded as provisional, since the samples were not sufficiently large and the time intervals should be taken larger. The finite-size effects restrict the structural development beyond the size of the MD simulation cell. It is important to note that dissociation of the SO_3H groups in PEMs occurs at water contents of less than $3\,H_2Os/SO_3H$, as shown by Paddison at al. [10, 11, 65] who used ab initio modeling. Due to hydrogen bonding between the neighboring sulfonate groups, both the existence of a continuous network of water formed among the sulfonate groups and the proton transport in PEMs are possible even under minimal hydration conditions [10, 11].

15.3.2. Field-Theoretic Simulation

In the field-theoretic SCMF method [46,47], the external mean field acting on a polymer chain is calculated self-consistently with the composition profile. The utilization of the SCMF simulation scheme demands several important assumptions. Polymers are represented by phantom chains (thread-like space curves) of statistical segments that pass through the spatial domain in the mean field produced by other components of the system (polymers and solvents). The interactions between pairs of segments are determined by phenomenological Flory-Huggins parameters χ.

To specify their configurations, each (macro)molecule is parameterized by a variable s that increases continuously from 0 to 1 along its length. The configuration of the μth molecule is then represented by a simple mathematical space curve, $\mathbf{r}_\mu(s)$. Each chain has independent statistics in average chemical potential fields, $\omega_\alpha(\mathbf{r})$, conjugate to the volume fraction fields, $\phi_\alpha(\mathbf{r})$, of monomer species α. The free energy per chain F is related to the statistical weight (pass integral) $q(\mathbf{r},s)$ that a segment of a chain, originating from the free end of the α block and with contour length s, has at its terminus at point $\mathbf{r}$. The free energy is to be minimized subject to the constraint of local incompressibility and the constraint that $q(\mathbf{r},s)$ satisfies a dynamical trajectory given by modified diffusion equations. Minimization of F with respect to $\omega_\alpha(\mathbf{r})$ and $\phi_\alpha(\mathbf{r})$ leads to a self-consistent set of equations that are solved iteratively. The propagators $q(\mathbf{r},s)$, together with their conjugate propagators $q^*(\mathbf{r},s)$, which propagate from the opposite end of the chain, are used to calculate the density fields. The calculations result in the space distribution of the volume fractions $\phi_\alpha(\mathbf{r})$.

Let us consider a system of n comb-shaped copolymers in volume V. Under the mean field approximation, each copolymer can be divided into a set of linear subchains which are jointed at the corresponding junction points and which are statistically independent. To specify the subchains having N_k segments, an index k is introduced. Each subchain contains $f_\alpha N_k$ monomer species of type α. In this case, the contribution from the k-th subchain to the monomer density field at point $\mathbf{r}$ is given by

$$\phi_\alpha^{(k)}(\mathbf{r}) = \frac{n_\alpha}{Q_\alpha^{(k)}} \int_{(s)} q_\alpha(\mathbf{r},s) q_\alpha^*(\mathbf{r},s) ds \qquad (15.9)$$

where n_α is the total number of the subchains of type α in the system and $Q_\alpha^{(k)}$ is the partition function of the single subchain subject to the mean field $w_\alpha(\mathbf{r})$, which can be written as

$$Q_\alpha^{(k)} = \int_{(\mathbf{r})} q_\alpha(\mathbf{r},s) q_\alpha^*(\mathbf{r},s) d\mathbf{r} \qquad (15.10)$$

The polymer segment probability distribution functions $q_\alpha(\mathbf{r},s)$ and $q_\alpha^*(\mathbf{r},s)$ for species α are obtained by solving the Schrödinger type time evolution equations

$$\begin{aligned}\partial q_\alpha(\mathbf{r},s)/\partial s &= (b^2/6)\nabla^2 q_\alpha(\mathbf{r},s) - w_\alpha(\mathbf{r}) q_\alpha(\mathbf{r},s) \\ \partial q_\alpha^*(\mathbf{r},s)/\partial s &= -(b^2/6)\nabla^2 q_\alpha^*(\mathbf{r},s) + w_\alpha(\mathbf{r}) q_\alpha^*(\mathbf{r},s)\end{aligned} \qquad (15.11)$$

with the following initial conditions: $q_\alpha(\mathbf{r},0) = q_0(\mathbf{r})$ and $q_\alpha^*(\mathbf{r},1) = q_1(\mathbf{r})$, where $q_0(\mathbf{r})$ and $q_1(\mathbf{r})$ are the statistical weights of the remaining parts of the whole copolymer connected to each end segment of the subchain. In Eq. (15.11), b is the Kuhn length of the polymer segment. Having $\phi_\alpha^{(k)}(\mathbf{r})$, one can calculate the segment density at position $\mathbf{r}$ contributed by all subchains that have α-type segments

$$\phi_\alpha(\mathbf{r}) = \sum_{k \in \alpha} \phi_\alpha^{(k)}(\mathbf{r}) \qquad (15.12)$$

With the above description, the free energy of the system can be written as

$$F/nk_BT \propto -\ln\left[\sum_k Q^{(k)}/V\right] + (N/V) \int_{(\mathbf{r})} \left[\sum_\alpha \sum_\beta \chi_{\alpha\beta} \phi_\alpha(\mathbf{r}) \phi_\beta(\mathbf{r}) \right. \\ \left. - \sum_\alpha w_\alpha(\mathbf{r}) \phi_\alpha(\mathbf{r}) - \xi + \xi \sum_\alpha \phi_\alpha(\mathbf{r})\right] d\mathbf{r} \qquad (15.13)$$

where $\xi(\mathbf{r})$ is the potential field that ensures the incompressibility of the system, also known as a Lagrange multiplier. Minimizing the free energy with respect to ϕ_α, ω_α and ξ leads to the following self-consistent field equations that, together with Eqs. (9)–(13), describe the equilibrium morphology

$$\omega_\alpha(\mathbf{r}) = \sum_\beta \chi_{\alpha\beta} \phi_\beta(\mathbf{r}) + \xi(\mathbf{r}) \qquad (15.14)$$

$$\sum_\alpha \phi_\alpha(\mathbf{r}) = 1 \qquad (15.15)$$

$$\xi(\mathbf{r}) = \lambda \left[1 - \sum_\alpha \phi_\alpha(\mathbf{r}) \right] \qquad (15.16)$$

Here, $\chi_{\alpha\beta}$ are Flory-Huggins parameters between different species and λ is large enough to enforce the incompressibility of the system.

The SCMF simulations are generally performed according to the following scheme [47]. After the initial guess of density distribution (which is usually random) and calculation of the initial mean fields the following steps are made: (i) from the current mean fields, the new path integrals are calculated; (ii) the new density distributions and new mean fields are calculated from the path integrals; (iii) the new mean fields are efficiently mixed with the old ones. This procedure is performed repeatedly in a self-consistent manner until the difference between the new and the old mean fields becomes less than a certain error level, and the incompressibility condition (15) is fulfilled within the same error level.

The first application of a lattice-based variant of the SCMF theory to perfluorosulfonated ionomers has been reported by Krueger et al. [66]. Their one-dimensional model predicted segregation of ionomer-water mixtures into water-rich and ionomer-rich microphases as the fluorocarbon-water interaction or the hydration level increased. However, it should be kept in mind that the one-dimensional model [66] cannot conclusively identify the preferred equilibrium morphology of the system. With this model, an unrealistic lamellar organization of the microphase-separated system can only be observed. To date, no such morphology has been found for Nafion. Another dubious point is that the ether groups in the side chains were hydrophilic [66], which may be questioned by some investigators. Indeed, quantum-chemical studies have revealed the hydrophilic nature of only sulfonic acid groups and hydrophobicity of vynil ethers and fluorocarbon sections [67].

Very recently, three-dimensional field-theoretic SCMF calculations have been employed for modeling hydrated Nafion membranes [68]. The concepts and techniques described in this study are relevant with a view toward mesoscopic simulation of swollen ionomer membranes by manipulating temperature, type of solvent and its composition, polymer architecture, and other variables.

Following a coarse-grained representation of Nafion, the three different segment types were defined: F, E, and S. Segment F represented fluorocarbon groups, E corresponded to ether groups, and S represented sulfonic acid groups. In addition, W stands for water. Varying the number of F groups between the side chains, one can change the equivalent weight of Nafion. The volume fraction of W-type interaction sites, ϕ_w, determined hydration level. In the SCMF calculation [68], S-type groups were assigned hydrophilic, while F- and E-type segments were assigned hydrophobic, F being more hydrophobic than E. Interactions between species were characterized by the Flory-Huggins parameters $\chi_{\alpha\beta} = (\tilde{z}/k_B T)\Delta E_{\alpha\beta}$, where $\Delta E_{\alpha\beta} = \varepsilon_{\alpha\beta} - (\varepsilon_{\alpha\beta} + \varepsilon_{\alpha\alpha\beta\beta})/2$ is the heat of mixing associated with site-site interaction energies $\varepsilon_{\alpha\beta}$, and $\tilde{z}$ is the effective coordination number. Therefore, the χ parameter between the segments of the same type is equal to zero, by definition. Fluorocarbon and ether groups are likely to mix with each other, and thus the choice $\chi_{EF} = 0$ seems to be reasonable. The interaction between water and sulfonate groups was assigned strongly attractive, $\chi_{WS} = -4$, while hydrophobic fluorocarbon and ether groups repelled water ($\chi_{WF} = 4$ and $\chi_{WE} = 2$) and sulfonate groups ($\chi_{FS} = \chi_{ES} = 1$) [68]. Since fluorocarbon and water are the two most prevalent species, their interaction plays a dominant role. A variation of the temperature changed simultaneously both the hydrophilic and hydrophobic interactions. This means that, similar to real experiments, a change in temperature should influence all relevant quantities of the system in a rather complex manner.

15.3.2.1. Phase Diagram

As an example, Figure 15.5 presents phase diagram calculated for water-containing Nafion with EW = 900 g/equiv [68]. The theoretical phase diagram in Figure 15.5 is nothing more than a map showing the morphology of lowest free energy plotted as a function of temperature and composition φ_w. As seen, even small variations of the parameters T and φ_w may result in transition from homogeneous to inhomogeneous, microphase-separated structures (below and above the solid line, respectively). The transition point shifts toward higher hydration level as the temperature is increased.

15.3.2.2. Effect of Hydration

When the temperature was fixed at $T/k_B = 1$, microphase separation into hydrophilic and hydrophobic domains took place in the model [68] at $\varphi_w \gtrsim 0.1$ (i.e., at $h \gtrsim 4$–5 for EW = 900 g/equiv). With hydration of the sample, distinct microstructural changes were observed. Figure 15.6 illustrates the evolution of these morphologies. In the case of relatively low water content,

15. Morphology of Nafion Membranes   473

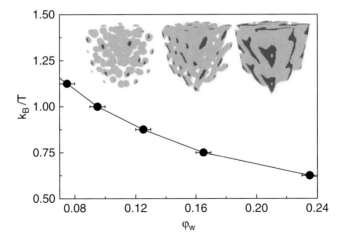

FIGURE 15.5. Equilibrium phase diagram calculated for water-containing Nafion (EW = 900 g/equiv) using self-consistent field theory [68].

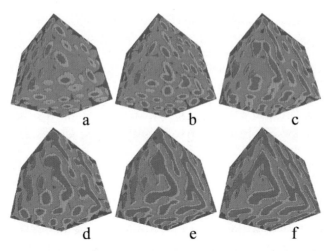

FIGURE 15.6. (a-f) Sequence of images, which illustrates the evolution of equilibrium morphology of hydrated Nafion with increasing water volume fraction, φ_w: (a) 0.1, (b) 0.15, (c) 0.2, (d) 0.25, (e) 0.3, and (f) 0.35. Regions occupied by polar (hydrophilic) species are shown in blue; hydrophobic regions are colored in red; interphase regions are shown in green. Adapted from Ref. [68].

$\varphi_w = 0.1$, the water simply collects in the nearly spherical domains shielded by the side chains and embedded into the hydrophobic matrix (15.6a). The domains grow in size with increasing φ_w and coalescence when $\varphi_w \gtrsim 0.15$ (Figure 15.6b). Further increase in hydration leads to the formation of a bicontinuous structure (Figure 15.6c–f). This picture is consistent with the results predicted by CA-based simulations [27] (see Figure 15.1).

At relatively low water contents, the observed morphologies support the spherical cluster-network model [6, 33], while at higher hydration, the morphology is in better agreement with the model of Falk [32] and Kreuer [69], who supposed that the hydrated ionic clusters are highly elongated (nonspherical) with frequent local intrusions of the fluorocarbon matrix. Also, as in the above-mentioned DPD [20] and CA-based [27] simulations, the SCMF modeling predicts a sponge-like structure, where hydrophilic clusters and channels cannot be distinguished from each other. The percolation threshold shifts toward lower hydration level with increasing equivalent weight. The solvent-filled regions provide energetically favorable pathways through the nonpolar fluorocarbon interior of the Nafion membrane. In fact, these regions are conductive nanopores, being responsible for the ionic flow and the transport of water from anode to cathode during the operation of a fuel cell.

In the range of water volume fractions $0.10 \leq \varphi_w \leq 0.35$, the hydration of Nafion membrane is accompanied by growth of both average diameter of channels and average spacing between them. In order to quantify these structural transformations, the structure factors, $S_w(q)$, and the pair correlation functions, $g_{ww}(r)$, were calculated for water component [68]. In particular, for Nafion with EW = 1200 g/equiv, it was found that the $S_w(q)$ function demonstrates a typical ionomer peak at $q \approx 0.2$ Å^{-1} that shifts toward lower values of the wave number and becomes more pronounced as the hydration level increases. Similar behavior was observed for the systems with other equivalent weights. Furthermore, the ionomer peak was found to increase in intensity and shift to lower angles with a decrease in equivalent weight. Such a behavior has been observed in scattering studies (see, e.g., Refs. [6, 33]).

The average size of hydrophilic domains, R, and the average spacing between them, d, were estimated from the first and second peaks of the water-water pair correlation function $g_{ww}(r)$. These quantities are shown in Figure 15.7 as a function of water content. In agreement with experimental observations [6, 33, 34] and the theoretical predictions discussed above, the average size of hydrophilic domains lies in the range from 20 to 40 Å and grows with hydration (15.7a). For a given hydration level, the higher equivalent weight of Nafion molecules results in the larger average cluster size. The average domain spacing d shows similar behavior, namely it grows with hydration and equivalent weight (Figure 15.7b). The fact that increasing the equivalent weight produces an increase in the R and d values is explained by decreasing the branch density, $\gamma = n_s/N_b$. Such behavior is similar to that known for self-assembling grafted comb-like copolymers with attractive side chains [70]. In this case, there is also similarity with usual linear copolymers for which the "crushing" of the blocks leads to an expected decrease in the average size of microdomains [15]. Thus, the field-theoretic SCMF

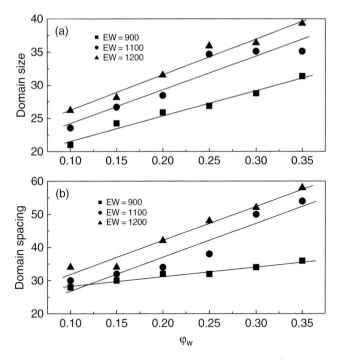

FIGURE 15.7. (a) Average size of hydrophilic domains, R (in Å), and (b) average spacing between them, d (in Å), as a function of water content, φ_w, in the range from 0.1 to 0.35 ($h \approx$ 4–20) for Nafion isomers with different equivalent weights. Adapted from Ref. [68].

calculations demonstrate the sensitivity of aggregation processes on local Nafion architecture.

It is important to note that both R and d values increase linearly with water content at solvent levels up to 35 vol% (Figure 15.7) as opposed to the 1/3 power behavior expected for isotropic swelling of spherical-shaped aggregates. Recently, Young et al. [71] performed detailed SANS investigations of Nafion-117 membranes swollen with a variety of absorbed solvents, including water. They found that shift in the position of the ionomer maximum was linearly related to the solvent volume fraction in the Nafion membrane for solvent volume fractions up to approximately 50 vol%. Correspondingly, the real-space dimensions or Bragg spacings corresponding to this maximum increased linearly with solvent content in the membranes. The swelling behavior of Nafion predicted from the SCMF calculation [68] is consistent with these experimental data, as well as with CA-based [27] and SC-MC/RISM [53] simulations discussed above. It should be noted that the average cluster separation increased with a concurrent decrease in the number density of clusters. The distribution of clusters was also shown to be nonuniform. These

476   D. Galperin et al.

findings, in agreement with other mesoscopic simulations [20, 27, 53], further support the conclusion that a considerable agglomeration of clusters occurs during the swelling process. Of course, at sufficiently high water contents, when the formation of network-like bicontinuous morphology is observed, the value of R must be considered as an "apparent cluster size", which determines, together with intercluster spacing d, the characteristic scale of the structural inhomogeneities, not the size of isolated, discrete clusters.

15.3.2.3. Microscopic Structure from Mesoscale Modeling

In order to study the internal structure of the hydrophilic channels and the spatial distribution of components, the local volume fractions of different components along an arbitrary axis was calculated (Figure 15.8) [68]. Obviously, the separation into hydrophilic and hydrophobic domains becomes more pronounced as the hydration level of the membrane is increased. The local volume fraction of water reaches almost 100% in the middle of hydrophilic channels of strongly hydrated membrane ($\varphi_w = 0.35$,

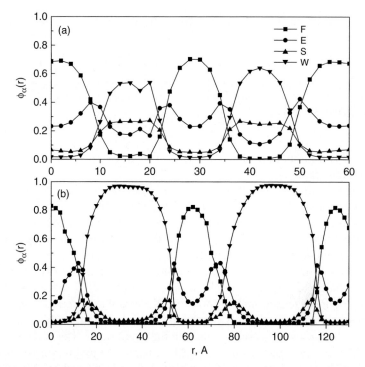

FIGURE 15.8. Volume fraction profiles of fluorocarbon groups of backbone (F), ether groups of side chains (E), sulfonate groups (S), and water (W) for hydrated Nafion with EW = 900 g/equiv, at (a) $\varphi_w = 0.1$ and (b) $\varphi_w = 0.35$. The distance r is given in Å. Adapted from Ref. [68].

Figure 15.8b), while for lower hydration levels ($\varphi_w = 0.1$, Figure 15.8a) this value only slightly exceeds 50%. In both cases, the side chains were shown to form the interphase between hydrophilic and hydrophobic domains. From the location of local maxima of E and S density profiles, one can conclude that this interphase has a layer structure: The ether groups are located closer to hydrophobic phase and sulfonic acid groups are turned toward water-filled domains. The interphase becomes more pronounced at higher water content, $\varphi_w = 0.35$, while at relatively low hydration, $\varphi_w = 0.1$, sulfonic acid groups can be found with equal probability anywhere in the hydrophilic phase, and ether groups substantially penetrate into the hydrophobic matrix. This allows suggesting that the fluorocarbon matrix is more amorphous at low hydration, since the side chains abundant in the matrix in this case are not capable of forming crystallites, due to their specific chemical structure. All these observations are in line with the transmission electron microscopy data of Hue et al. [34].

To extract further information from the field-theoretic modeling, a hierarchical procedure was employed to bridge the gap between microscopic and mesoscopic length scales. To this end, atomistic structures of Nafion and water molecules were first mapped onto the volume fraction fields, $\phi_\alpha(\mathbf{r})$, obtained in the SCMF calculation, using a MC technique. Then, these structures were utilized as input structures for detailed all-atom MD simulations. In short, the mapping procedure can be outlined as follows. A certain atom of the desired (macro)molecule was randomly generated at a sample position. According to the standard Metropolis algorithm, this trial step was either accepted or rejected depending on the volume fraction fields of the

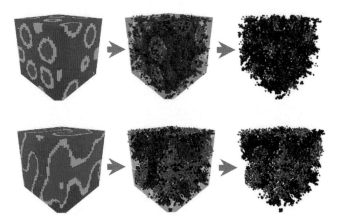

FIGURE 15.9. Snapshots illustrating "reverse-mapping" procedure in which the atomistic structure is mapped onto the average concentration fields of components, at (top) $\varphi_w = 0.1$ and (bottom) $\varphi_w = 0.35$. Regions occupied by polar (hydrophilic) species are shown in blue; hydrophobic regions are colored in red; interphase regions are shown in green.

FIGURE 15.10. Snapshots from full atomistic MD simulations showing microphase-separated structure of hydrated Nafion, at hydration levels of (a) $h = 4$ and (b) $h = 20$. Spherical domains were "cut out" from the cubic MD simulation box. For visual clarity, three-dimensional Connolly surfaces were generated for the subsystem of hydrophobic atoms. The equilibrium Nafion structure consists of water-filled pores (diameter ~ 40 Å) that are connected by channels having diameter of order 10 Å. This pore-channel nanostructure provides energetically favorable pathways through the nonpolar fluorocarbon matrix of the membrane for water and other mobile species.

atoms of this type at the sample position. Then, the polymer chain or small (water) molecule was generated step-by-step starting from this position by trial MC moves, which were also accepted or rejected so as to be consistent with the corresponding SCMF density profiles and intramolecular geometry constraints. In this way, the equilibrium atomistic structures, which fit well the density profiles, were generated. Some of them are shown in Figure 15.9 (for $\varphi_w = 0.1$ and 0.35). They can be compared to those obtained in extensive all-atom MD simulations. As an example, Figure 15.10 shows two well-equilibrated samples at different water contents.

15.4. Concluding Remarks

The main question addressed in this review was what role do the competing interactions play for the structural properties of hydrated ionomer membranes at low water contents. The behavior of such systems, in particular, their nanoscale organization, is dictated by the relative strength of the competing hydrophobic/polar interactions and can be tuned by varying parameters such as temperature, hydration level, and molecular architecture.

Due to up-to-date demands on the membrane materials with high proton conductivity over a wide range of water content, closer connections between the polymer morphology and microstructure with physical properties are needed. Because the proton mobility can hardly become higher than in bulk water, the task is to increase the proton mobility by optimizing the morphology of PEM. The challenges are in the effort to improve membranes presently used and to search for new materials exhibiting rapid exchange of the ionic species, optimum fluid permeability, and a long life (where composition, structure, and morphology of the materials are directly involved). Equally important is the development of a general understanding of the morphology-property relationships for mechanical and thermal response of these systems. In addition, understanding of electrochemical reactions and degradation processes is also important because chemical reactions occur at all stages of PEFC operation and lead to materials modifications that must be carefully identified in order to optimize the materials. These are fundamentally multiscale materials science problems.

As we have stressed throughout this article, ionomer membranes, including hydrated Nafion membranes, are complex nanostructured materials. In addition, they are multicimponent and multilength-scale materials. Furthermore, they are also interactive nanocomposites, where the amount and properties of one subphase (hydrophilic domains that conduct protons) affects the properties of other subphases (amorphous and semicrystalline perfluorocarbon regions that provide structural stability) and of interfacial regions. Composite interactions and percolation processes play key roles in the performance of PEFCs and determine the overall dependence of transport properties on the microstructure. Because of this complexity and multiscale organization, for the foreseeable future, many of physical phenomena occurring in hydrated ionomer membranes are not computationally tractable within a single numerical approach even with terascale computing facilities. In fact, they should be described within hierarchical, multilevel numerical models combining the individual advantages of atomistic and continuum statistical-mechanical approaches. Therefore, the development of coupled atomistic-continuum modeling strategies is one of the primary challenges associated with hierarchical simulation and accurate prediction of physical/chemical processes and behavior from the quantum level to nanoscale to mesoscale and beyond without loss of intrinsic structural information.

Several attempts have been made to couple microscopic simulations with statistical-mechanical theories. We have demonstrated that the hybrid MC/RISM technique combining atomistic/coarse-grained MC simulations with integral equation RISM theory is a very effective tool in the computational treatment of equilibrium properties and structural reorganizations in the weak segregation limit, when atomic level information is passed on to mesoscale level. Of course, RISM theory does not predict types of resulting nanostructures or their symmetries. It appears possible, and in fact desirable, to combine RISM or DF formalism with molecular dynamics – both classical

MD and first-principles MD. The development in this direction is very interesting and could result in a powerful multiscale modeling approach spanning several orders of magnitude over multiple time and length scales, from the quantum mechanical domain to the mesoscopic domain. Such hybrid computational schemes appear to be especially promising for studying chemical reactions. Although the modeling of transport properties is much more problematic, that possibility cannot be ruled out.

One of the main advantages of the field-theoretic SCMF approach [46, 47] is that any arbitrary morphology can be described easily by choosing appropriate molecular and macroscopic parameters. There is no a priori assumption on transient and nonequilibrium structures. However, despite the remarkable success of field-theoretic simulations in providing fundamental understanding of various inhomogeneous multiphase systems, existing field-theoretic simulation results are largely qualitative. Integrating field-theoretic methods with atomistic simulations is thus the key to *quantitatively* predicting the equilibrium structure and thermodynamics of complex heterogeneous systems that exhibit nanoscale and macroscale phase separation. We have illustrated the use of such "reverse-mapping" procedure in which the atomistic structure is mapped onto the average concentration fields of components, thus providing essential input information for more detailed (MD or MC) simulations. This computational strategy can be used to study the local organization and molecular level processes, taking into account the global organization of the microphase-separated material.

Certainly, the primary function of computer modeling is to explain and suggest experiments, much like the use of theory in physics. As was stated above, the relationships between microstructure and properties for PEMs are too complex for analytical methods to be very useful. Computational materials science provides the theory that is needed to understand current experiments and help plan new experiments. However, once the validity of a computer model is unequivocally demonstrated, one can then progress to replacing some experiments with models.

Acknowledgments

The financial support from the E.I. Du Pont de Nemours Company, RFBR and the Deutsche Forschungsgemainschaft (SFB 569, Projects A11 and B13) is highly appreciated.

References

[1] Yeager HL, Eisenberg A (1982) Perfluorinated Ionomer Membranes, Eisenberg A, Yeager HL (eds.), ACS Symp. Series 180; Amer. Chem. Soc., Waschington, DC.

[2] Gottesfeld S, Zawodzinski TA (1997) Polymer Electrolyte Fuel Cells. In AdVances in Electrochemical Science and Engineering, Alkire RC, Gerischer H, Kolb DM, Tobias CW (eds.), Wiley-VCH: Weinheim, Germany, Vol. 5, pp 195–301
[3] Ionomers: Synthesis, Structure, Properties and Applications (1997) Tant MR, Mauritz KA, Wilkes GL (eds.) Blackie Academic & Professional: London, 514 pp.
[4] Ionomers: Characterization, Theory and Applications (1996) Schlick S (ed.), CRC Press, Boca Raton, FL
[5] Roche EJ, Pineri M, Duplessix R, Levelut AM (1981) J Polym Sci Polym Phys Ed 19: 1–11
[6] Gierke TD, Munn GE, Wilson FC (1981) J Polym Sci Polym Phys Ed 19: 1687
[7] Gebel G, Moore RB (2000) Macromolecules 33: 4850–4859
[8] Elliott JA, Hanna S, Elliott AMS, Cooley GE (2000) Macromolecules 33: 4161–4171
[9] Gebel G. Polymer (2000) 41: 5829–5838
[10] Paddison SJ (2003) Annu Rev Mater Res 33: 289–319
[11] Paddison SJ (2003) Handbook of Fuel Cells – Fundamentals, Technology and Applications. Volume 3 – Fuel Cell Technology and Applications, Vielstich W, Lamm A, Gasteiger H (eds.), J Wiley & Sons, Chichester, UK
[12] Mauritz KA, Moore RB (2004) Chem Rev 104: 4535–4585
[13] Kreuer K-D, Paddison SJ, Spohr E, Schuster M (2004) Chem Rev 104: 4637–4678
[14] Weber AZ, Newman J (2004) Chem Rev 104: 4679–4726
[15] Hamley IW (1998) The Physics of Block Copolymers, Oxford University Press, Oxford
[16] Leibler L (1980) Macromolecules 13: 1602
[17] Glotzer SC, Paul W (2002) Annu Rev Mater Res 32: 401–436
[18] Hoogerbrugge PJ, Koelmann JMVA (1992) Europhys Lett 19: 155
[19] Espanol P, Warren P (1995) Europhys Lett 30: 191
[20] Yamamoto S, Hyodo S (2003) Polymer J 35: 519–527
[21] Hyodo S (2004) Molec Simul 30: 887–893
[22] Carmesin I, Kremer K (1988) Macromolecules 21: 2819
[23] Khalatur PG, Khokhlov AR, Prokhorova SA, Sheiko SS, Möller M, Reineker P, Shirvanyanz DG, Starovoitova NY (2000) Eur Phys J E 1: 99–103
[24] Khalatur PG, Shirvanyanz DG, Starovoitova NY, Khokhlov AR (2000) Macromol Theory Simul 9:141–155
[25] Shirvanyanz DG, Pavlov AS, Khalatur PG, Khokhlov AR (2000) J Chem Phys 112: 11069–11079
[26] Koelmann JMVA (1990) Phys Rev Lett 64: 1915
[27] Mologin DA, Khalatur PG, Khokhlov AR (2002) Macromol Theory Simul 11: 587–607
[28] Fried H, Binder K (1991) J Chem Phys 94: 8349
[29] Gauger A, Wayersberg A, Pakula T (1993) Macromol Chem Theory Simul 2: 531
[30] Litt MH (1997) Polymer Preprints 38: 80
[31] Roche EJ, Pinéri M, Duplessix R (1982) J Polym Sci, Polym Phys Ed 20: 107
[32] Falk M (1980) Can J Chem 58: 1495
[33] Hsu WY, Gierke TD (1982) Macromolecules 15: 101
[34] Hue T, Trent JS, Osseo-Asare K (1989) J Membr Sci 45: 261

[35] Gebel G, Lambard J (1997) Macromolecules 30: 7914
[36] Rollet A-L, Diat O, Gebel G (2002) J Phys Chem B 106: 3033
[37] Chen S, Lin YC (1994) Polym Mater Sci Eng 71: 702–703
[38] Khalatur PG, Khokhlov AR, Mologin DA, Zheligovskaya EA (1998) Macromol Theory Simul 7: 299–316
[39] MacMillan B, Sharp AR, Armstrong RL (1999) Polymer 40: 2471, 2481
[40] Dreyfus B, Gebel G, Aldebert P, Pineri M, Escoubes M, Thomas M (1990) J Phys (Paris) 51: 1341
[41] Schweizer KS, Curro JG (1994) Adv Polym Sci 116: 319
[42] Schweizer KS, Curro JG (1997) Adv Chem Phys 98: 1
[43] Chandler D, Andersen HC (1972) J Chem Phys 57: 1930
[44] Chandler D, McCoy JD, Singer SJ (1986) J Chem Phys 85: 5977
[45] Fraaije JGEM (1993) J Chem Phys 99: 9202
[46] Matsen MW, Schick M (1994) Phys Rev Lett 72: 2660
[47] Drolet F, Fredrickson GH (1999) Phys Rev Lett 83: 4317
[48] Croxton CA (1974) Liquid State Physics. A Statistical Mechanical Introduction. Cambridge University Press, Cambridge
[49] Schweizer KS, Yethiraj A (1993) J Chem Phys 98: 9053
[50] Chandler D, Singh Y, Richardson DM (1984) J Chem Phys 81: 1975
[51] Khalatur PG, Khokhlov AR (1998) Molec Phys 93: 555
[52] Khalatur PG, Zherenkova LV, Khokhlov AR (1998) Eur Phys J B 5: 881
[53] Khalatur PG, Talitskikh SK, Khokhlov AR (2002) Macromol Theory Simul 11: 566–586
[54] Flory PJ (1969) Statistical Mechanics of Chain Molecules. Wiley, New York
[55] Ostrowska J, Norebska A (1983) Colloid Polym Sci 261: 93
[56] Elliott JA, Hanna S, Elliott AMS, Cooley GE (1999) Phys Chem Chem Phys 1: 4855–4863
[57] Urata S, Irisawa J, Takada A, Shinoda W, Tsuzuki S, Mikami M (2005) J Phys Chem B 109: 4269–4278
[58] Hsu WY, Barkley JR, Meakin P (1980) Macromolecules 13: 198
[59] Safran SA, Grest GS, Webman I (1985) Phys Rev 32: 506
[60] Netemeyer SC, Glandt ED (1986) J Chem Phys 85: 6054
[61] Zawodzinski Jr. TA, Neeman M, Sillerud LO, Gottesfeld S (1991) J Phys Chem 95: 6040–44
[62] Weber AZ, Newman JJ (2004) Electrochem Soc 151: A311
[63] Edmondson CA, Fontanella JJ (2002) Solid State Ionics 152–153: 355
[64] Jang SS, Molinero V, Cagin T, Goddard WA (2004) J Phys Chem B 108: 3149–3157
[65] Eikerling M, Paddison SJ, Pratt LR, Zawodzinski TA Jr (2003) Chem Phys Lett 368:108–114
[66] Krueger JJ, Simon PP, Ploehn HJ (2002) Macromolecules 35: 5630–5639
[67] Paddison SJ, Zawodzinski TA (1998) Solid State Ionics 113–115: 333
[68] Galperin D, Khokhlov AR (2006) Macromol Theory Simul 15: 137–146.
[69] Kreuer KD (2003) Handbook of Fuel Cells – Fundamentals, Technology and Applications. Volume 3 – Fuel Cell Technology and Applications, Vielstich W, Lamm A, Gasteiger H (eds.), J Wiley & Sons, Chichester, UK
[70] Khalatur PG, Khokhlov AR (2000) J Chem Phys 112: 4849–4861
[71] Young SK, Trevino SF, Tan NCB (2002) J Polym Sci, Part B: Polym Phys 40: 387–400

The Catalyst

16
Molecular-Level Modeling of Anode and Cathode Electrocatalysis for PEM Fuel Cells

MARC T.M. KOPER

16.1. Introduction

Molecular-level modeling of heterogeneously catalyzed reactions is playing an increasingly important in the understanding of existing catalysts and the rational design of new catalysts. The progress in theoretical and computational modeling of heterogeneous catalysis, both in the gas and liquid phase, has been reviewed in several recent texts [1,2,3,4]. Although many of the conceptual aspects of modeling are similar, theoretical descriptions of catalytic reactions at the solid–liquid interface feature some important complications, mainly related to the presence of the liquid (electrolyte) phase and the electrical polarizability of the interface. This has consequences at various levels of theoretical and computational approaches.

The purpose of this chapter is to selectively summarize recent advances in the molecular modeling of anode and cathode electrocatalytic reactions employing different computational approaches, ranging from *first-principles* quantum-chemical calculations (based on density functional theory, DFT), ab initio and classical molecular dynamics simulations to kinetic Monte Carlo simulations. Each of these techniques is associated with a proper system size and timescale that can be adequately treated and will therefore focus on different aspects of the reactive system under consideration.

In the next section, I will briefly summarize key aspects of the most popular computational methods. Section 16.3 will then describe applications of these methods to relevant electrocatalytic processes such as the adsorption of CO on Pt and its alloys, the adsorption of OH, the oxidation of CO, and briefly also methanol oxidation. Section 16.4 deals with the computational modeling of oxygen reduction.

16.2. Methods

16.2.1. Ab Initio Quantum-Chemical Calculations

Quantum-chemical electronic structure calculations deal with obtaining a numerical solution to the electronic Schrödinger equation. An important advancement in electronic structure calculations in the last decades has been the development of density functional theory (DFT). DFT makes it possible to treat large many-electron systems with significantly reduced computational time and effort compared to "conventional" wave-function-based Hartree-Fock methods. There are many excellent texts on the principles and applications of DFT (see, e.g., Ref. [5]). Briefly, DFT is based on a theorem first proved by Hohenberg and Kohn [6] that states that the electronic energy is uniquely determined by the electronic density. Therefore, the exact electronic energy of a quantum-chemical system can be calculated if the exact electronic density of the system is known. However, the "functional" linking the electronic density to the electronic energy is not known and has to be approximated. There is an entire industry in computational chemistry involved in trying to establish such functionals, and the most successful functionals currently available come under the name of Generalized Gradient Approximation (GGA) and have an estimated accuracy of 10–20 kJ/mol. This may not be very satisfactory to many a "hardcore" quantum chemist, but is still very acceptable when a semi-quantitative or qualitative understanding is required. As qualitative trends produced by DFT tend to be very reliable, and the numbers are quite close to the experimental estimates, DFT–GGA calculations have become an extremely popular tool for both theoretical and experimental chemists. Many codes and packages are now available to perform DFT calculations routinely, even on a single-processor personal computer if systems are not too large.

In setting up a quantum-chemical calculation, one has to choose a geometrical model for the system under consideration. This is especially important for large systems, such as a solid, a solid surface, or a macromolecule, for which treating all atoms in the system would be prohibitively time consuming. For the interaction of molecules with surfaces, a situation relevant to (electro)catalysis, one usually resorts to either "cluster" or "slab" calculations. In cluster calculations, one models the surface by a cluster of 10–100 atoms having the solid-surface structure, and studies the interaction of the molecule with that cluster. Especially when the cluster is small, this type of calculation can be very efficient. However, small clusters do not have the electronic structure of an extended surface, and their properties tend to vary significantly with cluster size. Energies obtained from cluster calculations, such as adsorption energies, tend to be unreliable. More local structural information, such as bond distances and vibrational properties, on the other hand, are generally well described by cluster calculations. However, the preferred geometry for treating surface-molecule interactions is the slab geometry. Slab calculations make use of periodic boundary calculations such as they are used for bulk solids. The solid is repeated by the periodic boundary conditions in two dimensions, but along the third axis of the periodic "supercell," the solid is truncated into a slab of 3–10 layers, with the remaining part of the supercell either empty (i.e., vacuum) or occupied by solvent molecules. The thickness of the slab and the width of the vacuum or solvent region are chosen sufficiently large such that the properties of the system become independent of their extension.

An extensive review (with references) of the application of ab initio quantum-chemical calculations to electrochemistry can be found in Ref. [7].

16.2.2. Ab Initio and Classical Molecular Dynamics

A very prominent development in DFT has been the coupling of electronic structure calculations (which, when the ground state is concerned, apply to zero temperature) with finite-temperature molecular dynamics simulations. Carr and Parrinello published the founding paper in this field in 1985 [8]. Carr and Parrinello formulated effective equations of motion for the electrons to be solved simultaneously with the classical equations of motion for the ions. The forces on the ions are calculated from first principles by use of the Hellman-Feynman theorem. An alternative to the Carr-Parrinello method is to solve the electronic structure self-consistently at every ionic time step. Both methods are referred to as ab initio Molecular Dynamics (AIMD).

AIMD is still a very time-consuming simulation method and is as yet limited to small system sizes and "real" simulation times of not more than a few picoseconds. However, some first applications of this technique to interfacial systems of interest to electrochemistry have appeared. There is no doubt that AIMD simulations of electrochemical interfaces will become increasingly important in the future

Classical molecular dynamics (MD) simulations rely on model potentials describing the interactions between the particles in the system, rather than a first-principles calculation of the system energy at each time step. Obviously, the quality of the outcome depends on the quality of the model potential. Model potentials are normally either fit to experimental results, or based on first-principles calculation. Since the majority of the model potentials is based on two-body interactions only, such potentials generally ignore the multi-body character of real interactions. The time evolution of the system is computed by solving numerically Newton's laws of motion. Molecular dynamics or molecular simulation is a vast field with many fields of applications. For detailed discussions, see Refs. [9, 10].

16.2.3. Kinetic Monte Carlo Simulations

The methods described in the previous sections allow one to obtain information on the energetics of single atoms or molecules, or a small ensemble of molecules, interacting with a surface. However, a simulation of the overall dynamic behavior of an extended catalyst with many adsorbates interacting and reacting with each other is currently beyond the scope of any quantum chemical or molecular dynamics calculation. Still, such behavior is very relevant for understanding the macroscopic properties of model and real (electro-)catalysts.

One way to bridge this gap is to use a "coarse-graining" approach in which the surface reaction is modeled in terms of elementary steps on a lattice-like surface. The lattice points correspond to the adsorption sites on the catalyst surface. Adsorption energies and interaction energies can be in principle obtained from a DFT–GGA calculation, and rate constants can be estimated by using transition state theory. Alternatively, rate constants and energetics can be estimated by comparison with experiment or used as adjustable parameters to study the influence of their variation on the overall behavior. The overall dynamic behavior is computed from a dynamic Monte Carlo (DMC) simulation. DMC simulations are similar to Metropolis-type Monte Carlo simulations [10], with the main difference that it involves an exact introduction of time into the Monte Carlo method by providing a numerical solution to the so-called dynamic Master Equation. In my own work, I have mainly used a very versatile and user-friendly DMC code called "CARLOS" developed at Eindhoven University of Technology, which incorporates various methods for solving the Master Equation [11].

A further reason for doing Dynamic Monte Carlo simulations is a more conceptual one. The standard approach to kinetic modeling of surface-catalyzed reactions is to express all reaction rates in terms of average coverages. Such an approach is always an approximation as it ignores possible local correlations that may exist on the surface. Basically, it amounts to assuming a perfect mixing of all reaction partners and hence neglects effects of ordering, island formation, or inhomogeneous surface properties. In the statistical mechanics literature, this approximation is known as the mean-field approximation. In general, little is known about the accuracy of the mean-field approximation and presumably the best way to test its validity is by comparison with Monte Carlo simulations, which always give the exact outcome provided a good statistical sampling of all the possible configurations is carried out.

16.3. Anode Electrocatalysis

16.3.1. Adsorption of Carbon Monoxide

Understanding the quantum-chemical aspects that determine the mode and strength of adsorption of carbon monoxide to metal surfaces is crucial to issues that relate to CO tolerance and methanol oxidation, in which surface-bonded CO features as an intermediate. The adsorption of carbon monoxide to metal surfaces can be qualitatively understood using a model originally formulated by Blyholder [12]. A simplified molecular orbital picture of the interaction of CO with a transition metal surface is given in Figure 16.1. The CO frontier orbitals 5σ and $2\pi^*$ interact with the localized d metal states by splitting into bonding and antibonding hybridized metal-chemisorbate orbitals, which are in turn broadened by the interaction with the much more delocalized sp metal states.

16. Molecular-Level Modeling of Anode and Cathode Electrocatalysis

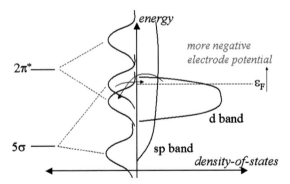

FIGURE 16.1. Energy level density-of-states diagram of the interaction of the CO frontier orbitals 5σ and $2\pi^*$ with a transition metal surface. The arrows indicate the donation of charge from the 5σ and the back donation of charge into the $2\pi^*$. The effect of the electrode potential is to shift the metal energy levels up with respect to the adsorbate and the solution as the potential becomes more negative.

Hammer, Morikawa, and Nørskov [13] have shown how this simple picture emphasizing the interaction of the CO frontier orbitals with the metal d states can explain trends in the binding energies of CO to various (modified) metal surfaces. Their model for CO chemisorption is based on an earlier, more general model for chemisorption suggested by Hammer and Nørskov [14]. The Hammer-Nørskov model singles out three surface properties contributing to the ability of the surface to make and break adsorbate bonds: (i) the center ϵ_d of the d band, (ii) the degree of filling f of the d band, and (iii) the coupling matrix element V_{ad} between the adsorbate states and the metal d states. These quantities have been calculated from extensive quantum-chemical calculations for a significant portion of the Periodic Table [15]. The basic idea of the Hammer-Nørskov model is that trends in the interaction and reactivity are governed by the coupling of the adsorbate states with the metal d states, since the coupling with the metal sp states is essentially the same for the transition and noble metals, and mainly acts to renormalize (i.e, shift and broaden) the energy of the adsorbate orbital.

In the Blyholder model, CO interacts with the metal states through two different states, the 5σ and $2\pi^*$ orbitals, but due to their different symmetry they interact with different metal d orbitals, and the two interactions can be treated independently. Hammer, Morikawa, and Nørskov [13] used the following expression to model the d contribution to the CO chemisorption energy:

$$E_{d-hyb} = -4\left[f\frac{V_\pi^2}{\varepsilon_{2\pi}-\varepsilon_d}+fS_\pi V_\pi\right] - 2\left[(1-f)\frac{V_\sigma^2}{\varepsilon_d-\varepsilon_{5\sigma}}\right] + (1+f)S_\sigma V_\sigma$$

(16.1)

where 2 is for spin, $\epsilon_{2\pi}$ and $\epsilon_{5\sigma}$ are the positions in energy of the (renormalized) adsorbate states, and S is the overlap integral. The first term on the right-hand side represents the back donation contribution, the second term the donation contribution, and the last term the contribution due to Pauli repulsion. Figure 16.2 compares the model expression and the full DFT–GGA chemisorption energies for a number of metal systems. It is seen that the model gives a good qualitative and even reasonable quantitative description of the CO chemisorption system. Hammer et al. [13] also find that the 5σ donation contribution is quite small, whereas the $2\pi^*$ back donation interaction is strongly attractive and dominates the variations among the different substrates. Generally, CO binds stronger to the lower transition metals (i.e., toward the upper left in the Periodic Table), mainly due to the center of the d band ϵ_d moving up in energy leading to a stronger back donation.

The same model may also be used to explain why on metals in the upper-right corner of the Periodic Table (Pd, Ni) CO prefers multifold coordination, whereas toward the lower left corner (Ru, Ir) CO preferentially adsorbs atop [16]. Pd and Ni combine a high d band filling factor f, which is favorable for a strong back donation, with a low steric repulsion as the d states on these metals are more contracted. Since the back donation interaction is a bonding interaction, it will prefer to coordinate to as many surface atoms as possible. A low steric repulsion is also a favorable condition for multifold coordination. Ru and Ir, on the other hand, combine a relatively lower d band filling with much more diffuse d band states leading to a strong steric Pauli repulsion. Both factors will favor atop coordination on these metals. Pt and Rh are borderline cases, and indeed on these metals there is no strong energetic preference for either atop, bridge or hollow coordination of CO.

One experimental aspect of CO chemisorption for which quantum-chemical calculations have provided particularly useful insight is the redshift

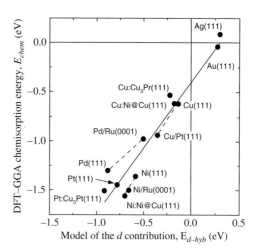

FIGURE 16.2. Comparison of the full DFT–GGA and model chemisorption energies for CO on a number of metal and metal-overlayer systems.

observed in the C–O intramolecular vibrational frequency upon adsorption. Depending on the substrate and the type of coordination, the C–O intramolecular vibration frequency in the chemisorbed state may be 100–400 cm^{-1} lower than its uncoordinated vacuum value (2170 cm^{-1}) (for a detailed comparison of UHV and electrochemical data, see Refs. [17, 18]). A simple but incomplete explanation for this observation refers to the Blyholder model. In its chemisorbed state, CO mainly interacts through its $2\pi^*$ orbital. This antibonding orbital is unoccupied for uncoordinated CO but becomes partially occupied in its chemisorbed state. Note that this back donation interaction is *bonding* in terms of the substrate-adorbate interaction but *antibonding* for the intramolecular C–O bond. This weakening of the C–O bond due to the metal to CO back donation is accompanied by a lowering of the C–O stretch frequency. This picture suggests a simple relationship between C–O stretching frequency and its adsorption strength: A lower C–O stretch frequency would imply a stronger back donation and hence a more favorable adsorption energy. This relationship is simple and attractive and hence very popular in the literature, but it is misleading and generally wrong.

Detailed ab initio-based analyses of the origin of C–O redshift have been pioneered by Bagus, Pacchioni and Illas [19,20,21]. We will discuss here a similar analysis of the origin of the vibrational shift of the CO and NO stretching frequency on a series of transition metal surfaces by my own group, carried out using DFT–GGA cluster calculations [16]. It was found in these calculations that the DFT-computed C–O and N–O stretching frequencies were in good agreement with the experimental frequencies, the DFT values being slightly lower (ca.20–60 cm^{-1}), as shown in Figure 16.3 Table 16.1 gives some representative and illustrative results of a frequency decomposition analysis for CO adsorbed in the atop and hollow site of a 13-atom Pt(111) cluster, and at the atop site of a 13-atom Ru(0001) cluster. The

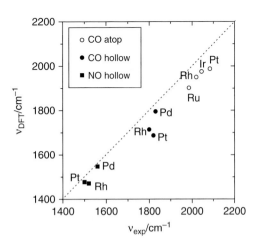

FIGURE 16.3. Comparison of the DFT–GGA calculated C–O and N–O internal stretching frequencies on a number of metals compared to the experimental values obtained at low coverage in UHV.

TABLE 16.1 Decomposition of the zero-field vibrational frequency shifts (in cm^{-1}) compared to the calculated vacuum values into steric and orbital components for CO chemisorbed in the atop and hollow site of Pt(111) and CO adsorbed in the atop site of Ru(0001)[a].

System	Steric	donation	back donation	rest	Final
Pt–CO (atop)	+743	−328(−486)	−569(−411)	+46	1987
Pt–CO (hollow)	+645	−198(−371)	−910(−737)	+82	1714
Ru–CO (atop)	+670	−349(−471)	−562(−440)	+47	1901

[a]The far-left column in the cluster-adsorbate system. Adjacent three columns (steric/donation/back donation) give the change in frequency due to each contribution, calculated with respect to the uncoordinated DFT frequency of 2095 cm^{-1} for CO. The main entries have been calculated in the order as given in the Table, i.e., steric-donation-back donation, whereas the entries within parentheses refer to the order steric-back donation-donation. The column "rest" refers to the residual frequency shift upon adsorption not accounted for by the sum of steric/donation/back donation. The far-right column gives the coordinated C–O vibrational frequency.

frequency decomposition analysis is based on the idea that the electronic energy can be decomposed in several contributions by constraining the electron reorganization in various steps in the calculation procedure [22]. In a first step of bond formation, no orbital hybridization is allowed, such that only the steric Pauli repulsion and electrostatic effects are included. This generally leads to a positive (repulsive) bond energy. In a next step, orbital hydrization is allowed, according to the different symmetry groups of the orbitals involved. This permits a decomposition, for instance, into the contribution due to s or σ-type orbitals and p or π-type orbitals. These contributions can be either negative (bonding) or positive (antibonding). A similar decomposition procedure can be introduced for analyzing changes in vibrational frequencies as the chemical environment changes [16]. Comparing the frequency decomposition in Table 16.1 for CO adsorbed atop and hollow on the Pt(111) cluster shows that the lower frequency in the hollow site is primarily due to the more negative back donation contribution there. Comparing CO adsorbed atop on Pt(111) and Ru(0001) shows that the C–O stretch frequency on Ru(0001) is lower due to the smaller steric contribution to the overall frequency change. This is in contrast with the common interpretation in the literature that a lower frequency is related to a stronger back donation, and hence a stronger chemisorption bond. A detailed discussion of the absence of a clear-cut relation between the C–O stretching frequency and the chemisorption energy was recently given by Wasileski and Weaver [23].

The interaction of CO with alloy or bimetallic surfaces is of special interest because of the importance of bimetallic catalysts in both the electrochemical and gas-phase oxidation of CO. Platinum-ruthenium alloys have long been known to be superior catalysts for the electrochemical CO oxidation, but the details of their catalytic action are still disputed.

Recent DFT–GGA calculations based on the slab geometry show that on the surface of a bulk alloy the CO binding to the Pt site weakens, whereas that to the Ru site gets stronger [24, 25]. For example, on the surface of a homogeneous $Pt_2Ru(111)$ alloy, the CO atop binding to Pt is weakened by about 0.2–0.3 eV, and that to Ru strengthened by about 0.1 eV. However, real catalytic surfaces are not homogeneous but may show the tendency to surface segregate. Figure 16.4 shows the binding energy and vibrational properties of CO coordinated to an atop Pt site on a series of surfaces with a pure Pt top layer, but for which the bulk composition changes from pure Pt, to Pt:Ru 2:1 to Pt:Ru 1:2 to pure Ru. It is observed that a higher fraction of Ru in the bulk causes a weakening of the CO bond to the Pt overlayer. This electronic alloying effect can be understood on the basis of the Hammer-Norskov d band shift model. Alloying Pt with Ru causes a flow of electrons from Ru to Pt, as evidenced from a charge analysis of the slab calculations [24], which pushes the local d band center on Pt downward with respect to the Fermi level. Interestingly, 16.4 also illustrates that there is no correlation between the binding energy and the C–O stretching frequency, even though such a

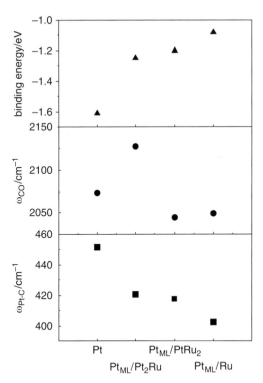

FIGURE 16.4. The binding energy, C–O stretching frequency, and Pt–C stretching frequency on four different surface, all with a Pt surface layer but different bulk compositions (Pt, Pt_2Ru, $PtRu_2$, Ru).

correlation is often assumed in the literature. As explained above, the frequency decomposition analysis suggest that there is no quantum-chemical basis for such a correlation, as both the back donation and steric contribution have an apparently equally strong influence of the C–O stretching frequency. On the other hand, the Pt–C stretching mode correlates well with the binding energy: A stronger bond causes the expected increase in the Pt–C frequency. We note that the weakening of Pt–CO bond on Pt/Ru(0001) in ultra-high vacuum was indeed observed experimentally by Buatier de Mongeot et al. using temperature-programmed desorption [26].

The opposite effect is observed when a surface with a pure Ru overlayer is considered, and the bulk is stepwise enriched with Pt. The CO binding to the Ru overlayer is strengthened with a higher content of Pt in the bulk. In fact, the Ru/Pt(111) surface is the one that shows the strongest CO binding [25]. A similar effect is also observed when the top layer of a Pt(111) surface is mixed with Ru, leading to a surface alloy [24]. As the Ru content of the surface layer increases, the binding to the Pt surface sites is weakened and the binding to the Ru surface sites is strengthened [24].

Under electrochemical conditions, the applied electrode potential has a significant influence on the binding properties of CO. We consider here the field-dependent binding energetics in the atop and hollow site of a 13-atom Pt(111) cluster [16]. Figure 16.5 shows the total binding energy $E_b(F)$ (upper left panel) and its field-dependent decomposition into steric (lower left panel) and orbital components (right panels), where the latter is further decomposed into contributions of A_1 symmetry (donation contribution, upper right panel) and E symmetry (back donation contribution, lower right panel). The orbital components have been set to zero for vanishing electric field, as their absolute values are not of interest. In addition to the calculations in which the C–O bond length was fully relaxed (solid lines), Figure 16.5 also shows the field-dependent binding parameters pertaining to a chemisorbed CO with its C–O bond length fixed at its uncoordinated vacuum value (dashed lines). This bond constraining is useful especially in analyzing the field-dependent steric contribution to the binding energy but will not be discussed here.

A significant result illustrated in Figure 16.5 is that atop coordination is preferred at positive fields, whereas at negative fields multifold hollow coordination is most favorable. This field-driven site switch on Pt(111) is in qualitative accordance with experiment [17, 18]. It is seen that the back donation component gets more negative (i.e., more stabilizing) toward negative field, whereas the opposite is the case for the donation component. This field-dependent donation and back donation effects are expected on the basis of Figure 16.1, and in agreement with the earlier ab initio calculations of Head-Gordon and Tully [27] and Illas and co-workers [28,29,30]. The back donation component is more strongly field dependent for the multifold coordination, i.e., more stabilizing toward negative fields. This can be understood in terms of the bonding character of the back donation interaction, which will prefer to bind to as many surface atoms as possible. By contrast, the donation

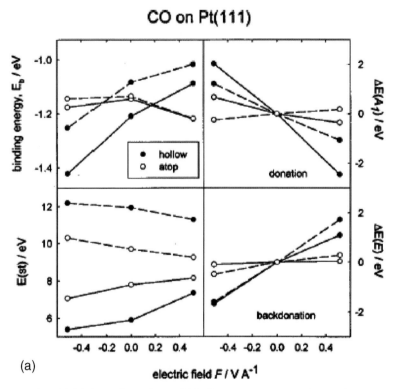

FIGURE 16.5. Field-dependent plots of binding energy, E_b, and constituent steric, E(st), and donation (A_1) and back donation (E) orbital components for chemisorbed CO in atop (*open circles*) and hollow (*filled circles*) sites on Pt(111) surface. Note that the orbital components are plotted as differences with respect to the zero-field values. Dashed plots refer to "bond-constrained" values, as described in the text.

interaction mainly occurs through an antibonding resonance and should therefore prefer to bind to as few surface atoms as possible, as already alluded to above. This explains the more destabilizing effect of a more negative field on the back donation component for multifold coordination compared to atop coordination. For the C–O "bond-constrained" case, the increase in the back donation interaction seen toward more negative fields is substantially larger than the opposite field dependence observed for the donation term. Since there is no significant difference in the field dependence of the steric term for atop and hollow coordination (although their absolute values are obviously different), the increased back donation interaction toward negative field is the dominant reason for CO to switch site from atop to multifold coordination on Pt(111) under the influence of a more negative electric field. Also note that the steric component gets more destabilizing toward negative field, which is due to a combination of the increased electron spill over at negative fields and the shorter Pt–CO bond length.

We now turn our attention to the field-dependent C–O stretching frequency. Stark tuning slopes were calculated from the slope of the $\nu - F$ curve at $F = 0$, for both CO and NO (nitric oxide) on the four different transition-metal surfaces Rh, Ir, Pd, and Pt [16]. The experimental Stark tuning slopes $d\nu/dE$, expressed in wave numbers per Volt, were converted into field units, i.e. wave numbers per Volt/Angstrom, by using the relation

$$\frac{d\nu}{dF} = d_{dl} \frac{d\nu}{dE} \quad (16.2)$$

where d_{dl} is the effective thickness of the double layer, taken to be ca. 3 Å [31]. The $d\nu_{DFT}/dF$ values were found to be in rough accordance with the $d\nu_{exp}/dF$ estimates, although the variations of the former are smaller. These discrepancies are probably due to the role of co-adsorbed solvent in affecting the local electrostatic field. Nevertheless, the larger Stark tuning slopes computed for NO vs. CO are in accordance with experimental observations.

The field-dependent changes in the C–O and N–O frequencies were analyzed following the decomposition method [16]. As reference state, the frequency at negative field $F = -0.514\,\text{V/Å}$ was taken, whereas the final state was the system at $F = +0.514\,\text{V/Å}$. To the negative field PES, the steric, donation, and back donation components calculated for the positive field were added sequentially and the corresponding frequency changes calculated. Table 16.2 shows the results for CO adsorbed in the atop and hollow site of the Pt(111) cluster. Both the steric and the donation contributions are seen make a negative contribution (this was also found for CO and NO on other metals). The negative steric contribution likely arises from the lower metal electron surface electron density toward positive fields, diminishing the extent of steric repulsion with the chemisorbate electrons and hence yielding a shallower PES. The negative donation contribution is also readily understandable given that this orbital interaction should lessen toward more positive fields. Note, however, that the absolute values of this contribution are small, indicating an interesting insensitivity of the donation interaction to the

TABLE 16.2 Decomposition of field-induced frequency shift (in cm^{-1}) for CO adsorbed in the atop and hollow site on Pt(111)[a].

System	Steric	donation	back donation	rest	overall
Pt–CO atop	−131	−7	+264	−16	+110
Pt–CO hollow	−72	−20	+201	+22	+131

[a] The far-left column is the cluster/binding site configuration. The far-right column is the frequency shift computed for a change in the applied field from −0.514 V/Å to +0.514 V/Å. The middle three columns give the corresponding field-induced frequency shift due to each contribution. The column "rest" refers to the residual field-induced frequency shift not accounted for by the sum of these components.

electrostatic field. Table 16.2 also illustrates that the dominant contribution to the field-induced frequency changes is the back donation component, being a large positive term, making the overall Stark tuning slope for the intramolecular vibration a positive quantity.

However, the decomposition analysis does not suggest any overriding trends that could be responsible for the variations observed in the Stark tuning slope as a function of adsorbate (CO or NO), coordination (atop vs. hollow), or metal substrate. For instance, Table 16.2 shows that the higher Stark tuning slope for CO bound in the hollow vs. the atop site (a result which itself is in agreement with experiment) cannot be attributed to a higher back donation component in the hollow site, as one might initially expect, at least not according to this method of analysis. The statement also holds for the higher Stark tuning slopes observed for chemisorbed NO vs. CO. The overall picture is rather one that invokes the offsetting influences of two or more components on the overall field-dependent behavior.

The field dependence of the metal-adsorbate $\nu_{\text{M-CO}}$ was also studied in some detail [32]. Figure 16.6 shows the $\nu_{M-CO} - F$ curves compared to the $\nu_{C-O} - F$ for CO adsorbed at the atop and hollow site on a Pt(111) cluster. For CO in the hollow site and for CO in the atop site at sufficiently positive potential, the $\nu_{M-CO} - F$ curve has a negative Stark tuning slope, in

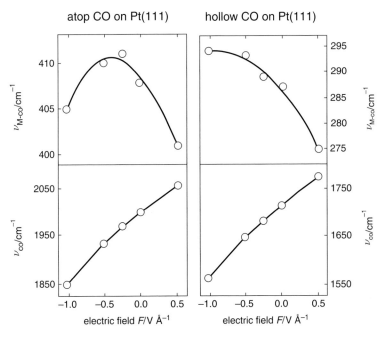

FIGURE 16.6. Stretching frequencies for metal-carbon ($\nu_{\text{M-CO}}$) and internal CO (ν_{CO}) bonds plotted against external field F for atop and hollow-site CO on Pt(111).

agreement with Raman spectroscopic results on polycrystalline Pt [33]. However, the curve for atop CO shows a well-pronounced maximum, as at negative potentials/fields the $\nu_{M-CO} - F$ curve has a positive slope. Such behavior is related, at least qualitatively, to a change in the sign of the dynamic dipole of the M–CO bond as a function of field. For hollow site CO, the maximum is not observed. This is most likely related to the much more negative chemisorption bond at negative fields for this geometry (see Figure 5), which favors progressively larger ν_{M-CO} values under these conditions [32]. The combination of the two effects yields the nearly constant ν_{M-CO} values seen at negative fields in the DFT calculations.

16.3.2. Electro-Oxidation of Carbon Monoxide

The mechanism for CO oxidation on metal electrodes is substantially different from that on metal surfaces in the gas phase. This is primarily related to a different oxygen donor: water in the aqueous phase vs. di-oxygen in the gas phase. It is now generally accepted in the experimental literature that on platinum water needs to be activated in order to react with chemisorbed CO. This "activated water" is most likely adsorbed hydroxyl. The reaction path between adsorbed CO and OH on Pt can be studied using DFT calculations. Figure 16.7 shows the energy profile as the CO and OH approach each other and react on Pt(111) slab [34]. An activation energy of ca. 0.6 eV is observed, which would agree with the experimental suggestion that the reaction between CO and OH to form adsorbed carboxyhydroxyl (COOH) may be the rate-determining step. Dunietz et al. [35] have looked at the influence of

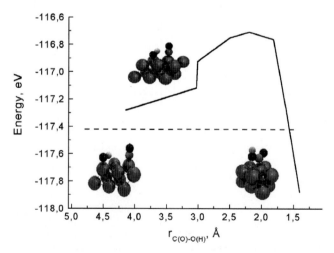

FIGURE 16.7. Energy profile as a function of the reaction coordinate for CO + OH on Pt(111). The reaction coordinate is defined as a distance between C(O) and O(H). The dashed line shows the energy at infinite separation of CO and OH.

an applied field and co-adsorbed water on the same reaction (though using a cluster instead of a slab geometry) and showed how at positive field the hydrogen is transferred to a water molecule, leading to the final formation of CO_2 and hydronium in solution.

The CO oxidation on PtRu is enhanced primarily because of the so-called bifunctional effect, first suggested by Watanabe and Motoo [36]. Ru is supposed to act as active sites to activate water at reduced overpotential, as OH binds stronger to Ru than to Pt (see Section 16.3.3), and CO adsorbed on Pt then reacts with OH adsorbed on Ru. It is anticipated that CO mobility will play an important role in the catalysis of this reaction. In order to test this presumption, we have employed DMC simulations to model CO diffusion on PtRu bimetallic electrodes, with the specific aim to assess the role of CO diffusion in CO electrooxidation on these surfaces [37].

The model for CO oxidation on PtRu electrodes was based on the bifunctional mechanism, in which OH is formed preferentially on Ru sites, where it may react with a CO on either Pt or Ru. The simulations explicitly take into account the mobility of CO on the surface, by introducing a "reaction" in which a CO may exchange places with an "empty" (ie, water-adsorbed) neighboring site. The rate of this process is proportional to the diffusion coefficient D.

The surfaces consisted of a random mixture of Pt and Ru sites in a square lattice, at different mixing ratios. Our initial simulations were designed to mimic as closely as possible the experiments carried out by Gasteiger et al. [38]. Therefore, we simulated the stripping voltammetry of a 0.99 monolayer of CO on the surface as a function of both the Ru content of the surface and the CO diffusion coefficient. A plot of the CO stripping peak potential as a function of the Ru content for three different diffusion coefficient is shown in Figure 16.8. In fact, the plot for large D reproduces a similar plot by

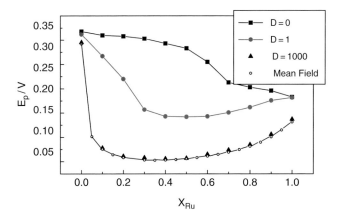

FIGURE 16.8. CO stripping peak potential E_p from kinetic Monte Carlo simulations as a function of the Ru fraction of the Pt–Ru model surface, for three different surface diffusion rates $D = 0$, 1, and 1000 s^{-1}. For details, see Ref. [78].

Gasteiger et al. very well: Especially, the experimentally observed fact that only a small amount of Ru is needed to see a ca. 200 mV shift in peak potential is also found in the simulation. However, at these Ru contents, the stripping peak is very broad, as found both in the simulations and the experiment. Interestingly, if CO is not mobile on these PtRu surfaces, the simulations predict that PtRu would not show any catalytic enhancement at all with respect to pure Ru. These results clearly demonstrate the importance of CO surface mobility in explaining the electrocatalytic activity of Pt–Ru alloys.

Our simulations also suggested that when Ru is present in/on Pt as islands, two CO stripping peaks may be observed if CO diffusion is sufficiently slow [37]. The first peak would correspond to CO stripping from Ru and the surrounding Pt, and the second to CO stripping from Pt. Experiments by the Baltruschat and Wieckowski groups [39, 40] indeed found two stripping peaks for CO stripping from a Ru-modified Pt(111) surface, on which Ru is known to form small islands. The common consensus at the moment regarding this observation is that even if CO diffusion on Pt is very fast, mobility near the Ru islands is strongly hampered as CO binding to a Pt site which is coordinated to Ru is (much) weaker than to pure Pt [25], as indeed suggested by the DFT calculations discussed above. Since the second peak observed experimentally is unlikely to be CO stripping from Pt(111) (as the difference in peak potential with pure Pt(111) would be too large), it is most likely to be interpreted as CO oxidation at the PtRu sites, but with the Pt sites being more difficult to reach as the CO binding energy there is unfavorable.

16.3.3. Adsorption of OH

The first product formed from water dissociation at positive electrode potentials is surface-bonded hydroxyl OH. Detailed calculations of the properties of chemisorbed OH suggest that it should be viewed upon as a surface hydroxide $OH^{\delta-}$ [41], with δ close to 1. This anionic character is related to the 1π orbital which is occupied by 1 electron in the uncoordinated OH, and whose energy lies below the Fermi level of the metal. Hence electronic charge is transferred from the metal to the OH upon adsorption. The bonding of OH is generally weaker than that of atomic O because of the lower degeneracy of the 1π orbital compared to the $2p$ orbital on oxygen [41].

The preferred binding site of OH on Pt(111) and Ru(0001) has been studied by DFT–GGA slab calculations by Michaelides and Hu [42] and Koper et al. [25]. Both groups find that on Pt(111) the atop adsorption site is preferred, with a binding energy of ca. 223 kJ/mol. Surface-bonded OH is tilted in the on-top and bridge sites but adsorbs upright in the hollow site. These binding geometries in the different sites can be understood qualitatively by the sp^3 hybridization of the atomic orbitals on oxygen, which prefers a tetrahedral coordination. Contrary to Pt(111), OH prefers the upright position in the fcc hollow site on Ru(0001) [25]. In agreement with the higher oxophilicity of Ru, the binding of OH to Ru(0001) is stronger than to Pt(111), i.e., 340 vs. 223 kJ/mol.

Vassilev et al. [43] studied the chemisorption of OH on Rh(111) in the presence of co-adsorbed water molecules by ab initio molecular dynamics. The OH was "created" on the surface by starting with a configuration from a water bilayer on Rh(111), and then removing the two hydrogens from a water molecule in the first layer of the bilayer. The resulting oxygen species is not stable and reacts quickly in the simulation with a neighboring water molecule to produce two surface-bonded OH species. It was found that the OH species is highly mobile on the surface due to a fast proton hopping between neighboring water molecules and the OH, implying an OH hopping in the opposite direction, as illustrated in Figure 16.9. This is similar to the Grotthus mechanism for proton and hydroxyl mobility in bulk water. Interestingly, in the simulation, the surface mobility of OH is higher than that calculated for OH^- in liquid water from similar simulations by Tuckerman et al. [44]. Also, from the simulations, it is concluded that surface-bonded OH should be considered a $OH(H_2O)_2$ complex composed of the OH itself and two neighboring water molecules. The complexing water molecules are characterized by shorter hydrogen bonds with the OH than with other water molecules.

16.3.4. Methanol

Another important molecule in low-temperature PEM fuel cells is methanol. The dehydrogenation of methanol on various surfaces has been studied extensively using DFT calculations [45,46,47]. The influence of co-adsorbed water was recently studied by a number of different groups [48,49,50,51]. One interesting aspect of catalytic methanol dehydrogenation is that the energetics of dissociation seems to depend on whether water is present or not. In the presence of water, the formation of hydroxymethyl CH_2OH by

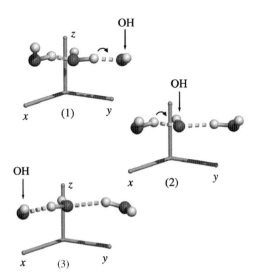

FIGURE 16.9. Subsequent configurations in the chain of proton exchange reactions between neighboring OH and H_2O molecules leading to an effective "shift" in the position of the surface hydroxyl. The Rh(111) surface is in the x–y plane.

C–H bond cleavage seems to be favored. Wasileski and Neurock found that, in general, the presence of water favors the formation of surface-bonded intermediates, but hydroxymethyl is stabilized more than methoxy. In a subsequent study, the influence of the electrode potential was included, suggesting that at more positive potentials, the difference in stability becomes smaller again. Interestingly, the stability of methoxy vs. hydroxymethyl also depends sensitively on the available binding site. Introducing a (100) step in the Pt(111) surface leads to almost equal binding energies of CH_2OH and CH_3O on the step. These results suggest that the mechanism of methanol oxidation is very structure sensitive, in particular the intermediates formed during the reaction.

16.4. Cathode Electrocatalysis

16.4.1. Mechansim of Oxygen Reduction

The electrochemical reduction of oxygen has been of interest to electrochemists and fuel cell chemists for more than a century [52]. Molecular models of oxygen reduction, incorporating the quantum-chemical interaction of oxygen with the metal surface and with the solvent (water), have been of much more recent vintage. These models aim at explaining the high overpotential observed for oxygen reduction, even on the most active electrocatalyst (platinum) and at identifying and characterizing the intermediates and relevant steps of oxygen reduction at the molecular level. In this section, I will review our attempts at treating oxygen reduction at increasing levels of sophistication, ultimately leading to a reaction scheme for oxygen reduction that simplifies existing reaction schemes by excluding certain intermediates and steps on theoretical grounds. Other workers in this field have dealt with identifying the molecular origin of the high overpotential of oxygen reduction. Their efforts will be briefly dealt with in Section 16.4.2

In many experimental papers on oxygen reduction, the rate-determining step in oxygen reduction is often considered to be the first electron transfer step (for a review, see [53]). The exact nature of this step remains elusive. In its simplest form, one could write:

$$O_2 + e \rightarrow O_2^-, \quad (16.3)$$

leaving the adsorbed or dissolved nature of either oxygen or the superoxide species unspecified. If the interaction of both species with the electrode can be neglected, reaction 3 can be regarded as a Marcus-type electron transfer reaction. We have constructed free energy curves for the solvent reorganization around the oxygen and the superoxide species using extensive molecular dynamics simulations [54]. These simulations employed a classical water interaction potential (SPC/E model), and the interaction between the

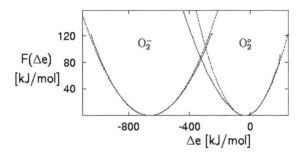

FIGURE 16.10. Free energy surfaces calculated for reaction 3 from molecular dynamics simulations. Dashed lines are parabolic fits to the minima of the curves.

O_2/O_2^- species and the water was considered a superposition of Lennard-Jones interactions between the constituting atoms and classical electrostatic interactions between the charges on the atoms. The resulting free energy curves shown in Figure 16.10 depict the free energy of solvent reorganization around the O_2/O_2^- species as a function of the so-called generalized solvent coordinate. This coordinate is essentially the electrostatic potential generated by the corresponding solvent configuration as generated at the center of the redox species. Where the curves for O_2 and O_2^- cross, the energies of the oxygen and the superoxide species are equal, and the electron may be exchanged radiationless. Two relevant conclusions were drawn from these calculations. First, the dominant barrier for reaction 4.1 is the outer-sphere solvent reorganization. The contribution of the inner-sphere reorganization, i.e. the O–O bond length changes, is much smaller. Second, the solvent response is significantly non-linear, in contrast to what is usually assumed in the Marcus theory for outer-sphere electron transfer reactions. According to linear response theory, the solvent reorganization energy λ, which is classical parameter in Marcus theory, can be estimated from the curvature in the minima of the potential energy surfaces. For the oxygen molecule, a value of 162 kJ/mol is obtained; for the superoxide molecular ion, $\lambda = 317$ kJ/mol. This significant difference (almost a factor of two) is due to the fact that effective radius of the hydration shell of the superoxide ion is smaller than of the oxygen molecule. The smaller radius is the result of the ion-dipole interaction, which is absent for the neutral molecule. We have shown elsewhere that these strong deviations from linear response theory are especially important for reactions involving a change from 0 to −1 in charge [55].

Although the above computations demonstrate the importance of solvent reorganization in activating the oxygen molecule, the interaction with the electrode surface is not included. To this end, we have performed DFT calculations of the interaction of O_2, HO_2 and H_2O_2 with Pt and Au surfaces [56, 57]. For gold, we have also considered the co-adsorption of oxygen with water.

The interaction of O_2 with Pt in ultra-high vacuum has been studied extensively, both experimentally and computationally [58, 59]. Oxygen adsorbs most strongly on Pt(110), and weakest on Pt(111). On all surfaces, the most favorable configuration is a flat adsorption mode with both oxygens coordinating with a metal surface atom, forming a bridge. Chemisorption of oxygen leads to electron transfer from the metal to the oxygen. Interestingly, the extent of this charge transfer depends on the adsorption mode. On Pt(111), the most favorable adsorption mode is the bridge mode. The O–O stretching frequency (ca 950 cm^{-1}) suggest that O_2 is in a superoxo O_2^- state. A slightly less favorable mode is one in which the center of the molecule is above a hollow site, with one oxygen coordinating with an atop Pt surface atom, and the other with a bridge site. In this case, the O–O stretching frequency (ca 740 cm^{-1}) suggest that O_2 is in an a peroxo O_2^{2-} state. Charge analyses also confirm that the latter adsorption is more strongly charged. Remarkably, the field dependence of both modes is also very different (see Figure 16.11). Although the peroxo-state is more charged, it is less sensitive to changes in the applied field than the superoxo-state. We have ascribed this effect to charge saturation. In the peroxo-state, the oxygen is almost fully saturated with electrons, and applying an electric field will not change the electron occupation much. In the superoxo-state, the $2\pi^*$ state is still only partially occupied. Therefore, this state is more polarizable (though less polar) than the peroxo-state and exhibits a stronger field dependence.

On gold, we found that O_2 adsorbs very weakly but always in the above-mentioned bridged position. When O_2 is co-adsorbed with a layer of water molecules on Au(100) in an AIMD simulation, the O_2 reacts rapidly with a neighboring water molecule to form $HO_{2,ads}$ and OH_{ads}. This strongly suggests that once oxygen has reached the surface, concerted electron and proton transfer take place, and the key first step in oxygen reduction is :

$$O_2 + H^+ + e^- \rightarrow HO_{2,ads} \tag{16.4}$$

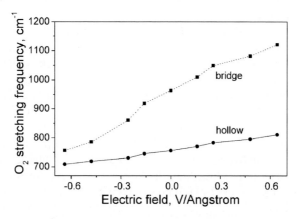

FIGURE 16.11. O–O stretching frequency of O_2 adsorbed at the bridged and hollow position on a Pt(111) surface as a function of the applied field.

or the corresponding reaction for neutral and alkaline solutions. The further fate of the reaction would then be determined by the stability and reactivity of the HO_2 intermediate. On Pt, we found that (in the absence of water) this species is (meta)stable only on Pt(111) – on Pt(100) and Pt(110) it dissociates immediately. On Au, it is more stable and does not show activationless dissociation. Hydrogen peroxide, H_2O_2, is never stable in adsorbed form; it dissociates on both Pt and Au. Therefore, we believe adsorbed H_2O_2 is not a stable intermediate in oxygen reduction.

Comparing the above results to the experimental tendency to form (dissolved) hydrogen peroxide, which is strong on gold [50] and stronger on Pt(111) than on the other two low-index platinum surfaces [60], suggests that the ability to make hydrogen peroxide may depend on the stability of the $HO_{2,ads}$ intermediate. If it is relatively stable toward dissociation—it may desorb as H_2O_2. If it dissociates rapidly, the final product of the oxygen reduction will be water.

16.4.2. Theoretical Volcano Plot for Oxygen Reduction

Nørskov and colleagues [61] have recently formulated a model for oxygen reduction based on the stability of intermediates, calculated using DFT methods, following a mechanism similar to that presented in the previous section. The attractive feature of this model is that it explains the high overpotential for oxygen reduction and why certain alloys of platinum, such as PtCo and PtNi, show a (slightly) higher activity than pure platinum. Figure 16.12 shows the

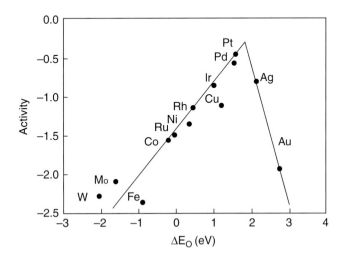

FIGURE 16.12. Trends in oxygen reduction activity (for details see original Ref. [61]) plotted as a function of the oxygen binding energy.

estimated catalytic activity for the oxygen reduction as a function of the adsorption energy of atomic oxygen. Although the involvement of atomic oxygen in the oxygen reduction mechanism may be disputed, a very similar "volcano plot" is obtained when the OH adsorption energy is plotted on the horizontal axis. On the left side of the volcano, oxygen binds so strongly that proton transfer to yield water becomes strongly activated, rendering oxygen reduction very slow. On the right side, the low oxygen adsorption energy is correlated with a high oxygen dissociation barrier, explaining the low activity of Ag and Au for oxygen reduction. On alloys such as PtNi and PtCo, Pt will segregate to the surface. Such "Pt-skin" surfaces have a slightly lower oxygen binding energy, and hence the oxygen reduction activity will be closer to the maximum in the volcano of Figure 16.12.

16.5. Concluding Remark

In this chapter, I have described some recent results illustrating the application of computer calculations and simulations to elucidate factors that are important in understanding catalytic reactions for PEM fuel cells. These methods, comprising ab initio DFT calculations, ab initio and classical molecular dynamics, and dynamic Monte Carlo simulations, have reached a level that they constitute a crucial if not indispensable tool in interpreting, understanding, and ultimately predicting experimental data. There is no question that computational chemistry will only become more important in understanding and developing fuel cell catalysts.

Acknowledgments

The work described in this chapter was done while I was at Eindhoven University of Technology. I would like to thank all my colleagues and co-authors who contributed to this work. Financial support from the Royal Netherlands Academy of Arts and Sciences (KNAW), the Netherlands Foundation for Scientific Research (NWO), and the Energy Research Centre of the Netherlands (ECN) and the European Union is also gratefully acknowledged.

References

[1] R.A. van Santen, M. Neurock, Catal. Rev. Sci. Eng. **37** 357 (1995).
[2] B. Hammer, J.K. Nørksov, Adv. Catal. **45** 71 (2001).
[3] J. Greeley, J.K. Nørksov, M. Mavrikakis, Annu. Rev. Phys. Chem. **53** 319 (2002).
[4] M.T.M. Koper, R.A. van Santen, M. Neurock, in Catalysis and Electrocatalysis at Nanoparticle Surfaces, E.R. Savinova, C.G. Vayenas, A. Wieckowski, eds., Marcel Dekker, New York, p.1 (2003).

- [5] F. Jensen, Introduction to Computational Chemistry, John Wiley & Sons, Chicester, (1999)
- [6] P. Hohenberg, W. Kohn, Phys. Rev. **136** (1964) B864
- [7] M.T.M. Koper, in Modern Aspects of Electrochemistry, Eds. C.G. Vayenas, B.E. Conway, R.E. White, Kluwer Academic/Plenum Press, New York, Vol. 36, p. 51 (2003)
- [8] R. Carr, M. Parrinello, Phys. Rev. Lett. **55** 2471 (1985)
- [9] M.P. Allen, D.J. Tildesley, Computer simulation of liquids, Clarendon, Oxford, (1987)
- [10] D. Frenkel, B. Smit, Understanding Molecular Simulation, Academic Press, London, (2002)
- [11] J.J. Lukkien, J.P.L. Segers, P.A.J. Hilbers, R.J. Gelten, A.P.J. Jansen, Phys. Rev. E **58** 2598 (1998)
- [12] G. Blyholder, J. Phys. Chem. **68** 2772 (1964)
- [13] B. Hammer, Y. Morikawa, J.K. Nørskov, Phys. Rev. Lett. **76** 2141 (1996)
- [14] B. Hammer, J.K. Nørskov, Surf. Sci. **343** 211 (1995)
- [15] A. Ruban, B. Hammer, P. Stoltze, H.L. Skiver, J.K. Nørskov, J. Mol. Catal. A **115** 421 (1997)
- [16] M.T.M. Koper, R.A. van Santen, S.A. Wasileski, M.J. Weaver, J. Chem. Phys. **113** 4392 (2000)
- [17] M.J. Weaver, S. Zou, C. Tang, J. Chem. Phys. **111** 368 (1999)
- [18] M.J. Weaver, Surf. Sci. **437** 215 (1999)
- [19] P.S. Bagus, C.J. Nelin, K. Hermann, M.R. Philpott, Phys. Rev. Lett. **36** 8169 (1987)
- [20] P.S. Bagus, G. Pacchioni, Electrochim. Acta **36** 1669 (1991)
- [21] F. Illas, S. Zurita, J. Rubio, A.M. Márquez, Phys. Rev. B **52** 12372 (1995)
- [22] T. Ziegler, A. Rauk, Theor. Chim. Acta **46** 1 (1977)
- [23] S.A. Wasileski, M.J. Weaver, Faraday Disc. **121** 285 (2002)
- [24] Q. Ge, S. Desai, M. Neurock, K. Kourtakis, J. Phys. Chem. B **105** 9533 (2001)
- [25] M.T.M. Koper, T.E. Shubina, R.A. van Santen, J. Phys. Chem. B **106** 686 (2002)
- [26] F. Buatier de Mongeot, M. Scherer, B. Gleich, E. Kopatzki, R.J. Behm, Surf. Sci. **411** 249 (1998)
- [27] M. Head-Gordon, J.C. Tully, Chem. Phys. **175** 37 (1993)
- [28] F. Illas, F. Mele, D. Curulla, A. Clotet, Electrochim. Acta **44** 1213 (1998)
- [29] D. Curulla, A. Clotet, J.M. Ricart, F. Illas, Electrochim. Acta **45** 639 (1999)
- [30] D. Curulla, A. Clotet, J.M. Ricart, F. Illas, J. Phys. Chem. B **103** 5246 (1999)
- [31] M.J. Weaver, Appl. Surf. Sci. **67** 147 (1993)
- [32] S.A. Wasileski, M.T.M. Koper, M.J. Weaver, J. Phys. Chem. B **105** 3518 (2001)
- [33] P. Gao, M.J. Weaver, J. Phys. Chem. **90** 4057 (1986)
- [34] T.E. Shubina, C. Hartnig, M.T.M. Koper, Phys. Chem. Chem. Phys. **6** 4215 (2004)
- [35] B.D. Dunietz, N.M. Markovic, P.N. Ross, M. Head-Gordon, J. Phys. Chem. B **108** 9888 (2004)
- [36] M. Watanabe, S. Motoo, J. Electroanal. Chem. **60** 259 (1975)
- [37] M.T.M. Koper, J.J. Lukkien, A.P.J. Jansen, R.A. van Santen, J. Phys. Chem. B **103** 5522 (1999)
- [38] H.A. Gasteiger, N.M. Markovic, P.N. Ross, E.J. Cairns, J. Phys. Chem. **98** 617 (1994)

[39] H. Massong, H.S. Wang, G. Samjeske, H. Baltruschat, Electrochim. Acta **46** 701 (2000)
[40] G.-Q. Lu, P. Waszczuk, A. Wieckowski, J. Electroanal. Chem. **532** 49 (2002)
[41] M.T.M. Koper, R.A. van Santen, J. Electroanal. Chem. **472** 126 (1999)
[42] A. Michaelides, P. Hu, J. Chem. Phys. **114** 513 (2001)
[43] P. Vassilev, M.T.M. Koper, R.A. van Santen, Chem. Phys. Lett. **359** 337 (2002)
[44] M. Tuckerman, K. Laasonen, M. Sprik, M. Parrinello, J. Chem. Phys. **103** 150 (1995)
[45] S.K. Desai, M. Neurock, K. Kourtakis, J. Phys. Chem. B **106** 2559 (2002)
[46] J. Greeley, M. Mavrikakis, J. Am. Chem. Soc. **124** 7193 (2002)
[47] J. Greeley, M. Mavrikakis, J. Am. Chem. Soc. **126** 3910 (2004)
[48] M. Neurock, S.A. Wasileski, D. Mei, Chem. Eng. Sci. **59** 4703 (2004)
[49] D. Cao, G.-Q. Lu, A. Wieckowski, S.A. Wasileski, M. Neurock, J. Phys. Chem. B **109** 11622 (2005)
[50] Y. Okamoto, O. Sugino, Y. Mochizuki, T. Ikeshoji, Y. Morikawa, Chem. Phys. Lett. **377** 236 (2003)
[51] C. Hartnig, E. Spohr, Chem. Phys. 319, 185 (2005)
[52] K. Kinoshita, Electrochemical Oxygen Technology, Wiley, New York, (1992)
[53] R.R. Adzic, in Electrocatalysis, Eds. J. Lipkowski,. P.N. Ross, Wiley, New York, p. 197 (1998)
[54] C. Hartnig, M.T.M. Koper, J. Electroanal. Chem. **531** 165 (2002)
[55] C. Hartnig, M.T.M. Koper, J. Chem. Phys. **115** 8540 (2001)
[56] A. Panchenko, M.T.M. Koper, T.E. Shubina, S.J. Mitchell, E. Roduner, J. Electrochem. Soc. **151** A2016 (2004)
[57] P. Vassilev, M.T.M. Koper, J. Phys. Chem. C 111 2607 (2007)
[58] A.C. Luntz, J. Grimblot, D.E. Fowler, Phys. Rev. B **39** 12903 (1989)
[59] A. Eichler, J. Hafner, Phys. Rev. Lett. **79** 4481 (1997)
[60] N.M. Markovic, T.J. Schmidt, V. Stamenkovic, P.N. Ross, Fuel Cells **1** 105 (2001)
[61] J.K. Norksov, J. Rossmeisl, A. Logadottir, L. Lindqvist, J.R. Kitchin, T. Blygaard, H. Jonsson, J. Phys. Chem. B **108** 17886 (2004)

17
Reactivity of Bimetallic Nanoclusters Toward the Oxygen Reduction in Acid Medium

Perla B. Balbuena, Yixuan Wang, Eduardo J. Lamas, Sergio R. Calvo, Luis A. Agapito and Jorge M. Seminario

17.1. Introduction

The oxygen reduction reaction (ORR) on Pt and Pt-alloys, the slowest of the two electrode reactions of low temperature fuel cells, has been studied for a long time in an effort to fully understand its mechanism and therefore be able to develop improved catalyst materials which may significantly contribute to enhance the overall fuel cell efficiency [1, 2].

The standard potential for the four-electron reduction of oxygen in acid medium is 1.23 V with respect to the standard hydrogen electrode. However, a negative overpotential – for the ORR – of about 0.3–0.5 V is needed to start the reaction on a Pt electrode, Pt being the best catalyst known so far for this reaction [3]. This overpotential is usually attributed both to kinetic and mass transport limitations at the cathode electrode. Nanoscale proton-exchange membrane (PEM) electrocatalysts have been used since the 1960s [1]; however they are in most cases the result of lucky trial and error experimentation, and there is no assurance that they correspond to the optimal materials [2]. There are advantages and challenges due to the nature of the nanometer-scale regime for the electrocatalytic system; nanocatalysts not only provide enhanced reaction rates with respect to those obtained from catalysis on extended surfaces [4, 5, 6, 7, 8, 9], but most importantly, they may be suitable for alternative reaction paths that are available only because of the electronic characteristics at nanodimensions. Besides, the feature size of nanoscale systems allows a theory-guided and a controlled atomic manipulation that should enable the fabrication of nanosystems with very precise characteristics.

Currently used electrode-catalysts (anode and cathode) consist of an assembly of metallic nanoparticles usually deposited on an electronic conducting substrate and embedded in a hydrated membrane [10, 11], which is the polymer electrolyte proton-conductive material (Figure 17.1). What differs between cathode and anode is the catalyst material, and also the significantly slow kinetics of the cathode oxygen reduction reaction compared to that of the anode hydrogen oxidation reaction. For this reason, several

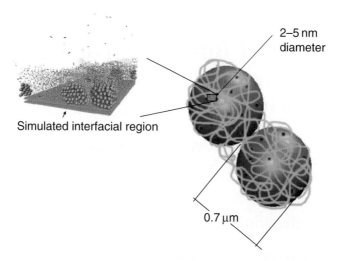

FIGURE 17.1. Carbon particles surrounded by a polymer electrolyte membrane. Inset: the catalyst/hydrated polymer interface in the nanoscale regime.

experimental [3, 11, 12, 13, 14] and theoretical studies [15, 16, 17, 18, 19, 20, 21, 22, 23, 24, 25] are devoted to improving the catalysis at the cathode, primarily pointed to the discovery of alternative catalytic materials [24, 26, 27, 28, 29, 30, 31, 32, 33, 34, 35, 36, 37, 38, 39].

Oxygen (air) is reduced at the cathode catalytic surface. It is generally accepted that the first electron transfer is the rate determining step (*rds*) in a four-electron oxygen reduction mechanism in acid medium, yielding water as the final product [3, 13, 28, 39, 40], but some disagreement still exists with respect to the reaction mechanism. The overall reduction reaction is:

$$O_2 + 4e^- + 4H^+ \xrightarrow{catalyst} 2H_2O$$

However, some authors propose a *direct* pathway where four electrons are sequentially added, others suggest a *series* of two electron reductions with the production of H_2O_2 as an intermediate, and finally a *parallel* (direct and series) mechanism is also suggested based both on experimental and theoretical evidences [3].

A second controversy exists with respect to the mechanism of the first step. One view is developed on the basis of the early proposition by Damjanovic et al. [40, 41, 42] that proton and electron are simultaneously transferred to a weakly adsorbed molecule in the *rds*: $(O_2)_{adsorbed} + H_3O^+ + e^- \xrightarrow{catalyst} (OOH)_{adsorbed} + H_2O$. A second view, advocated by Yeager [43, 44, 45], suggested that the most likely mechanism of the direct four-electron oxygen reduction on Pt involves dissociative chemisorption of the O_2 molecule on a Pt surface probably accompanied by electron transfer, $O_2 + e^- \xrightarrow{catalyst} (O_2^-)_{adsorbed}$. They also concluded that the dissociative adsorption of O_2 is the *rds* and that the proton transfer follows rather than being involved in the *rds* [43].

Several reports indicate that Pt-based alloys are at least as good as pure Pt, and in many cases the alloyed material shows a better performance for the oxygen electroreduction (OER) [26, 27, 33, 46, 47, 48]. Experimental data on the catalytic activity of bimetallic surfaces regarding the OER is controversial. It is clear that certain bimetallic catalysts (Pt-Cr, Pt-Fe, Pt-Co, Pt-Ni) yield a slightly enhanced oxygen reduction current, but the reported degree of enhancement differs among researchers [2].

Theoretical and experimental studies have investigated the reasons for such enhancement [24, 26, 27, 31, 32, 33, 46, 49], whose magnitude is not yet significant enough to induce a huge step forward in fuel cell technology, which requires reducing four times the current amount of Pt used, maintaining maximum power density and fuel cell efficiency [2].

Thus, a strong motivation exists to find alternative catalysts with at least a four-fold increase in their activity. But characterization of the experimental catalysts is far from being complete, mainly because of uncertainties about the structure of the nanoparticles (usually in the range of 2–5 nm, but sometimes with a much wider size distribution) and about the overall atomic distribution of the two components, since those aspects are extremely dependent on the nanoparticle fabrication method [50, 51, 52, 53, 54, 55]. Recent studies attributed the catalytic activity enhancement to the effect induced by the amorphous structure of the nanoparticle [47]. This is certainly a good point since particles in such size range have melting points much lower than their bulk counterparts [56, 57, 58], and both morphology changes and solid-solid structural transitions, which can take place at relatively low temperatures, are extremely dependent on the overall composition and on the nature of the substrate [56, 57, 59, 60, 61, 62]. Such variable morphology characteristic of nanostructures may play an important role in the output of the catalytic process and should be accounted in the analysis of catalytic activity. Others attribute the activity enhancement to differences between Pt and Pt-alloys in relation to their ability to form an oxide layer, which is found to be dependent on the ratio water/acid, i.e., in the water activity [48].

Some aspects of the current catalysts – in many cases not properly accounted when discussing catalytic activity – include the facts that under the PEM fuel cell conditions of temperature ($\sim$353 K) and pressure (1–2 atm), and under the effect of the surrounding medium (hydrated polymer), nanocatalysts are not necessarily in their lowest energy geometries, and their structures can adopt local minima configurations, including amorphous structures [63, 64, 65]. The substrate can also significantly alter the conformation and chemistry of the active sites [66, 67, 68].

The problem we would like to address is schematically shown in Figure 17.1: A catalytic interface (inset) is surrounded by a proton-carrier hydrated membrane; the reactants, intermediates, and products diffuse in and out to reach or leave the catalyst surface, and the substrate is the electronic conductor material which may also influence the reactive system given the nanoscale of the actual catalysts. When a new material is proposed and tested theoretically or experimentally, it is of paramount importance to understand the reaction and mass transfer mechanisms; for complex systems, such as that in Figure 17.1, the theoretical approach that we illustrate in this chapter becomes essential to aid and complement the design of costly experimental synthesis and testing efforts.

The structure of our chapter is as follows: We first discuss the oxygen reduction mechanism on Pt(111) surfaces based on our recently reported Car-Parrinello molecular dynamics (CPMD) simulations [21, 69], making emphasis on the effects of degree of proton solvation, proximity proton-molecular oxygen, and surface charge on the first step of the reaction, in relation to the catalyst/hydrated membrane interfacial structure that might be expected in the actual membrane electrode assembly of low-temperature fuel cells. In this context, we also analyze the effect that alloy catalyst ensembles may have on the first reduction step based on our own density functional theory (DFT) calculations and other theoretical and experimental results. Second, we discuss the subsequent electron reduction steps, emphasizing the differences between the adsorption of intermediates, such as O, OH, and water, and the production, adsorption, and dissociation of hydrogen peroxide, on Pt and Pt-alloy surfaces and clusters. Finally, we discuss the current status and perspectives for the design of more efficient and less costly fuel cell electrocatalysts.

17.2. Oxygen Reduction Mechanism in Acid Medium on Pt and Pt-Alloys

As recently reviewed by Gasteiger et al. [2], a variety of experimental techniques have been applied to investigate the origins of the enhancement on the catalytic activity of some Pt alloys with respect to ORR currents in acid medium. Experimental studies suggested that the enhanced activity originates in the inhibition of OH formation in alloy surfaces where the second element (such as Ni or Co) is easily oxidized [46]. On the theoretical side, Norskov et al. [39] introduced a relatively simple approach to investigate the potential energy surface for the ORR on Pt and on other metal surfaces, based on DFT calculations of the adsorbed energies of OH and O on those surfaces. They found a volcano-type description that correlates ORR activity with oxygen-binding energy concluding that Pt and Pd would be the metals producing the smallest overpotentials. Experimental studies [27, 28, 34, 46, 70] had suggested that Pt-skin surfaces consisting of a monolayer of Pt deposited over an alloy of Pt with Ni, Co, Cr, or Fe yielded various degrees of catalytic enhancements depending on the electrolyte, catalyst preparation, and temperature. The same idea has been tested theoretically [24, 49], where the strain effect caused by the lattice mismatch has been analyzed. In the same spirit of Norskov et al., Xu et al. [24] introduced correlations between transition state energies and final energies for O_2 dissociation, suggesting that reasons for the enhanced activity of the Pt-Co skin surfaces may include a lower affinity of

Pt-skin by atomic oxygen and the presence of a few Co atoms that would easily dissociate O_2 and bind O more strongly. In agreement with these findings, Kitchin et al. [49] have suggested that conservation of the d-band filling causes a d-band broadening and lowering of the average energy as a result of the interactions of Pt with the subsurface atoms, inducing weaker dissociative energies of adsorbates such as H_2 and O_2 on these surfaces. Moreover, DFT calculations [71] have also shown that both strain and the presence of adsorbates [72] are at least partially responsible for the displacement of the d-band center inducing changes in reactivity. Some of these concepts have also been proved by recent experiments from Adzic's group [73] that correlate ORR currents with DFT-calculated oxygen-binding energies and also with the energy corresponding to the alloy d-band center, suggesting that most of the Pt-skin surfaces should have an intermediate value of such energy, providing a compromise between the ability to dissociate O_2 and that to avoid poisoning of the surface with oxygenated intermediates or products. Regarding the ORR mechanism, experimental evidence suggests that the mechanism remains unaltered when Pt is alloyed with other transition metals such as Co, Ni, Cr, and Fe [46, 48, 74].

Finally, the most important concerns regarding alloys as substitutes for Pt in fuel cell electrodes include the potential leach and contamination of the electrolyte membrane with cations coming from the dissolution of the base-metal; therefore, the design of new catalysts requires not only optimizing the catalytic activity but also analyzing the stability of the Pt and non-Pt elements under proton exchange fuel cell conditions.

17.2.1. The First Reduction Step

As outlined in the Introduction, a couple of suggested pathways have been proposed for the first electron transfer step: (a) dissociative chemisorption of O_2 (rds) probably accompanied by e-transfer and followed by proton transfer; (b) simultaneous proton and electron transfer to a weakly adsorbed O_2 molecule. We have recently shown through CPMD [21, 69] and DFT [75] results that both pathways may take place under different conditions of the interfacial structure; i.e., proton transfer may be involved in the first reduction step depending on the relative location of the O_2 molecule with respect to the surface and to the proton, on the degree of proton hydration, and on the surface charge which is dependent on the electrode potential. Moreover, it was shown that proton transfer may precede or follow the first electron transfer, but in most cases the final product of the first step is an adsorbed HOO*.

These points are illustrated in a series of snapshots from CPMD simulations. The initial configurations are displayed by Schemes 17.1 and 17.2 First we analyze the effect of the distance d_{OH} between the oxygen molecule and the closest H of a hydronium ion solvated by two water molecules (Scheme 17.1) for various values of the distance d_{OS} between the oxygen molecule and the surface.

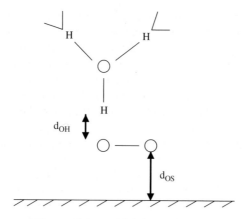

FIGURE Scheme 17.1. low solvation.

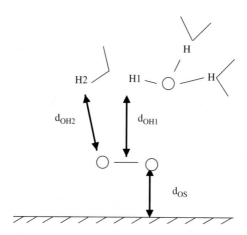

FIGURE Scheme 17.2. intermediate solvation.

At the shortest initial d_{OH} distance (2.2 Å), Figure 17.2 (left), a hydrated proton, still H-bonded by the water molecules, gradually migrates toward O_2. The O_2 molecule, initially located parallel to the surface (Scheme 17.1), rotates from the symmetric bridge site to a canted top-bridge site, as found in previous work [76, 77], and is partially polarized by the surface, which in this case carries a negative charge due to the addition of one electron to the unit cell to counterbalance the positive charge, as discussed in a later section [19, 20, 75]. At this point there is no clear chemisorption of O_2, which is still far from the surface. After 0.25 ps, the next image (Figure 17.2, center) shows the formation of an end-on chemisorbed species *OOH; analysis of the CPMD trajectory [69] shows that the O–O distance elongates whereas the

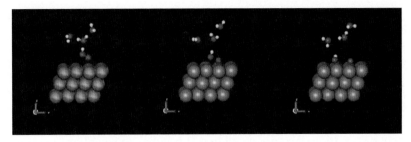

FIGURE 17.2. Snapshots from CPMD simulations [69] of the first ORR reduction step on Pt (111) for initial distances $d_{OH} = 2.2$ Å and $d_{OS} = 3$ Å, at a low degree of proton solvation (see Scheme 1). Left: at 0.11 ps, Center: at 0.36 ps, Right: at 0.46 ps. The sequence illustrates that when O_2 is sufficiently close to a weakly solvated proton, proton transfer precedes chemisorption.

distance from the OH group to the surface (HO–S) becomes shorter, as a consequence of the increasing interaction between the OH group and the surface, while the electronic energy remains approximately constant. At 0.46 ps (Figure 17.2, right), the chemisorbed *OOH is decomposed into adsorbed O and OH, after a lifetime of approximately 0.15 ps. DFT studies on small Pt_n clusters (n = 3, 4, 6, and 10) carried out to investigate the adsorption and dissociation of *OOH [75] confirmed the existence of an end-on adsorbed *OOH state which can be decomposed with an activation barrier of ∼0.25 eV for the case of Pt_3.

The effect of increasing the initial OH distance between O_2 and the hydrated proton (still under low solvation, Scheme 17.1) tested by CPMD simulations [69] shows that these conditions may revert the sequence of events for the first reduction: As illustrated by Figure 17.3, O_2 chemisorption takes place first (Figure 17.3, left), followed by proton transfer and formation of *OOH (Figure 17.3, center), and *OOH decomposition (Figure 17.3, right).

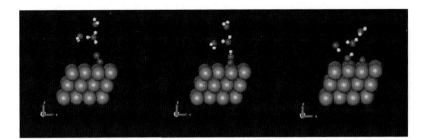

FIGURE 17.3. Snapshots from CPMD simulations [69] of the ORR first reduction step on Pt (111) for initial distances $d_{OH} = 3$ Å and $d_{OS} = 3$ Å, at a low degree of proton solvation (see Scheme 1). Left: at 0.22 ps, Center: at 0.38 ps, Right: at 0.87 ps. The sequence illustrates that longer separation between the O_2 molecule and the solvated proton causes chemisorption to take place first, followed by OOH formation, chemisorption, and dissociation on the surface.

The initial step (Figure 17.3, left) is characterized by a strong chemisorption, accompanied by O–O bond elongation to ~ 1.5 Å, and a drastic decrease of the electronic Kohn-Sham energy. However, the chemisorbed O_2 is not dissociated and instead a proton transfer takes place, forming adsorbed *OOH, suggesting that the O_2 dissociation has a higher barrier than the *OOH formation under these conditions of low proton solvation. At 0.87 ps (Figure 17.3, right) finally *OOH decomposes, with the resulting adsorbed O* adsorbed on top site, instead of being at the most favorable hollow site location. We speculated that H-bonding to neighbor water molecules may be responsible for this type of adsorption [69].

Second, we examined the effect of solvation on the barriers for proton transfer. As found in previous work [20], CPMD simulations revealed that using an intermediate degree of solvation, such as $(H_3O)^+(H_2O)_3$, O_2 chemisorption precedes proton transfer even if the distance between the O_2 and the closest H atom is initially relatively low (Scheme 17.2). Figure 17.4 illustrates the sequence of events: O_2 chemisorption takes place first (Figure 17.4 left), followed by proton transfer from the closest water molecule (Figure 17.4, center), and proton transfer from the hydronium ion to water, and finally *OOH dissociation. Thus, the net effect of higher solvation is the delay of proton transfer.

In order to assess further solvation effects, new CPMD simulations were carried out using a high degree of proton solvation which includes a first and a second shell of water molecules surrounding the hydronium ion $(H_3O)^+(H_2O)_3(H_2O)_6$. Snapshots of the dynamic sequence are depicted in Figure 17.5. In the initial configuration, the O_2 molecule is located at about 3 Å from the surface, with the closest H atoms located at 1.76 Å (Figure 17.5, top left). During the initial 0.7 ps, proton transfer is observed from the hydronium ion to one of the water molecules (Figure 17.5, top center), whereas the O_2 molecule approaches the surface and locates over a bridge position (Figure 17.5, top right). Note also the rotation of the water molecule

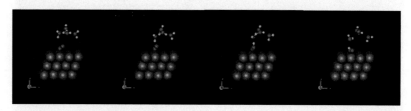

FIGURE 17.4. Snapshots from CPMD simulations [69] of the ORR first reduction step on Pt(111) for initial distances $d_{OH} = 2.5$ Å and $d_{OS} = 3$ Å, at an intermediate degree of proton solvation: $(H_3O)^+(H_2O)_3$ (Scheme 2 shows the initial configuration). Left: at 0.12 ps, Center: at 0.19 and 0.22 ps, Right: at 0.30 ps. The net effect of higher proton solvation is the delay of proton transfer; thus chemisorption takes place first, followed by OOH formation, chemisorption, and dissociation on the surface. Proton transfer from hydronium to a water molecule is also detected (third image from the left).

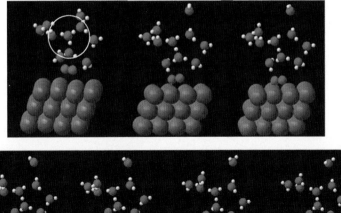
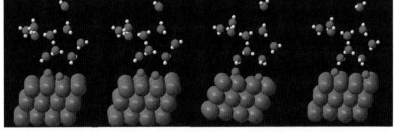

FIGURE 17.5. Snapshots from CPMD simulations of the ORR first reduction step on Pt(111) for initial distances $d_{OH} = 2.5$ Å and $d_{OS} = 3$ Å, at a high degree of proton solvation: $(H_3O)^+(H_2O)_3(H_2O)_6$. First row: Initial configuration (left), at 0.7 ps (center), and at 0.8 ps (right). Second row: At 1.1, 1.2, 1.6, and 2 ps. Chemisorption takes place followed by O_2 decomposition, whereas OOH formation was not observed during 5 ps.

close to one of the oxygen atom, favoring H-bond formation, while the O–O bond is elongated from an initial value of 1.21 Å, to 1.43 Å in Figure 17.5, top right. In the next images (Figure 17.5, bottom), the dissociation of O_2 is observed, the O–O bond length is 2.6, 3.12, 4.62, and 6.36 Å, respectively, in the four bottom images of Figure 17.5. The adsorbed oxygen atoms are located on top-bridge-top (at 0.7 and 0.8 ps, top center and right) locations, diffusing to top and hollow upon dissociation at 1.1 ps (bottom left), the O adsorbed in hollow site diffuses to bridge site at 1.2 ps (bottom center) and then both adsorbed O diffuse to two-fold bridge sites at 1.6 and 2 ps (bottom right). Although the closest water molecules become very close to the adsorbed O atoms (Figure 17.5, bottom right), formation of adsorbed OH was not observed in the total duration of the CPMD simulation, 5 ps. Thus, the strong solvation effect causes O_2 chemisorption to take place first, followed by O_2 dissociation, but proton transfer is much delayed.

To analyze the possible influence of alloys on the dissociative pathway – where O_2 is dissociated previous to protonation – we discuss the results of a combined density functional theory and Green function approach to determine electron transfer through an interface that was applied to investigate the effect of alloying on O_2 dissociation [32]. Such procedure yields the variation of the density of states of the complex adsorbate/adsorbent, as illustrated by

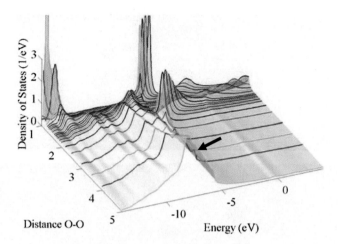

FIGURE 17.6. Effect of adsorbed O_2 on the density of states of a Pt surface calculated by a DFT/Green function method[32], as a function of the O-O distance. For all curves, the center of the O-O bond is located at 2.10 Å from the center of the Pt-Pt bond. The Pt-Pt (2.47 Å) and the Pt-O bond distances are kept constant, whereas the O-O bond length is varied progressively from the gas phase O-O bond length, 1.2 Å, to 5 Å.

Figure 17.6 for O_2 adsorbed on a bridge site as a function of the O–O distance. The dissociation limit – density of states (DOS) curves located at the front of the graph in Figure 17.6 – is represented by the atomic orbitals of the oxygen atoms (indicated by an arrow in Figure 17.6), each being doubly degenerated since there are two atoms; on the other hand, when the O–O distances become close to the value of the O–O bond length (1.2 Å), the DOS shows that the single peak of O is substituted by two peaks corresponding to the molecular orbitals of the oxygen molecule, usually one above and one below the Fermi level of Pt (-5.93 eV). The splitting of the single peak becomes clear as the O–O distance decreases, forming the O–O bond (Figure 17.6). We found that this feature, the separation between peaks corresponding to the frontier crystal orbitals, is an indication of the ability of the system toward oxygen dissociation [32], the closest the two peaks the greater the tendency of the system to split into two adsorbed oxygen atoms. Figure 17.7 illustrates the effect of the adsorbate on the DOS as a function of adsorption geometry and nature of the neighbor atoms for bimetallic Pt-based systems alloyed with Co.

The density of states in Figure 17.7 were calculated based on DFT optimizations of O_2 adsorbed on Co_2Pt (a and c) and Pt_2Co (b and d); the density matrix information from such calculations is fed to an integrated DFT-Green function procedure [78] where the system is defined as the cluster embedded in a Pt bulk matrix, yielding the effect of the adsorbate on the density of electronic states of bulk Pt. Figures 17.7 (a) and (c) correspond to a Co_2Pt cluster located on the surface of Pt bulk, but O_2 is adsorbed on bridge Co-Co in (a) and on the Pt-Co bridge in (c). In Figures 17.7 (b) and (d), the system is a

17. Reactivity of Bimetallic Nanoclusters 519

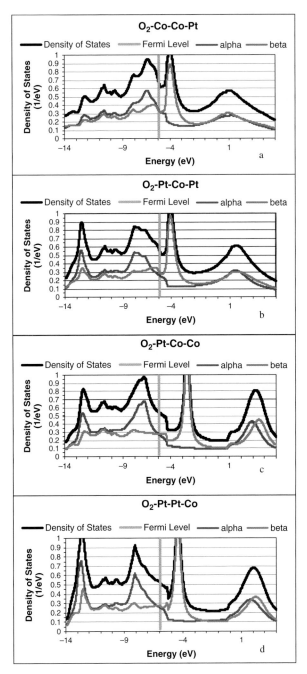

FIGURE 17.7. DOS of O adsorbed on bridge Co-Co (a), Pt-Co (b and c) and Pt-Pt (d) sites of a Pt(111) surface, calculated on the basis of Co_2 Pt (a and c) and Pt_2 Co (b and d) between the peaks of the unoccupied (left of Fermi level) and unoccpied (right of Fermi level is shown as a vertical line) is an indication of the degree of O_2 dissociation.

Pt$_2$Co cluster located on the surface of Pt bulk, O$_2$ is adsorbed on Pt-Co bridge site in (b), whereas is adsorbed on bridge Pt-Pt in (d). The O–O distances are 1.49, 1.45, 1.40, and 1.38 Å in (a) to (d), and their corresponding O$_2$ binding energies are −2.93, −1.50, −1.44, and −0.62 eV, respectively. Adsorption on the Co–Co bridge (Figure 17.7, a) appears the most favorable to O$_2$ dissociation; note also the proximity of the band of occupied states to the Fermi level of the metal and to the peak of lowest unoccupied states. The advantage of the present procedure compared to the "extended periodic systems" is that the method allows active sites to be treated with larger basis sets using the best theoretical methods available for these systems. Experimental results suggested that Ni, Co, and Cr would become oxidized and act as "sacrificial sites" where other species may adsorb, leaving the Pt sites available for O$_2$ dissociation [29, 30]. Although this may be the case for Ni, we showed that alternative mechanisms are possible, especially for Co and Cr, which act as active sites for O$_2$ dissociation. Using this theoretical screening technique, we found that XXPt (X = Co and Cr) are the best active sites to promote O$_2$ dissociation. On the other hand, ensembles involving Ni atoms produce similar degrees of O$_2$ dissociation as those of pure Pt.

17.2.2. The Catalyst/Hydrated Membrane Interface

The next question is which of the above-discussed scenarios is most likely to be found at the catalyst/hydrated membrane interface. It is generally accepted that the proton conductivity of the membrane depends on the characteristics of ionic clusters formed surrounding the polymer hydrophilic sites, both within the bulk polymeric structure and at the interface with the catalyst [79]. The ionic clusters located at the membrane/catalyst interface are the ones that close the circuit of this electrochemical system. That is, these ionic clusters act as bridges through which protons and other hydrophilic reactants and products may pass from the membrane to the catalyst surface and vice versa during fuel cell operation. To get some insights into the possible formation of ionic clusters, we have analyzed the conformation of a hydrated model nafion membrane over Pt nanoparticles deposited on a carbon substrate via classical MD simulations [80] at various degrees of hydration.

Usually a λ parameter is defined as the number of water molecules per sulfonic site to characterize the degree of membrane hydration. Our MD simulations were carried out at various values of λ to analyze the possible water structure in such environment [80]. As shown by Figure 17.8, at low water contents ($\lambda = 5$), small water clusters are nucleated near the sulfonic hydrophilic sites, which are barely connected with each other. Thus, at these conditions, proton transfer could be severely limited because of the poor contact between water clusters. At a much higher water concentration ($\lambda = 45$, not shown), water forms a continuous phase [80] occupying almost all accessible sites on the surface; thus oxygen diffusion will be reduced because of the low O$_2$ solubility in water. When λ is 24 (Figure 17.8), a value close to the amount

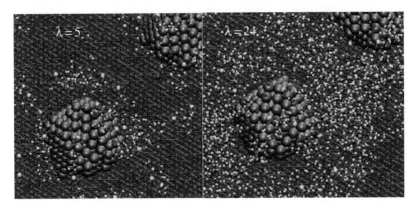

FIGURE 17.8. MD simulations of water distribution near platinum nanoclusters on a graphite substrate surrounded by nafion at different λ (water:S ratio) values. Nafion atoms other than the sulfonic (SO_3^-) groups are not shown in order to clearly illustrate the water clusters. At $\lambda = 24$ the clusters are interconnected. S atoms are intermediate spheres, O atoms and H atoms are the smallest spheres [80].

found experimentally in fully hydrated PEM membranes [79], the interface is defined by a semicontinuum water film where water clustering can be identified. According to the classical MD simulation results [80], these clusters have diameters of 1 to 1.5 nm and are interconnected by multiple water bridges. The average water density in this phase is estimated to be about 0.682 g/cm^3, much lower than that of the bulk water phase at 353 K, as a consequence of the reduced water density present in the cluster bridging areas.

Trying to reproduce events at the electrochemical interface, it is essential to elucidate the effects of the potential difference between the two electrodes. The existence of a charged electrochemical double layer is generally recognized, and it is included in several microscopic [81] and macroscopic models [82] In our CPMD simulations [69], such effect was incorporated either by adding one electron to the cell (in all cases shown in Figures 17.2 to 17.5) or through a background charge [83]. In the first case, the average charge on the Pt surface is initially negative (Figure 17.9) and becomes increasingly positive as charge is transferred to the adsorbates, as indicated by the time evolution of the net charge of the OOH group. Comparing the net charges in the slab and those on the first surface layer shown in Figure 17.9, it is seen that most of the negative charge is carried by the first layer, whereas no significant changes are observed on the charge carried by the Pt atom initially close and then bonded to an adsorbed O atom.

Using the second approach of counterbalancing the positive charge of the hydronium ion by a negative background charge, the net charge on the Pt surface is initially positive and becomes even more positive after electron transfer to the adsorbates [69]. CPMD simulations done at $d_{HO} = 2.2$ and 2.5 Å (Scheme 17.1) using this alternative approach yielded an opposite

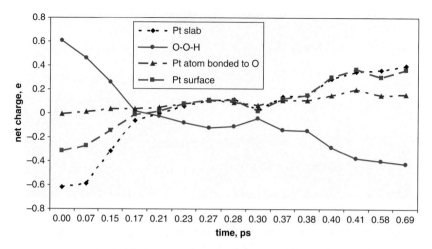

FIGURE 17.9. Time evolution of the net charge (in e) of the Pt slab, Pt first surface layer, OOH group, and Pt atom bonded to an adsorbed O atom. The results were obtained from CPMD trajectories [69] corresponding to Scheme 1 and Figure 2.

sequence for the first reduction step to that shown in Figure 17.2, i.e., O_2 chemisorption takes place first, the proton is repelled and oxygen is attracted by the more positive surface, suggesting that there is competition between the positively charged surface and the proton for the high electronic density available in O_2. A less positive surface charge as shown in Figure 17.9 causes less attraction between O_2 and the surface, and a proton reacts with O_2 previous to chemisorption (Figure 17.2). Thus, changes in the electrode potential will be manifested as different distribution of surface charges which would also alter the sequence of events. The first step equations can be summarized as:

$$O_2 + H^+ \rightarrow OOH^+ \tag{17.1}$$

$$OOH^+ +^* + e^- \rightarrow H\text{-}O\text{-}O^* \tag{17.2}$$

Or

$$O_2 +^* + e^- \rightarrow OO^{-*} \tag{17.3}$$

$$OO^{-*} + H^+ \rightarrow H\text{-}O\text{-}O^* \tag{17.4}$$

$$H\text{-}O\text{-}O^* +^* \rightarrow HO^* + O^* \tag{17.5}$$

17. Reactivity of Bimetallic Nanoclusters   523

Except in the highly solvated case shown in Figure 17.5, the final product of the first reduction step is the formation of adsorbed *OOH.

A related question is how the presence of a foreign element on the surface or in subsurface layers may affect the first reduction step. According to Norskov et al. [39] the origin of the overpotential is independent of the details of the first step, and the strength of the interactions with intermediates such as OH and O determine a volcano-type dependence for both the "associative" mechanism (forming *OOH) and the dissociative mechanism. Panchenko et al. [25] emphasized the structure-sensitive character of this reaction, in particular in relation to nanocatalysts where crystallographic facets other than (111) may be present. Thus, they made a careful evaluation of the adsorption of relevant ORR intermediates and products, such as O, OH, OOH, H_2O_2, and H_2O on small clusters and surfaces representing the Pt (111), (110), and (100) surfaces.

A comparison of OOH adsorption on elements different than Pt obtained from DFT calculations of OOH adsorbed on Pt-based alloy 3-atom clusters [84] revealed a stronger *OOH adsorption to foreign elements (such as Ni, Co, Cr) compared to Pt, with characteristics that depend on the specific adsorption site, cluster composition, and nature of the foreign element. As shown by Figure 17.10, the end-on adsorption on top is preferred in Pt_3 clusters, and similar configurations, but with stronger binding energies, are found when one of the Pt atoms is substituted by a Cr, Co, or Ni (not shown). Ensembles containing two foreign atoms can stably adsorb OOH in bridge configuration with two (Figure 17.10, top center) or one (Figure 17.10, top right) of the oxygen atoms connected to the two metal atoms. Cr has the strongest ability to bind OOH and some Cr-containing ensembles are able to dissociate *OOH without barrier [84]. Therefore the presence of Cr, Co, and Ni would strongly bind OOH but not always favoring *OOH dissociation and the continuation of the reduction process.

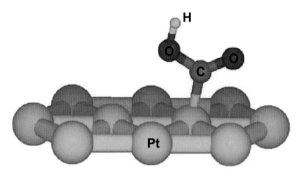

FIGURE 17.10. DFT optimized geometries of OOH adsorption to selected mono and bimetallic clusters [84].

17.2.3. The Subsequent Reduction Steps

17.2.3.1. The Second Electron Transfer

Two alternative pathways may follow the formation of adsorbed *OOH, as described by the equations [69]:

$$HOO^* + H^+ + e^- \rightarrow HOO^*H \qquad (17.6)$$

$$HOO^*H + ^* \rightarrow HO^* + HO^* \qquad (17.7)$$

and

$$H\text{-}O\text{-}O^* + ^* \rightarrow HO^* + O^* \qquad (17.5)$$

$$O^* + HO^* + H^+ + e^- \rightarrow H_2O^* + O^* \qquad (17.8)$$

$$H_2O^* + O^* \rightarrow 2HO^* \qquad (17.9)$$

Equation (6) represents the formation of the one-end adsorbed HOO*H generated via reduction of HOO*, i.e., the *series* pathway, which may be followed by equation (7) because of the high instability of the adsorbed HOO*H. Electrochemical experiments done on Pt and Pt-alloy surfaces [46] indicated that the amount of H_2O_2 found at potentials below 0.8 V (with respect to the standard hydrogen electrode) depends on the strength and nature of the acid medium, surface composition, and temperature. Similar experiments carried out on nanoparticles (3 to 4 nm) deposited on carbon substrates [30] showed that alloys with compositions Pt_3X (X = Ni and Co) yield no significant differences with respect to the production of H_2O_2, in all cases considered negligible. However, higher contents of Ni (~50%) make the catalysts unstable and large amounts of H_2O_2 are detected. Earlier work by Arvia et al. [85, 86] found that the (100) facets would favor the *series* reduction pathway. Similar studies were reported recently by Mukerjee's group [48], pointing again to the complexity of the nanoparticle environment.

DFT studies of the adsorption and dissociation of H_2O_2 on Pt_n clusters (n = 3, 4, 6, and 10) [75] showed the existence of a one-fold weakly adsorbed state, with binding energies ranging from –0.67 eV (on Pt_3) to –0.36 eV on Pt_{10} [75]. Similar weak adsorbed states were identified on bimetallic clusters, whereas spontaneous H_2O_2 decomposition has been observed in all cases where H_2O_2 is located on a bridge position both on Pt and Pt-alloy clusters and surfaces (Figure 17.11) [31, 75]

Equation (5) is the alternative 2nd electron transfer path, where HOO* is not reduced but dissociates first forming adsorbed O* and HO*, which is followed by reaction (8) whereby an electron and proton are transferred to HO*, probably the

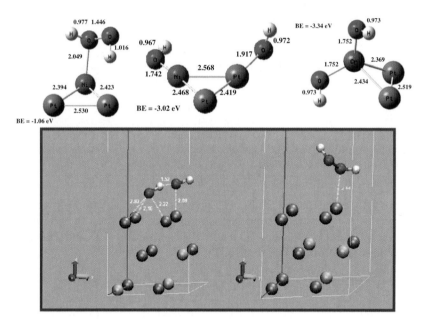

FIGURE 17.11. DFT calculated adsorption of H_2O_2 on bimetallic clusters (top) showing one-end weak adsorption on Ni, and on a Pt-skin (Pt_3Co) surface (bottom, right). Spontaneous dissociation into two adsorbed OH is observed both in small clusters (top, center and right), and surfaces (bottom left) [31].

preferred site because of the highest negative charge carried by the oxygen atom of HO* compared to that of O* [69]. The adsorbed water formed by reaction (6) is observed to be unstable, rotating from its initial position and becoming parallel to the surface, where it forms H-bond with the adsorbed oxygen atom, transfers a proton, and the end product are two adsorbed *OH radicals.

Table 17.1 provides a comparison of the various binding energy strengths of the relevant oxygenated compounds on mono and bimetallic clusters, which may be used to evaluate the relative affinities. The range of energies shown under each of the bimetallic clusters refers to adsorption on the various corresponding bridge or top sites. For example, adsorption on Pt_2Ni can occur on bridge Pt-Pt or Pt-Ni, on top Pt or Ni, or in the hollow site. On average, the strongest affinities are found on the PtX_2 compounds, where the X element becomes easily oxidized. Ni and Co show similar behavior, and Cr has the strongest affinities for the oxygenated compounds and in many cases is able to spontaneously dissociate peroxide compounds such as OOH or H_2O_2. In the presence of the alloy elements, Pt binds less strongly to the oxygenated compounds. Thus, this is in agreement with the cooperative effect where one of the elements acts to bind specific intermediates, whereas the Pt sites are used to dissociate O_2 suggested on the basis of other theoretical and experimental results.

TABLE 17.1 DFT calculated binding energies (BE) of oxygenated reactants and intermediates of the ORR to Pt_3 and Pt-based alloy clusters.

	-BE (eV)						
	Pt_3	Pt_2Ni	$PtNi_2$	Pt_2Co	$PtCo_2$	Pt_2Cr	$PtCr_2$
O_2 [32]	1.08	0.92–1.55	1.28–2.14	0.62–1.5	1.44–2.93	0.38–1.58	1.25–3.09
O [87]	3.24	3.23–3.36	3.60–3.99	3.03–3.87	1.88–3.90	4.09	4.15–4.98
OH [87]	2.10–2.92	3.07–3.13	2.74–3.88	2.94–3.56	2.88–4.32	2.90–3.58	3.32–4.89
H_2O [87]	1.02	1.25	0.12–0.97	0.86–1.06	0.69–0.97	0.81–1.02	0.39–1.05
OOH [84]	0.86–1.72	1.48–1.84	1.49–2.65	1.70–2.20	1.65–2.92	1.69–2.40	1.43–4.90
H_2O_2 [31]	0.75–0.88	0.75–1.06	0.58–0.93	0.84–1.09	0.62–0.81	0.71	0.31

Based only on energetics, Table 17.1 suggests that adsorption of two *OH radicals (as in equations (7) and (9)) is more favorable than the alternative products of the second step, *O and *H_2O (equation (8)) and HOO* H (equation (6)) not only for Pt but for Pt alloys.

17.2.3.2. The Third and Fourth Electron Transfers

The last two steps consist of successive formation of water molecules by electron and proton transfer to the adsorbed OH species, given by equations (10) and (11).

$$HO^* + HO^* + H^+ + e^- \rightarrow H_2O^* + HO^* \qquad (17.10)$$

$$H_2O^* + HO^* + H^+ + e^- \rightarrow 2H_2O^* \qquad (17.11)$$

Here the effect of the applied field, as well as a realistic representation of the surrounding environment, should be crucial as it will determine the specific orientations of water on the surface, influencing adsorption or desorption of water or H_2O_2, as observed experimentally [46].

17.3. Summary and Conclusions

Recent experiments by Bard et al. [38] reported ORR polarization curves measured for catalysts designed on the basis of an alloy model where one of the metal elements had the function of dissociating O_2 and the other metal would easily reduce adsorbed atomic oxygen. Carbon-supported bi- and trimetallic catalysts were prepared with such selection rule, and it was concluded that for the metals examined (Ag, Au, and Pd), adding 10 to 20% of Co showed a good catalytic activity, which however was inferior to that of Pt electrocatalysts. Better degree of success was achieved with non-Pt catalysts reported by Savadogo et al. [37] and the Pt-skin catalysts by Adzic et al. [73] In most of these catalytic designs, it is encouraging that first principles calculations are at the point of contributing useful insights into developing new understanding of the reaction mechanisms [17, 69, 88], helping in the interpretation of experimentally found enhanced activities [36, 49, 71, 73], and even providing theory-based guidelines [39, 72, 89].

We would like to emphasize, however, that ab initio electrochemistry is still in its infancy. Besides the complexity of the problem of assessing the true interfacial structure, including hydrated nafion, O_2 and catalyst actual nanostructure, and surface segregation effects, there are other challenges such as the need of a proper description of the applied potential. We expect that advances in theory, computational implementation, and faster computers will help us achieve less expensive and more efficient fuel cell catalysts in the near future.

Acknowledgments

We gratefully acknowledge the financial support of the Department of Energy, Basic Energy Sciences (Grant DE-FG02-04ER15619), and of the Army Research Office (DURIP grant W911N F-04-1-0098).

References

[1] Steele, B. C. H.; Heinzel, A. Materials for fuel-cell technologies *Nature* **2001**, *414*, 345–352.

[2] Gasteiger, H. A.; Kocha, S. S.; Sompalli, B.; Wagner, F. T. Activity benchmarks and requirements for Pt, Pt-alloy, and non-Pt oxygen reduction catalysts for PEMFCs. *Appl. Cat. B: Environmental* **2005**, *56*, 9–35.

[3] Adzic, R. Recent advances in the kinetics of oxygen reduction. In *Electrocatalysis*; Lipkowski, J., Ross, P. N., Eds.; Wiley-VCH: New York, **1998**, pp. 197–242.

[4] Poirier, J. A.; Stoner, G. E. Microstructural effects on electrocatalytic oxygen reduction activvity of nano-grained thin-film Platinum in acid media. *J. Electrochem. Soc.* **1994**, *141*, 425–430.

[5] Takasu, Y.; Ohashi, N.; Zhang, X.-G.; Murakami, Y.; Minagawa, H.; Sato, S.; Yahikozawa, K. Size effects of platinum particles on the electroreduction of oxygen. *Electrochim. Acta* **1996**, *41*, 2592–2600.

[6] Peuckert, M.; Yoneda, T.; Betta, R. A. D.; Boudart, M. Oxygen reduction on small supported platinum particles. *J. Electrochem. Soc.* **1986**, *133*, 944–947.

[7] Choi, K. H.; Kim, H. S.; Lee, T. H. Electrode fabrication for proton exchange membrane fuel cells by pulse electrodeposition. *J. Power Sources* **1998**, *75*, 230–235.

[8] Joo, S. H.; Choi, S. J.; Oh, I.; Kwak, J.; Liu, Z.; Terasaki, O.; Ryoo, R. Ordered nanoporous arrays of carbon supporting high dispersions of platinum nanoparticles. *Nature* **2001**, *412*, 169–172.

[9] Wilson, M. S.; Valerio, J. A.; Gottesfeld, S. Low platinum loading electrodes for polymer electrolyte fuel cells fabricated using thermoplastic ionomers. *Electrochim. Acta* **1995**, *40*, 355–363.

[10] Sheppard, S. A.; Campbell, S. A.; Smith, J. R.; Lloyd, G. W.; Ralph, T. R.; Walsh, F. C. Electrochemical and microscopic characterization of platinum-coated perfluorosulfonic acid (Nafion 117) materials. *Analyst* **1998**, *123*, 1923–1929.

[11] Markovic, N. M.; Ross, P. N. Electrocatalysts by design: from the tailored surface to a commercial catalyst. *Electrochim. Acta* **2000**, *45*, 4101–4115.

[12] Markovic, N. M.; Gasteiger, H. A.; Grgur, B. N.; P. N. Ross, J. Oxygen reduction reaction on Pt(111): effects of bromide. *J. Electroanal. Chem.* **1999**, *467*, 157.

[13] Markovic, N. M.; Ross, P. N. Electrocatalysis at well-defined surfaces: Kinetics of oxygen reduction and hydrogen oxidation/evolution on Pt(hkl) electrodes. In *Interfacial Electrochemistry. Theory, Experiment and Applications*; Wieckowski, A., Ed.; Marcel Dekker: New York, **1999**, pp. 821–841.

[14] Adzic, R. R.; Wang, J. X. Configuration and site of O_2 adsorption on the Pt(111) electrode surface. *J. Phys. Chem. B* **1998**, *102*, 8988–8993.

[15] Anderson, A. B.; Albu, T. V. Ab initio approach to calculating activation energies as functions of electrode potential. Trial application to four-electron reduction of oxygen. *Electrochem. Comm.* **1999**, *1*, 203–206.

[16] Anderson, A. B.; Albu, T. V. Ab initio determination of reversible potentials and activation energies for outer-sphere oxygen reduction to water and the reverse oxidation reaction. *J. Am. Chem. Soc.* **1999**, *121*, 11855–11863.

[17] Anderson, A. B.; Albu, T. V. Catalytic effect of platinum on oxygen reduction: An ab initio model including electrode potential dependence. *J. Electrochem. Soc.* **2000**, *147*, 4229–4238.

[18] Anderson, A. B. O_2 reduction and CO oxidation at the Pt-electrolyte interface. The role of H_2O and OH adsorption bond strengths. *Electrochim. Acta* **2002**, *47*, 3759–3763.

[19] Li, T.; Balbuena, P. B. Computational studies of the interactions of oxygen with platinum clusters. *J. Phys. Chem. B* **2001**, *105*, 9943–9952.

[20] Li, T.; Balbuena, P. B. Oxygen reduction on a platinum cluster. *Chem. Phys. Lett.* **2003**, *367*, 439–447.

[21] Wang, Y.; Balbuena, P. B. Ab initio-molecular dynamics studies of O_2 electro-reduction on Pt (111): Effects of proton and electric field. J. Phys. Chem. B **2004**, *108*, 4376–4384.

[22] Xu, Y.; Mavrikakis, M. Adsorption and dissociation of O_2 on Cu(111): thermochemistry, reaction barrier and the effect of strain. *Surf. Sci.* **2002**, *505*, 369.

[23] Xu, Y.; Mavrikakis, M. Adsorption and dissociation of O_2 on Ir(111). *J. Chem. Phys.* **2002**, *116*, 10846–10853.

[24] Xu, Y.; Ruban, A. V.; Mavrikakis, M. Adsorption and dissociation of O_2 on Pt-Co and Pt-Fe alloys. *J. Am. Chem. Soc.* **2004**, *126*, 4717–4725.

[25] Panchenko, A.; Koper, M. T. M.; Shubina, T. E.; Mitchell, S. J.; Roduner, E. Ab Initio calculations of intermediates of oxygen reduction on low-index platinum surfaces. *J. Electrochem. Soc.* **2004**, *151*, A2016–A2027.

[26] Mukerjee, S.; Srinivasan, S. Enhanced electrocatalysis of oxygen reduction on platinum alloys in proton-exchange membrane fuel cells. *J. Electroanal. Chem.* **1993**, *357*, 201–224.

[27] Mukerjee, S.; Srinivasan, S.; Soriaga, M. P. Effect of preparation conditions of Pt alloys on their electronic, structural, and electrocatalytic activities for oxygen Reduction-XRD,XAS, and electrochemical studies. *J. Phys. Chem.* **1995**, *99*, 4577–4589.

[28] Markovic, N. M.; Ross, P. N. Surface science studies of model fuel cells electrocatalysts. *Surf. Sci. Rep.* **2002**, *45*, 117–229.
[29] Paulus, U. A.; Vokaun, A.; Scherer, G. G.; Schmidt, T. J.; Stamenkovic, V.; Markovic, N. M.; Ross, P. N. Oxygen reduction on high surface area Pt-based alloy catalysts in comparison to well defined smooth bulk alloy electrodes. *Electrochim. Acta* **2002**, *47*, 3787–3798.
[30] Paulus, U. A.; Vokaun, A.; Scherer, G. G.; Schmidt, T. J.; Stamenkovic, V.; Radmilovic, V.; Markovic, N. M.; Ross, P. N. Oxygen reduction on carbon-supported Pt-Ni and Pt-Co alloy catalysts. *J. Phys. Chem. B* **2002**, *106*, 4181–4191.
[31] Balbuena, P. B.; Calvo, S. R.; Lamas, E. J.; Seminario, J. M. Adsorption and dissociation of H_2O_2 on Pt_3, Pt_2M, PtM_2(M = Cr, Co, and Ni), Pt(111), and Pt_3Co(111). *J. Phys. Chem. B* **2005**, *110*, 17452–17459.
[32] Balbuena, P. B.; Altomare, D.; Agapito, L. A.; Seminario, J. M. Adsorption of oxygen on Pt-based clusters alloyed with Co, Ni, and Cr. *J. Phys. Chem. B* **2003**, *107*, 13671–13680.
[33] Drillet, J. F.; Ee, A.; Friedemann, J.; Kotz, R.; Schnyder, B.; Schmidt, V. M. Oxygen reduction at Pt and $Pt_{70}Ni_{30}$ in H_2SO_4/CH_3OH solution. *Electrochim. Acta* **2002**, *47*, 1983–1988.
[34] Toda, T.; Igarashi, H.; Watanabe, M. Enhancement of the electrocatalytic O_2 reduction on Pt-Fe alloys. *J. Electroanal. Chem.* **1999**, *460*, 258–262.
[35] Zhang, J.; Mo, Y.; Vukmirovic, M. B.; Klie, R.; Sasaki, K.; Adzic, R. R. Platinum monolayer electrocatalysts for O-2 reduction: Pt monolayer on Pd(111) and on carbon-supported Pd nanoparticles. *J. Phys. Chem. B* **2004**, *108*, 10955–10964.
[36] Anderson, A. B.; Roques, J.; Mukerjee, S.; Murthi, V. S.; Markovic, N. M.; Stamenkovic, V. Activation energies for oxygen reduction on platinum alloys: Theory and experiment. *J. Phys. Chem. B* **2005**, *109*, 1198–1203.
[37] Savadogo, O.; Lee, K.; Oishi, K.; Mitsushima, S.; Kamiya, N.; Ota, K.-I. New palladium alloys catalyst for the oxygen reduction reaction in an acid medium. *Electrochem. Comm.* **2004**, *6*, 105–109.
[38] Fernandez, J. L.; Walsh, D. A.; Bard, A. J. Thermodynamic Guidelines for the Design of Bimetallic Catalysts for Oxygen Electroreduction and Rapid Screening by Scanning Electrochemical Microscopy. M-Co (M: Pd, Ag, Au). *J. Am. Chem. Soc.* **2005**, *127*, 357–365.
[39] Norskov, J. K.; Rossmeisl, J.; Logadottir, A.; Lindqvist, L.; Kitchin, J. R.; Bligaard, T.; Jonsson, H. Origin of the overpotential for oxygen reduction at a fuel-cell cathode. *J. Phys. Chem. B* **2004**, *108*, 17886–17892.
[40] Damjanovic, A.; Brusic, V. Electrode kinetics of oxygen reduction on oxide-free platinum electrodes. *Electrochim. Acta* **1967**, *12*, 615–628.
[41] Damjanovic, A.; Sepa, D. B.; Vojnovic, M. V. New evidence supports the proposed mechanism for dioxygen reduction at oxide free platinum electrodes. *Electrochim. Acta* **1979**, *24*, 887–889.
[42] Sepa, D. B.; Vojnovic, M. V.; Vracar, L. M.; Damjanovic, A. Different views regarding the kinetics and mechanisms of oxygen reduction at platinum and palladium electrodes. *Electrochim. Acta* **1987**, *32*, 129–134.
[43] Yeager, E.; Razaq, M.; Gervasio, D.; Razaq, A.; Tryk, D. "The electrolyte factor in oxygen reduction electrocatalysis."; Proc. Workshop Struct. Eff. Electrocatal. Oxygen Electrochem. **1992**.

[44] Yeager, E.; Razaq, M.; Gervasio, D.; Razaq, A.; Tryk, D. Dioxygen reduction in various acid electrolytes. *J. Serb. Chem. Soc.* **1992**, *57*, 819–833.

[45] Clouser, S. J.; Huang, J. C.; Yeager, E. B. Temperature dependence of the Tafel slope for oxygen reduction on platinum in concentrated phosphoric acid. *J. Appl. Electrochem.* **1993**, *23*, 597–605.

[46] Stamenkovic, V.; Schmidt, T. J.; Ross, P. N.; Markovic, N. M. Surface composition effects in electrocatalysis: Kinetics of oxygen reduction on well-defined Pt_3Ni and Pt_3Co alloy surfaces. *J. Phys. Chem. B* **2002**, *106*, 11970–11979.

[47] Yang, H.; Vogel, W.; Lamy, C.; Alonso-Vante, N. Structure and electrocatalytic activity of carbon-supported Pt-Ni alloy nanoparticles toward the oxygen reduction reaction. *J. Phys. Chem. B* **2004**, *108*, 11024–11034.

[48] Murthi, V. S.; Urian, R. C.; Mukerjee, S. Oxygen reduction kinetics in low and medium temperature acid environment: Correlation of water activation and surface properties in supported Pt and Pt alloy electrocatalysts. *J. Phys. Chem. B* **2004**, *108*, 11011–11023.

[49] Kitchin, J. R.; Norskov, J. K.; Barteau, M. A.; Chen, J. G. Modification of the surface electronic and chemical properties of Pt(111) by subsurface 3 d transition metals. *J. Chem. Phys.* **2004**, *120*, 10240–10246.

[50] Gai, P. L.; Roper, R.; White, M. G. Recent advances in nanocatalysis research. *Curr. Op. Sol. St. Mat. Sci.* **2002**, *6*, 401–406.

[51] Dassenoy, F.; Casanove, M.-J.; Lecante, P.; Verelst, M.; Snoeck, E.; Mosset, A.; Ely, T. O.; Amiens, C.; Chaudret, B. Experimental evidence of structural evolution in ultrafine colbalt particles stabilized in different polymers-From a polytetrahedral arrangement to the hexagonal structure. *J. Chem. Phys* **2000**, *112*, 8137–8145.

[52] Tadaki, T.; Koreeda, A.; Nakata, Y.; Kinoshita, T. Structure of Cu-Au alloy nanoscale particles and the phase transformation. *Surf. Rev. and Lett.* **1996**, *3*, 65–69.

[53] Sra, A. K.; Schaak, R. E. Synthesis of atomically ordered AuCu and $AuCu_3$ nanocrystals from bimetallic nanoparticle precursors. *J. Am. Chem. Soc.* **2004**, *126*, 6667–6672.

[54] Liz-Marzan, L. M. Nanometals: formation and color. *Materials Today* **2004**, 26–31.

[55] Guo, Z.; Kumar, C. S. S. R.; Henry, L. L.; Doomes, E. E.; Hormes, J.; Podlaha, E. J. Displacement synthesis of Cu shells surrounding Co nanoparticles. *J. Electrochem. Soc.* **2005**, *151*.

[56] Huang, S.-P.; Balbuena, P. B. Melting of bimetallic Cu-Ni nanoclusters. *J. Phys. Chem. B* **2002**, *106*, 7225–7236.

[57] Huang, S.-P.; Balbuena, P. B. Platinum Nanoclusters on Graphite Substrates: A Molecular Dynamics Study. *Mol. Phys.* **2002**, *100*, 2165–2174.

[58] Huang, S.-P.; Mainardi, D. S.; Balbuena, P. B. Structure and dynamics of graphite-supported bimetallic nanoclusters. *Surf. Sci.* **2003**, *545*, 163–179.

[59] Sato, K.; Kajiwara, T.; Fujiyoshi, M.; Ishimaru, M.; Hirotsu, Y.; Shinohara, T. Effects of surface step and substrate temperature on nanostructure of $L1_0$-FePt nanoparticles. *J. Appl. Phys.* **2003**, *93*, 7414–7416.

[60] Parravicini, G. B.; Stella, A.; Tognini, P.; Merli, P. G.; Migliori, A.; Cheyssac, P.; Hofman, R. Insight into the premelting and melting processes of metal nanoparticles through capacitance measurements. *Appl. Phys. Lett.* **2003**, *82*, 1461–1463.

[61] Wang, Z.; Sasaki, T.; Shimizu, Y.; Kirihara, K.; Kawaguchi, K.; Kimura, K.; Koshizaki, N. Effect of substrate position on the morphology of boron products by laser ablation. *Appl. Phys. A* **2004**, *79*, 891–893.

[62] Qi, W. H.; Wang, M. P. Size and shape dependent melting temperature of metallic nanoparticles. *Mat. Chem. and Phys.* **2004**, *88*, 280–284.

[63] Rossi, G.; Rapallo, A.; Mottet, C.; Fortunelli, A.; Baletto, D.; Ferrando, R. Magic polyicosahedral core-shell clusters. *Phys. Rev. Lett.* **2004**, *93*, 105503.

[64] Koga, K.; Ikeshoji, T.; Sugawara, K. Size- and temperature-dependent structural transitions in gold nanoparticles. *Phys. Rev. Lett.* **2004**, *92*, 115507.

[65] Bas, B. S. D.; Ford, M. J.; Cortie, M. B. Low energy structures of gold nanoclusters in the size range 3–38 atoms. *J. Mol. Struct.-Theochem.* **2004**, *686*, 193–205.

[66] Chen, M. S.; Goodman, D. W. The structure of catalytically active gold on titania. *Science* **2004**, *306*, 252–255.

[67] Min, B. K.; Wallace, W. T.; Goodman, D. W. Synthesis of a sinter-resistant, mixed-oxide support for nanoclusters. *J. Phys. Chem. B* **2004**, *108*, 14609–14615.

[68] Khanra, B.; Sarkar, A. D. Impurity and support effects on surface composition and CO plus NO reactions over Pt-Rh/CeO_2 nanoparticles: A comparative study. *Int. J. Mod. Phys. B* **2003**, *17*, 4831–4839.

[69] Wang, Y.; Balbuena, P. B. Ab initio Molecular Dynamics Simulations of the Oxygen Electroreduction Reaction on a Pt(111) Surface in the Presence of Hydrated Hydronium $(H_3O)+(H_2O)_2$: Direct or Series Pathway? *J. Phys. Chem. B* **2005**, *109*, 14896–14907.

[70] Toda, T.; Igarashi, H.; Uchida, H.; Watanabe, M. Enhancement of the electroreduction of oxygen on Pt alloys with Fe, Ni, Co. *J. Electrochem. Soc.* **1999**, *146*, 3750–3756.

[71] Kitchin, J. R.; Norskov, J. K.; Barteau, M. A.; Chen, J. G. Role of strain and ligand effects in the modification of the elctronic and chemical properties of bimetallic surfaces. *Phys. Rev. Lett.* **2004**, *93*, 156801.

[72] Greeley, J.; Mavrikakis, M. Alloy catalysts designed from first principles. *Nat. Mat.* **2004**, *3*, 810–815.

[73] Zhang, J.; Vukmirovic, M. B.; Xu, Y.; Mavrikakis, M.; Adzic, R. R. Controlling the catalytic activity of platinum-monolayer electrocatalysts for oxygen reduction with different substrates. *Angew. Chem. Int. Ed.* **2005**, *44*, 2132–2135.

[74] Stamenkovic, V.; Grgur, B. N.; Ross, P. N.; Markovic, N. M. Oxygen reduction reaction on Pt and Pt-bimetallic electrodes covered by CO – Mechanism of the air bleed effect with reformate. *J. Electrochem. Soc.* **2005**, *152*, A277–A282.

[75] Wang, Y.; Balbuena, P. B. Potential Energy Surface Profile of the Oxygen Reduction Reaction on a Pt Cluster: Adsorption and Decomposition of OOH and H_2O_2. *J. Chem. Theory and Comp.* **2005**, *1*, 935–943.

[76] Eichler, A.; Hafner, J. Molecular precursors in the dissociative adsorption of O_2 on Pt (111). *Phys. Rev. Lett.* **1997**, *79*, 4481–4484.

[77] Eichler, A.; Mittendorfer, F.; Hafner, J. Precursor-mediated adsorption of oxygen on the (111) surfaces of platinum-group metals. *Phys. Rev. B* **2000**, *62*, 4744–4755.

[78] Derosa, P. A.; Seminario, J. M. electron transport through single molecules: scattering treatment using density functional and green function theories. *J. Phys. Chem. B* **2001**, *105*, 471–481.

[79] Paddison, S. J. Proton conduction mechanisms at low degrees of hydration in sulfonic acid-based polymer electrolyte membranes. *Ann. Rev. Mat. Res.* **2003**, *33*, 289–319.

[80] Balbuena, P. B.; Lamas, E. J.; Wang, Y. Molecular modeling of polymer electrolytes for power sources. *Electrochim. Acta* **2005**, *50*, 3788–3795.

[81] Cao, D.; Lu, G. Q.; Wieckowski, A.; Wasileski, S. A.; Neurock, M. Mechanisms of methanol decomposition on platinum: A combined experimental and ab initio approach. *J. Phys. Chem. B* **2005**, *109*, 11622–11633.

[82] Weber, A. Z.; Newman, J. Modeling transport in polymer-electrolyte fuel cells. *Chem. Rev.* **2004**, *104*, 4679–4726.

[83] Marx, D.; Sprik, M.; Parrinello, M. Ab initio molecular dynamics of ion solvation. The case of Be^{2+} in water. *Chem. Phys. Lett.* **1997**, *273*, 360.

[84] Seminario, J. M.; Agapito, L. A.; Yan, L.; Balbuena, P. B. Density functional theory study of adsorption of OOH on Pt-based bimetallic clusters alloyed with Cr, Co, and Ni. *Chem. Phys. Lett.* **2005**, *410*, 275–281.

[85] Zinola, C. F.; Luna, A. M. C.; Triaca, W. E.; Arvia, A. J. The influence of surface faceting upon molecular-oxygen electroreduction on platinum in aqueous solutions. *Electrochim. Acta* **1994**, *39*, 1627–1632.

[86] Zinola, C. F.; Triaca, W. E.; Arvia, A. J. Kinetics and mechanism of the oxygen electroreduction reaction on faceted platinum-electrodes in trifluoromethanesulfonic acid solutions. *J. Appl. Electrochem.* **1995**, *25*, 740–754.

[87] Balbuena, P. B.; Altomare, D.; Vadlamani, N.; Bingi, S.; Agapito, L. A.; Seminario, J. M. Adsorption of O, OH, and H_2O on Pt-based bimetallic clusters alloyed with Co, Cr, and Ni. *J. Phys. Chem. A* **2004**, *108*, 6378–6384.

[88] Sidik, R. A.; Anderson, A. B. Density functional theory study of O_2 electroduction when bonded to a Pt dual site. *J. Electroanal. Chem.* **2002**, *528*, 69–76.

[89] Wang, Y.; Balbuena, P. B. Design of oxygen reduction bimetallic catalysts: Ab-initio derived thermodynamic guidelines. *J. Phys. Chem. B* **2005**, *109*, 18902–18906.

18
Multi-Scale Modeling of CO Oxidation on Pt-Based Electrocatalysts

CHANDRA SARAVANAN, N. M. MARKOVIC, M. HEAD-GORDON
AND P. N. ROSS

18.1. Introduction

A serious limitation of modern low-temperature fuel cells is the use of highly purified H_2 as fuel [1, 2]. While using H_2/CO mixtures from steam-reforming hydrocarbon fuels considerably improves fuel flexibility, trace amounts of CO in the feed stream poison electrode surface. One solution to this problem is the development of temperature-stable membranes that allow high- temperature (e.g. 200°C) fuel cell operations (for e.g. see Refs. [3, 4]), thereby reducing CO adsorption on electrode surfaces. An alternate approach is the development of CO-tolerant electrocatalysts. The search for new electrocatalysts can be performed using a purely combinatorial screening, where an alloy catalyst is chosen from a library of possible combinations by a rapid screening procedure [5]. Alternatively, the search can proceed in a materials-by-design approach, where new electrocatalysts can be tailor-made to overcome the CO poisoning problem. This process is guided by developing a working hypothesis on how best known CO electro-oxidation catalysts function. In this regard, there has been significant effort in the last few decades to understand electro-oxidation of CO on Pt-based electrodes. Surface X-ray scattering, scanning tunneling microscopy, IR spectroscopy, sum frequency generation, NMR, ex-situ low energy electron diffraction, voltammetry, and other experiments on clean single crystal surfaces have provided valuable information on structure, thermodynamics, and kinetics of the complex CO electro-oxidation phenomenon (for example see Refs. [6, 7, 8, 9, 10, 11, 12, 13, 14]. However, a complete molecular picture of the CO oxidation mechanism, and a deep understanding of the complex kinetics on electrode surfaces can only be obtained using an interplay of theory/simulations with experiment.

Multi-scale modeling provides a hierarchical computational approach to describe macroscopic catalytic processes. In this approach, atomistic methods (first principle quantum chemistry calculations and classical molecular dynamics) are used which reveal microscopic insight into the mechanisms and molecular-scale dynamics of reactions at electrode surfaces (for e.g. see [15,

16, 17, 18, 19, 20, 21, 22, 23, 24]). In a staged multi-scale approach, the energetics and reaction rates obtained from these calculations can be used to develop coarse-grained models for simulating kinetics and thermodynamics of complex multi-step reactions on electrodes (for example see [25, 26, 27, 28, 29, 30]). Varying levels of complexity can be simulated on electrodes to introduce defects on electrode surfaces, composition of alloy electrodes, distribution of alloy electrode surfaces, particulate electrodes, etc. Monte Carlo methods can also be coupled with continuum transport/reaction models to correctly describe surfaces effects and provide accurate boundary conditions (for e.g. see Ref. [31]). In what follows, we briefly describe density functional theory calculations and kinetic Monte Carlo simulations to understand CO electrooxidation on Pt-based electrodes.

18.2. Quantum Chemical Calculations of CO Oxidation on Pt-Based Catalysts

Gas-phase quantum chemistry (QC) calculations of CO and OH adsorption on Pt-based anodes provide valuable information on structure and energetics of adsorbates (for e.g. see [15, 19, 21]). A detailed review of CO adsorption calculation was presented by Fiebelman and co-workers [15]. Detailed CO and OH adsorption calculations on Pt-based electrodes have also been reported (for e.g. see [19] and references therein). Potential effects on CO binding energy and frequency have been discussed in detail by Koper and co-workers [20]. However, these calculations do not attempt to investigate the mechanism of the CO electrooxidation. Anderson and co-workers have used first-principle QC chemistry and semi-empirical calculations to understand the effect of potential on fuel cell electrochemistry in general, and CO oxidation electrochemistry in particular [16, 32, 33].

In what follows, we briefly review our density functional theory (DFT) calculations that investigate the mechanism of CO electrooxidation reaction. A commonly accepted mechanism for CO electrooxidation is the Langmuir-Hinshelwood mechanism, where oxygen-containing species, OH, formed on the Pt surface reacts with adsorbed CO to form CO_2 [16, 34, 35]. In this model, OH is formed on the electrode surface by dissociative adsorption H_2O. [26, 35] OH adsorption is assumed to be reversible and is given by the following two reactions:

$$H_2O_{ads} \xrightarrow{k^+_{OH}} OH_{ads} + H^+(solution) + e^- \qquad (2.1)$$

and

$$OH_{ads} + H^+(solution) + e^- \xrightarrow{k^-_{OH}} H_2O_{ads}, \qquad (2.2)$$

where OH$_{ads}$ denotes the adsorbed OH species. The reaction between OH$_{ads}$ and adsorbed CO (CO$_{ads}$) is irreversible and is given by:

$$CO_{ads} + OH_{ads} \xrightarrow{k_{CO_2}} CO_2 + H^+ + e^-. \qquad (2.3)$$

This bifunctional mechanism is typically invoked to explain CO oxidation on Pt-based alloys as well. While this model has been successful in describing several experimental results on Pt-based electrodes, a complete first-principle theoretical evidence to this mechanism does not exist.

Quantum calculations can test reaction mechanisms and also lend support to these mechanisms. We show that the formation of a 'COOH' complex between CO and OH (adsorbed on neighboring one-fold sites) plays an important role in the oxidation process [16, 18, 36]. We also see that oxidation occurs under appropriate potentials through a proton transfer from the complex to a neighboring water molecule. A similar complex formation was suggested earlier by Anderson and Grantscharova based on semi-empirical calculations [16]. Recently, Anderson and Neshev published a DFT study, which reconfirmed their earlier findings by ab initio methodology [17]. However, their cluster models were too small to describe the surface chemistry reliably and thus could not be subjected to geometry optimizations. Our results reviewed here support most of their findings and place them on firm ground by performing geometry optimization of relevant structures.

Applications of QC calculations to study electrochemistry face two major challenges: modeling the electrode potential and modeling the solvent at the electrode electrolyte interface [33, 37, 38, 39]. The problem of introducing an electrode (chemical) potential to first-principle QC calculations has been discussed by several authors [38, 39, 40]. Insights into electrochemical reactions at the interface can be obtained by simulating a double-layer electric field formed near the electrode surface [20, 41, 42, 43]. In this approach, an electric field is imposed at the interface to simulate the field resulting from applying an electrode potential [44, 45]. This provides a model to simulate the near-linear electric fields ($10^7 - 10^8 V cm^{-1}$) present in the inner Helmholtz double layer, which is consistent with experimental observations [45].

In this chapter, we review results of a simplified cluster model of the electrode electrolyte system that contains just the minimal features necessary to describe electrooxidation [18]. Our cluster model consists of a rigid piece (of sufficient size) of the electrode surface, one of each adsorbed species (CO and OH), and a single electrolyte (H$_2$O) molecule, together with an applied field to simulate the potential bias. Within the model used, a positive field applied to the surface corresponds to charging the metal atoms positively. This also corresponds to applying a positive electrode potential in an experiment.

Our DFT calculations employ the B3LYP exchange-correlation functional and Pople 6-31G basis set along with the LANL2DZ effective core potential

(ECP) for the Pt atoms [46, 47, 48]. All calculations were performed using Q-Chem suite of QC programs [49]. In order to simulate the surface electrochemistry reliably on Pt(111) electrode surfaces, we find that the surface atoms adjacent to atop adsorption sites (of CO and OH) must be included. This corresponds to a total of 10 surface atoms. We have also verified that the adsorption energy of the COOH complex is well behaved with respect to cluster size, DFT model, and size of basis functions. With no potential bias, we find negligible geometrical differences of the complex with a Pt_{10} model cluster (with no sub-surface atoms) and a Pt_{15} cluster (with five sub-surface atoms on the three-fold sites of the Pt_{10} cluster). All calculated adsorption energies are within a reasonable range of values, indicating that the cluster models are acceptable for the systems investigated. In all our geometry optimizations, the metal atoms were fixed, while the adsorbed organic species were allowed to move. This is in accordance with surface reconstruction measurements, where Pt atoms on the electrode surface are shown to move by less than 4%(0.1Å) during the CO oxidation process [50]. For details of these calculations the reader is referred to Ref. [18]. Cluster models of this type have previously been qualitatively successful in describing the essential features of chemisorption and potential dependent Stark shifts [20, 42].

Our results show the formation of a stable planar complex of CO and OH on Pt surface as shown in Figure 18.1. The structure of this intermediate complex is obtained by optimizing the geometry of co-adsorbed CO and OH on the two adjacent surface atoms using the model described above. This complex plays a crucial role in the CO electrooxidation process as a stable intermediate structure.

Next, we show solvent and electrode potential effects on the structure of COOH intermediate. As explained previously, this is modeled using a single H_2O molecule and applying electric fields. We performed DFT geometry optimizations of the bound COOH intermediate and H_2O solvent molecule at different field strengths. The fields range from -0.03 to 0.03 au in 0.01

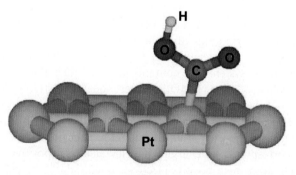

FIGURE 18.1. Figure showing a stable COOH intermediate on pure Pt when no electric field is applied.

18. Multi-Scale Modeling of CO Oxidation on Pt-Based Electrocatalysts 537

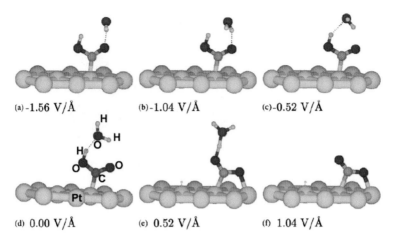

FIGURE 18.2. Showing potential dependence of the intermediate as a function of electric field. The atoms are labeled for the intermediate at zero field. At 1.04 VÅ^{-1} the preferred structure is the product itself i.e. CO_2 and H_3O^+ (H_3O^+ is not shown for this particular structure).

increments. This increment corresponds to 0.52 VÅ^{-1}. The optimized minimum energy structures are shown in Figure 18.2. As discussed in Ref. [18], the figure indicates that a proton transfer from the COOH complex to the water molecule occurs easily under more positive fields. We note that for fields below 1.04 VÅ^{-1} CO, oxidation proceeds through a stable intermediate. However, at 1.04 VÅ^{-1}, the preferred structure is the product itself (CO_2 and H_3O^+), i.e. activation energy for CO oxidation decreases with increasingly positive fields.

Ru has a higher oxophilicity compared to Pt (for e.g. see Ref. [19]). In Ref. [18], we perform a preliminary DFT study to understand the effect of the higher oxophilicity on electrooxidation of CO on PtRu alloy electrode surfaces. Our results indicate the formation of a stable COOH intermediate on PtRu surfaces. However, our DFT geometry optimizations show several structural differences between the stable complexes formed on these surfaces. As shown in Figure. 18.3, oxygen atoms of the OH_{ads} species are bound to the cluster surface through the Ru metal. This suggests that the proton transfer to the water molecule is easier, and hence implies a lower barrier for CO oxidation reaction. For a detailed description of energetics and electrode potential dependence of CO oxidation on Pt and Pt-Ru alloys, the reader is referred to Ref. [18].

For details on the potential energy surface (PES) for CO oxidization, the reader is referred to Ref. [18]. In what follows, we describe the second stage of our Multi-scale approach. We show results of our Monte Carlo simulations, to elucidate the CO electrooxidation kinetics on Pt-based electrodes.

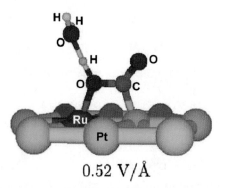

FIGURE 18.3. Showing stable intermediate on PtRu alloys at 0.52 V Å^{-1}. A similar complex is obtained at 0.52 V Å^{-1} for Pt surface (see Figure 18.2).

18.3. Monte Carlo Simulations

In the last section, we discussed the use of QC calculations to elucidate reaction mechanisms. First-principle atomistic calculations offer valuable information on how reactions happen by providing detailed PES for various reaction pathways. Potential energy surfaces can also be obtained as a function of electrode potential (for example see Refs. [16, 18, 33, 38]). However, these calculations do not provide information on the complex reaction kinetics that occur on timescales and lengthscales of electrochemical experiments. Mesoscale lattice models can be used to address this issue. For example, in Refs. [25, 51, 52] kinetic Monte Carlo (KMC) simulations were used to simulate voltammetry transients in the timescale of seconds to model Pt(111) and Pt(100) surfaces containing up to 256×256 atoms. These models can be developed based on insights obtained from first-principle QC calculations and experiments. Theory and/or experiments can be used to parameterize these models. For example, rate theories [22, 24, 53, 54] can be applied on detailed potential energy surfaces from accurate QC calculations to calculate electrochemical rate constants. On the other hand, approximate rate constants for some reactions can be obtained from experiments (for example see Refs. [25, 26]). This chapter describes the later approach.

In this section, we review the use of MC simulations to understand the complex reaction kinetics of CO electrooxidation near Pt(111) surfaces in sulfuric acid solutions. We address the role of specific adsorption of anions on electrode surfaces, and its effect on CO electrooxidation. We also review some results of CO oxidation in alloy surfaces [55]. This section Is Arranged as follows. In Section 18. 3.1, we provide a brief background onanion adsorption and CO oxidation. Section 18. 3.2 outlines the model used. In Section 18. 3.3, we briefly outline the KMC methodology. In Section 18. 3.4 we review the effect of competitive adsorption on base voltammograms (i.e. under CO-free conditions) and CO electrooxidation. We also briefly review CO diffusion effects and CO electrooxidation on PtRu alloys [55].

18.3.1. Background

Studies on thermodynamics and kinetics of anion adsorption on single-crystal metal surfaces have provided experimental data on microstructure of anion adlayers and revealed fascinating phenomena such as phase transitions on electrode surfaces [8, 56, 57, 58, 59, 60, 61]. These studies have also provided a strong basis for understanding CO oxidation on Pt single-crystal surfaces [9, 62, 63, 64, 65, 66, 67]. Experiments of CO oxidation with different supporting electrolytes help in understanding the effect of anions on this oxidation process. However, these experiments do not provide a complete picture of the complex reaction processes on the electrode surface. A better understanding of the surface kinetics can be obtained by the interplay of theory and simulation with experiment. Monte Carlo simulations have made significant contributions toward elucidating the structure of adlayers and their response to an applied potential [28, 29, 30, 52]. For example, KMC simulations reveal disorder-order transitions during anion adsorption in voltammetry experiments [25, 52]. These simulations also elucidate the effect of anion adsorption and CO diffusion on CO oxidation. Monte Carlo simulations have also been used to understand CO oxidations on alloy surfaces [55]. In what follows, we briefly describe a model used to understand CO oxidation on Pt(111)-sulfuric acid solution interface.

18.3.2. Model

As described previously, our model for CO electrooxidation on Pt(111) surface is based is on a Langmuir-Hinshelwood mechanism, where oxygen-containing species, OH, formed on the surface reacts with adsorbed CO to form CO_2 [16, 34, 35]. We have modeled adsorption and desorption of anions from sulfuric acid solution as reversible electrochemical reactions, i.e.,

$$AN_{sol} \overset{k_{AN}^+}{\rightarrow} AN_{ads} + e^- \quad (3.1)$$

and

$$AN_{ads} + e^- \overset{k_{AN}^-}{\rightarrow} AN_{sol}, \quad (3.2)$$

where AN_{ads} and AN_{sol} are the adsorbed anion and anion in solution, respectively.

The adsorption (Eqs. 2.1, 3.1) and desorption (Eqs. 2.2, 3.2) of X_{ads} (X = OH, AN) on Pt surface occur at finite rates that depend on potential (E) and temperature (T). These adsorption and desorption rates are modeled using Butler-Volmer type rate coefficients as:

$$k_X^+ = k_X^{+0} \exp[\alpha \beta e E] \quad (3.3)$$

and

$$k_X^- = k_X^{-0} \exp[-(1-\alpha)\beta eE], \quad (3.4)$$

respectively, where α is assumed to be 0.5, $\beta = (k_B T)^{-1}$ and k_B is the Boltzmann's constant [45]. In these expressions, k_X^{+0} and k_X^{-0} are chosen such that reversible voltammograms are produced at a scan rate of 50 mV/s. The CO oxidation rate constant is also modeled using a Butler-Volmer expression as $k_{CO_2} = k_{CO_2}^0 \exp[\alpha\beta eE]$. In our model, CO mobility is included as the rate of site-to-site hops, k_D. Table 18.1 summarizes the parameters in our model. In our voltammetry simulations, we fix the ratios $\kappa = k_{OH}^{0-}/k_{OH}^{0+}$ and $\chi = k_{AN}^{0-}/k_{AN}^{0+}$, in order to fix the OH and anion peak positions. For example, we set $\kappa = 10^{13}$ and $\chi = 10^6$ to obtain peak potentials for OH adsorption and HSO$_4$ adsorption near 0.77 V and 0.35 V, respectively. For complete details on parameter ranges, the reader is referred to Ref. [25]. All our voltammetry simulations have sweep rates between 1 and 100 mV/s.

In our model, OH adsorbs on atop sites of the Pt(111) surface. This is in accordance with DFT calculations (for e.g. see Ref. [19]). Experiments results suggest atop sites as preferred sites for CO adsorption (see Ref. [15] and references therein). However, DFT calculations show an apparent contradiction by predicting higher binding energies for high-coordination sites. [15] While this concern must be addressed by the DFT community, our model assumes that CO adsorbs on atop sites. We do not expect to see a change in the qualitative predictions of our model based on preferred adsorption sites of CO. Anions (SO$_4$ or HSO$_4$) require an 'ensemble' of sites to adsorb the Pt surface. [68, 69, 70] This is modeled by allowing anions to adsorb on the three-fold hcp and fcc sites, where sulfate adsorbs with three oxygen atoms coordinated to atop Pt surface atoms. The model is simplified by allowing a single type of discharged anion (eg. HSO$_4$) to adsorb on the three-fold sites. Our model also includes a difference in binding energy, $\Delta\varepsilon$, between the fcc and hcp sites. In our model, the intermolecular interactions are introduced as follows. If an atop site is occupied by OH or CO, none of the neighboring three-fold sites is occupied. if a three-fold site is occupied, none of the nearest-neighbor three-fold sites is occupied. We will show that the present model is sufficient to describe the 'butterfly' feature observed in base voltammograms

TABLE. 18.1. Showing parameter values of rate constants (in sec^{-1}).

Model parameters	Parameter values
k_{OH}^{0+}	$10^{-5} - 10^{-1}$
k_{OH}^{0-}	$10^8 - 10^{12}$
$k_{HSO_4}^{0+}$	$10^{-2} - 1$
$k_{HSO_4}^{0-}$	$10^3 - 10^6$
$k_{CO_2}^0$	$10^{-8} - 10^{-3}$
k_D	$0 - 10^3$

of sulfuric acid solutions [52, 57, 71, 72, 73]. In what follows we briefly describe the simulation method.

18.3.3. Simulations

The program CARLOS [74] was used to perform KMC simulations of voltammetry and potential step experiments. The details of the MC algorithms are available elsewhere [51, 74]. Therefore, we make only brief comments about the method. The simulations are performed on a 64×64, 128×128 or 256×256 Pt(111) lattice with periodic boundary conditions. In stripping voltammetry simulations, the simulation is started by fixing the initial concentration θ_{CO} on the Pt surface. The electrode potential is changed linearly with time causing a time dependence of the activation energy and hence the time dependence of electrochemical rate. [90] At a particular configuration γ and a given time t, a process list is prepared by compiling a list of possible reactions for the system. CARLOS uses a variable time step Monte Carlo (VTSMC) to update the time and configurations at each MC step. In a VTSMC algorithm, a time step, $\Delta t(\gamma)$, is chosen such that exactly one event occurs during that time step. The chosen event occurs with a probability proportional to its rate coefficient. Further, the time step is obtained as: $\Delta t(\gamma) = -\ln(1-x)/k_{tot}(\gamma)$, where $x \in (0,1)$ is a uniform random number, and k_{tot} is total rate of all possible processes available at time t. Once Δt is picked and the chosen event is performed, the process list is updated for choosing an event in the next VTSMC *step*. Many such VTSMC steps are performed before a single VTSMC *run* is complete. For example, during a voltammetry simulation, a VTSMC run is complete when the potential scan reaches its final preset value. The current transient, during a VTSMC run, is determined by calculating the net rate of electrons flowing into the electrode as a funtion of time. Several VTSMC runs are averaged to obtain a voltammogram.

18.3.4. Results of KMC Simulations

In this section, we discuss anion and OH coadsorption on Pt(111) surface to elucidate the effect of competitive adsorption on CO oxidation electrocatalysis. We first perform coadsorption simulations to understand base voltammograms, i.e., in the absence of CO oxidation. Next, we show the effect of anions on CO electrooxidation by performing simulations of CO stripping voltammetry, where a monolayer of CO is oxidized by a potential sweep.

18.3.4.1. Coadsorption Effects on Base Voltammograms

We understand the effect of HSO_4 and OH coadsorption by (i) varying the binding energy difference of anions ($\Delta \varepsilon$) on the fcc and hcp sites, and (ii) changing the relative adsorption rates of OH and anions ($\lambda = k_{AN}^{0+}/k_{OH}^{0+}$).

In order to model the effect of the difference in binding energy of the hcp and fcc sites, $\Delta\varepsilon$, we assume that the difference in activation energy for anion desorption from these sites is equal to the difference in binding energy $\Delta\varepsilon$ i.e. $\kappa(\text{fcc}) = \exp(\beta\Delta\varepsilon) \cdot \kappa_{\text{AN}}^-(hcp)$. Figure 18.4(a). shows the voltammogram as a function of $\Delta\varepsilon$. The voltammogram twin-peak near 0.35 V and the peak near 0.77 V are there because of HSO_4 adsorption and OH adsorption, respectively. This is understood from Figure 18.4(b) which shows an increase in HSO_4 adsorption and OH adsorption near 0.35 V and 0.77 V, respectively.

An interesting effect to note is that the twin-peak near 0.35 V occur only for values of $\Delta\varepsilon \gg k_B T$, and the peak separation is essentially unchanged as we increase the difference in site energies [75]. This is the result of a disorder-order phase transition near this potential. As described in Ref. [25], a delicate balance between the low-energy, low-entropy ordered state and the high-energy, high-entropy disordered state results in a disorder-order phase-transition. Such potential-induced transitions have previously been reported by several groups [8, 76, 77]. In the present case, this transition results in the formation of an ordered $(\sqrt{3} \times \sqrt{3})R30°$ adlayer. Ex-situ investigations of anion adsorption from sulfuric acid solutions indicate the formation of this structure [78, 79]. However, in-situ experiments show the formation of an ordered $(\sqrt{3} \times \sqrt{7})$ adlayer [58, 61, 80, 81]. It has been suggested that the loss of water during ex-situ measurements may result in these different structures. Similar structures are also observed in Au(111) and Rh(111) surfaces [82, 83]. Zhang et al. have modeled these $(\sqrt{3} \times \sqrt{7})$ structures using three body interactions [28].

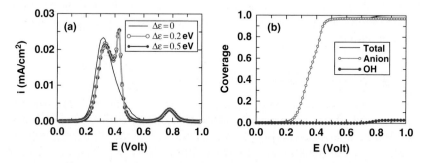

FIGURE 18.4. Effect of binding energy difference between the hcp and fcc sites, $\Delta\varepsilon$. Anion and OH adsorption peak positions are fixed with $\lambda = k_{\text{AN}}^{0+}/k_{\text{OH}}^{0+} = 10^3, \kappa = k_{\text{OH}}^{0-}/k_{\text{OH}}^{0+} = 10^{13}$, and $\chi = k_{\text{AN}}^{0-}/k_{\text{AN}}^{0+} = 10^6$. The voltammograms are shown as a function of $\Delta\varepsilon$ in (a). The figure shows the twin-peak near 0.35 V and a single peak near 0.77 V because of HSO_4 adsorption and OH adsorption, respectively. Adsorption isotherms of HSO_4, OH and total coverage for $\Delta\varepsilon = 0.2$ eV are shown in (b). The figure shows an increase in HSO_4 adsorption and OH adsorption near 0.35 V and 0.77 V, respectively.

Ordered anion domains formed are incommensurate with each other. This results in the formation of domain walls. The second sharp peak is the result of this disorder-order transition which occurs around 0.4 V in our model. Almost complete coverage of the Pt surface results in near-zero currents because it results in the lack of space for further adsorption of anions. However, near 0.77 V, OH adsorbs on the atop sites along the trenches between the incommensurate domains of anion islands. This is clear from the snapshot of the simulation at 0.95 V in Figure 18.5. For details of this phenomena, the reader is referred to Ref. [25]. Also since anions do not adsorb near 0.77 V, the predominant processes in this potential region are OH adsorption and OH desorption resulting in an OH peak. Experimentally, such a small peak near 0.77 V is observed and has indeed been ascribed to OH adsorption [56]. We do note, however, that the experimental peak is not reversible, in contrast with our simulation. Although we are not sure that the experimental peak is caused by OH adsorption in domain walls of the (bi)sulfate structure, we do believe the agreement between experiment and simulation is very suggestive of the qualitative correctness of our model.

Now we discuss the effect of $\lambda (= k_{AN}^{0+}/k_{OH}^{0+})$ on base voltammograms. In Figure 18.6, we show base voltammograms as a function of λ for a fixed $\Delta \varepsilon$ and T. We note that as we decrease λ (i.e. increase k_{OH}^{+}), we find an increase in the amount OH adsorption (see Figure 18.6a) and hence an increase in the OH peak height. We also note that the height of the second peak near 0.43 V, corresponding to the disorder-order transition, decreases with decreasing λ. This is because this phase transition is quenched by the competitive adsorption of OH in this potential range. For low concentrations of sulfuric acid solutions

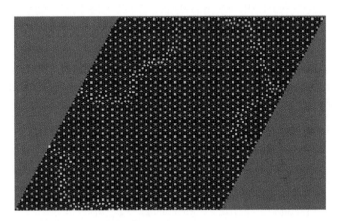

FIGURE 18.5. Snapshot of voltammetry simulation at 0.95 V. The simulation is performed at a scan rate of 50 mV/s for $\lambda = k_{AN}^{+}/k_{OH}^{+} = 10^3, \kappa = 10^{13}$ $\chi = 10^6, \Delta \varepsilon = 0.2$ eV. Light blue features indicate adsorbed anions and yellow features indicate adsorbed OH. The snapshot shows OH adsorption on atop sites along the domain walls between incommensurate anion domains.

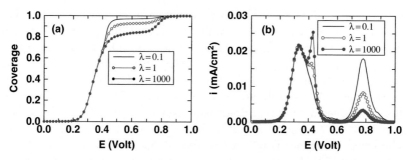

FIGURE 18.6. Showing the effect of λ with $\kappa = 10^{13}$, $\Delta E = 0.2 eV$ and $\chi = 10^6$. The total coverages and voltammograms are shown as a function of λ in (a) and (b), respectively.

(less than 0.01 M), voltammetric transients with single anion peaks are observed, suggesting competitive OH adsorption at these concentrations [84].

18.3.4.2. CO Electrooxidation on Pt-Based Electrodes

In this section, we show the effect of anions on CO electrooxidation by performing simulations of CO stripping, where a monolayer (initial coverage of $\theta_{CO} = 0.99$) of CO is oxidized by a potential sweep. The electrocatalytic activity is studied as a function of $\gamma = k_{CO_2}/k_{OH}^+$ and CO diffusion, k_D.

Here CO oxidation studied as a function of $\gamma = k_{CO_2}/k_{OH}^+$ is shown in Figure 18.7. Near-zero currents are observed until around 0.7 V followed by a prewave and a sharp peak. These features are explained as follows. Near-zero currents are obtained for potentials until 0.7 V because no net OH adsorption occurs in this region. Once OH adsorption begins, it results in the oxidation of CO (Eqs. 2.3) and the formation of empty sites, thus creating a 'hole' in the CO adlayer. When this hole is 'small' CO oxidation is slow. This is because of the small number of sites available for correlated events of OH and anion competitive adsorption. As a result in this slow CO oxidation regime a

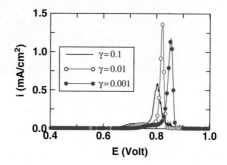

FIGURE 18.7. Showing CO oxidation function of $\gamma = k_{CO_2}/k_{OH}^+$. Near-zero currents are observed until around 0.7 V followed by a CO oxidation regime, which corresponds to a prewave and a sharp peak.

prewave is observed in the voltammogram. The holes formed in this regime are further enlarged by adsorption of OH and oxidation of CO. This process of CO oxidation can be modeled as a nucleation-growth phenomenon [26, 85, 86, 87, 88]. It is useful to note that the sharp peak shift to more positive potentials with decreasing γ. This is because CO oxidation is shifted to positive potentials with decreasing decreasing k_{CO_2}.

Diffusion of CO plays an important role in the oxidation of CO on Pt surfaces [26, 51]. We find that for high surface mobility of CO, the prewave vanishes and the peak shifts to more negative potentials. In the limit of fast diffusion, CO moving rapidly on the electrode surface washes out correlation effects resulting in a voltammogram with no prewave.

Koper and coworkers have performed KMC simulations to simulate CO stripping voltammetry on PtRu alloy electrodes in the absence of specific anion adsorption [55]. Their results show that for a randomly dispersed alloy of Pt and Ru, their model provides good agreement to experiments of Gasteiger and coworkers [89].

18.4. Summary and Future Work

This chapter discusses a staged multi-scale approach for understanding CO electrooxidation on Pt-based electrodes. In this approach, density functional theory (DFT) is used to obtain an atomistic view of reactions on Pt-based surfaces. Based on results from experiments and quantum chemistry calculations, a consistent coarse-grained lattice model is developed. Kinetic Monte Carlo (KMC) simulations are then used to study complex multi-step reaction kinetics on the electrode surfaces at much larger lengthscales and timescales compared to atomistic dimensions. These simulations are compared to experiments. We review KMC results on Pt and PtRu alloy surfaces.

In our quantum calculations, electrochemistry is modeled by applying electric fields. Our calculations show that the CO electrooxidation proceeds through a stable intermediate. The structure of the intermediate optimized with DFT under increasingly positive potentials shows decreasing activation energy for CO oxidation. Our calculations on PtRu alloy surfaces also show the formation of stable intermediate for CO oxidation. However, this intermediate has significant structural differences from that on Pt surface suggesting lower activation energy for CO oxidation. Our KMC simulations for base voltammograms (i.e. under CO free conditions) for Pt(111) in sulfuric acid solutions show disorder-order transitions resulting in 'butterflies' observed in experimental voltammograms. We also show that this transition is quenched by increase coadsorption of OH at low sulfuric acid concentrations. Our calculations of CO stripping voltammetry on Pt surface show excellent qualitative agreement to experiment. We also show that CO oxidation on Pt surfaces fits a nucleation-growth model. We also briefly review KMC results on PtRu alloy surfaces.

A molecular perspective of reactions from quantum chemistry calculations is the first step toward a theoretical design of new electrodes (e.g. binary or even ternary alloys). While reaction mechanisms on Pt and Pt alloy surfaces are getting clearer, details of these mechanism still remain elusive. For example, bifunctional mechanism of CO oxidation on PtSn and PtMo has received very little theoretical attention. Loading effects (CO, OH or specifically adsorbed anions) on CO oxidation is also poorly understood and requires further investigation. Theoretical calculations are also required to understand catalyst reorganization. Details of these calculations are required for accurately modeling the macroscopic kinetics on well-defined electrode surfaces and ultimately designing nanocatalyst particles.

References

[1] B. C. H. Steele and A. Heinzel, Nature **414**, 345 (2001).
[2] N. M. Markovic and P. N. Ross, Cat. Tech. **4**, 110 (2000).
[3] J. S. Wainright, J. T. Wang, D. Weng, R. F. Savinell, and M. H. Litt, J. Electrochem. Soc. **142**, L121 (1995).
[4] R. F. Savinell, M. H. Litt, J. T. Wang and H. Yu, Electrochim. Acta **41**, 193 (1996).
[5] E. Reddington, A. Sapienza, B. Gurau, R. Viswanathan, S. Sarangapani, E. S. Smotkin, and T. E. Mallouk, Nature **280**, 1735 (1998).
[6] N. M. Markovic, C. A. Lucas, A. Rhodes, V. Stamankovic, and P. N. Ross, Surf. Sci. **499**, L149 (2002).
[7] I. M. Tidswell, N. M. Markovic, and P. N. Ross, Phys. Rev. Lett **71**, 1601 (1993).
[8] B. M. Ocko, J. X. Wang, and T. Wandlowski, Phys. Rev. Lett **79**, 1511 (1997).
[9] I. Villegas and M. J. Weaver, J. Chem. Phys. **101**, 1648 (1994).
[10] I. Villegas, X. Gao, and M. J. Weaver, Electrochim. Acta **40**, 1267 (1995).
[11] S. Baldelli, N. Markovic, P. N. Ross, Y. R. Shen, and G. A. Somorjai, J. Phys. Chem. B **103**, 8920 (1999).
[12] S. Hoffer, S. Baldelli, K. Chou, P. N. Ross, and G. A. Somorjai, J. Phys. Chem. B **106**, 6473 (2002).
[13] Y. Y. Tong, H. S. Kim, P. K. Babu, P. Waszczuk, A. Wieckowski, and E. Oldfield, J. Am. Chem. Soc. **124**, 468 (2002).
[14] W. F. Lin, M. S. Zei, and G. Ertl, Chem. Phys. Lett. **312**, 1 (1999).
[15] P. J. Feibelman, B. Hammer, J. K. Norskov, F. Wagner, M. Scheffler, R. Stumpf, R. Watwe, and J. Dumesic, J. Phys. Chem. B **105**, 4018 (2001).
[16] A. B. Anderson and E. Grantscharova, J. Phys. Chem. **99**, 9149 (1995).
[17] A. B. Anderson and N. M. Neshev, J. Elec. Chem. Soc. **149**, E383 (2002).
[18] C. Saravanan, B. Dunietz, N. Markovic, M. Head-Gordon, and P. Ross, J. Electroanal. Chem. **554–555**, 459 (2003).
[19] M. T. M. Koper, T. E. Shubina, and R. A. van Santen, J. Phys. Chem. B **106**, 686 (2002).
[20] M. T. M. Koper, R. A. van Santen, S. A. Wasileski, and M. J. Weaver, J. Chem. Phys. **113**, 4392 (2000).

[21] E. Christofferson, P. Liu, A. Ruben, H. L. Skriver, and J. K. Norskov, J. Catal. **199**, 123 (2001).
[22] A. Calhoun and G. A. Voth, J. Electroanal. Chem. **450**, 253 (1998).
[23] J. W. Halley, A. Mazzolo, Y. Zhou, and D. Price, J. Electroanal. Chem. **450**, 273 (1998).
[24] J. Hautman, J. W. Halley, and Y. J. Rhee, J. Chem. Phys. **91**, 467 (1989).
[25] C. Saravanan, M. T. M. Koper, N. Markovic, M. Head-Gordon, and P. Ross, Phys. Chem. Chem. Phys. **4**, 2660 (2002).
[26] M. T. M. Koper, A. P. J. Jansen, R. A. van Santen, J. J. Lukkien, and P. A. J. Hilbers, J. Chem. Phys. **109**, 6051 (1998).
[27] A. V. Petukhov, W. Akemann, K. A. Friedrich, and U. Stimming, Surf. Sci. **404**, 182 (1998).
[28] J. Zhang, Y. Sung, P. A. Rikvold, and A. Wieckowski, J. Chem. Phys. **94**, 6887 (1996).
[29] P. A. Rikvold, J. Zhang, Y.-E. Sung, and A. Wieckowski, J. Chem. Phys. **94**, 6887 (1996).
[30] L. Blum and D. A. Huckaby, J. Electroanal. Chem. **375**, 69 (1994).
[31] D. G. Vlachos, AIChE J. **43**, 3031 (1997).
[32] A. B. Anderson and E. Grantscharova, J. Phys. Chem. **99**, 9144 (1995).
[33] A. B. Anderson and T. V. Albu, J. Am. Chem. Soc. **121**, 11855 (1999).
[34] P. Gilman, J. Phys. Chem. **68**, 70 (1964).
[35] H. A. Gasteiger, N. Markovic, P. N. Ross, and E. J. Cairns, J. Phys. Chem. **98**, 617 (1994).
[36] T. E. Shubina, C. Hartnig, and M. T. M. Koper, Phys. Chem. Chem. Phys. **6**, 4215 (2004).
[37] D. L. Price and J. W. Halley, J. Chem. Phys. **102**, 6603 (1995).
[38] A. B. Anderson and D. B. Kang, J. Am. Chem. Soc. **102**, 5993 (1998).
[39] X. Crespin, V. M. Geskin, C. Bureau, R. Lazzaroni, W. Schmickler, and J. L. Bredas, J. Chem. Phys. **115**, 10493 (2001).
[40] H. Nakatsuji, J. Chem. Phys. **87**, 4995 (1987).
[41] P. S. Bagus, C. J. Nelin, W. Muller, M. R. Philpott, and H. Seki, Phys. Rev. Lett. **58**, 559 (1987).
[42] M. Head-Gordon and J. C. Tully, Chem. Phys. **175**, 37 (1993).
[43] P. Liu, A. Logadottir, and J. K. Norskov, Electrochim. Acta **48**, 3731 (2003).
[44] W. Schmickler, *Interfacial Electrochemistry* (Oxford University Press, New York, 1996).
[45] A. J. Bard and L. R. Faulkner, *Electrochemical Methods : Fundamentals and Applications* (Wiley, New York, 1980).
[46] A. Becke, J. Chem. Phys. **98**, 1372 (1993).
[47] A. Becke, J. Chem. Phys. **98**, 5648 (1993).
[48] W. R. Wadt and J. P. Hay, J. Chem. Phys. **82**, 299 (1985).
[49] J. Kong, C. A. White, A. I. Krylov, C. D. Sherrill, R. D. Adamson, T. R. Furlani, M. S. Lee, A. M. Lee, S. R. Gwaltney, T. R. Adams, et al., J. Comp. Chem. **21**, 1532 (2000).
[50] N. M. Markovic and P. N. Ross, Surf. Sci. Rep. **45**, 121 (2002).
[51] C. Saravanan, N. Markovic, M. Head-Gordon, and P. Ross, J. Chem. Phys. **114**, 6404 (2001).
[52] M. T. M. Koper and J. J. Lukkien, J. Electroanal. Chem. **485**, 161 (2000).

[53] D. Chandler, *Introduction to Modern Statistical Mechanics* (Oxford University Press, Oxford, 1987).
[54] B. B. Smith and Halley, J. Chem. Phys. **101**, 10915 (1994).
[55] M. T. M. Koper, J. J. Lukkien, A. P. J. Jansen, and R. A. van Santen, J. Phys. Chem. **103**, 5522 (1999).
[56] N. Markovic and P. N. Ross, J. Electroanal. Chem. **330**, 499 (1992).
[57] W. Savich, S. G. Sun, J. Lipkowski, and A. Wieckowski, J. Electroanal. Chem. **388**, 233 (1995).
[58] A. M. Funtikov, U. Stimming, and R. Vogel, J. Electroanal. Chem. **428**, 147 (1997).
[59] C. A. Lucas, N. Markovic, and P. N. Ross, Phys. Rev. B **55**, 7964 (1997).
[60] N. Li and J. Lipkowski, J. Electroanal. Chem. **491**, 95 (2000).
[61] A. Kolics and A. Wieckowski, J. Phys. Chem. B **105**, 2588 (2001).
[62] F. Kitamura, M. Takahashi, and M. Ito, Surf. Sci. **223**, 497 (1989).
[63] M. J. Weaver, S.-C. Chang, L.-W. Leung, X.Jiang, M. Rubel, M. Szklarczyk, D. Zurawski, and A. Wieckoski, J. Electroanal. Chem. **327**, 247 (1992).
[64] J. M. Feliu, J. M. Orts, A. Fernandez-Vega, A. Aldaz, and J. Clavilier, J. Electroanal. Chem. **296**, 191 (1990).
[65] J. M. Orts, A. Fernandez-Vega, J. M. Feliu, A. Aldaz, and J. Clavilier, J. Electroanal. Chem. **327**, 261 (1992).
[66] J. Clavilier, R. Albalat, R. Gomez, J. M. Orts, J. M. Feliu, and A. Aldaz, J. Electroanal. Chem. **330**, 489 (1992).
[67] H. Kita, H. Narumi, S. Ye, and H. Naohara, J. Appl. Electrochem. **23**, 589 (1993).
[68] N. Markovic, N. S. Marinkovic, and R. Adzic, J. Electroanal. Chem. **214**, 309 (1988).
[69] P. Faguy, N. Markovic, , R. Adzic, C. Fierro, and B. Yeager, J. Electroanal. Chem. **289**, 245 (1990).
[70] N. Markovic, N. S. Marinkovic, and R. R. Adzic, J. Electroanal. Chem. **314**, 289 (1991).
[71] J. Clavilier, J. Electroanal. Chem. **107**, 211 (1980).
[72] D. A. Scherson and D. M. Kolb, J. Electroanal. Chem. **176**, 353 (1984).
[73] E. Herrero, J. M. Feliu, and A. Wieckowski, Surf. Sci. **325**, 131 (1995).
[74] J. J. Lukkien, J. P. L. Segers, P. A. J. Hilbers, R. J. Gelten, and A. P. J. Jansen, Phys. Rev. E **58**, 2598 (1998).
[75] C. G. M. Hermse, A. P. van Bavel, M. T. M. Koper, J. J. Lukkien, R. A. van Santen, A. P. J. Jansen Surf. Sci. **572**, 247 (2004).
[76] B. E. Conway, Prog. Surf. Sci. **16**, 1 (1984).
[77] M. T. M. Koper, J. Electroanal. Chem. **450**, 189 (1998).
[78] S. Thomas, Y.-E. Sung, H. S. Kim, and A. Wieckowski, J. Phys. Chem. **100**, 11726 (1996).
[79] H. Ogasawara, Y. Sawatari, I. Inukai, and M. Ito, J. Electroanal. Chem. **358**, 337 (1993).
[80] A. M. Funtikov, U. Linke, U. Stimming, and R. Vogel, Surf. Sci. **324**, L3243 (1995).
[81] K. Itaya, Prog. Surf. Sci. **58**, 121 (1998).
[82] H. Angerstein-Kozlowska, B. E. Conway, A. Hamelin, and L. Stoicoviciu, J. Electroanal. Chem. **228**, 429 (1987).
[83] Y.-E. Sung, S. Thomas, and A. Wieckowski, J. Phys. Chem. **99**, 13513 (1995).

[84] K. Jaaf-Golze, D. M. Kolb, and D. A. Scherson, J. Electroanal. Chem. **200**, 353 (1986).
[85] A. Bewick, M. Fleischmann, and H. R. Thirsk, Trans. Faraday. Soc. **58**, 2200 (1962).
[86] M. Avrami, J. Chem. Phys. **7**, 1103 (1939).
[87] M. Avrami, J. Chem. Phys. **8**, 212 (1940).
[88] C. Saravanan, P. Sunthar, and E. Bosco, J. Electroanal. Chem. **375**, 59 (1994).
[89] H. A. Gasteiger, N. Markovic, and P. N. Ross, J. Phys. Chem. **98**, 617 (1994).
[90] For a theory of variable time step Monte Carlo algorithm for time dependent rate constants the reader is referred to Ref. 51

19
Modeling Electrocatalytic Reaction Systems from First Principles

SALLY A. WASILESKI, CHRISTOPHER D. TAYLOR, AND
MATTHEW NEUROCK

19.1. Introduction

Electrocatalytic reaction systems demonstrate markedly different behavior than those carried out in the vapor phase or under ultrahigh vacuum conditions. The differences in reactivity can be attributed to the significant difference between the reaction environment of the electrocatalytic system which includes the presence of solution, electrolyte, and intrinsic as well as extrinsic potentials, in addition to the vapor phase system. The solution environment and the applied potential can stabilize or destabilize charge transfer events, thus influencing many of the physicochemical processes that occur at the surface of a working electrode and strongly impacting the activity, as well as the selectivity of the active catalyst. For example, the hydrogen-bonding character of water stabilizes charge on reactant, product, and transition states to varying extents and thereby influences both the thermodynamic, as well as the intrinsic kinetics. Such charge stabilization will enhance heterolytic bond activation involving electron and/or proton transfer. Similarly, the potential difference that results at the metal/solution interface can also significantly influence the kinetics. The ultimate goal of the design of electrocatalysts will therefore likely depend on establishing a critical understanding of the atomic- and molecular-level processes that occur at the active catalytic sites that exist at the electrochemical interface and the specific influence of their local reaction environment.

The tremendous advances that have occurred in the spectroscopic analysis of the electrode/electrolyte interface have begun to provide a fundamental understanding of the elementary processes and the influence of process conditions. Surface-sensitive spectroscopic and microscopic analyses such as surface-enhanced Raman scattering (SERS) [1], potential-difference infrared spectroscopy (PDIRS) [2], surface-enhanced infrared spectroscopy (SEIRS) [3], sum frequency generation (SFG) [4], and scanning tunneling microscopy (STM) [5,6] have enabled the direct observation of potential-dependent changes in molecular structure [2,7] chemisorption [8,9], reactivity [10], and surface reconstruction [11].

In principle, the application of ab initio quantum mechanical methods can provide synergistic guidance for those experiments aimed at elucidating the controlling surface chemistry. Theory has, for example, greatly enhanced our understanding of chemisorption, surface diffusion, and surface reactivity on model single crystal substrates and under ultrahigh vacuum conditions [12,13]. It has been used to elucidate chemical bonding and the reactivity of adsorbates on metal and metal oxide surfaces. Theory has also been used in a more predictive manner to calculate binding energies [13,14], reaction energies [15,16,17], activation barriers [15,16,17], and vibrational frequencies [18,19] for UHV and vapor-phase systems that are in good accordance with experiment. The complexity of the solution phase, as well as applied potentials present in the electrochemical environment, however, have hindered most previous theoretical efforts. The rapid and dramatic increases in computational power along with development of more efficient and robust computational algorithms have significantly enhanced our ability to simulate complex electrochemical environments, including aspects of aqueous-phase solvation [20,21,22,23,24,25,26] and surface potential [20,21,25,26,27,28,29,30,31,32,33,34,35]. These theoretical investigations begin to provide a closer complement to the conditions realized experimentally and thus provide a more fundamental basis toward understanding electrochemical and electrocatalytic reactions.

In this chapter, we provide a succinct review of some of the advances in the development and application of ab initio methods toward understanding the intrinsic reactivity of the metal and the influence of the reactive site and its environment. We draw predominantly from some of our own recent efforts. More specifically we describe (a) the chemistry of the aqueous-phase on transition metal surfaces and its influence on the kinetics and thermodynamics within example reaction mechanisms, and (b) computational models of the electrode interface that are able to account for a referenced and tunable surface potential and the role of the surface potential in controlling electrocatalytic reactions. These properties are discussed in detail for an example reaction of importance to fuel cell electrocatalysis: methanol dehydrogenation over platinum(111) interfaces [24,25].

19.2. The Role of the Aqueous Phase

The presence of solution or solvent can appreciably perturb the chemistry of surface-catalyzed reactions compared to their ultra-high vacuum or vapor-phase counterparts. Polar solvents, such as water, are able to stabilize charged intermediate and transition-state species at the surface that are unstable (or less stable) as gas-phase adsorbates, thus altering both the thermodynamics (i.e., reaction energy) and kinetics (i.e., activation barrier) for specific reaction steps. This can influence the activity, as well as the selectivity of the overall catalytic system, and thus control aqueous-phase electrocatalysis. Thiel and Madey [36] and Henderson [37] present exceptional reviews that describe in detail the interactions of water at transition-metal and metal oxide surfaces. We present only some of the salient features of metal–water interactions herein.

Water Adsorption. The adsorption of water on metal surfaces is, in general, rather weak and controlled by competing influences from both metal–water and water–water interactions [36–39]. Metal–water interactions are comprised of electron donation and back-donation between the frontier orbitals of water and the states at or near the Fermi level of the metal, whereas water–water interactions are dominated by H–O–H$\cdots$OH$_2$ hydrogen bonds. At low coverage, H$_2$O typically adsorbs weakly atop a single metal atom site through its oxygen atom in order to enhance the donation of electrons from oxygen into non-bonding surface states. Water takes on a tilted configuration with metal–O bond distances around 2.1–2.3 Å [39]. Michalaelides et al. [38–39] have shown that for monolayers and bilayers of water on Ru, the predominant overlap is between the $2p_x$-type $3a_1$ molecular orbital on water and the d_{z^2} state on the surface Ru atom. In addition, there is overlap between the lone-pair $1b_1$ orbital on water and the Ru d_{z^2} state. The overlap and mixing of the metal surface state with this orbital is significantly stronger. A more detailed charge analysis indicates that there is a charge depletion from the water $1b_1$ orbital and the Ru d_{z^2} states with a charge accumulation on the lower lying d_{xz} and d_{yz} states. Water prefers to adsorb in an atop configuration so as to minimize Pauli repulsive interactions between the lone pair of electrons on water and the filled states of the metal.

Depending on the strength of the metal–water vs. water–water interactions, low-coverage H$_2$O can adsorb as an isolated species if the metal–water interaction is greater than the water–water interaction, as two- and three-dimensional clusters if the metal–water interactions are less than or comparable to the water–water interactions, or dissociate into adsorbed hydroxy or hydride intermediates if the metal–water interaction is much greater than the water–water interaction [37]. Adsorption energies for vapor-phase (low-coverage) H$_2$O have been calculated by DFT to be -0.24 eV on Cu(111), -0.38 eV on Ru(0001), -0.42 eV on Rh(111), -0.33 eV on Pd(111), -0.18 eV on Ag(111), -0.35 eV on Pt(111), and -0.13 eV on Au(111) [39]. The binding energies generally increase (become stronger) as we move from right to left across the periodic table and especially from the noble to Pt-group metals. At higher coverages, adsorbed water can form mono-layers, bi-layers, and three-dimensional water clusters and overlayers in crystalline (ice-like) or amorphous (liquid-like) structures [37], depending on the strength of metal–water vs. water–water bonding interactions. On most close-packed surfaces, H$_2$O adsorption forms the well-known bilayer structure in which the H$_2$O molecules form a hydrogen-bonding network with hexagonal rings that match the registry of the metal surface [37,38].

Water Dissociation. Water dissociation (i.e., water activation) involves a fundamentally different mechanism at low coverage (vapor-phase) compared to high-coverage (liquid-phase) systems. At low coverage, H$_2$O dissociates homolytically to form radical-like intermediates that adsorb as neutral hydroxide and hydrogen species:

$$H_2O_{(ad)} \rightarrow OH_{(ad)} + H_{(ad)} \qquad (19.1)$$

This process involves a metal atom insertion into the O–H bond. We calculated this reaction to be endothermic by $+90\,kJ/mol$ over Pt(111) surfaces with an activation barrier of $+130\,kJ/mol$ [40]. Under UHV conditions, water remains intact. At higher temperatures water will preferentially desorb before it reacts. Water activation can proceed, however, at higher temperatures and higher pressures. A schematic diagram of the homolytic pathway for water dissociation is given on the left-hand-side of Figure 19.1. For high-coverage aqueous-phase systems, however, favorable hydrogen-bonding interactions with nearby H_2O molecules weaken the metal–water bond, yet stabilize the adsorbed OH and H surface intermediates. Under certain conditions, the aqueous phase facilitates water activation by enabling the heterolytic activation of water, where the aqueous phase stabilizes the formation of a solvated proton and the hydroxide intermediate that adsorbs to the metal surface:

$$H_2O_{(ad)} \rightarrow OH^-_{(ad)} + H^+_{(aq)} \rightarrow OH_{(ad)} + H^+_{(aq)} + e^-_{(metal)} \qquad (19.2)$$

The excess charge on the hydroxide is somewhat delocalized over the infinitely polarizable metal surface, hence providing further charge stabilization that enhances this reaction. A schematic diagram for the heterolytic water activation pathway is presented on the right-hand-side of Figure 19.1. For Pt(111), the heterolytic pathway is endothermic by $+75\,kJ/mol$ with an activation barrier of $+90\,kJ/mol$ which is $40\,kJ/mol$ less endothermic (i.e., more favorable) than the homolytic path [40]. The dissociation barrier is significantly lower due to the ability to activate the OH bond without metal insertion, whereby the hydrogen-bonded water network directly accepts the proton that forms as $H_3O^+_{(aq)}$ or $H_5O_2^+_{(aq)}$ (Figure 19.1). Furthermore, it

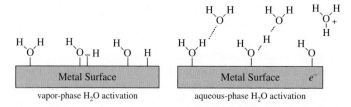

FIGURE 19.1. Left: Schematic diagram depicting the homolytic dissociation mechanism for vapor-phase water activation over a metal surface. Right: Corresponding diagram for the heterolytic dissociation mechanism for aqueous-phase water activation.

has been shown both experimentally [41,42] and from ab initio calculations [40] that adsorbed hydrogen on Pt(111) will be released as $H^+_{(aq)}$ in the presence of liquid water.

The ability of a metal surface to generate protons (as hydronium ions) from adsorbed hydrogen in the presence of water, i.e.,

$$H_{(ad)} + H_2O_{(aq)} + \text{Metal} \rightarrow H_3O^+_{(aq)} + [\text{Metal} + e^-] \quad (19.3)$$

can be explained by a Born-Haber cycle such as presented in Figure 19.2 for this system. The cycle here is comprised of the following steps: (1) the desorption of $H_{(ad)}$ into the gas phase as $H\cdot$ [$\Delta E = E_{b(H\ \text{gas phase})}$], (2) the ionization of $H\cdot$ to $H^+ + e^-$ [$\Delta E = E_{(H\cdot\ \text{ionization})}$], (3) the solvation of H^+ [$\Delta E = E_{(H+\ \text{solvation})}$], and (4) the capture of the electron by the metal surface [$\Delta E = -\Phi$, where Φ is the metal work function]. The energy of proton desorption, $\Delta E_{(H+\ \text{desorption})}$, is

$$\Delta E_{(H+\ \text{desorption})} = E_{b(H\ \text{gas phase})} + E_{(H\cdot\ \text{ionization})} + E_{(H+\ \text{solvation})} - \Phi \quad (19.4)$$

Contributions from $E_{(H\cdot\ \text{ionization})}$ and $E_{(H+\ \text{solvation})}$ are independent of the metal surface. As a result, the prevalence for proton formation is dependent on the strength of the metal–H bond [$E_{b(H\ \text{gas phase})}$] and the ability of the metal surface to accept an electron (i.e., the work function of the metal, Φ) (Figure 19.2). Although $E_{b(H\ \text{gas phase})}$ does vary with the metal surface, the changes are small compared to changes in Φ. Therefore, one can predict whether a metal will form protons from adsorbed hydrogen atoms by the magnitude of the work function. Kizhakevariam and Stuve estimated that proton formation would be exothermic for metals with work functions greater than 4.88 eV [42]. Desai and Neurock confirmed this result with a first principles analysis and determined that water-solvated Pt and Pd which have work functions of 5.08 and 4.88 eV, respectively, lead to

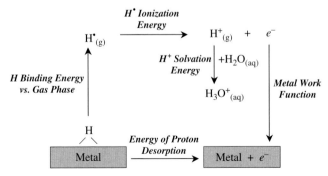

FIGURE 19.2. Born-Haber cycle for proton formation from adsorbed hydrogen in the aqueous phase.

exothermic proton formation energies of -57 and $-14\,\text{kJ/mol}$, respectively, and that water-solvated Ru with a work function of 4.54 eV gave an endothermic proton formation energy of $+40\,\text{kJ/mol}$ [40]. This suggests that both the Pt(111) and the Pd(111) will likely release hydrogen as a proton. Hydrogen adsorbed to Ru, however, is likely to remain on the surface as a hydride.

Following a similar approach, a Born-Haber cycle can be used to approximate the ability of other transition metal surfaces to activate water in the aqueous phase from the energetics of water activation in the vapor phase. This is quite useful since the vapor phase calculations are much less computationally intensive. The Born-Haber cycle for such a reaction scheme is given in Figure 19.3. The heterolytic activation of water over a metal surface is directly tied to the homolytic dissociation of water (Eq. 19.1) on that surface and the ease with which it can form protons from adsorbed hydrogen (Eq. 19.3). The specific steps in the Born cycle presented in Figure 19.3 include (1) the dissociation of H_2O in the vapor phase to form $OH_{(ad)}$ and $H_{(ad)}$ [$\Delta E = \Delta E_{rxn(\text{vapor phase})}$], (2) the desorption of $H_{(ad)}$ into the gas phase as H· [$\Delta E = E_{b(\text{H gas phase})}$], (3) the ionization of H· to form $H^+ + e^-$ [$\Delta E = E_{(\text{H· ionization})}$], (4) the solvation of H^+ [$\Delta E = E_{(\text{H+solvation})}$], and (5) the capture of the electron by the metal surface [$\Delta E = -\Phi$]. The overall reaction energy for heterolytic aqueous-phase water activation, $\Delta E_{rxn(\text{aqueous phase})}$, is:

$$\Delta E_{rxn(\text{aqueous phase})} = \Delta E_{rxn(\text{vapor phase})} + E_{b(\text{H gas phase})} + E_{(\text{H·ionization})} \\ + E_{(\text{H+solvation})} - \Phi \quad (19.5)$$

As for proton formation from adsorbed hydrogen, $E_{(\text{H· ionization})}$, $E_{(\text{H+solvation})}$ and $E_{b(\text{H gas phase})}$ are essentially metal independent (vide supra).

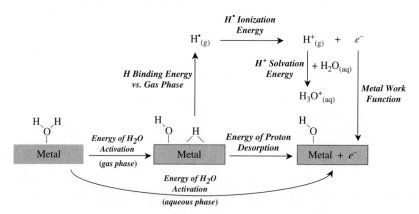

FIGURE 19.3. Born-Haber cycle for aqueous-phase water activation from vapor-phase water activation and proton formation from adsorbed hydrogen.

19. Modeling Electrocatalytic Reaction Systems from First Principles   557

Therefore, the ability of a metal to heterolytically cleave water depends principally on the vapor-phase dissociation energy and the work function of the metal surface (Figure 19.3). The energies for the vapor-phase H_2O dissociation over various (111) surfaces calculated by DFT [43] are listed in periodic fashion in Table 19.1 along with corresponding calculated metal work functions [44]. As anticipated, the vapor-phase dissociation energy becomes less endothermic (more exothermic) as we move from right-to-left across a row and/or as we move up a particular column in the periodic table. The most active metals follow the order of Ni > Ru > Rh > Cu (Table 19.1). The metal work function tends to increase from left-to-right across a row (except for the noble metals), requiring that the optimum metals for aqueous-phase cleavage to have a unique balance between the calculated vapor-phase surface reaction energy ($\Delta E_{rxn(\text{vapor phase})}$) and metal work function (Φ). The combination of these two trends leads to a reversal of Rh and Ru in the ability to activate water in the condensed liquid phase: Ni > Rh > Ru > Cu (Figure 19.3, Table 19.1). The higher work function of Rh ascompared to Ru changes the relative ordering for the aqueous-phase compared to the vapor phase.

Actual ab initio calculations which explicitly examined the reactants and the products at the metal/solution interface yield overall energies for the dissociation of water on Cu and Pt in an aqueous medium to be 0.31 eV and 0.93 eV which are in reasonable agreement with the Born cycle estimates of 0.44 eV and 0.97 eV, respectively.

TABLE 19.1. Periodic trends calculated from DFT for the dissociation energies of water in the vapor-phase over the metal [43] [$\Delta E_{rxn(\text{vapor phase})}$], the work function of the metal [44] [Φ] and estimated water dissociation energies at the metal solution interface.

		Ni (111)	Cu(111)
		ΔE_{rxn} (vap) = -0.47 eV	ΔE_{rxn} (vap) = $+0.28$ eV
		$\Phi = 5.1$ eV	$\Phi = 4.8$ eV
		ΔE_{rxn}(aq) = -0.07 eV	ΔE_{rxn}(aq) = $+0.44$ eV
Ru(0001)	Rh(111)	Pd(111)	Ag(111)
ΔE_{rxn} (vap) = -0.36 eV	ΔE_{rxn} (vap) = $+0.03$ eV	ΔE_{rxn} (vap) = $+0.73$ eV	ΔE_{rxn} (vap) = $+0.93$ eV
$\Phi = 5.0$ eV	$\Phi = 5.2$ eV	$\Phi = 6.0$ eV	$\Phi = 4.6$ eV
ΔE_{rxn} (aq) = $+0.27$ eV	$\Delta E_{rxn} = +0.16$ eV	ΔE_{rxn} (aq) = $+0.85$ eV	ΔE_{rxn} (aq) = $+1.63$ eV
	Ir(111)	Pt(111)	Au(111)
	$\Delta E_{rxn} = +0.74$ eV	ΔE_{rxn} (vap) = $+1.10$ eV	ΔE_{rxn} (vap) = $+1.58$ eV
	$\Phi = 5.6$ eV	$\Phi = 6.0$ eV	$\Phi = 5.2$ eV
		ΔE_{rxn} (aq) = $+0.97$ eV	ΔE_{rxn} (aq) = $+1.10$ eV

In addition to stabilizing charge states, water can also directly participate in elementary physicochemical processes such as surface reaction and diffusion. Water enhances the activation of an X-H bond by solvating the resulting H^+ that forms, thereby reducing the barrier for O–H bond cleavage and facilitating a pathway for proton transfer. To a much weaker degree, water also helps to stabilize the anion that forms. The electron, however, can delocalize through the metal surface. The aqueous phase can also facilitate the apparent transport of adsorbed surface hydroxyl intermediates across a metal surface by providing a conduit by which protons can be conducted. The facile proton transfer between coadsorbed water and hydroxyl intermediates gives rise to the rapid diffusion of hydroxyl intermediates across the surface. This, however, is not due to OH hopping but instead the result of proton transfer via the following symmetric surface reaction:

$$OH_{(ad)[site1]} + H_2O_{(ad)[site2]} \rightarrow H_2O_{(ad)[site1]} + OH_{(ad)[site2]} \quad (19.6)$$

Proton transfer in one direction appears as OH diffusion in the opposite direction which is effectively the Grotthus mechanism. Such behavior has been shown by Vassilev for enhanced OH diffusion in the presence of additional H_2O on Rh(111) [45] and by Desai for the activation of water over the aqueous-phase PtRu alloy [40].

19.2.1. Case Study: Aqueous- vs. Vapor-Phase Methanol Dehydrogenation Over Pt(111)

We have seen that water present in the aqueous solution phase can considerably alter the chemical behavior of water adsorbed at an electrode surface by stabilizing particular charge transfer reactions. In addition, water can significantly influence the pathways and energetics of different electrocatalytic reactions via direct participation. Herein, we examine methanol dehydrogenation over Pt(111) as a case study since it is a central reaction in the electrocatalytic direct methanol fuel cell.

Methanol oxidation over platinum and other platinum-group metals begins with methanol adsorption to the metal surface. This is followed by a sequence of dehydrogenation steps. Two possible pathways have been suggested for the initial dehydrogenation of methanol. The first involves the activation of one of the C–H bonds of methanol (CH_3OH) to form an adsorbed hydroxymethyl ($CH_2OH_{(ad)}$) and adsorbed hydrogen or a solvated proton. The second reaction involves the activation of the O–H bond to form adsorbed methoxy ($CH_3O_{(ad)}$), along with an adsorbed hydrogen or a proton in solution. Both Desai and Neurock [15] and Greeley and Mavrikakis [16,17] have investigated the pathways and energetics of methanol dehydrogenation for dry, vapor-phase methanol over platinum(111) using periodic DFT calculations. A graphical representation of an example super-cell structure for vapor-phase methanol adsorption system is shown on the left-hand-side of Figure 19.4. C–H activation over Pt(111) was found to be significantly more favored than O–H activation. The overall reaction energy for methanol to form an adsorbed hydroxymethyl and a surface hydrogen intermediate, as shown in Eq. (19.9), was found to be exothermic at -16 kJ/mol. The barrier for this reaction was $+90$ kJ/mol.

$$CH_3OH_{(ad)} \rightarrow CH_2OH_{(ad)} + H_{(ad)} \qquad (19.6)$$

The activation of the O–H bond, however, was significantly higher in energy than that for the C–H bond activation. The overall reaction energy for the dehydrogenation of methanol to an adsorbed methoxy and hydrogen intermediates was found to be $+64$ kJ/mol, which is 80 kJ/mol less favorable than that to form the hydroxymethyl intermediate. The activation barrier was also found to be 50 kJ/mol higher than that for the C–H activation path. The

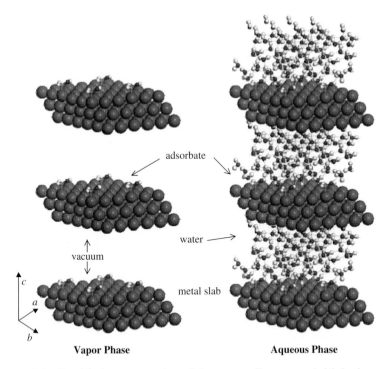

FIGURE 19.4. Graphical representation of the super cell structure (with lattice vector directions, a, b, and c indicated) for vapor-phase (left) and aqueous-phase (right) models of the reaction environment. Both diagrams depict the CH_2OH intermediate with a 1/9 surface coverage in a (3×3) adsorption structure.

most favorable energetics for subsequent $CH_2OH_{(ad)}$ dehydrogenation involve thermoneutral C–H cleavage to form the hydroxymethylene ($CHOH_{(ad)}$) intermediate and a second adsorbed hydrogen. This is followed by the C–H cleavage of the CHOH intermediate to form $COH_{(ad)}$ and $H_{(ad)}$ which was found to be 46 kJ/mol exothermic. The COH intermediate then undergoes a final O–H bond activation in order to form CO. This final reaction was calculated to be exothermic by 51 kJ/mol.

$$CH_3OH_{(ad)} \rightarrow CH_2OH_{(ad)} + H_{(ad)} \rightarrow CHOH_{(ad)}$$
$$+ 2H_{(ad)} \rightarrow COH_{(ad)} + 3H_{(ad)} \rightarrow CO + 4H_{(ad)} \quad (19.7)$$

Wasileski and Neurock [24,25] expanded the work on platinum(111) to include the influence of the aqueous phase by incorporating explicit water molecules within the unit cell to approximate a bulk density of water within the DFT calculations. An example of the aqueous-phase super-cell structure used in these simulations is depicted in the figure on the right-hand-side of Figure 19.4. In these calculations the entire vacuum layer from the vapor-phase cell system was filled with H_2O molecules in order to approximate the density of water between the two metal slabs. The structure of the metal–solution interface for Pt(111) [24,25] was adapted from a previously-determined structure from an ab initio molecular dynamics simulation for water on Pd(111) [46] and involves a hexagonal hydrogen-bonded network of H_2O molecules forming a bilayer structure (vide supra). The inclusion of the aqueous phase expectedly stabilized both the $CH_3OH_{(ad)}$ and $CH_xO_{(ad)}$ species via hydrogen bonding, in addition to enabling the formation of protons over platinum (as discussed for Figure 19.2 above). Therefore, it is expected that methanol dehydrogenation over platinum prefers the heterolytic pathway, as depicted here in Eq. (19.8):

$$CH_3OH_{(ad)} + H_2O_{(aq)} \rightarrow CH_2OH_{(ad)} + H_3O^+_{(aq)} + e^-_{(metal)} \quad (19.8)$$

Okamoto et al. [47] examined the homolytic dehydrogenation of methanol in an aqueous media and found that only the C–H activation step was favorable. They had not, however, considered the heterolytic path. We calculate that the heterolytic activation of methanol is exothermic at −43 kJ/mol [24,25] (Eq. 19.8) and is expected to have a lower activation barrier as shown for heterolytic vs. homolytic water activation [40] (vide supra).

The aqueous-phase dehydrogenation mechanism [24] was found to follow similar paths and relative ordering of the C–H and O–H activation steps as that found for the vapor-phase case [15,16,17], despite the fundamental differences that result from the heterolytic versus homolytic activation. The optimized structures [24,25] and reaction energies [24,25] (including vapor-phase reaction energies [15,16,17] in parentheses) for the aqueous phase methanol decomposition over Pt(111) are presented in Figure 19.5.

The initial C–H activation of methanol to form the $CH_2OH_{(ad)}$ intermediate is favored over the O–H activation to form $CH_3O_{(ad)}$. The path to form the hydroxymethyl ($CH_2OH_{(ad)}$) intermediate is exothermic by $-49\,kJ/mol$, whereas the path to form methoxy ($CH_3O_{(ad)}$) is endothermic by $+48\,kJ/mol$, a difference of $97\,kJ/mol$. The activation of the C–H bond in the aqueous-phase ($-49\,kJ/mol$) is considerably favored over its vapor-phase activation on Pt(111) ($-16\,kJ/mol$) due to the favorable solvation energy of H^+ to form $H_3O^+_{(aq)}$, as well as the stabilization of the $CH_2OH_{(ad)}$ intermediate from hydrogen bonding to the solution phase. The subsequent heterolytic dehydrogenation of hydroxymethyl intermediate ($CH_2OH_{(ad)}$) follows a second C–H activation step to form the hydroxymethylene ($CHOH_{(ad)}$) intermediate ($-46\,kJ/mol$), followed by O–H cleavage to form the formyl species ($CHO_{(ad)}$) ($-60\,kJ/mol$). The formyl intermediate undergoes one last C–H activation step in order to form $CO_{(ad)}$ ($-80\,kJ/mol$). This results in the following preferred dehydrogenation pathway of:

$$CH_3OH_{(ad)} \rightarrow CH_2OH_{(ad)} + H^+_{(aq)} \rightarrow CHOH_{(ad)}$$
$$+2H^+_{(aq)} \rightarrow CHO_{(ad)} + 3H^+_{(aq)} \rightarrow CO + 4H^+_{(aq)} \quad (19.9)$$

The activation of the C–H bond of $CHOH_{(ad)}$ to form COH ($-48\,kJ/mol$), followed by O–H cleavage to $CO_{(ad)}$ ($-100\,kJ/mol$), is also a viable pathway.

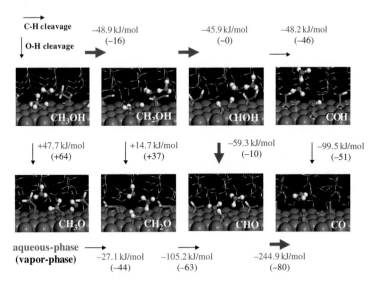

FIGURE 19.5. Optimized structures of CH_xO species, as indicated, over aqueous-solvated Pt(111) as determined by DFT [24,25]. Horizontal and vertical arrows indicate C-H and O-H cleavage steps, respectively, with indicated reaction energies for the aqueous phase [24,25] and vapor phase [15,16,17] (given in parentheses). The preferred aqueous-phase pathway is indicated by bold arrows.

In general, both the vapor phase as well as the aqueous phase reaction pathways indicate that C–H bond activation is preferred over O–H activation for most of the surface intermediates. The crossover to O–H activation appears to require the activation of some of the C–H bonds first. This case study demonstrates the strong influence of the aqueous phase on determining both the fundamental pathway and energetics of electrocatalytic processes.

19.3. The Role of the Surface Potential

A primary factor that controls the structure and chemistry of the electrochemical interface is the surface potential and the corresponding potential drop across the double-layer. A schematic diagram of the electrochemical double-layer is given in Figure 19.6. The electrochemical interface is simply a metal–solution interface where ions or molecules diffuse from the solution and interact with the metal electrode (a more elegant and detailed description is given in the text by Bard and Faulkner [48]). The external or intrinsic potential of the electrode causes a charge polarization at the surface, resulting from an excess of electrons or electron holes at the surface. The magnitude of the excess charge depends on the magnitude of the potential, ϕ, relative to the potential of zero charge, ϕ_{pzc}, (an inherent property of the metal substrate in its environment). By convention, if ϕ is less than or more negative of ϕ_{pzc}, the excess surface charge is negative. Molecules or ions from the solution phase can specifically adsorb to the metal surface along with co-adsorbed solvent molecules (for example, carbon monoxide and water, respectively, in Figure 19.6). Just beyond the adsorption layer is a layer of solvated counter ions having the opposite charge to the surface polarization. These counter ions are located at the outer Helmholtz plane (OHP) and define the boundary of the inner layer (Figure 19.6). The thickness of the inner layer, x_{il}, depends on the size of the adsorbate, solvent molecules, and counter ion and is usually a few Ångstroms in magnitude. Beyond the OHP is the diffuse layer, composed of solvent and bulk concentrations of supporting electrolyte (anion and cationic) and other solvated molecular species. Essentially all of the excess charge at the surface is canceled by the counter charge at the OHP, generating an approximately linear potential drop across the double layer with an electric field on the order of 10^8 V cm^{-1}.

Modeling the electrochemical system from first principles presents a considerable challenge. First, the system potential is controlled by combined chemical interactions at the anode and cathode, each with their respective double layer and "connected" via macroscopic distances of metallic conductor and solution. Such a scale is not accessible to quantum-mechanical calculations. This dilemma is typically resolved, however, by modeling the anode and cathode separately as artificially charged half-cells or approximate models of half-cells. Second, quantum-mechanical calculations are usually performed using a canonical ensemble, in which the number of electrons is the

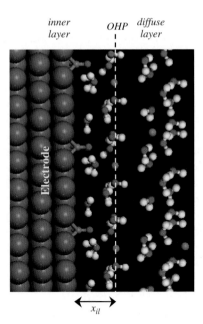

FIGURE 19.6. Schematic diagram of the electrochemical metal–solution interface, with the electrode, inner layer, diffuse layer, outer Helmholtz plane (OHP), and inner-layer thickness x_{il} indicated.

controlled variable and remains constant throughout the simulation. Electrochemical systems, however, involve a constant chemical potential (which is maintained by non-Faradaic current flow to the electrode surface within the chemical cell). A simulation that treats the chemical potential as the controlled variable requires a grand canonical ensemble description which cannot currently be achieved quantum-mechanically due to the size and complexity of the full electrochemical system that would be needed.

We can, however, begin to model the system instead as an electrochemical half-cell using the canonical ensemble. In order to do so, we must (1) enable the system potential to be extracted from the calculation such that it can be related to a known reference potential and (2) allow the system potential to be accurately controlled in order to access a range of electrochemical conditions. A variety of approximate quantum-mechanical approaches have been developed to extract the influence of the electrochemical potential on the surface chemistry (and reactivity) of adsorbed species. These methods, briefly reviewed below, typically differ from one another in the model of the electrode surface and how the potential is applied to the electrode surface.

Some of the first quantum-mechanical descriptions of the electrochemical interface were developed by Anderson, utilizing an atom superposition and electron delocalization molecular orbital method (ASED-MO) to generate potential surfaces for the reactant and product states using a cluster model of the electrode surface [27,32,34,49,50]. The activation barrier and equilibrium potential were extracted from the potential surfaces in a manner reminiscent of Marcus theory [51]. Recently, this approach has been elaborated and recast in terms of ab inito electronic structure theory to include influences of the solvent and electrolyte through the use of Madelung potentials and by adding explicit solvent molecules to the simulation [52–54]. Some of the shortcomings of this approach are its neglect of the electronic structure due to the extended electrode surface and the lack of potential-dependent effects on bond polarization and changes to the complex reaction environment.

Much of the theoretical effort to understand molecular structure at the metal–water interface is contained in the work of Halley [21,22,28,29,55]. A tight-binding molecular dynamics method was developed in which the electrons in the metal are treated via first principles calculations. The water phase was treated classically and coupled to the metal such that the electronic structure is matched at the metal–water interface and the double-layer was appropriately accounted for. The structure and orientation of water at the interface and the nature of the potential drop were investigated [21,22,55]. The results agree with experimental observations of the extent of water orientation at the electrode–electrolyte interface [20].

Another approach to investigate the influence of electrode potential is to simulate the potential drop across the double layer by applying an external homogeneous field (of a magnitude of $\sim 10^8$ V/cm) across the electrode surface (usually modeled as a finite cluster of metal atoms) to investigate the potential-dependent vibrational behavior of chemisorbates, such as carbon monoxide [30,31,33,56], and the adsorption behavior of H_2O and OH species [57–59]. The external electric field polarizes the surface, influencing surface adsorption. The potential is tuned with the strength of the applied field, but relating this potential to a known reference is not straightforward [60]. Illas et al. have coupled this approach with molecular dynamics to describe electron transfer through Marcus theory [61].

Nørskov et al. utilized the H_2/H^+ electrochemical couple to determine potential-dependent reaction and activation energies for electrochemical processes on Ru(111) [35]. This method is unique in that a reference potential is implicit within the calculation. The approach, however, does not take into consideration the potential-dependent changes in surface charge density. More recent calculations indicate that while the polarization is important in an absolute sense, the changes in reaction energies are not as sensitive provided the reactant and products have similar directional dipoles [62] Lozovoi and Alavi have developed a method to generate an internal potential reference from a homogeneous or arbitrarily shaped background charge within the electronic structure calculation using periodic calculations of the metal–vapor phase interface [63,64] In this approach, the potential is tuned by adding or removing electrons from the metal surface slab. Charge neutrality is maintained by the background charge which acts to terminate the electric field within the vacuum region of the periodic cell. The point of field termination acts as a reference to define the chemical potential of the surface.

A similar approach was developed by Filhol and Neurock and extended to metal–solution interfaces [26,46] through the implementation of two internal potential references within periodic density functional theory calculations. The first is a vacuum reference, which enables the surface potential to be related to the vacuum scale. The second is an aqueous reference state used to relate the potential to the vacuum scale in cases where the surface potential is varied by external charging. In this method, a homogeneous background charge is utilized, which subsequently polarizes the aqueous phase and metal slab to generate a pseudo-double layer. Chemical transformations can be investigated by calculating the free energy vs. potential profile for reactant, intermediate and product states, thereby enabling reaction energetics to be determined at constant surface potential.

In this section, the methodology adopted by Filhol and Neurock is discussed in greater detail, particularly with respect to establishing the reference potential and determining electrochemical potential-dependent structural and energetic transformations at the metal–solution interface. A more complete discussion of the approach can be found in the paper by Taylor et al. [26]. Methanol dehydrogenation over Pt(111) is used once again as a case study to examine potential-dependent chemical reactions.

19.3.1. Determining an Internal Reference

An integral part of exploring electrocatalytic behavior from computational methods is the ability to relate the surface potential to a known reference state. The Filhol and Neurock approach enables such a relation by incorporating a vacuum reference state within the aqueous-phase super cell. This is accomplished by bisecting the aqueous layer of a previously-optimized aqueous-phase system with the addition of a 15–20 Å layer of vacuum, as illustrated in the right-hand portion of Figure 19.7. Plotted in Figure 19.8 is the average electrostatic potential vs. distance normal to surface for the vacuum reference system (Figure 19.7, right-hand side), determined by averaging the three-dimensional potential map of electrostatic potential along the surface normal. The potential fluctuations appearing as "wells" in Figure 19.8 are due to the local electronic structure of the metal surface atoms, the CH_2OH intermediate and the H_2O molecules in the aqueous phase. The portion of the potential vs. distance across the vacuum layer is linear and has a non-zero slope due to the asymmetry of the periodic slab. The potential at the center of the vacuum layer, ϕ_{vacuum} (i.e., at the position bisecting the upper and lower surfaces in adjacent periodic cells) defines the vacuum reference (which is the potential of the free electron) and is defined to be 0 V. The potential at all other locations in the cell are shifted linearly, including the potential of the center slab layer, ϕ_{metal} (Figure 19.8).

The potential at the center of the slab, ϕ_{metal}, is now referenced with respect to vacuum and can be used as a reference point in the initial aqueous-phase optimization (illustrated in the left-hand-side of Figure 19.7). Given in Figure

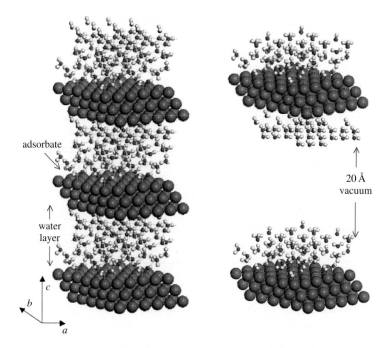

FIGURE 19.7. Graphical representation of the super cell structure (with lattice vectors, a, b, and c) for the internal vacuum reference (right) from the optimized aqueous-phase structure (left). Both diagrams depict the CH_2OH intermediate with a 1/9 surface coverage in a (3×3) adsorption structure.

19.9 is a plot of average potential vs. distance normal to the surface plane for the aqueous-phase system. The potential for the center slab layer has been set to the same value determined from the vacuum reference (i.e., from Figure 19.8), and all other potentials are shifted linearly by a value $\Delta\phi_{shift}$. The surface potential is the potential of the electrons at the highest energy state (i.e., at the Fermi level, E_{Fermi}), defined as the Fermi potential, ϕ_{Fermi}, where

$$\phi_{Fermi} = \frac{E_{Fermi}}{e} - \Delta\phi_{shift} \qquad (19.10)$$

and e is the charge of an electron. Since the ϕ_{metal} is referenced vs. the vacuum scale, ϕ_{Fermi} is also referenced vs. the vacuum scale and can be converted to the scale of the normal hydrogen electrode via the relationship [65]

$$\phi_{NHE} = -4.8V - \phi_{Fermi} \qquad (19.11)$$

This method enables the potential vs. vacuum for any aqueous-phase system (with no additional external charge) to be calculated directly from the electronic and atomic structure.

19. Modeling Electrocatalytic Reaction Systems from First Principles 567

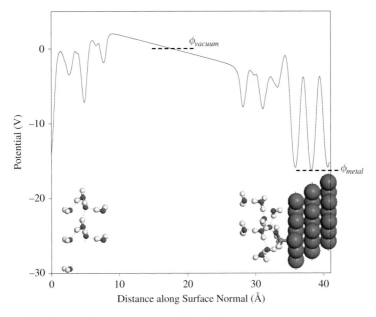

FIGURE 19.8. Plot of the average potential vs. distance along surface normal for the (CH_2OH + H) vacuum reference, with the vacuum reference potential (ϕ_{vacuum}) and potential of the center slab layer (ϕ_{metal}) indicated.

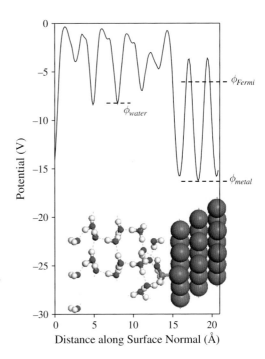

FIGURE 19.9. Plot of the average potential vs. distance along surface normal for the (CH_2OH + H) system, with the potential of the center slab layer (ϕ_{metal}), the Fermi potential (ϕ_{Fermi}) and the aqueous reference potential (ϕ_{water}) indicated.

The Filhol and Neurock methodology also enables the surface potential to be varied to enable a wide range of potential-dependent behavior to be investigated. The potential is varied by applying an external charge (q) to the system. The previous determination of a vacuum reference applies only to the $q = 0$ calculations (i.e., no external charge). Periodic boundary conditions allow q to be non-integer, which enables the potential to be tuned on a very fine scale. The corresponding background (with a total charge $q_{bg} = -q$) is applied as a homogeneous grid of point charges across the entire unit cell (similar to the vacuum-phase methods [63,64]) and simulates the electrochemical double layer [26]. Since ϕ_{metal} necessarily varies with q, an internal vacuum reference cannot be determined for cases when $q \neq 0$. Instead, a second reference is established within the aqueous phase. The water molecules in the center of the unit cell will experience little effect from charge polarization at the metal–solution interface, since this region experiences little to no electric field [26,63,64]. As a result, the potential in this region will typically not vary with q. This phenomenon is exploited such that an aqueous-phase potential reference (labeled ϕ_{water} in Figure 19.9) can be established and is referenced to the vacuum scale via the internal vacuum reference described in Figure 19.8.

For example for a calculation at $q \neq 0$, the potential at all positions in the cell is shifted by a value $\Delta\phi_{shift}(q)$ such that the potential of the water molecule(s) in the center of the cell are equal to ϕ_{water}. The calculated Fermi potential is also shifted by $\Delta\phi_{shift}(q)$ as in:

$$\phi_{Fermi(q)} = \frac{E_{Fermi(q)}}{e} - \Delta\phi_{shift(q)} \qquad (19.12)$$

so that the surface potential referenced to vacuum can be determined for any q and related to the scale of the normal hydrogen electrode via Eq. (19.14).

19.3.2. Establishing the Potential-Dependent Free Energy

Inclusion of the homogeneous background unavoidably influences the system free energy because of interactions between the charge of the background with the charge in the slab. As a result, the total energy calculated by DFT, E_{DFT}, includes a number of contributions for a system with an external charge q and background charge q_{bg} (where $q_{bg} = -q$)

$$E_{DFT}(q, q_{bg}) = E_{slab}(q) + E_{slab-bg}(q, q_{bg}) + E_{bg}(q_{bg}) \qquad (19.13)$$

where E_{slab} is the energy of the slab without the background, E_{bg} is the energy of the background without the slab and $E_{bg-slab}$ is the interaction energy between the slab and the background. To decouple the influence of the background from the energetics of the slab, $E_{slab-bg(q,qbg)}$ and $E_{bg}(q_{bg})$ must be subtracted from Eq. (19.13). In addition, corresponding influences from the additional external charge at the Fermi level must be removed in order to relate the calculated energies for various q on the same scale. The sum of $E_{slab-bg}(q,q_{bg})$ and $E_{bg}(q_{bg})$ is equal to the integrated average internal potential of the unit cell, $-\int_0^q \Delta\phi_{shift(Q)}dQ$. The energy of the excess charge at the Fermi level is equal to $q\phi_{Fermi}$ Therefore, the total free energy of the system can be determined from E_{DFT}, $\Delta\phi_{shift}(q)$, q and ϕ_{Fermi} as:

$$E_{Free} = E_{DFT} + \int_0^q \Delta\phi_{shift(Q)}dQ - q\phi_{Fermi} \qquad (19.14)$$

The Filhol and Neurock method [26,46] enables the system potential to be related to a known reference scale and varied to encompass a range of potential conditions for atoms, molecules, and ions in an aqueous-phase electrochemical environment. Although this formalism is a clear advancement in modeling electrochemical interactions on the surface (i.e., electrocatalysis), it only enables an estimate of the surface potential due to different approximations necessary. First, the estimated potential at $q = 0$ is dependent on the position of the 20 Å vacuum layer within the aqueous layer, which is an unavoidable consequence of the finite size of the aqueous layer. The orientation of the hydrogen-bonded H_2O network induces a propagating dipole moment perpendicular to the surface and across the entire unit cell, which results in a non-constant potential–distance profile along the vacuum region in the vacuum reference cell (Figure 19.8). Changes in the location of the vacuum layer cause a variation in the value of ϕ_{vacuum} and hence the relative potential vs. the vacuum scale, ϕ_{Fermi}. Comparisons of similar systems using identical reference conditions produce relative potential differences that have accuracies that are on the order of to 0.1–0.2 V. Second, the $q = 0$ potential refers to the potential of the system at zero external charge. However, this potential estimate is not comparable to the potential of zero charge, ϕ_{pzc}. The computations refer to a simulation (and therefore a potential estimate) at zero Kelvin. Experimentally measured values of ϕ_{pzc} refer to a dynamic aqueous system. Estimation of the experimental ϕ_{pzc} using the Filhol and Neurock method can be derived by time-averaging the calculated surface potential over a very long ab initio molecular dynamics run, as in the work by Taylor and Neurock [66].

19.3.3. Case Study: Potential-Dependent Methanol Dehydrogenation over Aqueous-Phase Pt(111)

The Filhol and Neurock method [26,46] for investigating the potential dependence of electrocatalytic reaction behavior was extended and used by Wasileski and Neurock [25] to examine the mechanism for the dehydrogenation methanol over Pt(111). The salient features of this study are presented here as a case study.

Electrocatalysis involves a catalytic reaction which takes place at a constant potential determined by the combined chemical interactions at the anode and cathode. In order to assess the reactivity at a particular potential, the energy vs. potential profile must be determined for the reactants and products at various external charges so as to be able to interpolate the energy at a particular potential. For example, the first step in the electrocatalytic dehydrogenation of methanol involves methanol adsorption followed by a series of C–H and O–H dissociation steps. Plotted in Figure 19.10A is the energy (determined by Eq. 19.17) as a function of estimated potential, where the potential is determined from aqueous-phase systems with adsorbed methanol ($CH_3OH_{(ad)}$, circles), hydroxymethyl ($CH_2OH_{(ad)}$, squares), and methoxy ($CH_3O_{(ad)}$, triangles) at various charges q (using the vacuum and aqueous reference states described in Figures 19.8 and 19.9). Since the adsorption structure is necessarily different for each adsorbate species, the corresponding potential at $q = 0$ is also different; the potential for the $q = 0$ systems are indicated as solid points in Figure 19.10A. This illustrates the complexity in using the canonical ensemble in that constant-potential behavior cannot be directly inferred from calculations at identical q.

Plotted in Figure 19.10B is the reaction energy that corresponds to the activation of the C–H or O–H bonds of methanol, as indicated, determined by subtracting the best-fit lines to the energy vs. potential plots in Figure 19.10A. Similar to the aqueous-phase behavior discussed above, C–H activation is thermodynamically favored over O–H activation in the potential range of -0.5 to $+1.0$ V vs. NHE (Figure 19.10B) [25]. However, the stability of the methoxy surface intermediate is much more potential dependent than that for the hydroxymethyl species (i.e., the energy vs. potential slope is greater for CH_3O), indicating that O–H activation may become a competing path toward higher potentials. This behavior is in agreement with experimental findings by Cao, Lu, and Wieckowski [25] which show that methanol dehydrogenation involves a dual-path mechanism on Pt(111) at potentials greater than $+0.35$ V vs. NHE.

Plotted in the remaining portions of Figure 19.10 are potential-dependent energies and reaction energies for O–H and C–H activation steps involved in the second, third, and fourth consecutive methanol dehydrogenation steps to form carbon monoxide. Following the most preferred pathway, methanol dehydrogenation involves [25]

$$CH_3OH_{(ad)} \to CH_2OH_{(ad)} + H^+_{(aq)} \to CHOH_{(ad)}$$
$$+2H^+_{(aq)} \to CHO_{(ad)} + 3H^+_{(aq)} \to CO_{(ad)} + 4H^+_{(aq)} \quad (19.15)$$

identical to the most favorable path determined in the aqueous phase (i.e, at a constant charge of $q = 0$ and not at constant potential). However, toward

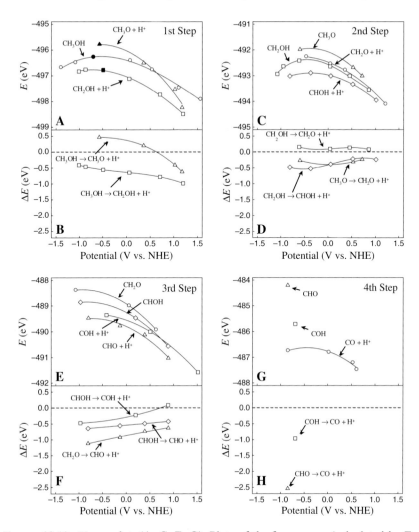

FIGURE 19.10. *Upper plots* (A, C, E, G): Plots of the free energy (calculated by Eq. 19.14 from DFT) vs. estimated potential for reactants and products involved the first, second, and fourth consecutive methanol dehydrogenation steps, as indicated, over Pt(111). Solid symbols in plot A refer to the energy and potential for the system at $q = 0$. *Lower plots* (B, D, F, H): Corresponding plots of the reaction energy for the first, second, third, and fourth consecutive methanol dehydrogenation steps, as indicated. Plots were extracted from Reference [25].

positive potentials, a parallel path involving initial O–H cleavage to form methoxy followed by C–H cleavage to form formaldehyde ($CH_2O_{(ad)}$) may become favorable; this path involves

$$CH_3OH_{(ad)} \rightarrow CH_3O_{(ad)} + H^+_{(aq)} \rightarrow CH_2O_{(ad)} + 2H^+_{(aq)} \quad (19.16)$$

Formaldehyde is a stable species in solution and may desorb or continue to dehydrogenate via

$$CH_2O_{(ad)} \rightarrow CHO_{(ad)} + H^+_{(aq)} \rightarrow CO_{(ad)} + 2H^+_{(aq)} \quad (19.17)$$

The variation in the preferred pathway(s) toward higher potentials is due to differences in the orbital interactions of the $CH_2OH_{(ad)}$ and $CH_3O_{(ad)}$ species (binding to the surface via the carbon and oxygen atoms, respectively) and the platinum surface. As the potential becomes more positive, the electronic band structure of the surface is depleted and the surface is more capable of accepting electron density from the adsorbate, thus resulting in a greater interaction energy. The electron-donation interactions of CH_3O are able to adapt more favorably with the changing potential than CH_2OH, thereby producing a steeper energy vs. potential curve toward higher potentials (Figure 19.10A).

19.4. Conclusions

Elucidating the mechanisms that govern electrocatalytic systems from first principles presents a tremendous challenge due to the complexity of the electrochemical environment, but is critical for the development of strategies for the rational design of new electrocatalytic materials with improved performance. The major difference between an electrocatalytic (or electrochemical) reaction environment and those occurring in the vapor phase or on a single crystal surface under UHV conditions is the influence of solvation (namely water) and the electronic character of the electrode surface (measured as a surface potential). Both of these features dictate the structure as well as the properties of the electrochemical double layer. The chemical interactions within the double layer subsequently influence the adsorption, desorption, diffusion and the mechanisms which control catalytic reactivity. Various ab initio methods have been introduced to treat the electrochemistry and begin to provide realistic representations of the atomic as well as the electronic structure of the double layer, appropriate electronic changes that result upon changes in the potential, and reactivity that follow experimental behavior.

In the case study for methanol dehydrogenation, ab initio electrochemical calculations identified the presence of a dual reaction pathway that proceeds through the formation of the hydroxymethyl intermediate as the dominant path to form CO and the formation of the methoxy intermediate as the path to form formaldehye. Differences in the electronic structure of the adsorbed methoxy and hydroxymethyl ultimately control the onset of the dual path at potentials lower than about 0.5 V. While these results are encouraging, theory has only just begun to tackle the complexity of the electrocatalytic systems.

The influence of the reaction conditions at which the fuel cell is operated; the dynamics of the solution phase present at the metal surface; the influence of the carbon support on the metal and the solution phase; and the size, shape and morphology of the supported nanoparticles that are used can all significantly alter the performance of these material. The influence from these extrinsic factors are critical and will likely be a significant focus of the future efforts from the theory community.

References

[1] Weaver, M. J., Zou, S., Chan, H. Y. H., *Anal. Chem.*, **71** 38A (2000).
[2] Iwasita, T., Nart, F. C., *Prog. Surf. Sci.*, **55** 271 (1997).
[3] Osawa, M., *Topics in Appl. Phys.*, **81** 163 (2001).
[4] Tadjeddine, A., Peremans, A., *J. Electroanal. Chem.*, **409** 115 (1996).
[5] Villegas, I., Weaver, M. J., *J. Chem. Phys.*, **101** 1648 (1994).
[6] Tao, N. J., Li, C. Z., He, H. X., *J. Electroanal. Chem.*, **492** 81 (2000).
[7] Bewick, A., Kunimatsu, K., *Surf. Sci.*, **101** 131 (1980).
[8] Cruickshank, B. J., Sneddon, D. D., Gewirth, A. A., *Surf. Sci. Lett.*, **281** 308 (1993).
[9] Kunze, J., Maurice, V., Klein, L. H., Strehblow, H.-H., Marcus, P., *Corrosion Sci.*, **46** 245 (2004).
[10] Niaura, G., *Electrochim. Acta*, **45** 3507 (2000).
[11] Wilms, M., Broekmann, P., Stuhlmann, C., Wandelt, K., *Surf. Sci.*, **416** 121 (1998).
[12] van Saten, R. A., Neurock, M., *Catal. Rev. Sci.-Eng.*, **37** 357 (1995).
[13] Hammer, B., Norksov, J. K., *Adv. Catal.*, **45** 71 (2000).
[14] Hammer, B., Morikawa, Y., Norksov, J. K., *Phys. Rev. Lett.*, **76** 2141 (1996).
[15] Desai, S. K., Neurock, M., Kourtakis, K., *J. Phys. Chem. B*, **106** 2559 (2002).
[16] Greeley, J., Mavrikakis, M., *J. Am. Chem. Soc.*, **124** 7193 (2002).
[17] Greeley, J., Mavrikakis, M., *J. Am. Chem. Soc.*, **126** 3910 (2004).
[18] Loffreda, D., Simon, D., Sautet, P., *Surf. Sci.*, **425** 68 (1999).
[19] Eichler, A., Hafter, J., *J. Chem. Phys.*, **109** 5585 (1998).
[20] Toney, M. F., Howard, J. N., Richer, J., Borges, G. L., *Nature*, **368** 444 (1994).
[21] Halley, J. W., Mazzolo, A., Zhou, Y., Price, D., *J. Electroanal. Chem.*, **450** 273 (1998).
[22] Halley, J. W., Smith, B. B., Walbren, S., Curtiss, L. A., Rigney, R. O., Sutjianto, A., Hung, N. C., Yonco, R. M., Nagy, Z., *J. Chem. Phys.*, **110** 6538 (1999).
[23] Desai, S. K., Pallassana, V., Neurock, M., *J. Phys. Chem. B*, **105** 9171 (2001).
[24] Neurock, M., Wasileski, S. A., Mei, D., *Chem. Eng. Sci.*, **59** 4703 (2004).
[25] Cao, D., Lu, G.-Q., Wieckowski, A., Wasileski, S. A., Neurock, M., *J. Phys. Chem. B*, **109** 11622 (2005).
[26] Taylor, C. D., Wasileski, S. A., Fanjoy, J. W., Filhol, J.-S., Neurock, M., *Phys. Rev. B*, **73** 165402 (2006).
[27] Anderson, A. B., *Surf. Sci.*, **105** 159 (1981).
[28] Halley, J. W., Johnson, B., Price, D., Schwalm, M., *Phys. Rev. B*, **31** 7695 (1985).
[29] Halley, J. W., Price, D., *Phys. Rev. B*, **35** 9095 (1987).
[30] Head-Gordon, M., Tully, M., *Electrochim. Acta*, **175** 37 (1993).

[31] Illas, F., Mele, F., Curulla, D., Clotet, A., Ricart, J. M., *Electrochim. Acta*, **44** 1213 (1998).
[32] Albu, T. V., Anderson, A. B., *Electrochim. Acta*, **46** 3001 (2001).
[33] Wasileski, S. A., Koper, M. T. M., Weaver, M. J., *J. Am. Chem. Soc.*, **124** 2796 (2002).
[34] Anderson, A. B., *Electrochim. Acta*, **48** 3743 (2003).
[35] Nørksov, J. K., Rossmeisl, J., Logadottir, A., Lindqvist, L., Kitchin, J. R., Bligaard, T., Jonsson, H., *J. Phys. Chem. B*, **108** 17886 (2004).
[36] Theil, P. A., Madey, T. E., *Surf. Sci. Rep.*, **7** 211 (1987).
[37] Henderson, M. A., *Surf. Sci. Rep.*, **46** 1 (2002).
[38] Michaelides, A., Alavi, A., King, D. A., *J. Am. Chem. Soc.*, **125** 2746 (2003).
[39] Michaelides, A., Ranea, V. A., de Andres, P. L., King, D. A., *Phys. Rev. Lett.*, **90** 216102 (2003).
[40] Desai, S. K., Neurock, M., *Phys. Rev. B*, **68** 75420 (2003).
[41] Wagner, F. T., Moylan, T. E., *Surf. Sci.*, **206** 187 (1988).
[42] Kizhakevariam, N., Stuve, E., *Surf. Sci.*, **275** 223 (1992).
[43] Desai, S. K., Neurock, M., unpublished results (2001).
[44] Wasileski, S. A., Neurock, M., *unpublished results*, (2005).
[45] Vassilev, P., Koper, M. T. M., van Saten, R. A., *Chem. Phys. Lett.*, **359** 337 (2002).
[46] Filhol, J.-S., Neurock, M., Angew. Chemie Int. Ed., accepted, 2005.
[47] Okamoto, Y., Sugino, O., Mochizuki, Y., Ikeshoji, T., Morikawa, Y., *Chem. Phys. Lett.*, **377** 236 (2003).
[48] Bard, A. J., Faulkner, L. R. *Electrochemical Methods: Fundamentals and Applications*; John Wiley & Sons: New York, (1980).
[49] Anderson, A. B., *J. Chem. Phys.*, **62** 1187 (1975).
[50] Anderson, A. B., *Int. J. Quant. Chem.*, **49** 581 (1994).
[51] Marcus, R. A., *J. Chem. Phys.*, **24** 966 (1956).
[52] Anderson, A. B., *J. Electrochem. Soc.*, **147** 4229 (2002).
[53] Anderson, A. B., Neshev, N. M., Sidik, R. A., Shiller, P., *Electrochim. Acta*, **47** 2999 (2002).
[54] Anderson, A. B., Sidik, R. A., Narayanasamy, J., *J. Phys. Chem. B*, **107** 4618 (2003).
[55] Price, D. L., Halley, J. W., *J. Chem. Phys.*, **102** 6603 (1995).
[56] Koper, M. T. M., van Saten, R. A., *J. Electroanal. Chem.*, **476** 64 (1999).
[57] Koper, M. T. M., van Saten, R. A., *J. Electroanal. Chem.*, **472** 126 (1999).
[58] Patrito, E. M., Paredes-Olivera, P., *Surf. Sci.*, **527** 149 (2003).
[59] Paredes-Olivera, P., Patrito, E. M., *J. Phys. Chem. B*, **105** 7227 (2001).
[60] Koper, M. T. M., *Modern Aspects of Electrochemistry*, **36** 51 (2003).
[61] Dominguez-Ariza, D., Hartnig, C., Sousa, C., Illas, F., *J. Chem. Phys.*, **121** 1066 (2004).
[62] Rossmeisl, J., J.K. Nørskov, C. Taylor and M. Neurock, *J. Phys. Chem. B*, **110** 21833.
[63] Lozovoi, A. Y., Alavi, A., Kohanoff, J., Lynden-Bell, R. M., *J. Chem. Phys.*, **115** 1661 (2001).
[64] Sanchez, C. G., Lozovoi, A. Y., Alavi, A., *Mol. Phys.*, **102** 1045 (2004).
[65] Wagner, F. T. In *Structure of Electrified Interfaces*; Lipkowski, J., Ross, P., Eds.; VDH Publishers: New York, (1993).
[66] Taylor, C. D., Neurock, M., *Curr. Opin. Solid State Mater. Sci.*, **9** 49 (2005).

Index

Ab initio molecular dynamics (AIMD), 406, 487
ACL, see Anode catalyst layer (ACL)
Active sites, 51, 53–54
Adiabatic connection principle, 441
Adsorption
 of carbon monoxide, 488–498
 energies for vapor-phase, 553
 of OH, 500–501
Adzic, R. R., 526–527
Agapito, L. A., 509
AIMD, see Ab initio molecular dynamics (AIMD)
Air entry pressure, 303
Analytical models of PEMFC, 199–200
 cell prototype and quasi-2D approach, 202–203
 1D + 1D model, humidified membrane, 213
  dynamics of cell performance degradation, 220–229
  finite oxygen stoichiometry, effect of, 214–219
  flow velocity in channel, 213–214
  local polarization curves, 219–220
 1D + 1D model, water management
  constant inlet flow rate of oxygen, 229–240
  constant oxygen stoichiometry, 240–243
 one-dimensional model of, 203
  cathode catalyst layer, performance of, 204–211
  voltage loss due to oxygen transport, 211–213
 voltage losses, 200–202
Andreaus, B., 41
Anode catalyst layer (ACL), 41
Anode electrocatalysis
 adsorption of carbon monoxide, 488–498
 adsorption of OH, 500–501
 electro-oxidation of carbon monoxide, 498–500
 methanol, 501–502
Aqueous-phase dehydrogenation mechanism, 560
Aqueous-phase solvation, 552–553
 water adsorption, 553
 water dissociation, 553–558
Aqueous- vs. vapor-phase methanol dehydrogenation over Pt(111), 558–562
Arrhenius dependence, 182
Arvia, A. J., 524
Atomistic structural modelling of ionomer membrane morphology, 413–433
 application to PFSI membrane morphology, 416–418
 atomistic computational studies of PFSIs, 418
 empirical valence bond models, 430–431

Atomistic structural (*continued*)
  molecular mechanics models
    monomeric sequence in polymeric models, effect of, 427–430
    oligomeric models, 420–421
    polyelectrolyte models, 418–420
    polymeric models, 423–427
  molecular orbital and density functional models, 432
  PFSI membranes, 413–416
Auhl, R., 422
Autonomous oscillations in STR PEMFC, 114–115

Back diffusion, 230
  flux, 158
Bagus, P. S., 491
Balbuena, P. B., 509
Bard, A. J., 526, 562
Benziger, J., 91, 109, 114
Berendsen, H. J. C., 421
Berg, P., 238, 254
Bernardi and Verbrugge's model, 29
BFM, *see* Bond fluctuation model (BFM)
BFM to PEM, specialization of, 136–137
Bifunctional effect, 499
Bimetallic nanoclusters and oxygen dissociation in acid medium, 509–512
Binary and ternary alloys, 31
Binary friction conductivity model, 142
  conductivity model, 142–143
  experimental data, 143–144
  fitting conductivity, 144–148
Binary friction membrane transport model (BFM2), 124, 136–137
  diffusion coefficients, 140–142
  in fuel cell model, implementation, 148
    membrane transport on fuel cell performance, effect of, 148–151
    PEM fuel cell model, 148
  magnitude of coefficients for driving force terms, 137–138
  model, 141–142
  mole numbers, volumes, porosities, 138–140
  specialization of BFM to PEM, 136–137

Binary friction model, 132–133
Bipolar plates, 321
Boltzmann's constant, 540
Bond fluctuation model (BFM), 425
Bontha, J. R., 365, 379, 416
Born-Haber cycle, 555–556
  for proton formation, 555
Born-Oppenheimer molecular dynamics, 444
Brooks-Corey function, 303
Brown, D., 422
Bruggemann formula, 61
Brunauer-Emmett-Teller (BET) sorption model, 134
Buatier de Mongeot, F, 494
Buchi, F., 254
Butler-Volmer equation, 47–48
Butler-Volmer kinetics, 148
Butler-Volmer type rate coefficients, 539–540

Cairns, E. J., 204
Calvo, S. R., 509
Campbell, C. T., 32
Capillary pressure, 302–303
Cappadonia, M., 176
Carbon corrosion, 32–33
Carbon monoxide (CO)
  diffusion of, 545
  electrooxidation on Pt nanoparticles, 52–53
  oxidation, 24
  oxidation on Pt-based electrocatalysts, modeling of, 533–533
    Monte Carlo simulations, 538–545
    Pt-based catalysts, 530–534
    quantum chemical calculations, 534–584
  oxidation on PtRu, 499
CARLOS, 488, 541
Carnes, B., 123
Car-Parrinello molecular dynamics, 444–445
  simulations, 512
Catalyst/hydrated membrane interface, 520–523

Catalyst layer(s), 27, 42
  modeling, 19–20
    of decay mechanisms, 29–30
    design, 21
    freeze, 34–35
    hydrogen oxidation, 24
    interfacial kinetics of supported
      catalysts, 22
    kinetic Limitations on, Implications
      of, 24–27
    loss of catalyst surface area, 30–33
    membrane degradation, 34
    oxygen reduction reaction, 22–23
    porous electrodes, 27–28
    porous-electrode theory, 28–29
    surface property changes, 34
  overpotential and reaction rate across,
    209–210
  transport and reactions in, 71–73
Catalyst layer operation
  catalyst utilization and performance at
    agglomerate level, 58–59
    spherical agglomerates, 59–62
    Tafel law, 63–64
    ultrathin planar catalyst layers,
      implications for, 65–66
  electrode processes
    exchange current density, 49–51
    faradaic current density, 46–49
  macrohomogeneous catalyst layer
    modeling, 66–67
    relations between structure and
      effective properties, 67–71
  macrohomogeneous catalyst layser
    modeling
    structural picture, 43–46
    transport and reactions in catalyst
      layers, 71–73
  nanoparticle reactivity
    active site concept, 53–54
    CO electrooxidation on Pt
      nanoparticles, 52–53
    kinetic Monte Carlo simulation with
      finite surface mobility, 57–58
    mean field model with active sites,
      54–56
    particle size effects in
      electrocatalysis, 51–52
  pacemakers, 41–43

porous structure and water
  management in CCLs, 77–79
critical liquid water formation,
  80–82
liquid-to-vapor conversion
  capability, 79–80
results of macrohomogeneous
  approach, 73–74
catalyst utilization, 76
composition effects, 75–76
optimum thickness and
  performance, 74–75
supremacy of ultrathin catalyst
  layers, 77
structure-function optimization,
  82–84
Catalyst utilization, 76
Cathode catalyst layer (CCL), 41–42,
  67, 202
Cathode electrocatalysis
  mechansim of oxygen reduction,
    502–505
  theoretical volcano plot for oxygen
    reduction, 505–506
Cathodic overpotential, 48
CCL, see Cathode catalyst layer (CCL)
Cell performance, 231
Cell polarization, 24
Cell potential, 225–227
Cellular-automaton-based simulation,
  457–461
Cell voltage and local current,
  282–284
CFD, see Computational fluid
  dynamics (CFD)
Channel condensation, 331
Channel flux equations, 326
Characteristics Times, 112
Charge-transfer reaction, 22
Cheddie, D., 3
Cho, E., 35
Chronoamperometric current
  transients, 56
Cluster-network model, 160–161
Cluster-network model of Hsu and
  Gierke, 419
Coarse-grained particle-based
  simulations, 455–456
Coarse-graining approach, 488

Complexity and proton conduction, 389
  cooperativity, 404–407
  diffusion, 392–400
  dissociation, 387–391
  transfer and separation, 391–392
Computational fluid dynamics (CFD), 199–200
  based models, 3, 199–200, 243
  codes, 318
Conductivity model, 142–143
  based on DFM, 134–136
Connectivity and proton conduction
  cooperativity, 408–410
  dry connectivity, 403–404
  through hydration, 404–407
Constant inlet flow rate of oxygen
  comparison with experiment, 238–241
  general case, 236–238
  limiting current density, 232–236
  model and governing equations, 230–232
  solution, 231–232
Constant oxygen stoichiometry
  condition of ideal membrane humidification, 241–243
  modification of model equations, 240–241
Contact porosimetry, 11
Coolant flow, 328
Cooperativity, 408–410
Coordinate system, 318–319
Core-shell, 427
CO-tolerant electrocatalysts, 533
Counter ions, solvated, 562
Coupled proton and water transport in PEM, 123–124
  BFM2 in fuel cell model, implementation, 148
    membrane transport on fuel cell performance, effect of, 148–151
    PEM fuel cell model, 148
  binary friction conductivity model, 142
    conductivity model, 142–143
    experimental data, 143–144
    fitting conductivity, 144–148
  binary friction membrane model
    diffusion coefficients, 140–142
    magnitude of coefficients for driving force terms, 137–138
    mole numbers, volumes, porosities, 138–140
    specialization of BFM to PEM, 136–137
  membrane conductivity models, 132
    binary friction model, 132–133
    conductivity model based on DFM, 134–136
    dusty fluid model, 133–134
  membrane families, 124–125
  membrane hydration, 126–128
  membrane transport models
    diffusion models, 131–132
    hydraulic models, 130–131
    transport mechanisms, 129–130
Critical liquid water formation, 80–82
Current density, 332

Dalton's law, 325
Damjanovic, A., 510
Darcy's Law, 300–302
Darling, R. M., 30, 32–33
Davydov "solitonic" mechanism, 351
Day, T. J. F., 431
$1D + 1D$ model, humidified membrane, 213–214
  dynamics of cell performance degradation, 220–229
  finite oxygen stoichiometry, effect of, 214–219
  flow velocity in channel, 214
  local polarization curves, 219–220
$1D + 1D$ model, water management
  constant inlet flow rate of oxygen, 229–240
  constant oxygen stoichiometry, 240–243
Decomposition analysis, 497
Degradation, 8–10
  wave, dynamics of, 222–225
Degree of blockiness, 428
Density functional theory (DFT), 439–441, 485–486, 512, 534
  approximate treatment, 441–443
  functionals in, 440–441
Density-functional tight-binding (DFTB) method, 442–443
Density of states (DOS), 518

Depleted-zone core-shell (DZCS), 427
DFM, see Dusty fluid model (DFM)
Differential PEMFC, 93–96
  STR PEM design, 96–97
Diffusion, 395–402
  coefficients, 140–142
  models, 131–132
Dimethylmethylphosphonate (DMMP), 420
Direct methanol fuel cell (DMFC), 6, 297, 349–350
  reaction kinetics
    methanol kinetics, 310–314
    oxygen kinetics, 308–310
    reaction models, 307–308
  two-phase flow
    alternative two phase flow formulations, 304–305
    capillary pressure, 302–303
    Darcy's Law to two-phase flow, extension of, 300–302
    motivation, 298
    numerical example, 306–307
    relative permeability, 305
    Stefan Maxwell model, 298–300
    two phase flow and Stefan Maxwell Model, combination, 305–306
Discretization, 331
Dissipative particle dynamics (DPD), 456–457
Dissociation, 389–394
Dissolution mechanism, 31
Distributed electrode theory, 48–49
Divisek, J., 254
Djilali, N., 123
DMFC, see Direct methanol fuel cell (DMFC)
DMMP, see Dimethylmethylphosphonate (DMMP)
1-D model of MEA transport, 322
1 + 1D models, 254
"Double counting terms", 441
Dow membranes, 125
DPD, see Dissipative particle dynamics (DPD)
DREIDING, 418–420, 427

Dry connectivity, 403–404
Dry regime, 265–267
Dunietz, B. D., 498–499
DuPont's Nafion, 386–387
Dusty fluid model (DFM), 133–135
Dynamic master equation, 488
Dynamic Monte Carlo (DMC) simulation, 488
Dynamics of cell performance degradation, 220–229

Edisonian design, 23
Eigen-like cation, 392
Eikerling, M., 41, 131, 159, 204, 244
Electrical and thermal coupling, 317–318
  channel flow for unit cell
    channel flux equations, 326
    gas concentrations in channels, 325–326
    operating conditions, 327
  computational results
    stack, 333–336
    unit cell, 333
  coordinate system, 318–319
  dimensional reduction, 320–322
  1-D model of MEA transport, 322
    electrochemistry, 325
    membrane transport, 322–324
    oxygen transport losses, 324–325
  fundamental variables, 319
  iterative solution strategy, 331
    current density update, 332
    temperatures and channel fluxes, 332
  model summary
    discretization, 331
    system description and important coupling, 330–331
  other quantities, 319–320
  stack level coupling
    coolant flow, 328
    electrical coupling, 329–330
    end cell conditions, 330
    thermal coupling, 327–328
  state of stack modeling, 336
Electrical coupling, 329–330
Electrocatalysis, particle size effects in, 51–52

Electrocatalysis for PEMFC, molecular-level modeling of, 485
  anode electrocatalysis
    adsorption of carbon monoxide, 488–498
    adsorption of OH, 500–501
    electro-oxidation of carbon monoxide, 498–500
    methanol, 501–502
  cathode electrocatalysis
    mechansim of oxygen reduction, 502–505
    theoretical volcano plot for oxygen reduction, 505–506
  methods
    ab initio and classical molecular dynamics, 487
    ab initio quantum-chemical calculations, 485–487
    kinetic Monte Carlo simulations, 487–488
Electrocatalysts, 341–342
Electrocatalytic reaction systems from first principles, 551–552
  aqueous phase, role of, 552–558
    aqueous- vs. vapor-phase methanol dehydrogenation over Pt(111), 558–562
  surface potential, role of, 562–565
    internal reference, 565–568
    potential-dependent free energy, 568–569
    potential-dependent methanol dehydrogenation over aqueous-phase Pt(111), 569–572
Electrochemistry, 325
Electrode potential, 46
Electrolytes, 341–342
Electronic Kohn-Sham energy, 516
Electronic transmission factor, 46
Electron transfer, 524–526
Electro-osmotic drag, 129
  coefficient, 230
  in Nafion, 230
Electro-oxidation of carbon monoxide, 498–500
Elliott, J. A., 392, 394, 403–404, 408, 413, 418–420, 432, 466
Empirical valence bond (EVB), 430

End cell conditions, 330
Ennari , J., 418–419
Equilibrium concentration of mobile platinum species, 31
Equilibrium rate constant, 47
Evaporative layer, 270–272
EVB, see Empirical valence bond (EVB)
Exchange current density, 49–51

Faradaic current density, 46–49, 64
Fermi potential, 566
Fick's equation, 204
Field-theoretic simulation, 469–478
Fimrite, J., 123
Finite oxygen stoichiometry, effect of, 214–219
First integral, 206
Fitting conductivity, 144–148
Flooded-agglomeratemodels, 28
Flory-Huggins parameters, 469, 471
Flow field, 199
  models, 8
Flow velocity in channel, 213–214
Flow visualization methods, 11
Fluoroalkanes, 420
Free water, 466
Freeze, 34–35
Freezing, 10–11
Freunberger, S., 254
Fuel cell
  electrocatalysis, 533
  electrode can, 46
  model, 331
  response to process control, 115–117
Fuhrmann, J., 297
Fundamental voltage equation, 329

Galperin, D., 453
Galvani potentials, 46
Gärtner, K., 297
Gas concentrations in channels, 325–326
Gas crossover, 183–185
Gas diffusion electrodes, 7
Gas-diffusion layers (GDL), 19, 29, 66, 188, 199, 229
Gas plenums, 281–282
Gasteiger, H., 21, 499–500, 512, 545
GAUSSIAN code, 417

GDL, *see* Gas diffusion layer (GDL)
Generalized gradient approximation (GGA), 441, 486
Generalized solvent coordinate, 503
Generic exchange current density, 49–50
GGA, *see* Generalized gradient approximation (GGA)
Gibbs-Duhem equation, 164
Gibbs free energy, 41
Gierke model, 353
Global pressure, 304
  saturation formulation, 304
Global velocity, 304
Grønbech-Jensen, N., 368
Grothuss-type proton transfer mechanism, 430
Grotthuss mechanism, 182
Grotthuss structural diffusion mechanism, 354
Grove, Sir William, 42
Gur, Y, 365–366
Guzman-Garcia, A., 365–366

Hamiltonian, 368
Hamiltonian matrix elements, 44
Hammer, B., 489–490, 493
Hanna, S., 418–420
Hartree-Fock exchange, 441
Hartree-Fock (HF) method, 438
Hazebroucq, S., 437
Head-Gordon, M., 494, 533
Heterolytic water activation pathway, 554–555, 560
HF and post-HF methods, 438–439
High-performance catalyst layers, 20
Hirschfelder, J. O., 164
Holt, D. B., 418
Hue. T, 475
Hybrid DFT (B3LYP), 432
Hydration, 404–474
Hydraulic models, 130–131
Hydrogen bonds, 127
Hydrogen fuel initiative, 3
Hydrogen oxidation, 24
  reaction, 13
Hydrophilic patch, 429
Hyodo, S., 456–457, 461
Hypernetted chain (HNC) approximation, 463

Hysteresis in, 275–276
  in counter-flowing PEMFC, 291–294
  STR fuel cells, 276–281
    cell voltage and local current, 282–284
    full model, 284–285
    gas plenums, 281–282
    in-plane direction, 280
    numerical hysteresis and transients, 285–290
    through-plane direction, 279–280

IEC, *see* Ion exchange capacity (IEC)
Ignition
  front migration in SAPC, 104–106
  parameter that affect, 106
  in STR, 101–104
Illas, F., 491, 494, 564
Imidazole, 446–448
Integral equation theory, 461–469
Interfacial charge-transfer reactions, 24
Interfacial kinetics of catalysts, 22
Inter-particle model, 354
Ion exchange capacity (IEC), 125
Ionomer density, 7
Ionomer membrane morphology, atomistic structural modelling of, 413
  application to PFSI membrane morphology, 416–418
  atomistic computational studies of PFSIs, 418
  empirical valence bond models, 430–431
  molecular mechanics models
    monomeric sequence in polymeric models, effect of, 427–430
    oligomeric models, 420–421
    polyelectrolyte models, 418–419
    polymeric models, 421–427
  molecular orbital and density functional models, 432
  PFSI membranes, 413–416
"Ionomer" peak, 454
Iterative solution strategy, 331

J_*, 211
Jang, S. S., 420, 427–429, 431–432
Janssen, M., 132, 159

Kelvin equation, 73
Khalatur, P. G., 354, 423–425, 453
Khokhlov, A. R., 354, 423–424, 453
Kinetic Monte Carlo simulations, 487–488
  with finite surface mobility, 57–58
Kinoshita, K., 22–23, 30, 52
Kinoshita's analysis, 25
Kohn-Sham energy expression, 442–443
Koper, M. T. M., 23, 50, 53, 485, 500, 534, 545
Kornyshev, A. A., 204, 365, 417, 430
Kreuer, K. D., 123, 125, 129, 160, 341, 351, 374, 474
Krueger, J. J., 471
Kulikovsky, A. A., 199, 254

Lack of smoothness, 330
Lagrange multiplier, 471
Lamas, E. J., 509
Lamellar model of Litt, 459
Langmuir-Hinshelwood mechanism, 539
Laplace equation, 73
Lattice molecular dynamics (LMD), 425, 457
LDA, *see* Local density approximation (LDA)
LDA SVWN functional, 440–441
Leverett function, 304
Liquid-to-vapor conversion capability, 79–80
Liquid water saturation, 44–45
LMD, *see* Lattice molecular dynamics (LMD)
Load, dynamics of changes in, 107–111
Local density approximation (LDA), 440–441
Local order model, 459
Local polarization curves, 219–220

Macrohomogeneous catalyst layer modeling, 66–67
Macroscopic transport in polymer-electrolyte membranes, model for, 157–158
  gas crossover, 183–185
  governing equations, 164–166

membrane structure
  experimental evidence of, 160
  as function of water content, 161–163
  model aspects, 186
    describing water management, 187–191
    validation, 186–187
  modeling approaches, 159–160, 166–168
  occurrence
    approach and governing equations, 174–175
    discussion and limitations, 178–179
    fraction of expanded channels, calculation, 175–178
  physical and transport properties
    thickness, 179–180
    transport properties, 180–183
  vapor-equilibrated transport mode, 168–169
    chemical model for determining water content, 169–172
    liquid-equilibrated transport mode, 172–173
Magnetic resonance imaging (MRI), 13
Magnitude of coefficients for driving force terms, 137–138
Marcus theory, 563–564
Markovic, N. M., 23, 533
Mauritz, K. A., 414, 416
MEA, *see* Membrane-electrode-assembly (MEA)
Mean-field approximation, 488
Mean field model with active sites, 54–56
Membrane, 349–351
Membrane conductivity models, 132
  based on DFM, 134–136
  binary friction model, 132–133
  dusty fluid model, 133–134
Membrane degradation, 34
Membrane density, 11
Membrane electrode assembly (MEA), 96, 317, 342–343
Membrane hydration, 126–128
Membrane transport, 322–324
Membrane transport models
  diffusion models, 131–132
  hydraulic models, 130–131

Membrane transport on fuel cell
  performance, effect of, 148–151
Mesoscopic continuum modeling, 461
  field-theoretic simulation, 469–478
  integral equation theory, 461–469
Meta-GGA methods, 441
Metal–water interactions, 553
Methanol, 501–502
Methanol kinetics, 310–314
Methanol oxidation, 558
Meyers, J. P., 19, 30–33, 159, 165,
  169–170, 176, 180
Michalaelides, 553
Model and nondimensionalization
  boundary conditions, 260–262
  GDL model, 257–260
  notation, 256–257
Molar fraction of bulk gas, 231
Molecular closures, 463
Molecular dynamics, 443–444
  Born-Oppenheimer molecular
    dynamics, 444
  Car-Parrinello molecular dynamics,
    444–445
  data analysis, 445
Molecular dynamics, ab initio and, 485
Molecular-level modeling, 485
  anode electrocatalysis
    adsorption of carbon monoxide,
      488–498
    adsorption of OH, 500–501
    electro-oxidation of carbon
      monoxide, 498–500
    methanol, 501–502
  cathode electrocatalysis
    mechansim of oxygen reduction,
      502–505
    theoretical volcano plot for oxygen
      reduction, 505–506
  methods
    ab initio and classical molecular
      dynamics, 485
    ab initio quantum-chemical
      calculations, 485–487
    kinetic Monte Carlo simulations,
      487–488
Molecular mechanics models
  monomeric sequence in polymeric
    models, effect of, 427–430

oligomeric models, 420–421
polyelectrolyte models, 418–420
polymeric models, 421–427
Molecular orbital and density functional
  models, 432
Mole numbers, volumes, porosities,
  138–140
Møller Plesset method, 439
Møller-Plesset perturbation theory
  (MP2), 432
Mologin, D. A., 425–426, 458
Monomeric sequence in polymeric
  models, effect of, 427–430
Monte Carlo simulations, 23, 54, 57,
  538, 539
  background, 539
  kinetic, 485
  KMC simulations, 541–545
  model, 539–541
Moore, R. A., 160, 414, 416
Morikawa, Y., 489–490
MRI, *see* Magnetic resonance imaging
  (MRI)
Multi-state empirical valence bond
  (MS-EVB) methodology, 402
  model, 431
Münch, W., 437
Munroe, N., 3

Nafion, 91, 125–126
  morphological model of inverted
    micelles in, 359–360
  pockets, 402
  polymer, 421–427
  water-nanostructure, 249–254
Nafion macromolecule, 465
Nafion membranes, morphology of,
  453–455
  cellular-automaton-based simulation,
    457–461
  coarse-grained particle-based
    simulations, 455–456
  dissipative particle dynamics (DPD),
    456–457
  mesoscopic continuum modeling, 461
    field-theoretic simulation, 469–478
    integral equation theory, 461–469
Neimark, A. V., 354, 420–423, 425,
  427, 430

Nernst-Planck equation, 61–62
Neurock, M., 502, 551, 555, 558, 560, 565, 569–570
Neutron imaging, 12–13
Newman, J., 28, 125, 132, 146–147, 157, 159, 169–170, 180, 204, 468
Noble-metal electrocatalysts, 22
Nondimensionalization, 262–263
Nonlocality, 330
Nørskov, J. K., 489, 564
Nosé-Hoover chain thermostat, 423
Numerical hysteresis and transients, 285–290

OER, see Oxygen electroreduction (OER)
OHP, see Outer Helmholtz plane (OHP)
Okamoto, Y., 560
Oligomeric models, 420–421
One-dimensional model of
  cathode catalyst layer, performance of, 204–211
  voltage loss due to oxygen transport, 211–213
ONIOM, 417–418
Ornstein-Zernicke equation, 401–402
Ornstein-Zernike-like (SSOZ) integral matrix equation, 461
ORR, see Oxygen reduction reaction (ORR)
Ostwald ripening, 32
Oszcipok, M., 35
Outer Helmholtz plane (OHP), 562–563
Oxygen and local current along the channel, 214–218
Oxygen electroreduction (OER), 511
Oxygen kinetics, 308–310
Oxygen-limiting current density, 230
Oxygen reduction
  mechanism in acid medium on Pt and Pt-alloys, 512–513
    catalyst/hydrated membrane interface, 520–523
    first reduction step, 513–520
    subsequent reduction steps, 524–526
  mechansim of, 502–505
  theoretical volcano plot for, 505–506
Oxygen reduction reaction (ORR), 19, 22–23, 202, 509

Oxygen starvation, 220
Oxygen transport losses, 324–325

Pacchioni, G, 491
Pacemakers, 41–43
Paddison, S. J., 368, 374, 385, 389, 392, 394, 403–404, 408, 416, 432, 46, 469
Pair-connectedness correlation functions, 468
Panchenko, A., 523
Parthasarathy, A., 60
Particle-size effect, 23
Paul, R., 365, 416
Pauli repulsive interactions, 553
PEMFC development, areas of, 4
PEM fuel cell model, 148
Percolation-type equation, 180
Percus-Yevick (PY) closure, 463
Perfluorosulfonic acid membranes (PFSA), 124
Perfluorosulfonic acid (PFSA), 123, 136, 385
Perfluorosulphonate ionomer (PFSI), 413–416
Permeation coefficients, 184
Perovskites, 448–450
Peroxide, 34
Perry, M. L., 204
Petersen, M. K., 431–432
Peuckert, M., 23
PFSA, see Perfluorosulfonic acid (PFSA)
PFSI membranes, 413–416
Phase change and hysteresis, 253–254
  hysteresis in, 275–276
    in counter-flowing PEMFC, 290–294
    STR fuel cells, 276–290
  two-phase flow in GDL, 254–256
    analysis of three regimes, 264–273
    model and nondimensionalization, 256–263
    resolution of front evolution, 273–275
Phase pressure saturation formulation, 304
Piela, P., 30–31
Pintauro, P. N., 365, 379, 416
Platinum electrocatalysts, 342

Poisson-Boltzmann theory, 357
Poisson-Nernst- Planck (PNP) theory, 61
Polarization curves, 310–311
Polarization voltages, 201
Polar solvents, 552–553
Polyelectrolyte models, 418–420
Polyelectrolyte system, 419–420
Polymer electrolyte membranes (PEM), functionality of, 123
Polymeric models, 421–427
  effect of monomeric sequence in, 427–430
Polymeric PFSI, 421–427
Polymer RISM (PRISM) theory, 461
Polytetrafluoroethylene (PTFE), 124, 385–387, 394–396, 403–404, 420, 428
Porosities, 44
Porous electrodes, 27–28
Porous-electrode theory, 28–29
Porous structure and water management in CCLs, 77–79
  critical liquid water formation in CCL, 80–82
  liquid-to-vapor conversion capability, 79–80
Porous transport layer (PTL), 297
Potential-dependent free energy, 568–569
Potential-dependent methanol dehydrogenation over aqueous-phase Pt(111), 569–572
Potential-difference infrared spectroscopy (PDIRS), 551
Potential energy surface (PES), 537
Pourbaix diagram for ruthenium stability, 31
Power density, 21
Pressure pressure formulatiom, 305
Promislow, K. S., 109, 192, 253
Proton conducting materials, quantum molecular dynamic simulation of, 437–450
  application, 445–446
    imidazole, 446–448
    perovskites, 448–450
    water, 446
  basic theory, 437–438

density functional theory, 439–440
  approximate treatment, 441–443
  functionals in, 440–441
HF and post-HF methods, 438–439
molecular dynamics, 443–444
  Born-Oppenheimer molecular dynamics, 444
  Car-Parrinello molecular dynamics, 444–445
  data analysis, 445
Proton conduction in PEMs, 385, 388–389
  complexity, 389
  diffusion, 395–402
  dissociation, 389–394
  transfer and separation, 394–395
  connectivity
    cooperativity, 408–410
    dry connectivity, 403–404
    through hydration, 404–405
  goals and challenges for modeling, 388
  motivation for modeling, 387–388
  types of, 386–387
Proton conductor, 357
Proton current across the catalyst layer, 207
Proton diffusion, 129
Proton hopping, 429
Proton surface diffusion, 7
Proton transfer, 357–359
Proton transport in PEM membrane, 349–351
  morphological model of inverted micelles in Nafion, 359–360
  Nafion-water-nanostructure, 249–254
  proton transfer, 357–359
PTFE, see Polytetrafluoroethylene (PTFE)
PTL, see Porous transport layer (PTL)

Quantum-chemical calculations, ab initio, 485–487
Quantum molecular dynamic simulation of proton conducting materials, 437
  application, 445–446
    imidazole, 446–448
    perovskites, 448–450
    water, 446

Quantum molecular (*continued*)
  basic theory, 437–438
  density functional theory, 439–440
    approximate treatment, 441–443
    functionals in, 440–441
  HF and post-HF methods, 438–439
  molecular dynamics, 443–444
    Born-Oppenheimer molecular dynamics, 444
    Car-Parrinello molecular dynamics, 444–445
    data analysis, 445
Quasi-Newton methods, 332

Raistrick, I. D., 19
Random phase approximation (RPA), 454
Reaction kinetics
  methanol kinetics, 310–314
  oxygen kinetics, 308–310
  reaction models, 307–308
Reaction models, 307–308
Reactor dynamics, 91–92
  autonomous oscillations in STR PEMFC, 114–115
  differential PEMFC, 93–96
    STR PEM design, 96–97
  dynamics of changes in load, 107–111
  dynamics of start-up of PEMFC, 99–100
    ignition front migration in SAPC, 104–106
    ignition in STR, 101–104
    parameters, affect ignition, 106–107
  fuel cell response to process control, 115–117
  segmented anode parallel channel (SAPC), 97–98
    fuel cell design, 98–99
  STR PEMFC, applications of, 117–118
  time constants for PEMFC, characteristic, 111–114
Realistic rotational- isomeric-state (RIS) model, 464
Real laboratory, 343
Reference molecular mean-spherical approximation (RMMSA), 463, 464

Reiser, C. A., 32
Relative humidity (RH), 149–150
Relative permeability, 301, 305
Rempel, A. W., 35
Representative elementary volume elements (REV), 67
REV, *see* Representative elementary volume elements (REV)
Rivin, D., 420
Ross, P. N., 22, 50, 52, 57, 529
Rotating ring-disk electrode (RRDE), 22
Rotational isomeric state (RIS), 423–424
Roughness factor, 25
RRDE, *see* Rotating ring-disk electrode (RRDE)

Saravanan, C., 533
Savadogo, O., 526–527
Scanning tunneling microscopy (STM), 533, 551
Schmitt, U. W., 430
Schröder paradox, 128, 151, 158, 276
Schrödinger equation, 437–438
Segmented anode parallel channel (SAPC), 97–98
  fuel cell design, 98–99
Seifert, G., 437
Self-consistent mean-field (SCMF) methods, 461
Seminario, J. M., 509
Separation, 394–395
Small angle x-ray scattering (SAXS), 343
Solution phase, 552, 558, 561, 562, 573
Sone, Y., 135, 143–144, 146
Specific reactivity for an electrode surface, 23
Spectroscopic analysis, 551
Spherical agglomerates, 59–62
Spillover effect, 52
Spohr, E., 374, 430, 431
Springer, T. E., 24, 29, 67, 131, 143–144, 148, 152, 159, 204
Springer model, 146, 149–150
Stack, 333–336
  modeling, 336
Stack level coupling
  coolant flow, 328
  electrical coupling, 329–330

end cell conditions, 330
thermal coupling, 327–328
Stack modeling, state of, 336
Start-up of PEMFC, dynamics of, 99–100
  ignition front migration in SAPC, 104–106
  ignition in STR, 101–104
  parameters, affect ignition, 106–107
State of water in PEM, 365–373, 376–380
  applications, 374–376
Stefan Maxwell model, 298–300, 305–306
St-Pierre, Jean, 3
STR fuel cells, 279–290
STR PEMFC, applications of, 117–118
Struchtrup, H., 123
Structure-function optimization, 82–84
Sub-layer and front motion, 272–273
Sum frequency generation (SFG), 533, 551
Superficial area, electrode, 24–25
Surface-enhanced infrared spectroscopy (SEIRS), 551
Surface-enhanced Raman scattering (SERS), 551
Surface potential, 552
SZG's ground-breaking model, 131

Tafel law, 63–64
Tafel-limit, 49
Tant, M. R., 413
Taylor, C. D., 551, 565, 569
Teflon, 161, 184–185, 298
  blocks, 96, 98
Temperatures and channel fluxes, 332
Thampan, T., 132, 134, 143–144, 146–147, 149, 152, 159, 170
Thermal and electrical coupling, 317–318
  channel flow for unit cell
    channel flux equations, 326
    gas concentrations in channels, 325–326
    operating conditions, 327
  computational results
    stack, 333–336
    unit cell, 333
  coordinate system, 318–319

dimensional reduction, 320–322
1-D model of MEA transport, 322
  electrochemistry, 325
  membrane transport, 322–324
  oxygen transport losses, 324–325
fundamental variables, 319
iterative solution strategy, 331
  current density update, 332
  temperatures and channel fluxes, 332
model summary
  discretization, 331
  system description and important coupling, 330–331
other quantities, 319–320
stack level coupling
  coolant flow, 328
  electrical coupling, 329–330
  end cell conditions, 330
  thermal coupling, 327–328
  state of stack modeling, 336
Thermal coupling, 327–328
Thiele modulus, 62
Three-phase interface, 19
Time constants for PEMFC, characteristic, 111–114
Transfer, 391–392
Transition state theory (TST), 46
Transport coefficient, 166
Transport mechanisms, 129–130
TST, *see* Transition state theory (TST)
Tuckerman, M. E., 497
Two-phase flow and DMFCs
  alternative formulations, 304–305
  capillary pressure, 302–303
  Darcy's Law, 300–302
  motivation, 298
  numerical example, 306–307
  relative permeability, 305
  Stefan Maxwell model, 298–300, 305–306
Two-phase flow in GDL, 254–256
  analysis of three regimes, 263–273
  model and nondimensionalization, 256–263
  resolution of front evolution, 273–275
Two-phase regime, 267–270

Ultrathin planar catalyst layers,
  implications for, 65–66
Unit cell, 333
Urata, S., 423, 432, 466, 469

Vacuum layer, 565
Validation, 11–14
Vapor-equilibrated membrane, 168
Vapor-equilibrated transport mode,
  168–169
  chemical model for determining water
    content, 169–172
  liquid-equilibrated transport mode,
    172–173
Variable time step Monte Carlo
  (VTSMC), 541
Vassilev, P., 501, 558
Verbrugge, M. W., 29, 130–131, 159,
  365–366
Virtual laboratory, 343
Vishnyakov, A., 352, 418–421, 423,
  425, 428
Volcano effect, 334
Voltage current curve, 208,
  218–219
Voltage losses, 200–202
Voth, G. A., 355, 417, 430
VTSMC, *see* Variable time step Monte
  Carlo (VTSMC)

Wang, Y., 509
Wasileski, S. A., 492, 502, 551, 560,
  569–570
Water, 446
  activation, *see* Water dissociation
  back flux, 158
  dissociation, 553–558
    heterolytic dissociation for aqueous-
      phase, 554
    homolytic dissociation mechanism
      for vapor-phase, 554
  management, 11, 187–191
  profiles, 189–191
Water adsorption, 553
Water dissociation, 553–558
Weber, A. Z., 29, 125, 132, 146–147, 157,
  468–469
Weber's model, 29
Wetting, 45
Wetton, B., 244, 317

Xu, G., 512–513

Yamamoto, S., 456, 461
Yao, K. Z., 3
Young, S. K., 475

Zawodzinski Jr., T. A., 129, 432, 468–469
"Zundel" ion, 129